Flow and Transport in Fractured Porous Media

P. Dietrich · R. Helmig · M. Sauter · H. Hötzl
J. Köngeter · G. Teutsch (Editors)

Flow and Transport in Fractured Porous Media

With 306 figures and 65 tables

Springer

Editors:

Dr. Peter Dietrich
University of Tübingen
Center for Applied Geoscience
Sigwartstr. 10
72076 Tübingen
Germany

Prof. Dr.-Ing. Rainer Helmig
University of Stuttgart
Institute of Hydraulic Engineering
Pfaffenwaldring 61
70569 Stuttgart
Germany

Prof. Dr. Martin Sauter
Geoscience Center of the
University of Göttingen
Department of Applied Geology
Goldschmidtstr. 3
37077 Göttingen
Germany

Prof. Dr. Heinz Hötzl
University of Karlsruhe
Department of Applied Geology
Kaiserstrasse 12
76128 Karlsruhe
Germany

Univ.-Prof. Dr.-Ing. Jürgen Köngeter
RWTH Aachen University
Institute of Hydraulic Engineering
Mies-van-der-Rohe-Str. 1
52056 Aachen
Germany

Prof. Dr. Georg Teutsch
Environmental Research Center (UFZ)
Permoser Str. 15
04318 Leipzig
Germany

ISBN 978-3-642-06231-5 e-ISBN 978-3-540-27012-6

Springer is a part of Springer Science+Business Media
springeronline.com

Printed in The Netherlands

Cover-Design: Erich Kirchner, Heidelberg
Printed on acid-free paper 89/3141/Yu - 5 4 3 2 1 0

Preface

Fractured porous aquifers are an important source of water for public, industrial and agricultural use, both in developed and developing countries. Recently, understanding, characterizing and modeling physical and chemical interactions within fractured porous aquifers has become increasingly important, particularly with regard to the question of water resources development and groundwater contamination. This book presents concepts and methods for characterizing flow and transport in fractured porous media from an experimental and modelling perspective on the basis of the aquifer-analogue concept.

A fundamental understanding of the response of a system to particular tests is a prerequisite, especially for the characterization of fractured porous aquifers, which are highly heterogeneous systems. This understanding is essential to identifying the governing flow and transport processes as well to calculating and predicting flow and transport in such systems.

In the classical concept of the aquifer analogue, which was established in the petroleum industry, the potential reservoir rocks are investigated in outcrops and characterized by detailed sedimentological studies. The properties of the system under investigation, i.e. the geometry and petrography of the sedimentological elements in the outcrop, are considered to be directly transferable to similar reservoir rocks. In the field of fractured porous rock hydrogeology, this approach has very rarely been applied. In the application of the aquifer-analogue approach presented in this book, accessible outcrops of the unsaturated fractured porous aquifer are examined. This concept allows sections of an aquifer to be investigated on several scales with field and laboratory techniques.

Employing appropriate investigation methods on each scale meant that a high resolution and precise data sets could be obtained, e.g. detailed information on the geometry of the fracture network and on the hydraulic characteristics of the porous matrix. The results of the investigations and the processes thus revealed are then assumed to be applicable to the inaccessible section of the saturated aquifer. While this approach offers considerable advantages, basic assumptions have to be made.

The choice of outcrop for the analogue investigation is crucial, because the samples chosen determine how representative the results are for the fractured porous aquifer system. Effects of the experiments themselves, for example a widening of the fracture apertures as a result of a pressure decrease following the removal of a sample, also have to be taken into account. In the investigation, the main focus is on basic experimental techniques for determining the effects of the heterogeneous system on flow and transport processes. The investigations that are based on the analogue concept usually cover small volumes of the aquifer or the aquifer analogue (ranging from centimeters to tens of meters) so that possible scale-dependency effects of the hydraulic properties need to be taken into account. On this assumption, the aquifer-analogue approach can be utilized in experimental and modelling investigations to characterize inaccessible fractured porous aquifers.

The work on the aquifer-analogue approach is divided into two parts. Firstly, flow and transport experiments are developed and applied to obtain highly resolved data sets enabling a detailed investigation of the system under consideration. Secondly, these data sets are used to develop reliable models, allowing the verification and interpretation of the observed results. Additionally, these models are simplified and verified in order to study the principle behavior of the fractured porous system. The proposed models are realized by using a discrete and a multi-continua modeling approach.

The experiments and model approaches used here allow the effects arising from heterogeneity to be analyzed and the heterogeneous nature of the investigated system to be characterized. In addition, the processes that can be expected to occur during flow and transport within fractured porous media are understood more clearly.

The aquifer analogue approach was also used as the basic concept of the research initiative "Fractured-Rock Aquifer-Analogue Approach". This research initiative was launched in 1996 and funded by the German Research Foundation (DFG). It was headed by the University of Tübingen and supported by research groups from the universities of Aachen, Karlsruhe and Stuttgart. While the emphasis of the groups from Tübingen and Karlsruhe was mainly on the design of appropriate and adapted experimental techniques on the laboratory and field scale, the Stuttgart and Aachen groups concentrated on the development of suitable modelling tools using discrete and continuum approaches. The major results of the research initiative "Fractured-Rock Aquifer-Analogue Approach"are presented in this book.

The work on the aquifer-analogue approach in the frame of the research initiative is divided into two parts. Firstly, flow and transport experiments are developed and applied to obtain highly resolved data sets enabling a detailed investigation of the system under consideration. Secondly, these data sets are used to develop reliable models, allowing the verification and interpretation of the observed results. Additionally, these models are simplified and verified in order to study the principle behavior of the fractured porous system. The proposed models are realized by using a discrete and a multi-continua modeling approach.

The experiments and model approaches used here allow the effects arising from heterogeneity to be analyzed and the heterogeneous nature of the investigated system to be characterized. In addition, the processes that can be expected to occur during flow and transport within fractured porous media are understood more clearly.

The book is divided into three main parts. In the first part *Assessment of Fractured Porous Media*, the aquifer-analogue approach and model concepts are introduced. The realization and results of flow and transport experiments in fractured porous media are described in the second part *Project Scale Studies*. The investigated objects range from small cores with a diameter of 10 cm to the field block with a dimension of approx. $10\,\mathrm{m} \times 8\,\mathrm{m} \times 2\,\mathrm{m}$. In the final part, scale-independent approaches and investigations are presented and illustrated by examples. the final part, *Scale-independent Approaches and Investigations* are presented and illustrated by examples.

The research initiative "Fractured-Rock Aquifer-Analogue Approach", funded by the German Research Foundation (DFG), was launched in 1996. It was headed by the University of Tübingen and supported by research groups from the universities of Aachen, Karlsruhe and Stuttgart. While the emphasis of the groups from Tübingen and Karlsruhe was mainly on the design of appropriate and adapted experimental techniques on the laboratory and field scale, the Stuttgart and Aachen groups concentrated on the development of suitable modelling tools using discrete and continuum approaches.

Our thanks are due to the German Research Foundation for funding the project. We are grateful to the firm Fauser, which supplied the block for the laboratory experiments, to Michel Lambert for his technical support and to the town of Pliezhausen where the quarry is situated. Thanks also to the publishing house Springer, particularly Dr. Thomas Ditzinger. Finally, we would like to thank all the numerous technicians, computer network administrators, student assistants and others for their indispensible contributions.

Tübingen and Stuttgart,
August 2004

Peter Dietrich
Rainer Helmig

Contents

Part I Assessment of Fractured Porous Media

Part III Scale-Independent Approaches and Investigations

List of Authors

Daniel Bachmann
RWTH Aachen University
Institute of Hydraulic Engineering
and Water Resources Management
Mies-van-der-Rohe-Str. 1
52056 Aachen
Germany
bachmann@iww.rwth-aachen.de

Roland Bäumle[2]
Ministry of Agriculture
Dept. of Water Affairs
Private Bag 13193
Windhoek
Namibia
baeumler@mawrd.gov.na

Salima Baraka-Lokmane[4]
Heriot-Watt University
Institute of Petroleum Engineering
Riccarton
Edinburgh EH14 4AS
United Kingdom
salima.lokmane@pet.hw.ac.uk

Ralf Brauchler[4]
Geoscience Center of the
University of Göttingen
Department of Applied Geology
Goldschmidtstr. 3
37077 Göttingen
Germany
rbrauch@gwdg.de

Peter Dietrich
University of Tübingen
Center for Applied Geoscience
Sigwartstr. 10
72076 Tübingen
Germany
peter.dietrich@uni-tuebingen.de

Hartmut Eichel
University of Stuttgart
Institute of Hydraulic Engineering
Pfaffenwaldring 61
70569 Stuttgart
Germany
Hartmut.Eichel@iws.uni-stuttgart.de

Rainer Helmig
University of Stuttgart
Institute of Hydraulic Engineering
Pfaffenwaldring 61
70569 Stuttgart
Germany
Rainer.Helmig@iws.uni-stuttgart.de

Heinz Hötzl
University of Karlsruhe
Department of Applied Geology
Kaiserstr. 12
76128 Karlsruhe
Germany
heinz.hoetzl@agk.uka.de

Martin Imig
Siegmundstr. 1
38106 Braunschweig
Germany
martin.imig@web.de

Dietmar Jansen[1]
Ingenieurgesellschaft
Dr. Ing. Nacken mbH
Valkenburger Str. 15
52525 Heinsberg
Germany
jansen@nacken-ingenieure.de

Jürgen Köngeter
RWTH Aachen University
Institute of Hydraulic Engineering
and Water Resources Management
Mies-van-der-Rohe-Str. 1
52056 Aachen
Germany
koengter@iww.rwth-aachen.de

Vincent Lagendijk[1]
Gewecke und Partner GmbH
Im Pesch 79
53797 Lohmar
Germany
v.lagendijk@bup-gup.de

Carsten Leven
University of Tübingen
Center for Applied Geoscience
Sigwartstr. 10
72076 Tübingen
Germany
c.leven@gmx.de

Rudolf Liedl
University of Tübingen
Center for Applied Geoscience
Sigwartstr. 10
72076 Tübingen
Germany
rudolf.liedl@uni-tuebingen.de

Christopher I. McDermott
University of Tübingen
Center for Applied Geoscience
Sigwartstr. 10
72076 Tübingen
Germany
chris.mcdermott@uni-tuebingen.de

Lina Neunhäuserer[3]
German Meteorological Service
Kaiserleistr. 42
63067 Offenbach
Germany
Lina.Neunhaeuserer@dwd.de

Yoram Rubin
University of California, Berkeley
Department of Civil and
Environmental Engineering
435 Davis Hall
Berkeley, CA 94720-1710
USA
rubin@ce.berkeley.edu

Martin Sauter[4]
Geoscience Center of the
University of Göttingen
Department of Applied Geology
Goldschmidtstr. 3
37077 Göttingen
Germany
martin.sauter@uni-goettingen.de

Annette Silberhorn-Hemminger
University of Stuttgart
Institute of Hydraulic Engineering
Pfaffenwaldring 61
70569 Stuttgart
Germany
Annette@iws.uni-stuttgart.de

Brett Sinclair[4]
Kingett
Mitchell Ltd.
Auckland
New Zealand

Mia Süß
University of Stuttgart
Institute of Hydraulic Engineering
Pfaffenwaldring 61
70569 Stuttgart
Germany
Mia.Suess@iws.uni-stuttgart.de

Georg Teutsch[4]
Environmental Research Center (UFZ)
Permoserstr. 15
04318 Leipzig
Germany
georg.teutsch@ufz.de

Christian Thüringer[2]
DB Verkehrsbau Logistik GmbH
Lammstr. 19
76133 Karlsruhe
Germany
christian.thueringer@bahn.de

Thomas Vogel
RWTH Aachen University
Institute of Hydraulic Engineering
and Water Resources Management
Mies-van-der-Rohe-Str. 1
52056 Aachen
Germany
vogel@iww.rwth-aachen.de

Matthias Weede
University of Karlsruhe
Department of Applied Geology
Kaiserstr. 12
76128 Karlsruhe
Germany
weede@agk.uka.de

Kai Witthüser[2]
University of Pretoria
Department of Geology
0002, Pretoria
South Africa
kai.witthueser@up.ac.za

[1]Formerly RWTH Aachen University
Institute of Hydraulic Engineering
and Water Resources Management
Mies-van-der-Rohe-Str. 1
52056 Aachen
Germany

[2]Formerly University of Karlsruhe
Department of Applied Geology
Kaiserstr. 12
76128 Karlsruhe
Germany

[3]Formerly University of Stuttgart
Institute of Hydraulic Engineering
Pfaffenwaldring 61
70569 Stuttgart
Germany

[4]Formerly University of Tübingen
Center for Applied Geoscience
Sigwartstr. 10
72076 Tübingen
Germany

Part I

Assessment of Fractured Porous Media

1

Aquifer-Analogue Approach

Characterizing of fractured rock systems is probably one of the most challenging problems that petroleum geologists as well as hydrogeologists have to face. The identification and assessment of fractures as barriers or hydraulic conductors in a geological environment, the understanding of flow and transport in fracture systems and the development of mathematical models are of utmost importance for the prediction of the hydraulic behavior of fractured porous geological systems.

Problems associated with the occurrence of fractures and their importance for flow and solute transport cover a wide area, ranging from subsurface oil, water and geothermal reservoirs, mining and quarrying, isolation of nuclear and hazardous wastes as well as geotechnical problems involved in tunnelling and cavern excavations.

Fractured rock systems display unique features compared to unconsolidated granular porous materials. The contrast in permeability between fractures and the porous matrix spans many orders of magnitude and can vary extremely in space. Fracture permeabilities can also change to a large extent, depending on whether a fracture is filled with, for example, sealing secondary calcite or wide open due to extensional stresses. Characterizing fractured systems therefore requires a careful examination of the interconnections in the fracture network as well as an evaluation of the fracture-matrix interaction.

Research into fracture flow and transport processes has mainly been motivated by the need to identify safe repositories for nuclear wastes. Comprehensive studies at the Yucca Mountains site in Nevada (e.g. Bodvarsson *et al.* (2003)), the ÄSPÖ underground laboratory (e.g. Wikberg *et al.* (1991)) and the STRIPA mine in Sweden (e.g. Olsson (1992)), the WIPP site in the United States (e.g. Ostensen (1998)), the Grimsel rock laboratory in Switzerland (e.g. Majer *et al.* (1990)), the Fanay-Augères mine in France (e.g. Cacas *et al.* (1990)), and the URL site in Canada (e.g. Everitt *et al.* (1994)) provided insight into the fundamental processes influencing fluid flow and (reactive) transport processes in fractured rock materials. Due to the nature of the problem, predominantly low-permeability environments such as granites and gneisses were investigated. Therefore, the results from the projects listed above cannot be directly transferred to fractured porous aquifer systems with a significant fracture permeability and storage in the rock matrix.

However, fractured rock systems forming relevant aquifers show some very varying characteristics. Frequently, consolidated sand-, silt- or claystones with multiple (fracture-) porosity features prevail. The hydraulic conductivity is generally much higher than in crystalline rocks and both the original sedimentary layering as well as the diagenetic cementation may affect the distribution of the effective porosity as well as the hydraulic conductivity. However, most of the existing subsurface investigation and hydrogeological assessment techniques were designed for highly conductive unconsolidated aquifers and a few techniques (e.g. fluid logging) were developed to work in low-permeability environments. There is in fact no appropriate concept or technique available for the characterization of materials with medium-range permeability, i.e. multiple-porosity systems commonly encountered within thick sedimentary beds.

The investigation of individual discrete flow paths with the volume-averaging techniques generally available, such as pumping tests or tracer tests, is almost impossible. This is mainly because these types of test provide integral information on system characteristics. In contrast, it is known from numerical modelling and laboratory experiments that the characterization of the fracture geometry (length, orientation, aperture, connectivity, etc.) as well as a knowledge of fracture/matrix interaction is a prerequisite for understanding transport in these systems. There is therefore a gap between the technical possibilities available for determining the large-scale spatially dependent flow and transport parameters of the fractured porous system on the large scale and the theoretical knowledge of factors influencing the system on the smaller scales. Furthermore, there are large discrepancies between hydraulic features in fractured porous rocks that are detectable and those that are actually relevant for process understanding and the prediction of flow and transport.

Below, the fundamentals of the "Fractured-Rock Aquifer-Analogue Approach"are presented as a concept to reconcile both a)an honoring of the effect of small-scale important hydraulic features, and b) the provision of integral spatially variable parameters on the large scale for prediction modelling.

An overview of the occurrence of fractured rock aquifers and a brief review of current methods employed in the investigation of flow and transport processes of fractured rocks is provided as a general background and framework, together with some new paradigms in the characterization of fractured porous systems.

1.1 Occurrence of Fractured Rock Aquifers

M. Weede

Fractured rock aquifers make up a significant part of the world's known aquifer systems. To be more precise: about 53.4 % (190,000 km^2) of Germany and about 75 % of the earth's surface consist of fractured or karstic fractured rock aquifers covered by negligibly small amounts of loose granular soil material (fig. 1.1 and fig. 1.2).

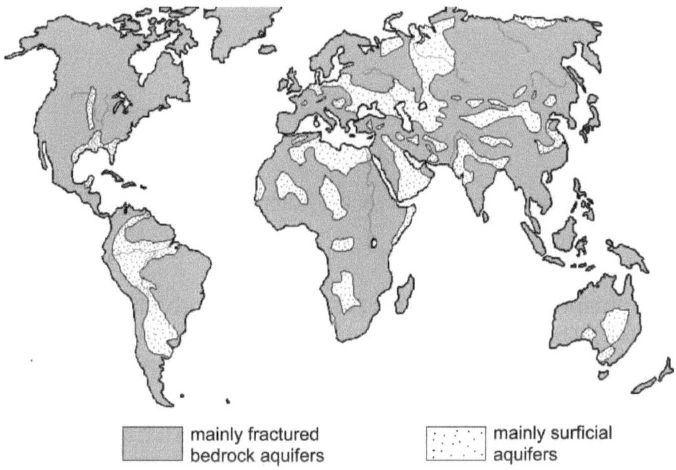

Fig. 1.1. Sketch-map of worldwide fractured rock aquifer distribution (interpreted after Plummer *et al.* (2002)).

The prevalence of rock-aquifer systems, even though they are less productive than porous media, makes their enormous potential for worldwide water resources evident. In addition, during the last few years the hydraulic characteristics of fractured rocks have gained increasing interest in the context of the search for storage of, for example, radioactive material.

1.2 Characterization of the Hydraulic Properties of Fractured Rock Aquifers

M. Sauter

There are a number of excellent reviews and summaries describing the current state of research into the characterization of fractured systems (e.g. Wang

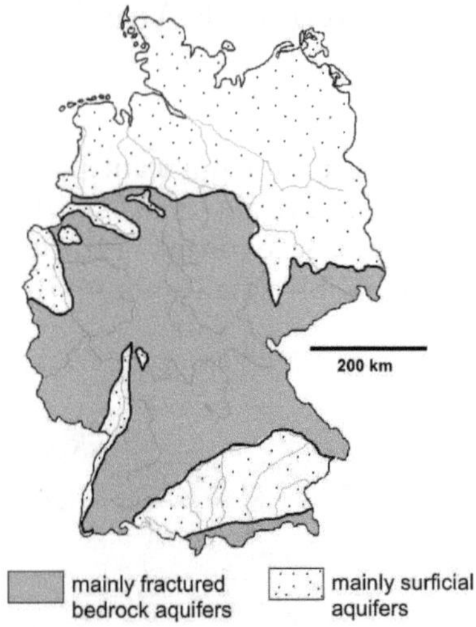

Fig. 1.2. Distribution of fractured rock aquifers in Germany (Hölting (1980), mod.).

(1991); Faybishenko *et al.* (2000); NRC (1996), Bodvarsson *et al.* (2003), Evans *et al.* (2001)). The review presented here is not intended to give a comprehensive and exhaustive overview on the subject but is meant to provide information on some trends in the investigation of fractured rock systems as well as to locate the "Fractured-Rock Aquifer-Analogue Approach"within the context of other approaches for parameterization of such systems.

1.2.1 Fracture Characterization and Flow Processes in Individual Fractures

The hydraulic properties of individual fractures as well as the transport parameters are determined by the fracture aperture, their variability in space (roughness), depending on the stress field. Many investigations focussed on the so-called "channelling effect". In natural joints only a small proportion of the total fracture face is actually open in contrast to the frequently adopted parallel plate assumption (Snow, 1965), i.e. a larger part of the fracture area is in contact with the opposite complementary fracture face. The field of rock mechanics introduced a number of techniques for describing fracture frequencies, fracture geometry, and fracture characterization (e.g. Priest (1993); Kendorski and Bindokas (1987)). A systematic compilation summarizing the current understanding of the formation of fractures is presented by Pollard and Aydin (1988). A direct determination of the spatial variability of fracture apertures can be achieved by casting techniques using epoxy resin or Wood's metal (Pyrak-Nolte *et al.* (1987); Gale (1987); Gentier and Billaux (1989); Cox *et al.* (1990)). The roughness of fracture faces is measured either mechanically or optically (Voos and Shotwell (1990); Miller *et al.* (1990); Brown and Scholz (1986)).The heterogeneous flow field in natural fractures was investigated in laboratory experiments (Moreno *et al.* (1985); Neretnieks *et al.* (1987); Bourke (1987); Abelin *et al.* (1987)). Tsang (1992) pointed out that the variability in experimentally determined equivalent apertures was often a result of different definitions by the authors. The theory of flow in individual fractures on a basis of fluid mechanics was recently reviewed by Zimmermann and Yeo (2000). The flow through individual fractures under variable stress was investigated by, for example, Raven and Gale (1985), Pyrak-Nolte *et al.* (1987), Sundaram *et al.* (1987), and Ryan *et al.* (1987).

Schrauf and Evans (1986) conducted laboratory experiments to investigate flow processes and parameters of individual fractures with pneumatic techniques. The authors were able to show that, especially at high flow velocities and with low-viscosity gases, non-linear flow laws apply. Kilbury *et al.* (1986) demonstrated in similarly designed experiments with adapted double-ring infiltrometers for fracture networks, that water-saturated parameters could be predicted with gas-flow methods. Using numerical experiments, Thunvik and Braester (1990) simulated flow through partially saturated fracture networks. According to Baehr and Hult (1991), permeabilities in the vadose zone can be predicted from air-injection measurements assuming homogeneous fracture distributions and good connectivity. Further useful information is available from the investigations in the fractured tuff at Apache Leap in Arizona (e.g. Guzman *et al.* (1996); Illman *et al.* (1998); Chen *et al.* (2000)). Systematic experiments were conducted in order to enable parameters found in the laboratory to be transferred to the field. Further knowledge regarding gas flow and transport in the unsaturated zone stems from the field of soil physics (e.g. Flury (1996); Jury and Wang (2000) and the investigations into air venting remediation at contaminated sites. Extensive reviews can be found in Croisé (1996), Wang and Dusseault (1991), and Brusseau (1991).

Fracture/matrix exchange processes, especially matrix diffusion, can play a major role in transport through a fractured porous system. Matrix diffusion is mainly determined by porosity. Skagius and Neretnieks (1986) and Birgersson and Neretnieks (1990) conducted experiments in the laboratory using different types of rock samples and demonstrated the long-term effect of matrix diffusion on radionuclide transport. Grader *et al.* (2000) investigated the multiphase flow of water and mineral oil in an induced fracture in a laboratory experiment using X-ray tomographic imaging techniques and demonstrated the importance of the fracture matrix interaction for flow through such systems.

1.2.2 Flow Processes in Fracture Networks

To investigate flow in fracture networks on the field scale, a number of different hydraulic and geophysical tests are available. Packer, drillstem and slug tests are employed to obtain hydraulic and geometric parameters at well-defined fracture intervals. Evaluation procedures for pumping and injection tests are generally based on a conceptual model assuming flow from a homogeneous matrix towards a single (equivalent) fracture, directly connected with the borehole. A review of the individual interpretation methods adopted from the petroleum-engineering field for individual conceptual models and well/fracture configurations is presented in Gringarten (1982). The effects of well storage and well skin as well as their influence on test interpretation are being discussed. More comprehensive summaries are presented by Matthews and Russel (1967), Earlougher (1977) and Streltsova (1988). The application of the Laplace inversion technique in the field of groundwater (Moench and Ogata, 1981) made it possible to consider more complex problems with elaborate conceptual models in the field. There are numerous further analytical solutions for more complex systems, partly based on double-porosity approaches (e.g. Naceur and Economides (1988); Karasaki *et al.* (1988); Barker (1988)). Barker (1988) first presented the concept of a fractional flow dimension for the analysis of hydraulic tests in fractured systems. The flow geometry towards the well can be derived from the fractional dimension. Heterogeneity, however, can be the reason for ambiguity in the interpretation. Chang and Yortsos (1990) introduced solutions for fracture networks which can be described by a fractal dimension. Hsieh (2000) presents a review of hydraulic testing and evaluation procedures in fractured and tight formations as well as illustrations of problems involved in rationalizing of local-scale measurements. The hydraulic properties of fracture networks can also be investigated with borehole-geophysical techniques, such as flow meter, temperature and fluid logging (e.g. Morin *et al.* (1988), Silliman (1989); Paillet *et al.* (1987); Tsang *et al.* (1990)).

Transport in fracture networks can be investigated with tracer experiments which require substantial effort for a differentiated measuring of tracer breakthrough. In Grimsel and STRIPA (Abelin *et al.*, 1987), sorbing and non-sorbing tracers were collected throughout the entire ceiling of former mine workings to obtain individual tracer-breakthrough curves. The interpretation demonstrated that only a few fractures dominate tracer transport and that matrix diffusion (Skagius and Neretnieks, 1986) and channelling (Tsang and Tsang (1989); Tsang *et al.* (1991)) play an important role in the distribution of the concentration measured. The development of specialized packer equipment by Novakowski and Lapcevic (1994) contributed largely to the advancement of the field testing of tracer experiments. The test and interpretation techniques whisch are currently generally available as well as the problems encountered are summarized by NRC (1996). Some experience in the investigation of single-fracture flow processes was also obtained from field experiments (e.g. Novakowski and Lapcevic (1994); Himmelsbach (1993); Kelley *et al.* (1987)).

Only recently, the investigation of crystalline basement rocks received more attention as to their potential as aquifers. In particular in the African

Shield and the Scandinavian countries, such resources can be very important locally (e.g. Krasny (2002); Olofsson (2002); Gudmundsson *et al.* (2002)). Several recent conferences were devoted to this topic, summarized in the proceedings of Rohr-Torp and Roberts (2002) and Krasny and Mls (1996).

1.2.3 Unsaturated Flow Processes in Fractured Systems

The research into flow through unsaturated fractured porous rocks, relevant for the aquifer-analogue experimental concept, received major attention within the context of determining recharge rates through the thick unsaturated zone of the tuff at the Yucca Mountains experimental site. Further contributions to the field result from investigations by the University of Arizona at the Apache Leap Research Site (Rasmussen *et al.*, 1993) where elaborate pneumatic experiments in fractured tuff were conducted and inversion techniques developed (Neuman and Federico, 2003). There are a number of excellent summaries highlighting major important processes as well as techniques for their identification and parameter quantification (Faybishenko *et al.* (2000); NRC (2001); Evans *et al.* (2001); Evans and Nicholson (1987); Bodvarsson *et al.* (2003)).

New developments in the understanding of flow and transport processes in the fractured vadose zone are emerging from the research at Yucca Mountains and from soil physics. The previous conceptual models of flow in the vadose fractured porous system predicted that fractures convey water only in a saturated state. These models rely on the capillary model (e.g. Wang and Narasimhan (1985); Peters and Klavetter (1988)) which implies that water preferentially remains in small pores and that fractures do not conduct water unless they reach saturation. Recently, isotope studies in the Yucca Mountains suggest that bomb-pulse radionuclides migrated over several hundred meters during the last 50 years. Pruess (1999) suggests limiting fracture-matrix interaction in the respective models because of fracture coating and Tokunaga and Wan (1997) propose the film flow model, which implies a flow mechanism that is supported by a continuous water supply and a water-solid contact angle of zero. Dragila and S. W. Wheatcraft (2001) provide an extensive review of unsaturated flow models and a mathematical model for the quantification of the flow due to a free-surface film. Another model by Doe (2001) proposed the drop-flow conceptual model, which takes into account variable contact angles between water and solid surfaces; according to the author, this is an important effect which is not considered in the film-flow models.

1.3 Aquifer Genesis Approaches for the Assessment of Hydraulic Characteristics of Geological Materials

M. Sauter

Determining aquifer heterogeneity constitutes a major challenge for hydrogeologists. Although the analysis of flow problems is more forgiving regarding the resolution of zones of variable hydraulic conductivity because of the averaging characteristics of diffusion-type processes, the investigation of solute transport demands a small-scale quantification of the variation in the hydraulic properties (Frind *et al.* (1988); Essaid *et al.* (1993)). In general, only point information on aquifer properties is available, e.g. from boreholes and core samples; this however, only allows flow and, in particular, transport prediction on a large scale for more or less homogeneous conditions. The uncertainty in prediction of contaminant migration as a result of limited information on aquifer characteristics is demonstrated by, for example, Whittaker and Teutsch (1996). Petroleum reservoir geologists and engineers face similar kinds of problems. The presence and the continuity of highly permeable features determines reservoir properties and their integration in reservoir performance models is therefore of utmost importance. Similar problems are incurred in the field of rock mechanics and slope-stability assessment as well as in the prediction of radionuclides from nuclear-waste repositories (Bodvarsson *et al.* (2003);Billaux *et al.* (1989)). Here, a few individual fractures, difficult to detect, might represent large failure risks.

In the following, the extent to which the understanding of sedimentary processes as well as the generation of fractures might contribute to the characterization problem in fractured aquifers is demonstrated briefly. This is reflected in some sessions at recent conferences of the Geological Society of America (GSA) held in 2001 and 2003 and devoted exclusively to this topic.

1.3.1 Aquifer Sedimentology

The investigation of aquifer sedimentology aims at the development of methods for quantifying the subsurface characteristics of aquifers and, in particular, the heterogeneities of hydraulic and transport parameters in a continuous manner on the basis of sedimentological information and principles.

Huggenberger and Aigner (1999) demonstrate the general approach employed in aquifer sedimentology. The geometry of the aquifer units is obtained from concepts employed in sequence stratigraphy and facies analysis, which identify sedimentary architectural elements and depositional units. These units are grouped on the basis of an understanding of the genesis (e.g. Miall (1985)) and the geological history of the depositional environment. From outcrop-analogue studies, further information can be obtained on spatial continuity and qualitative parameter distribution. Actual hydrogeological parameters are then attributed to the individual units, which can be specified by a single parameter value. Different approaches are taken for obtaining permeabilities, e.g. minipermeameter measurements in consolidated deposits (Hornung and Aigner, 1999) or gas-flow measurements in unconsolidated fluvial materials (Klingbeil *et al.*, 1999). Ground penetrating radar investigations have been shown to provide a spatial resolution of the geometry of individual units (Huggenberger *et al.*, 1997) as well as a semi-quantitative estimate of the geological characteristics of the material. Kleineidam *et al.* (1999) demonstrated how aquifer material properties as well as their spatial variability, which determine transport in the subsurface, can be obtained from sedimentary composition.

1.3.2 Genesis of Fracture Networks

In the last ten years, a development can be observed from a purely descriptive analysis of fracture patterns towards an understanding of fracture genesis caused by variable stress (e.g. Rawnsley *et al.* (1992), Rives *et al.* (1992); Narr and Suppe (1991); Gudmundsson *et al.* (2002); Brenner and Gudmundsson (2002)). A systematic summary of the processes relevant to the formation of fractures and fracture networks can be found in the review of Pollard and Aydin (1988). Gudmundsson *et al.* (2003) discuss the importance of understanding fracture genesis for permeability generation and groundwater flow in crystalline rocks in general. They also present a principle numerical model illustrating important factors for fracturing. Rutqvist and Stephansson (2003) present a review of the state of the art in hydromechanical coupling and highlight its importance for the fields of rock engineering, groundwater and geothermics. Gentier *et al.* (2000) show how a variation in the stress field changes the hydromechanical behavior of fractures and thus also the flow and transport.

Modelling of fracture formation, however, does not yet extend beyond the formation of individual fractures. It can, however, be expected that future progress in the field will contribute largely to the quantification of the parameter field in fractured systems.

1.4 The Aquifer Analogue Approach for Fractured Porous Aquifers

M. Sauter

The need to provide alternative approaches for characterizing fractured porous aquifer systems different from those applied in low-permeability crystalline or shaly rocks is outlined above. In sum, methodologies have to be developed which are capable of (a) distinguishing between the relative importance of the contributions of fractures and matrix to flow and transport processes, and (b) reconciliating hydraulic integral measurements from, for example, pumping tests with the hydraulic effect of discretely visible permeable features (fractures).

Therefore, the research initiative "Aquifer-Analogue Approach for Fractured Porous Aquifers", funded by the German Science Foundation (DFG), was launched in 1996. Groups from the universities of Aachen, Karlsruhe, Stuttgart and Tübingen collaborated in this research initiative.

1.4.1 Objectives

The objectives of the research initiative "Aquifer-Analogue Approach for Fractured Porous Aquifers"were to:

- provide integral large scale hydraulic information for model prediction, while still considering important small-scale features in characterization,
- provide criteria and techniques for differentiating between visibly discrete features and those that are actually hydraulically important,
- provide recommendations for minimal model requirements by combining both detailed discrete process models with integrating continuum models,
- contribute to the discussion on the extent to which upscaling techniques (e.g. Neuman and Federico (2003)) can assist in developing large models from small scale measurements.

1.4.2 Concept

The research concept of the "Fractured-Rock Aquifer-Analogue Approach" is based on the so called outcrop-analogue method (Fig. 1.3), in which well defined samples from sandstone outcrops (quarries) are used as a realistic representation of those sections of the (aquifer) system where access is limited to (a few) boreholes.

Outcrop analogue studies have mainly been used in the petroleum industry for reservoir characterization (Flint and Bryant, 1993). Based on a detailed sedimentary analysis of outcrops, assumed to represent the characteristics of the reservoir rocks, deductions are made about their hydraulic properties. The advantages of analogue studies are the accessibility of the material in a larger (than a borehole, for example) context and the opportunity to obtain detailed small-scale measurements in two and sometimes three dimensions.

For fractured systems the aquifer analogue approach has rarely been employed and then only in two dimensions (Kiraly (1969); Billaux *et al.* (1989)). In the mine adit of Fanay-Augères, Billaux *et al.* (1989) attempted to generate fracture networks statistically based on a 2D survey of fracture traces. However, this approach suffers from the problem of ambiguity, i.e. there is an

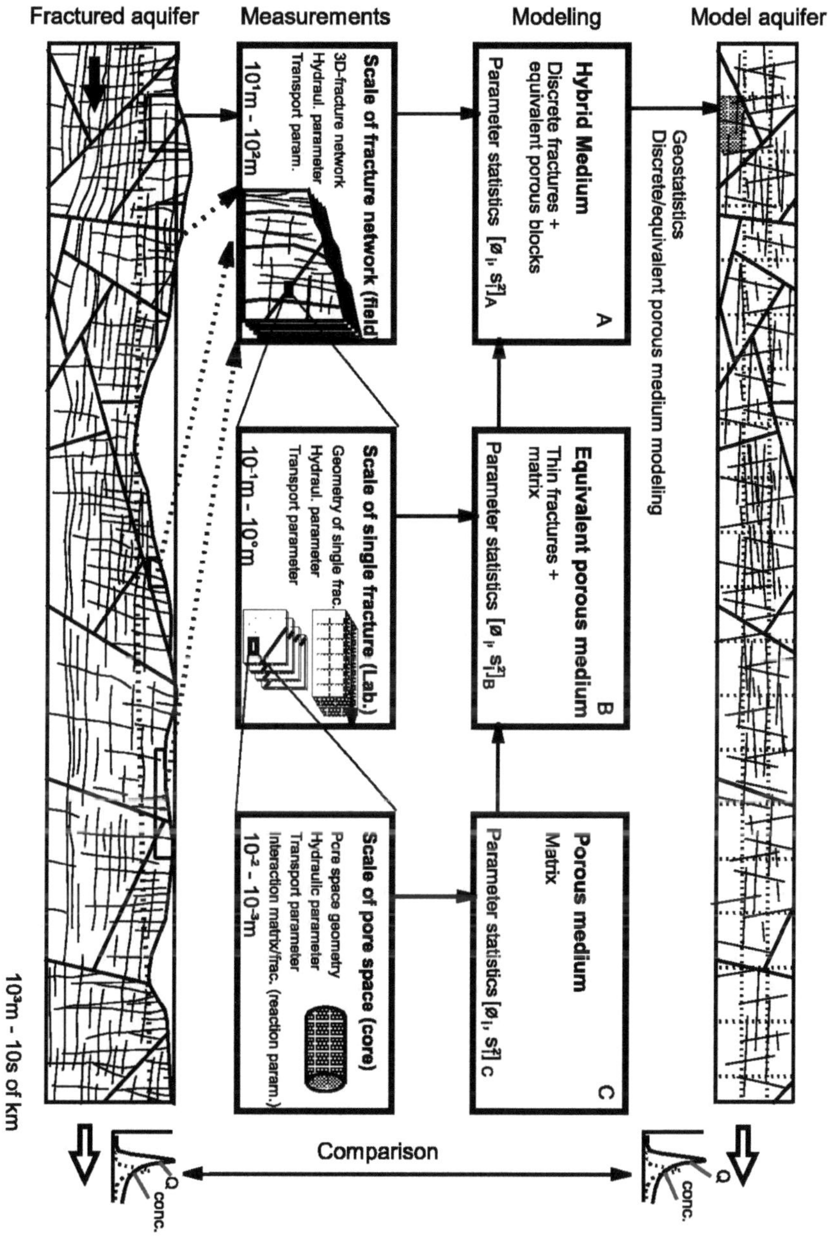

Fig. 1.3. Fractured rock aquifer analogue approach.

infinite number of possibilities that can satisfy the 2D observations. Applying various statistical and pattern-recognition techniques, La Pointe (2000) showed some relationship between increased flow rates measured by well tests and geological data (structure, lithology) and derived a regional hydrogeological model on the basis of 8 boreholes. Similar ambiguities apply here as well.

In the frame of the research initiative "Aquifer-Analogue Approach for Fractured Porous Aquifers", the flow and transport properties of fractured porous sandstone materials were determined on different scales and appropriate measuring techniques were developed. Samples with dimensions from approx. 0.05 m (cores) to 10 m (field block) of lengths were used to investigate the main hydraulic features of the system, i.e. the influence of the fractured system as compared to the properties of the permeable porous matrix on the flow and transport behavior. Various direct (e.g. hydraulic and pneumatic) and indirect (e.g. geophysical) investigation methods were designed in order to parameterize the system adequately. For example, resin casting techniques were employed to discretely determine fracture dimensions on the core scale, gas flow and gas transport tomographical experimental set-ups were used to evaluate the three-dimensional properties of large scale (1 m^3) laboratory samples. Diffusivity (K/S) investigations proved to be extremely useful to emphasize the contrast in hydraulic conductivity between fractures and permeable matrix and to categorize the samples. Inverse techniques allowed the determination of the equivalent continuum three-dimensional anisotropy tensor of the hydraulic conductivity from the experimental data, i.e. the estimation of equivalent parameters. The experimental methods were also employed on the field block, allowing estimates on the effect of the scale of investigations on the estimates of flow and transport properties. On the basis of these parameters, various physically based distributed and lumped parameter models were developed to analyze the experimental data and also to predict the outcome of future experiments (experimental design). These model approaches included a stream-tube model that proved to be very efficient and useful for the interpretation of the experimental results, but also multi-phase discrete fracture flow and transport models capable of simulating the actual physical processes. Double continuum model investigations assisted in identifying the simplest equivalent model that could represent the experimental data and phenomenological observations, such as fracture geometries.

For some scenarios, we were able to conduct independent model validation exercises and used those to rank the modelling concepts employed.

2

From Natural System to Numerical Model

Natural geological systems are generally highly complex, both as far as the geological structure and the physical processes occurring within them are concerned. In order to investigate such systems, it is necessary to understand the natural processes, their interaction with the geological structure and their relative importance on different problem scales.

Due to the geometrical complexity of the natural system and the large number of physical processes involved, it is not feisable to describe the exact system in great detail. The system structure and the processes occurring within it are therefore represented by conceptual models, designed to meet the requirements of certain types of problems on a given scale. This implies, for example, the introduction of parameters representing the material properties and of physical descriptions of the relevant flow and transport processes.

In order to solve a problem, a mathematical description of the model concept is required. Due to the degree of complexity of this type of problem, analytical solutions are not an option. The numerical solution involves the spatial and the temporal discretization of the problem and the use of efficient and stable numerical algorithms.

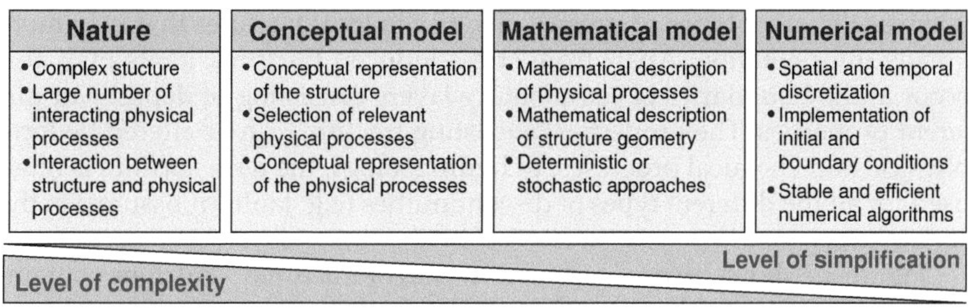

Fig. 2.1. Transformation from a complex natural system to a simplified numerical model (Süß, 2004).

The transformation of a natural system to a model inevitably leads to simplifications of the real system as visualized in Fig. 2.1. When model results are being interpreted, it is therefore essential to keep in mind that a model is merely an approximation of nature. One must always be aware of the assumptions and the concepts used for a specific model in order to assess its results correctly.

This chapter describes the transformation from nature to model of saturated fractured porous systems as well as the flow and the transport processes that are relevant for the investigations discussed in this book. The first part (Sect. 2.1) deals with the characteristics of natural fractured porous systems. This is followed by a discussion of different model concepts used on different scales and for different problem types (Sect. 2.2). Finally, the governing equations of the pertinent physical processes and the mathematical implementation of the discussed model concepts are presented (Sects. 2.3 – 2.5). It should be underlined that this chapter does not provide a complete overview of physical processes, theories, concepts and models. It is restricted to the scope of the research work presented in this book. For a more detailed discussion of the topic, we refer to the literature references given in the various sections.

2.1 Natural Fractured Porous Systems

A. Silberhorn-Hemminger, M. Süß, R. Helmig

Solid rock can be classified according to its diagenetic characteristics (Kolditz, 1997). Consolidated sedimentary rock evolves from the cementation of mineral grains, metamorphic rock is the result of recrystallization under high temperature and stress whereas igneous rock forms by the direct crystallization of minerals from magmaic melt.

In direct response to the stress applied, which may be lithostatic, tectonic, thermal or the result of high fluid pressures, joints, faults and systems of such discontinuities occur on different scales and with different geometries. Tectonic fractures tend to be oriented along stress fields on a regional scale, whereas the other types of stresses give rise to local fractures that vary more greatly in orientation. Apart from stress-induced fractures, joints may also occur at the boundaries of sedimentary layers consisting of deposits of different properties. The properties of existing fractures can be altered by local physical and chemical processes. In hydrogeology, the term *fractures* is often used for all the different types of discontinuities (e.g. faults or fissures) in the rock matrix.

Figure 2.2 shows an exposed vertical wall of fractured sandstone containing both vertical and horizontal fractures. In this case, the horizontal fractures are separation planes between different sedimentary layers, whereas the vertical fractures are the result of mechanical stress.

Fig. 2.2. Vertical exposed wall of fractured sandstone at the field site (Sect. 5).

The hydraulic properties of hard rock are to a large extent determined by the porosity of the rock. Table 2.1 presents typical ranges of porosity for different types of hard rock, considering the rock as a whole, including both fractures and matrix (see below). Depending on the type of rock, the contribution of these two components to the porosity varies. For example for granite, the porosity is almost exclusively determined by the fractures, whereas for sandstone, the matrix porosity is considerable. Table 2.1 shows that there is a large difference between the total n and the effective porosity n_e. The total porosity includes all the pores of the system whereas, for the effective porosity, only connected pores that are available to fluid flow are considered. It is important to point out that the effective porosity is not directly correlated with the hydraulic conductivity of the system. The hydraulic conductivity varies over a wide range for different rock types as well as for one single rock type.

This book is concerned with *fractured porous rock*, in which the rock matrix is considered to be permeable to flow. Fractured porous rock is generally divided into three different components:

- A *fracture network* is a system of partially intersecting single fractures. Its hydraulic properties are typically characterized by the distribution of fracture size, fracture permeability, fracture orientation, fracture distance, and fracture density. Due to its small volume relative to the volume of the total domain, the storage capacity of the fracture system is small.
- Within the fractures, *filling material* consisting of mineral deposits can be found. Open fractures can channel and speed up the transport of pollutants from disposal sites, e.g. leading to a locally high concentration of a

Table 2.1. Total porosity, effective porosity and hydraulic conductivity of selected hard rocks (Domenico and Schwartz (1990); Matthess and Ubell (1983)).

Rock	Total porosity n (%)	Effective porosity n_e (%)	Hydraulic conductivity K (m s^{-1})
Granite	0.1	0.0005	$0.5 \cdot 10^{-12} - 2.0 \cdot 10^{-12}$
Limestone	5 – 15	0.10 – 5	$1.0 \cdot 10^{-09} - 6.0 \cdot 10^{-06}$
Chalk	5 – 44	0.05 – 2	$6.0 \cdot 10^{-09} - 1.4 \cdot 10^{-07}$
Sandstone	5 – 20	0.5 – 10	$3.0 \cdot 10^{-10} - 6.0 \cdot 10^{-06}$
Shale	1 – 10	0.5 – 5	$1.0 \cdot 10^{-13} - 2.0 \cdot 10^{-09}$

pollutant at a great distance from the source, whereas filled fractures may inhibit the flow in otherwise highly permeable aquifers (Odling, 1995).

- The *matrix* blocks between the fractures have a spatially varying texture and porosity. The permeability contrast between fractures and matrix is decisive for the importance of the matrix for flow and transport processes. As opposed to the fracture system, the storage capacity of the matrix is often significant as a result of its large volume relative to the total domain.

The research results discussed in this book are obtained from flow and transport experiments in fractured porous systems on different scales. This means that the matrix is permeable to flow and plays a significant role for both flow and transport within the experimental domains. Open as well as filled fractures occur, in most cases acting as preferential flow paths. In some cases, however, the filling material inhibits the flow.

In order to characterize a fractured system, criteria and properties must be defined that can be qualitatively or quantitatively determined directly in the field or in laboratory investigations. Since the evaluation of fracture geometry and the generation of fracture systems for modeling are discussed in this book, some frequently used properties are now briefly presented.

Fracture Size

The fracture size, i.e. the lateral limitations of a fracture, can in most cases not be recorded and determined directly. Usually, fracture traces are detected at exposed walls (e.g. outcrops, quarries, tunnels). The fracture trace is the intersection line of a fracture with the exposed wall. In the field, the actual fracture sides are therefore often not determined, but the fracture traces are recorded and evaluated further using statistical methods. These difficulties are demonstrated in Fig. 2.3. The vertical two-dimensional section, showing the fracture intersection lines with the x-y-plane at $y = 1$ m, is only a limited representation of the actual three-dimensional system.

The approximation of the empirical fracture trace length distribution can be accomplished using different theoretical distribution functions, e.g. power

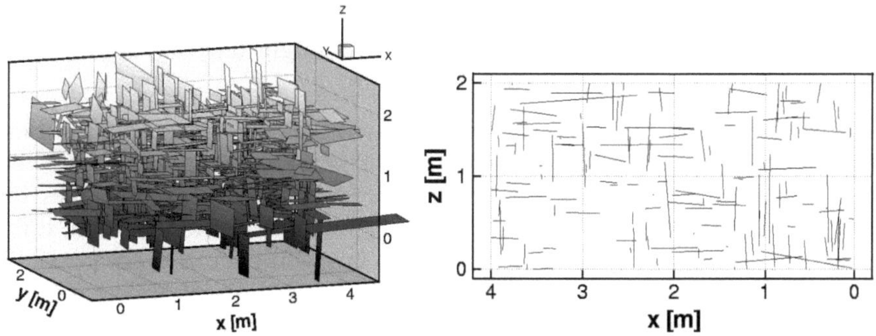

Fig. 2.3. Artificial three-dimensional fracture system (left) and a vertical two-dimensional section showing the fracture traces at $y = 1$ m (right).

law, log-normal, hyperbolic or gamma-1 distributions. References for these distributions are given by Dershowitz and Einstein (1988). They explain the diversity of possible distributions by the fact that fracture traces and not the actual fracture sides are detected, and by the processes that cause the formation of fractures. An interesting aspect is whether it is possible to reconstruct the actual three-dimensional fracture system based on the determined fracture-trace distribution. Considering fractures with circular shape, Baecher *et al.* (1977) show that a power law and a log-normal distribution of the fracture radii both lead to a log-normal distribution of the fracture traces. Hence, a unique reconstruction of the fracture system is not possible.

Fracture Distance

The fracture distance is defined as the distance measured between two directly neighboring fractures along a straight line. The fracture distance, or the distribution of the fractures distance controls the geometrical arrangement of the fractures. According to Meier and Kronberg (1989), the principle of the fracture distance is based on the idea that the formation of a fracture causes a tension decrease in the vicinity of the fracture. The next fracture can only be formed if the critical regional tension occurs again, causing the genesis of another fracture. In sedimentary rock, the fracture distance depends, for example, on the elastic properties of the individual layers of a sedimentary sequence, the thickness of the layers, the permeabilities and the deformation intensity.

According to Priest (1993), three different definitions of fracture spacing can be distinguished:

- *Total spacing*: Distance between two directly neighboring fractures with different orientation measured along a straight line.
- *Set spacing*: Distance between two directly neighboring fractures with equal orientation measured along a straight line.

- *Normal set spacing:* Distance between two directly neighboring fractures with equal orientation measured along a straight line that is parallel to the mean normal direction of the fractures.

Extensive field investigations have shown that a POISSON distribution, describing the arrangement of the fractures in space, corresponds to a power-law distribution of the fracture distances. Sachs (1997) circumscribes the POISSON distribution as follows: "The POISSON distribution is valid, if the average number of events is the result of a large number of event possibilities and a very small event probability. The POISSON distribution is used to solve problems that occur when counting relatively rare random and independent events in time, length, area or space domains".

The determination of the fracture distance is schematically represented in Fig. 2.4.

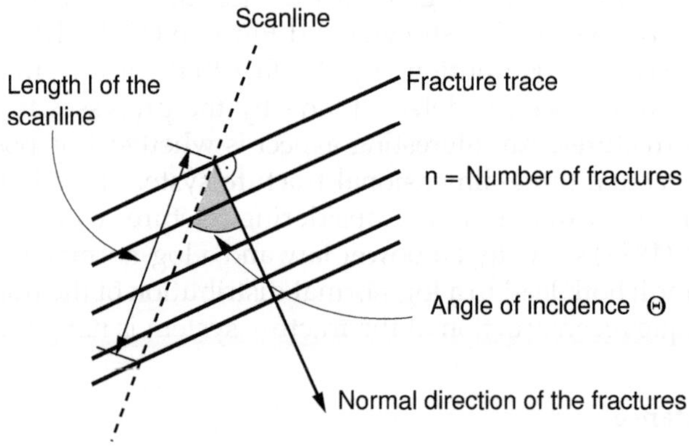

Fig. 2.4. Determination of fracture distances.

The mean fracture distance can be determined from the length l of the scanline, the number of fractures n and the angle of incidence θ between the scanline and the normal direction of the fractures:

$$\text{mean distance} = \frac{l}{n} \cdot \cos\theta \,. \tag{2.1}$$

Fracture Density

The fracture density is determined on the basis of core samples and scanline measurements. Scanlines are observation lines, positioned on an exposed wall. Along the scanline, the number of intersections with fractures, the angle of incidence etc. can be determined.

The one- and two-dimensional fracture density d_1 and d_2 (see below) are, except for isotropic systems, dependent on the orientation of the scanline and the exposed wall. The volumetric fracture density d_3, i.e. the mean fracture area per unit volume, is, according to Chilès and de Marsily (1993), determined by:

$$d_3 = \frac{1}{l} \sum_{i=1}^{n} \frac{1}{\sin \theta_i}, \tag{2.2}$$

where θ_i is the angle of incidence of the fracture, l is the length of the scanline and n is the number of fractures per scanline.

For randomly distributed fracture orientations, the following two relationships between the volumetric fracture density d_3 and the linear fracture density d_1 and the area-related fracture density d_2, respectively, are obtained. The linear fracture density d_1 describes the average number of fractures per unit length along a line:

$$d_1 = \frac{1}{2} d_3 . \tag{2.3}$$

The area-related fracture density d_2 defines the average fracture length per unit area:

$$d_2 = \frac{\pi}{4} d_3 . \tag{2.4}$$

Fracture Aperture

The fracture aperture is defined as the perpendicular distance between two directly neighboring fracture walls. The aperture can be increased by, for example, dissolution and erosion processes. This is mainly observed in the weathered zones close to the ground surface. Another reason for an increase in aperture is displacement due to external forces or subsidence. The aperture generally decreases with increasing depth due to the increasing thickness of the overlying rock.

Figure 2.5 shows two sides of a fracture. The roughness of the two surfaces can be clearly seen. If the two pieces are put together, the actual fracture is obtained as the space between the two surfaces. The aperture varies significantly throughout a fracture. It is therefore not possible to characterize the fracture aperture by a single measurement. In the case of single fractures, one possibility is to assume a heterogeneous fracture-aperture distribution; however, for the consideration of multiple fractures or fracture systems, this approach is made impossible by the degree of detail required.

In general, for the discrete representation of fractures in models, the fracture aperture is described using the parallel-plate concept. This concept is explained in detail in Sect. 2.4.1.

The fracture aperture has an essential influence on the flow and transport processes in a fractured system. However, the determination of the aperture is not trivial (Chilès and de Marsily, 1993):

- In nature, there is no constant aperture throughout a fracture plane. There are closed and open regions.

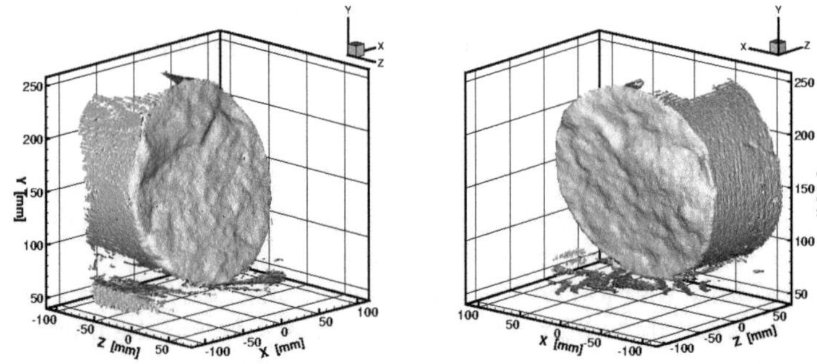

Fig. 2.5. Image of the two opposite walls of a single fracture from a sandstone core sample with a diameter of approximately 20 cm (created in cooperation with the Institute for Robotics and Process Control, Technical University of Braunschweig, Germany).

- The flow between two parallel plates (parallel-plate concept), separated by a constant aperture, has little in common with the actual flow through a natural fracture.
- Due to the pressure release that a sample experiences during sample extraction, the measured aperture deviates from the aperture that would be measured *in situ*.

Fracture Orientation

The orientation of geological formations in space is uniquely determined by the strike angle S_t or the azimuth A_z, and the dip D_i. According to Murawski (1998), the strike is the section boundary of a natural surface (e.g. layer or fracture) at an imaginary horizontal plane. The strike angle S_t is defined as the angle between the northerly direction and the section boundary. The projection of the line of greatest slope onto the horizontal plane is the dip direction and is always perpendicular to the strike. The inclination angle between the line of greatest slope and the dip direction is the dip D_i. The angle between the northerly direction and the dip direction is defined as the azimuth A_z. It is connected to the strike angle through the equation $A_z = S_t + 90°$. Figure 2.6 summarizes the relationship between strike, dip and azimuth.

If fractures or layers occur in a preferred direction, the statistical distribution of the orientation is often described by the FISHER distribution, also called the spherical normal distribution. The FISHER distribution is characterized by the fact that orientations are distributed around a certain main orientation with rotational symmetry. In Wallbrecher (1986) and Fisher *et al.* (1993), the distribution is discussed in detail. The probability density function of the FISHER distribution has the following form:

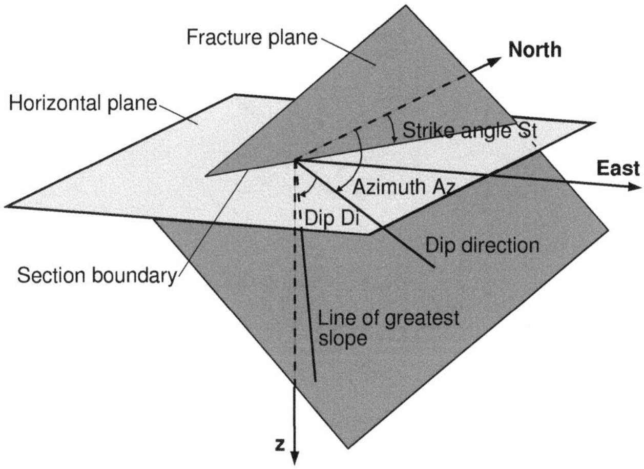

Fig. 2.6. Determination of the location of a geological surface: strike, dip, azimuth. The dip direction is the projection of the line of greatest slope onto the horizontal plane.

$$f(\Theta,\phi) = \frac{\kappa}{4\pi \cdot \sinh \kappa} \cdot exp[\kappa\,(\sin\Theta\ \sin\alpha\ \cos(\phi-\beta) + \cos\Theta\ \cos\alpha)] \cdot \sin\Theta\,. \tag{2.5}$$

Here, α is the Θ-pole coordinate (latitude) of the main direction, β is the ϕ-pole coordinate (longitude) of the main direction and κ is the concentration parameter. The concentration parameter κ is a measure of the distribution of the orientations around the main orientation. For $\kappa = 0$, the orientations are uniformly distributed. The larger κ is, the stronger the concentration around the main orientation. The *cone of confidence* is used to quantify the significance of the distribution. The cone of confidence yields a small circle around the main orientation R_i. The calculation of the cone of confidence is only permissible for $\kappa \geq 4$. For $\kappa < 4$, there is no spherical normal distribution or the sample size is too small. The measures for the cone of confidence are the spherical variance

$$S^* = \frac{n-|R_i|}{n} \tag{2.6}$$

and the spherical aperture

$$\omega = \arcsin\sqrt{2\frac{1-1/n}{\kappa}}\,. \tag{2.7}$$

Here, R_i is the main orientation and n is the sample size. Figure 2.7 shows the relationship between the main direction R_i and the spherical aperture ω. The spherical aperture for the FISHER distribution corresponds to the standard deviation of the GAUSSIAN normal distribution.

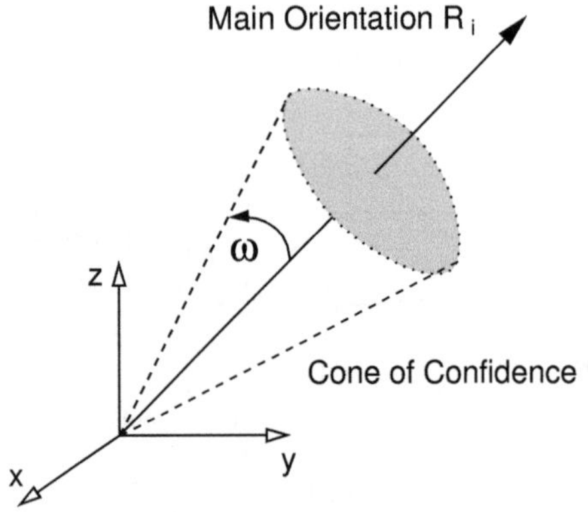

Fig. 2.7. Sketch of the cone of confidence with the spherical aperture ω.

Orientation data are represented graphically in pole diagrams. The intersection points of the normal vectors of the planes with one half of a globe are projected onto a circular area. Examples of such diagrams are shown and discussed in Sect. 5.3.

2.2 Model Concepts in Fractured Porous Systems

M. Süß, R. Helmig

As discussed in the introduction to this chapter, it is not possible to set up a model that is an exact representation of reality, but conceptual models are developed that are able to describe the relevant structures and physical processes of a problem. The choice of a model concept for the description of fractured media strongly depends on the scale of the problem, the geological characteristics of the area of investigation, and the purpose of the simulation. Bear (1993) classifies various problems of flow and transport in fractured porous media according to their scale. On these scales, different types and extensions of heterogeneities occur (Rats and Chernyashov, 1967).

- **Zone 1: The very near field.** Interest is focused on flow and transport processes within small-scale fractures (fissures) and the pore space. Single, well-defined fractures and the surrounding porous rock, which is possibly accessible to transport, are considered.
- **Zone 2: The near field.** The flow and transport processes are considered in a relatively small domain, which contains a small number of well-defined small and intermediate fractures. The location and shape of the individual fractures are either deterministically defined or can be generated stochastically, based on statistical information from the real system.
- **Zone 3: The far field.** On this scale, the flow and transport processes are regarded as taking place, simultaneously, in at least two continua. One

continuum is composed by the network of large scale fractures and the other one by the porous rock. Mass of the fluid phase and its components may be exchanged between the two continua.

- **Zone 4: The very far field.** The entire fractured medium is considered as one single continuum, possibly heterogeneous and anisotropic in order to account for large scale geological layers and fault zones.

In order to set up models of systems with such varying characteristics, different model concepts are necessary. These concepts are discussed on the basis of Fig. 2.8. Two principal approaches are possible (Helmig, 1993):

1. Assuming that the concept of the *representative elementary volume* (REV) (see Sect. 2.3.1) is valid and that the scale of the investigation area is sufficiently large, it is possible to describe the model area as a heterogeneous, anisotropic *continuum*. According to Bear (1993), this is possible on the very far scale. Kröhn (1991) also considers this to be a feasible approach for describing poorly fractured rock (type I) and rock with a very high fracture density (type II) on a smaller scale.
2. If the flow and transport processes in the fractured media are dominated by shear zones (spatially concentrated small-scale fractures) or fracture systems, it is feasible to describe these features specifically, neglecting the rock matrix, using a *discrete* fracture network model (consideration of each single fracture) (type III).

However, the rock matrix, filling the space between the fractures, is often not negligible, but plays an essential role in flow and transport processes. If the porous rock matrix can be idealized as a continuum with averaged material properties, a model can be set up where a continuum model, accounting for the matrix, is coupled with a discrete model considering the fractures (Fig. 2.8, type IV). This type of model is further discussed in Sect. 2.4.

Another widely used possibility for describing areas of type II (system with high fracture density) and IV (dominant single fractures + rock matrix) (Fig. 2.8) is to transform the matrix and the fracture systems on different scales into separate homogeneous equivalent continua. This approach is mainly applied on large scales. With the concept originally presented by Barenblatt *et al.* (1960), the flow and the transport processes between the continua can be represented by coupling them via exchange terms and in this way setting up a so-called double-continuum model. It is essential to the principle of homogenization for heterogeneous media to define equivalent model parameters and to find appropriate expressions for the interaction of the hydraulic components which are capable of describing the correct physical system behavior. Extending this model allows the consideration of more than two continua, a multi-continuum model, if required by the geological characteristics of the investigation area and by the nature of the problem. This type of model concept is further described in Sect. 2.5.

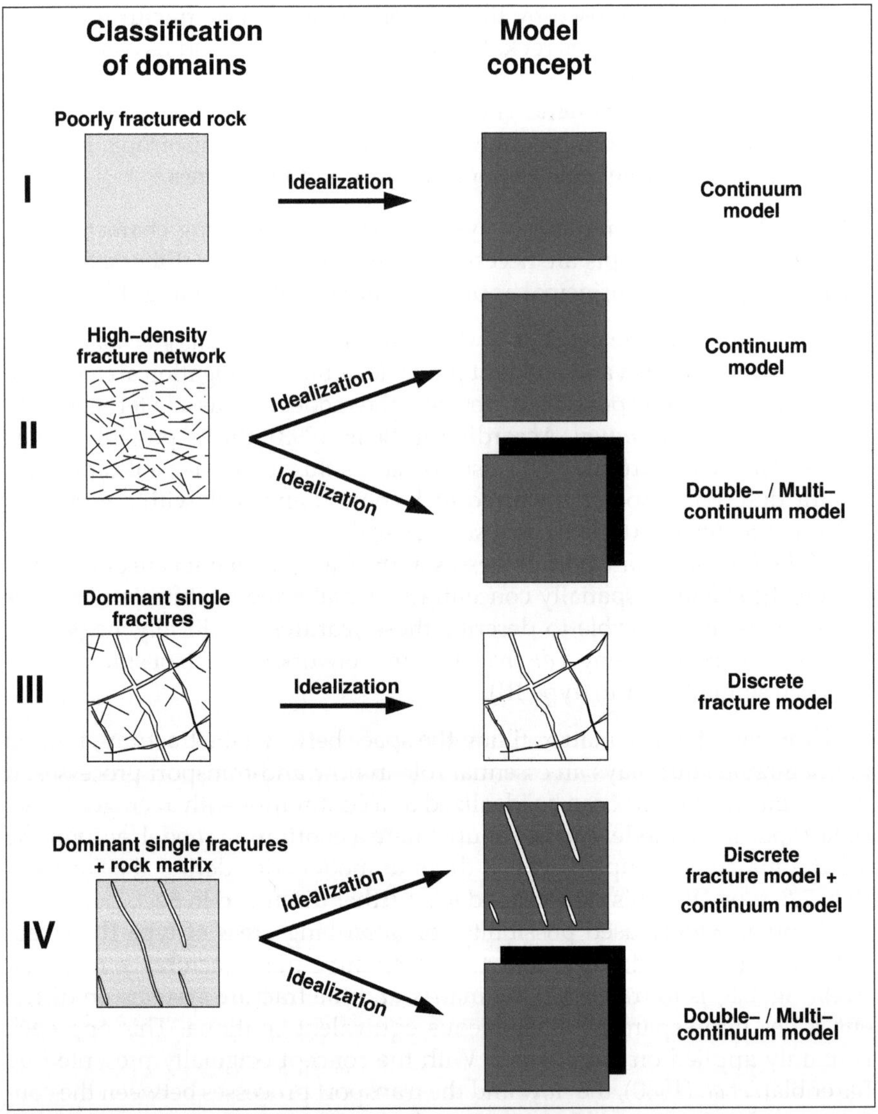

Fig. 2.8. Model concepts for the description of fractured porous media (based on Kröhn (1991) and Helmig (1993)).

2.3 Governing Equations of Flow and Transport in Porous Media

M. Süß, A. Silberhorn-Hemminger, R. Helmig

In this section, the mathematical description and the governing equations of flow and transport processes in *porous media in general*, i.e. without the consideration of fractures, are presented. The discussion focuses on the processes which are relevant to the research work presented in this book. The implementation of the discrete and the multi-continuum concepts for considering the fractures is discussed in Sects. 2.4 and 2.5 respectively.

2.3.1 Representative Elementary Volume

The concept of the *representative elementary volume* (REV) as defined by Bear (1972) is fundamental to the mathematical description of fluid flow and transport in porous media. By means of volume averaging, the micro-scale properties of the porous medium (grain-size and pore-space geometry) are represented by an equivalent continuum on a larger scale described by new properties. On the one hand, the REV must be large enough to avoid undesirable fluctuations of the averaged properties, and on the other hand, it must be small enough to render the spatial dependency of these properties. In Fig. 2.9, the definition of a suitable extent of the REV is visualized. The application of the REV approach in different model concepts for *fractured* porous media is discussed in Sects. 2.2, 2.4 and 2.5.

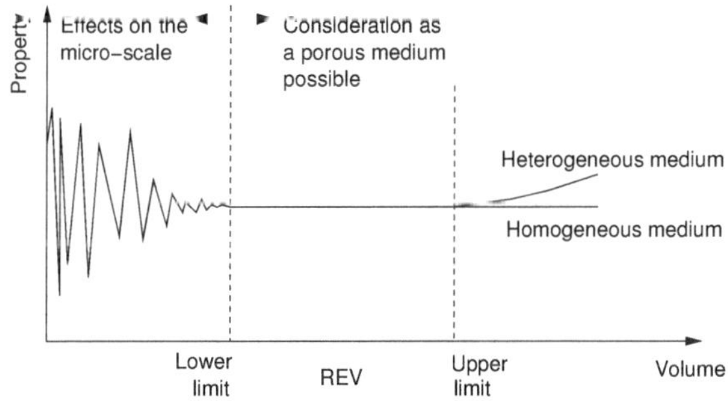

Fig. 2.9. Representative elementary volume (REV) (modified from Bear (1972)).

2.3.2 Flow Processes

2.3.2.1 DARCY Equation

The DARCY equation for laminar flow in porous media was defined by HENRY DARCY in 1856. In one-dimensional column experiments, DARCY found that the volume discharge Q is proportional to the hydraulic gradient $\Delta h/\Delta l$ as described by the following equation:

$$Q = -A\,K\,\frac{\Delta h}{\Delta l}\,. \tag{2.8}$$

Here, A is the cross-sectional area of the column, K is the hydraulic conductivity, Δh is the difference in hydraulic head and Δl is the distance between the measurement points. From (2.8) and the relationship

$$Q = q_i A \, , \tag{2.9}$$

the three-dimensional DARCY velocity q_i is determined:

$$q_i = -K_{ij} \frac{\partial h}{\partial x_j} \, . \tag{2.10}$$

The hydraulic conductivity tensor K_{ij} depends on the properties of the porous medium as well as of the fluid:

$$K_{ij} = \frac{\rho \, g \, k_{ij}}{\mu} \, , \tag{2.11}$$

where ρ is the fluid density, g is the gravitational acceleration, k_{ij} is the permeability tensor and μ is the dynamic viscosity. The permeability tensor k_{ij} represents the directional resistance of a porous medium and is independent of the fluid properties. Expressing the piezometric head h as a pressure p and inserting the permeability k_{ij} instead of the hydraulic conductivity K_{ij}, yields the following expression for the DARCY velocity q_i:

$$q_i = -\frac{k_{ij}}{\mu} \left(\frac{\partial p}{\partial x_j} + \rho g \frac{\partial z}{\partial x_{ij}} \right) \, . \tag{2.12}$$

In the case of gas flow, the gravity effect is often neglected. This is legitimate if the gravity effect, due to low density, is small compared to the effect of the pressure gradient. Neglecting the gravity effect, the equation can be simplified to

$$q_i = -\frac{k_{ij}}{\mu} \frac{\partial p}{\partial x_{ij}} \, . \tag{2.13}$$

This form of the equation may be applied for two-dimensional horizontal calculations as well.

The range of validity of the DARCY equation is expressed in terms of the REYNOLDS number *Re*. According to Bear (1972), the upper limit of the validity of the DARCY equation is at a value of *Re* between 1 and 10.

2.3.2.2 Continuity Equation

The continuity equation is based on the principle of conservation of mass, and states that the temporal change of mass in a control volume is the sum of the mass flux across the volume boundaries and the mass flux due to sources and sinks. The temporal change of the mass in the control volume is described as follows:

$$\frac{\partial(n\rho)}{\partial t} = -\frac{\partial(\rho q_i)}{\partial x_i} + q_s \,. \tag{2.14}$$

Here, q_s is the source and sink term, e.g. describing well withdrawal or injection, and n represents the total porosity, also including the pores through which there is no flow. The porosity is slightly pressure-dependent (Kinzelbach, 1992). However, this aspect is neglected here, i.e. the matrix is considered as inelastic. From this, the continuity equation is obtained in the following form:

$$n\,\frac{\partial \rho}{\partial t} = -\frac{\partial(\rho q_i)}{\partial x_i} + q_s \,. \tag{2.15}$$

If we introduce the piezometric head h as the independent variable instead of the DARCY velocity q_i, we may express the continuity equation as

$$n\,\frac{\partial \rho}{\partial t} = \frac{\partial}{\partial x_i}\left(\rho^2 g\,\frac{k_{ij}}{\mu}\,\frac{\partial h}{\partial x_i}\right) \,. \tag{2.16}$$

The source and sink term q_s is omitted for simplicity's sake. This general expression is valid for heterogeneous, anisotropic and compressible media. Bear (1972) rewrites (2.16) for the independent variable pressure p using the relationship

$$g\,\frac{\partial h}{\partial x_i} = g e_{x_3} z + \frac{1}{\rho}\,\frac{\partial p}{\partial x_i} \,, \tag{2.17}$$

where e_{x_3} is the unit vector in z-direction. Assuming that $k_{ij} = 0$ for $i \neq j$, this results in the expression

$$n\frac{\partial \rho}{\partial t} = \frac{\partial}{\partial x_1}\left(\rho\frac{k_1}{\mu}\frac{\partial p}{\partial x_1}\right) + \frac{\partial}{\partial x_2}\left(\rho\frac{k_2}{\mu}\frac{\partial p}{\partial x_2}\right) + \frac{\partial}{\partial x_3}\left(\rho\frac{k_3}{\mu}\left[\frac{\partial p}{\partial x_3} + \rho g\right]\right) . \tag{2.18}$$

In (2.18), the term ρg, expresses the gravity effect. In many cases, this effect is much smaller than the pressure gradient $\partial p/\partial x_3$, and is therefore neglected (Bear, 1972).

Depending on the fluid, the pressure dependency of the fluid density ρ is more or less significant, e.g. water is generally assumed to be incompressible whereas gas is highly compressible. The experiments and the numerical simulations discussed in this book are mainly concerned with gas-saturated media. Assuming an ideal gas, the relationship between the density and the pressure is described by the ideal gas law:

$$\rho = \frac{p}{R_i T} \,. \tag{2.19}$$

Here, R_i is the individual gas constant and T is the temperature.

Introducing (2.19) into (2.18) and neglecting the gravity term, the following diction of the continuity equation is obtained:

$$n\,\frac{\partial p}{\partial t} = \frac{\partial}{\partial x_i}\left(\frac{k_{ij}}{2\mu}\,\frac{\partial p^2}{\partial x_i}\right) \,. \tag{2.20}$$

2.3.3 Transport Processes

Assuming a conservative tracer, i.e. no adsorption and no reactions, in an isothermal system, three mechanisms determine the transport process, namely advection, dispersion and diffusion.

2.3.3.1 Advection

Advective transport comprises the movement of the tracer in the direction and with the average fluid velocity of a control volume (Kinzelbach, 1992). Here, the determining velocity is the seepage velocity v_i defined as:

$$v_i = \frac{q_i}{n_\mathrm{e}} . \tag{2.21}$$

Dividing by the effective porosity n_e, and not by the total porosity n, takes into account that the fluid can only flow through the connected pore space of the control volume, i.e. not through the dead-end pores. The seepage velocity v_i is a bulk property and is therefore only indirectly measurable. The advective mass flux is expressed as

$$J_{\mathrm{a},i} = c v_i , \tag{2.22}$$

where c is the solute concentration of the transported substance.

2.3.3.2 Hydrodynamic Dispersion

Dispersion describes the mixing of two miscible fluids due to fluctuations around the average velocity, caused by the morphology of the medium, the fluid flow condition and chemical or physical interaction with the solid surface of the medium (Sahimi, 1995). The concept generally used to describe this mixing process is based on FICK's law, assuming that there is a compensation of the concentration in the direction of the negative concentration gradient. The dispersive mass flux is expressed as

$$J_{\mathrm{d},i} = - D_{\mathrm{d},ij} \frac{\partial c}{\partial x_j} , \tag{2.23}$$

where $D_{\mathrm{d},ij}$ is the dispersion tensor. Dispersive mixing is assumed to take place in two principal directions, longitudinal and transversal to the direction of the seepage-velocity vector v_i. The dispersion tensor $D_{\mathrm{d},ij}$ is not constant, but depends on the seepage-velocity v_i. Provided that the system of coordinates is aligned with the direction of flow, the dispersion tensor $D_{\mathrm{d},ij}$ is diagonal:

$$D_{\mathrm{d},ij} = \begin{bmatrix} D_\mathrm{l} & 0 & 0 \\ 0 & D_\mathrm{t} & 0 \\ 0 & 0 & D_\mathrm{t} \end{bmatrix} = \begin{bmatrix} \alpha_\mathrm{l} v_1 & 0 & 0 \\ 0 & \alpha_\mathrm{t} v_2 & 0 \\ 0 & 0 & \alpha_\mathrm{t} v_3 \end{bmatrix} . \tag{2.24}$$

Here, D_l and D_t are the longitudinal and the transversal dispersion coefficients respectively, assuming that the vertical and the horizontal transversal dispersivity is equal. α_l and α_t are the longitudinal and the transversal dispersion lengths. The ratio between α_l and α_t is generally larger than 1.

Diffusion induces a mass flux between regions of different concentration. Mass flux occurs in the direction of the negative concentration gradient and is described by FICK's law:

$$j_{\mathrm{m,e},i} = -D_{\mathrm{m,e}} \frac{\partial c}{\partial x_i} . \tag{2.25}$$

Here, $D_{\mathrm{m,e}}$ is the effective diffusion coefficient which takes into account that the diffusion process is dependent not only on the combination of fluids, on the temperature and the pressure (Reid *et al.*, 1987) but also on the porous medium (Grathwohl, 1998). The diffusion coefficient for gases is higher than for liquids. Since the diffusion process is slow, the significance of this mass flux depends on its relative importance compared to the advective and the dispersive fluxes. In regions of high velocities, it may be neglected, whereas for low velocities, it is one of the essential processes that determine the shape of the tracer-breakthrough curve. Since, in this book, first, most of the cases discussed involve gas flow and, second, the investigations are concerned with flow in fractured low-permeable porous media, the diffusive processes may not be neglected.

From the above discussion, it is obvious that dispersive as well as diffusive processes are implemented according to the same concept, i.e. FICK's law. Consequently, these transport flux terms may be combined in one term, generally defined as the *hydrodynamic-dispersion* term, where $D_{\mathrm{d},ij}$ and $D_{\mathrm{m,e}}$ are summarized in the hydrodynamic-dispersion tensor D_{ij} (Scheidegger, 1961):

$$D_{ij} = \begin{bmatrix} D_{\mathrm{m,e}} + \alpha_\mathrm{l} v_1 & 0 & 0 \\ 0 & D_{\mathrm{m,e}} + \alpha_\mathrm{t} v_2 & 0 \\ 0 & 0 & D_{\mathrm{m,e}} + \alpha_\mathrm{t} v_3 \end{bmatrix} . \tag{2.26}$$

The mass flux due to hydrodynamic dispersion is expressed by:

$$j_{\mathrm{hd},i} = -D_{ij} \frac{\partial c}{\partial x_j} . \tag{2.27}$$

The concept of hydrodynamic dispersion accounts for the spreading of the tracer due to the irregular pore space. On a larger scale, the concept of *macro-dispersion* accounts for dispersion due to heterogeneities of the porous medium. In continuum models, a FICKIAN approach is often chosen for the macro-dispersion. In, for example, Cirpka (1997), the concept of macro-dispersion as well as different model approaches are discussed in detail.

2.3.3.3 Transport Equation

Following the same principle as for the continuity equation (2.14), the transport equation is derived by balancing all mass fluxes across the boundaries of a control volume:

$$J_i + \frac{\partial}{\partial x_i} \left(J_{\mathrm{a},i} + J_{\mathrm{hd},i} \right) = 0 . \tag{2.28}$$

If J_i is expressed as

$$J_i = \frac{\partial c}{\partial t} + q_{\mathrm{m}} , \tag{2.29}$$

where q_{m} is the tracer mass source/sink term, (2.28) can be written as

$$\frac{\partial c}{\partial t} + \frac{\partial}{\partial x_i}(v_i c) - \frac{\partial}{\partial x_i}\left(D_{ij} \frac{\partial c}{\partial x_j} \right) + q_{\mathrm{m}} = 0 . \tag{2.30}$$

A measure of the relative importance of advective and dispersive/diffusive transport is the PECLET number Pe. It is defined as

$$Pe = \frac{|v|\, L}{|D_l|} , \tag{2.31}$$

where v is the average seepage velocity in flow direction, L is a typical length scale of the problem and D_l is the hydrodynamic dispersion coefficient in the flow direction. Large PECLET numbers indicate that advection dominates the transport process.

2.4 The Discrete Model Concept

A. Silberhorn-Hemminger, M. Süß, R. Helmig

As discussed above, in situations where the fractures as well as the matrix play a significant role for the flow and transport processes, the model domain cannot be homogenized but a model concept that includes fractures as well as matrix is required. One approach is to use the discrete model concept, where the matrix and the fractures are locally idealized as continua and the fractures are implemented discretely at their actual location within the domain. It is obvious that the amount of data required to set up a discrete model of the actual domain is very large and to some extent not measurable. Consequently, the discrete model concept is preferably used for relatively small domains and is a suitable tool for principle studies of flow and transport processes.

In the previous sections (2.3), the physical-mathematical description of the flow and transport processes in porous media, i.e. in the porous matrix,

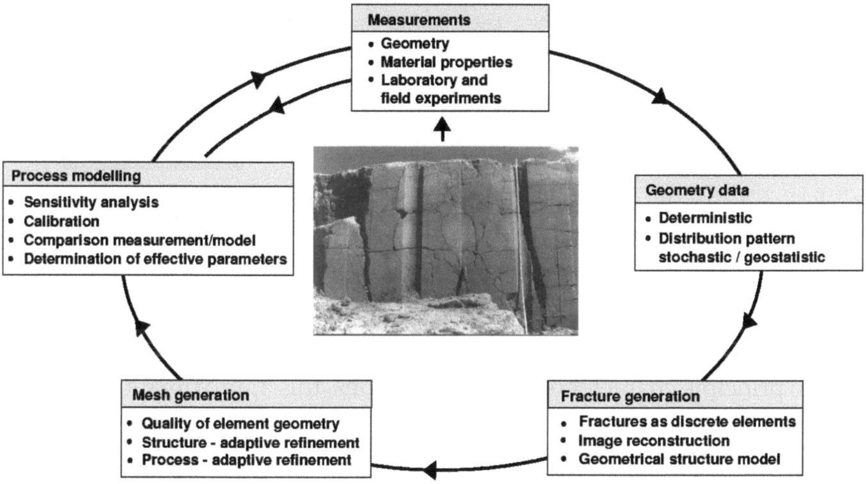

Fig. 2.10. Discrete modeling of flow and transport processes in fractured porous media - necessary steps.

is presented. This section deals with the model concept specific to the fractures and the mathematical description of the flow within a fracture as well as of its location and geometry. The stochastic fracture generator FRAC3D is presented and the implementation of the discrete model concept in the numerical model is briefly discussed. Figure 2.10 shows the steps from a natural system to a discrete process model.

2.4.1 Parallel-Plate Concept

A natural fracture is bounded on both sides by the rock surface (Fig. 2.5). The rough fracture walls do not have an identical profile and the normal tension is carried by contact zones between the walls. A model concept frequently used for a fracture consists of two plane parallel plates, representing the fracture walls. As illustrated in Fig. 2.11, it can be applied locally, maintaining a variation in fracture aperture throughout the fracture, or globally, assuming one constant aperture for the total fracture. It is a well-known fact that especially the latter approach is a strong simplification of nature. However, other methods proposed in the literature have not yet found general acceptance (Berkowitz, 2002).

Tsang and Tsang (1987) showed that preferential flow paths exist, hence *channeling* effects may have significant influence on the flow and therefore also on the transport processes. For multi-phase flow, the variation in entry pressure is strongly related to the distribution of the aperture; therefore channeling effects are particularly important for simulations including more than one fluid phase.

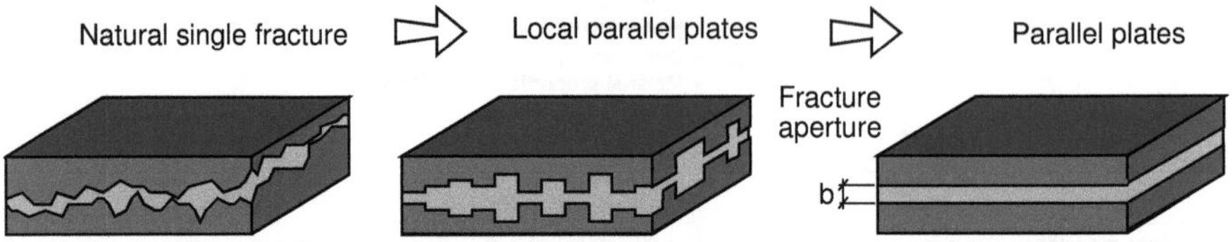

Fig. 2.11. From nature to parallel-plate concept.

For the numerical studies presented in this book, the parallel-plate concept is applied. The decision to use this simplified concept is justified by the fact that the simulations are concerned with single-phase flow only and that, for the principle character of the investigations, this approximation of nature is sufficient.

When the parallel-plate concept is applied, it is assumed that the length scale l of the plates is much larger than the distance between them b ($l \gg b$). Furthermore, hydraulically smooth walls and laminar flow are assumed, corresponding to the POISEUILLE fluid model (Wollrath, 1990) . Figure 2.12 shows the two parallel plates and the parabola-shaped velocity profile, indicating laminar flow. The NAVIER-STOKES equation for the laminar single-

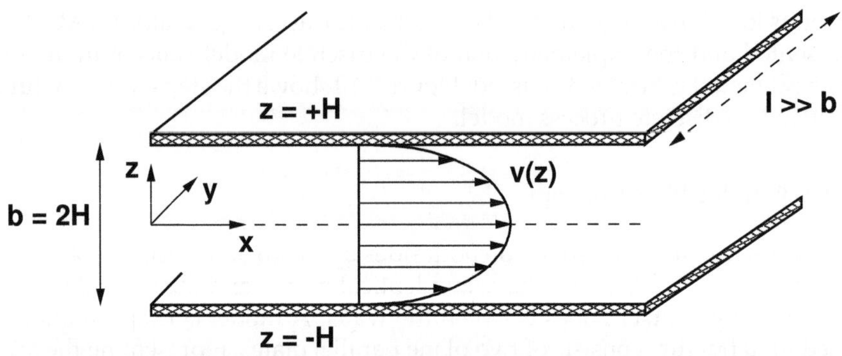

Fig. 2.12. Laminar flow between two parallel plates: parabola-shaped velocity profile.

phase flow of an incompressible NEWTONIAN fluid yields the following equation for the velocity profile between two parallel plates (Snow (1969); White (1999)):

$$v(z) = \frac{\rho g}{2\mu}\left[-\frac{d}{dx}\left(\frac{p}{\rho g} + z\right)\right](H^2 - z^2) . \qquad (2.32)$$

The maximum velocity $v_{\max}$ is reached at $z = 0$:

$$v_{\max} = v(z=0) = \frac{\rho g}{2\mu} H^2 \cdot -\frac{d}{dx}\left(\frac{p}{\rho g}\right) . \qquad (2.33)$$

For a parabola-shaped profile, the mean velocity $\overline{v}$ is derived from the maximum velocity v_{max}:

$$\overline{v} = \frac{2}{3}\, v_{\mathrm{max}} = \frac{\rho g}{\mu} \frac{H^2}{3} \cdot - \frac{d}{dx} \left(\frac{p}{\rho g} \right) . \tag{2.34}$$

From (2.34) and under consideration of the distance between the plates b ($b = 2H$), the mean three-dimensional velocity $\overline{v}_i$ can be written as:

$$\overline{v}_i = - \frac{b^2}{12} \frac{\rho g}{\mu} \frac{\partial h}{\partial x_i} = - K \frac{\partial h}{\partial x_i} . \tag{2.35}$$

Here, the hydraulic conductivity K and the permeability k have the following relationship (see (2.11)):

$$K = k \frac{\rho g}{\mu} \qquad \text{with} \qquad k = \frac{b^2}{12} . \tag{2.36}$$

From this, it can be concluded that the permeability of a fracture, approximated by the parallel plate concept, is proportional to the square of the fracture aperture b. The volume discharge Q is derived by integrating the velocity over the distance between the plates (assuming a constant depth l parallel to the y-axis):

$$Q = \int_{-H}^{+H} v(z)\, l\, dz . \tag{2.37}$$

Including (2.32) yields:

$$Q = - \frac{\rho g}{\mu} \frac{b^3}{12}\, l\, \frac{\partial h}{\partial x_1} \tag{2.38}$$

Due to the proportionality of Q to the third power of the aperture b, (2.38) is referred to as the *cubic law* (Romm, 1966).

2.4.2 Generation of Fracture Structural Models – FRAC3D

The use of the discrete modeling approach for simulating flow and transport processes in fractured porous media requires the discrete description of the fractures in space. Here, the fracture generator provides the geometrical description and the structural properties of the fracture system. The fracture generator represents a link between the natural and the numerical model.

The generating algorithm requires information on the geometrical characteristics of the fractures. This is obtained by investigating core samples or outcrop sites. The information gained from these samples and locations is of one- or two-dimensional character. Here, one problem of generating a stochastic fracture field, particularly a three-dimensional fracture field, becomes obvious: the information content of the data does not account for the clearly three-dimensional characteristics of fractured porous systems. The reliability of such stochastically generated geometric models depends strongly on the qualitative and quantitative description of the aquifer system: are there main fracture orientations, is the fracture density high or low, is there a single dominating fracture or fault zone, are the fractures open, are they filled, how wide and rough are the fractures? These questions are just a selection of many more that have to be answered or at least to be considered during the process of generating stochastic fracture systems. Additionally, we assume that the selected real aquifer from which the field data is obtained can be considered representative of the ensemble of all possible realizations. The stochastic properties of the ensemble are given by adapted theoretical distributions. Consequently, a stochastically generated fracture network based on the field data can be regarded as one possible realization out of the ensemble. The fracture-generating algorithm makes it possible to combine the available field information adequately in order to obtain structural geometric fracture models which are realizations of the real aquifer system.

In structural models, the natural fractures are represented by discrete elements. In a two-dimensional model, the fractures are one-dimensional elements. In a three-dimensional model, the fracture planes are two-dimensional elements. Additionally, distinctive flow channels may be represented by one-dimensional elements in the three-dimensional model. Figure 2.13 shows a two- and a three-dimensional stochastic structural fracture model.

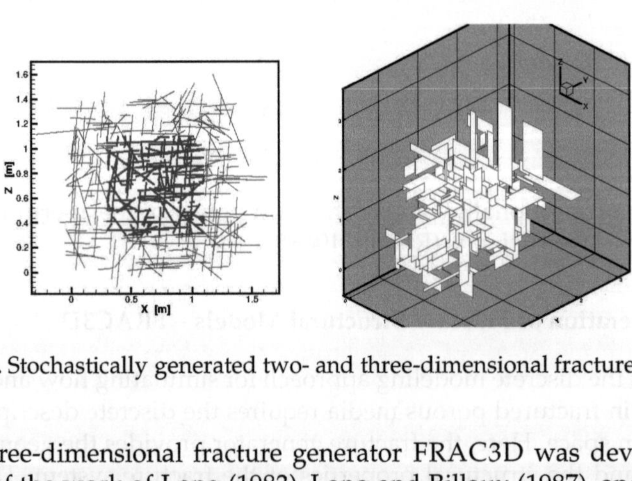

Fig. 2.13. Stochastically generated two- and three-dimensional fracture models.

The three-dimensional fracture generator FRAC3D was developed on the basis of the work of Long (1983), Long and Billaux (1987), and Wollrath (1990) A flow chart of the program algorithm can be seen in Fig. 2.14. Beside the fracture-generating routine itself, the program FRAC3D offers various methods for analyzing the quality of the generated fields, and for optimizing the generated fields. An interface for the mesh generation program ART and the flow and transport simulation program MUFTE-UG (see Sect. 2.4.4) is included. A detailed description of the fracture generation program can be found in Silberhorn-Hemminger (2002).

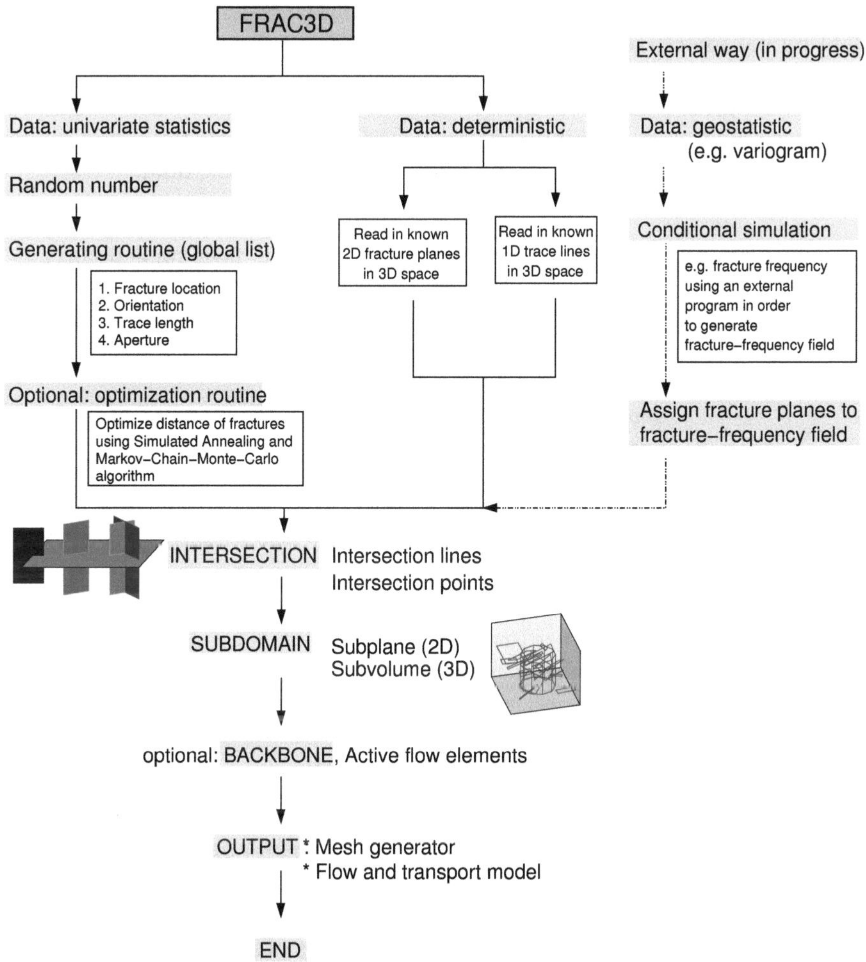

Fig. 2.14. Program algorithm of the fracture generator FRAC3D.

2.4.2.1 Generation Routine

As can be seen in Fig. 2.14, the algorithm for structure models for generating fractures is based on two different approaches: deterministic and stochastic. The choice of approach depends on the quantity and the quality of the input data available.

The deterministic approach requires exact information about a fracture network or a single fault zone. One of its main problems is that one has to generate a three-dimensional system out of one- and two-dimensional information. Often, this approach is not feasible due to lack of information.

The second basic approach is the stochastic one. A large sample of fracture data (e.g. length, orientation) is required and a description of these field data by parameterized theoretical distributions, such as the FISHER distribution for the spherical orientation or the exponential distribution for the fracture lengths, must be available. It is important to be aware of the fact that these theoretical distributions are based on linear statistics. They do not include any information about the spatial variability of the data.

The generating routine includes the following steps:

```
while (simulated_fracture_density < target_fracture_density)
{
    step 1: Fracture location
            Generation of the mid point of fracture [i]
    step 2: Fracture orientation
            Generation of the normal vector of fracture [i]
    step 3: Fracture extension / length
            Generation of the spatial extension of
            fracture [i] and calculation of the four edge
            points of the fracture [i]
    step 4: Inclusion of the new generated fracture
            element [i] into the global list of all
            fracture elements
    step 5: Calculation of the simulated_fracture_density

    ++i
    total number of fractures: nfrac=i
}
```

An optional optimization routine for the parameter *fracture distance* completes the fracture generation.

The statistical characteristics of the newly generated fracture field are analyzed. For this purpose, the distribution functions of several fracture parameters are calculated. The difference between the input distribution functions and the distribution functions of the new field indicates the quality of the new field. If the differences are too large, the newly generated field is either rejected and the generation routine run again or the optimization routine starts.

2.4.2.2 Optimization

As the optimization routine, a *Simulated Annealing* algorithm followed by a *Markov-Chain-Monte-Carlo* algorithm is implemented in the fracture generator FRAC3D. The Simulated Annealing optimization step serves as a preconditioning of the starting field for the subsequent Markov-Chain-Monte-Carlo optimization.

Because of the relatively smooth distributions of the fracture parameters *orientation* and *fracture length*, a good agreement between the input distributions and the distributions of the generated fields is generally achieved.

However, for the fracture parameter *fracture distance* an optimization is necessary since the information of the fracture distances cannot be taken into account in the generating routine described above. In order to include this important information in the generated fracture fields, a modified scanline technique is applied. The scanline technique allows the calculation of the fracture distance distribution and the optimization of the distribution.

2.4.2.3 Post-processing

In the next step, the intersection lines of the fracture planes and the intersection points of the intersection lines are determined (Fig. 2.15). Subsequently, the investigation domain is extracted from the generated domain (Fig. 2.16). Optionally, the inactive, disconnected fracture elements can be removed from the fracture network.

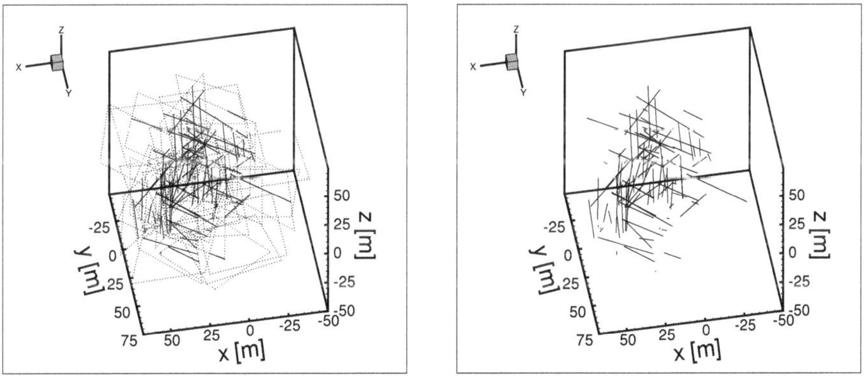

Fig. 2.15. Fracture network (left) and intersection lines (right).

Finally, the xyz-coordinates of the investigation domain, the fracture planes, the intersection lines, and the intersection points are converted into the data format required for the following mesh generation. Figure 2.17 shows a stochastically generated three-dimensional fracture network and two details of the finite element mesh of the fracture network. Fuchs (1999) gives detailed information about the mesh-generation program ART (Almost Regular Triangulation).

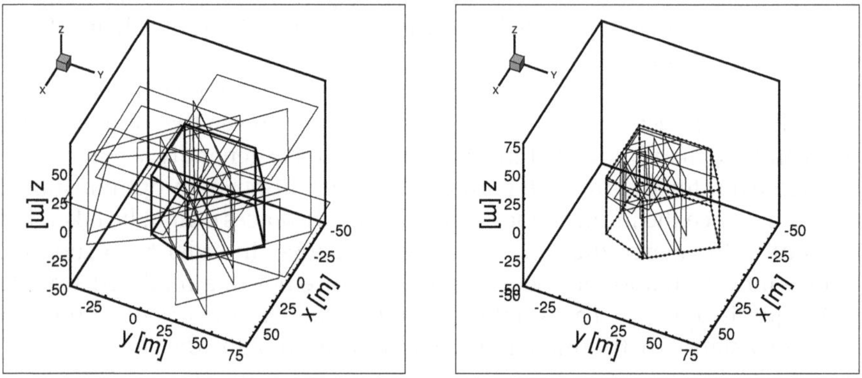

Fig. 2.16. Generated domain and extracted investigation domain.

2.4.2.4 Remarks

There is never a perfect match between a stochastic geometrical model and the real system on which it is based. It is possible to generate different realizations which are similar in their statistical description. However, one single realization can never exactly predict the behavior at a certain point of the real system. Therefore, one has to consider a large number of realizations. On the basis of these realizations, system properties such as the effective permeability and the effective dispersion of the system can be investigated and calculated (see e.g. Sect. 4.3.3). Additionally, one has to be aware of the fact that single fractures sometimes control the flow and transport processes in a fracture network completely by connecting different independent fracture clusters. If such dominating fractures are known and can be described by their orientation and extension, a combination of the deterministic and the stochastic approaches improves the reliability of the generated fracture network.

The generation approach discussed above incorporates deterministic and univariate stochastic information. A further improvement of the generation process is the implementation of routines for considering geostatistical information as well. Such algorithms are being developed as this book is being written. In Sect. 5.4, a geostatistical evaluation of the test site is discussed.

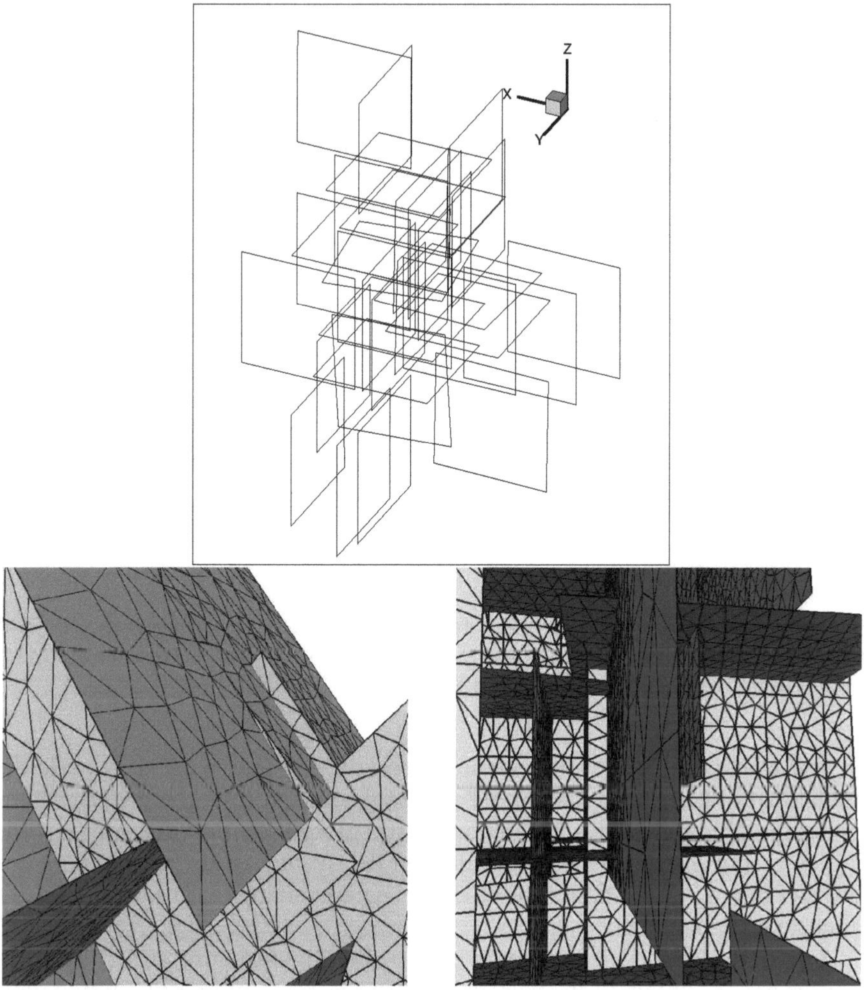

Fig. 2.17. Three-dimensional fracture network and details of the finite element mesh. Mesh generator ART (Almost Regular Triangulation) (Fuchs, 1999).

2.4.3 Spatial and Temporal Discretization

Due to the large discrepancy between the properties of the matrix and the fractures, large gradients occur in the vicinity of the fracture-matrix interface. To achieve an acceptable numerical accuracy, the mesh of the numerical model must have a high degree of refinement in these areas. Fractures may either be discretized with one dimension less than the matrix, i.e. as lines in a two-dimensional matrix or as a surface in a three-dimensional matrix, or with the same dimension as the surrounding matrix elements, i.e. equidimensionally (Fig. 2.18).

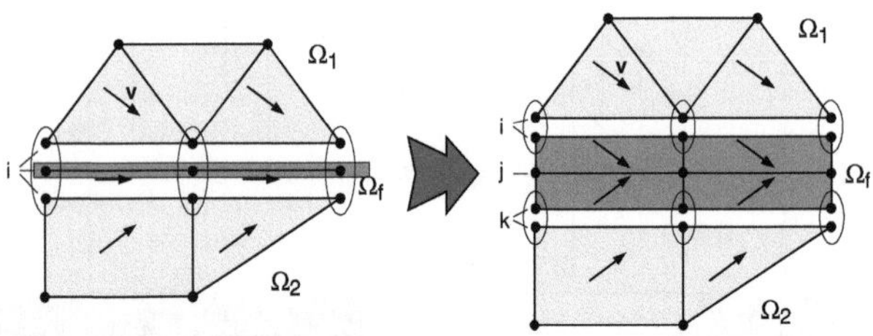

Fig. 2.18. Spatial discretization of a fracture embedded in a porous matrix in a two-dimensional domain. Left: Lower-dimensional discretization. Right: Equidimensional discretization. From Ochs *et al.* (2002).

For transport simulations, Neunhäuserer (2003) showed that the equidimensional approach yields a more accurate solution than the lower-dimensional one. However, the differences in the global solution are only significant if certain local effects accumulate in the system or if the processes are slow so that a relevant amount of tracer mass exchange occurs between the fracture and the matrix.

For time-dependent problems, sharp moving pressure and concentration fronts are obtained in the fractures, suggesting an adaptive refinement method in order to save computing time, especially for highly complex systems. The temporal discretization required is in general finer in the vicinity of the fractures, where very high flow velocities occur compared to the velocities in the surrounding matrix. An implicit temporal discretization is often chosen in order to achieve a stable solution, despite the wide range of velocities. The disadvantage of this approach is a significant influence of numerical dispersion.

Detailed discussions on spatial and temporal discretization methods can be found in, for example, Neunhäuserer (2003), Helmig (1997), Bastian *et al.* (1999), Kinzelbach (1992), and Hirsch (1984).

2.4.4 Applied Numerical Model – MUFTE-UG

The numerical model used for most discrete modeling investigation in this book is MUFTE-UG. It consists of the two parts: MUFTE (Multiphase Flow, Transport and Energy Model) and UG (Unstructured Grids). Figure 2.19 gives an overview of the features of the models. UG is a software toolbox providing techniques for the numerical solution of partial differential equations (PDEs) on unstructured grids (Bastian, 1997). For solving linear and nonlinear PDEs, several multi-grid solvers are available as well as adaptive and parallel techniques. MUFTE contains numerous discretization methods and applications for modeling non-isothermal multi-phase processes in porous and fractured media (Helmig (1997); Helmig *et al.* (1998)). Geometrically complex structures, such as fractures systems, can be simulated with MUFTE-UG due to the flexibility of the system and its compatibility with the powerful mesh generator ART (Fuchs, 1999).

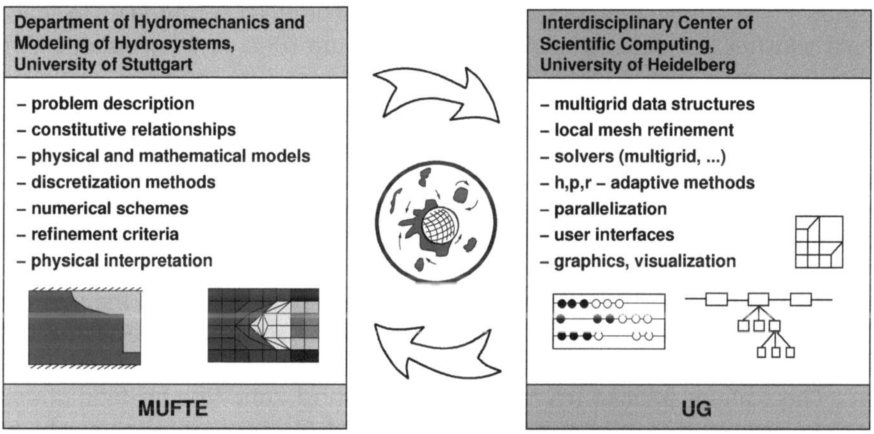

Fig. 2.19. Overview of the model system MUFTE-UG.

2.4.5 Summary

This section provides a rough overview of the discrete model concept as it is applied and implemented for the investigations within the framework of the research work presented in this book. It gives a basis for a better understanding of the simulation results presented and discussed later on.

Discrete modeling of fractured porous media requires not only an accurate approximation of the flow and transport processes in the matrix, within the fractures and at their interfaces but also the best possible description of the geometrical properties of the system.

The parallel-plate concept for the description of flow within the fractures is explained and the fracture generator FRAC3D presented. Possible spatial and temporal discretization methods are discussed briefly and the numerical model MUFTE-UG is introduced.

Typical numerical difficulties are not included here. Detailed discussions on this topic are given in, for example, Jakobs (2004), Reichenberger (2003), Neunhäuserer (2003) and Barlag (1997).

2.5 Implementation of the Multi-continuum Concept

T. Vogel, D. Jansen, J. Köngeter

The basic idea of multi-continuum modeling is to model separate, coupled hydraulic components of a heterogeneous aquifer. This principle of the multi-continuum model is illustrated in Fig. 2.20 for three identified continua. It is assumed that each component is distributed continuously in space and satisfies the conditions of a porous medium (Bear and Bachmat, 1990). For fracture matrix systems, this could be two fracture continua, such as a micro- and a macro-fracture system, and a matrix continuum with appropriate equivalent parameters (cf. Sect. 2.5.5).

A detailed modeling of such systems by a discrete model approach requires a high standard of modeling techniques and sufficient computer resources, as well as very detailed experimental investigations of the aquifer properties. Conversely, a representation of the aquifer by a single continuum model neglects the interactions between the components, which may be significant for the integral transport behavior of the aquifer. Multi-continuum models offer an efficient solution for this conflict.

The scale of interest (cf. Sect. 2.2) is the third zone (the far-field scale), where flow and transport may be considered to occur simultaneously, in overlapping continua. Single fractures are not observed to be dominant, as the mean length of the fractures is much smaller than the scale of interest.

It may seem that fewer input data are required for setting up a multi-continuum model than for the discrete approach (cf. Sect. 2.4). However, for high-quality multi-continuum modeling, a very good data basis is necessary. Data are needed so that hydraulically effective components and their interactions can be identified. An aquifer within fractured porous media with two components could be identified as double-porous and single-permeable (DPSP), double-porous and double-permeable (DPDP) or as single-porous and single-permeable (SPSP).

2.5.1 Governing Equations

The governing equations concerning flow and transport are formulated in the same way as for a porous medium (cf. Sect. 2.3) and are based on the

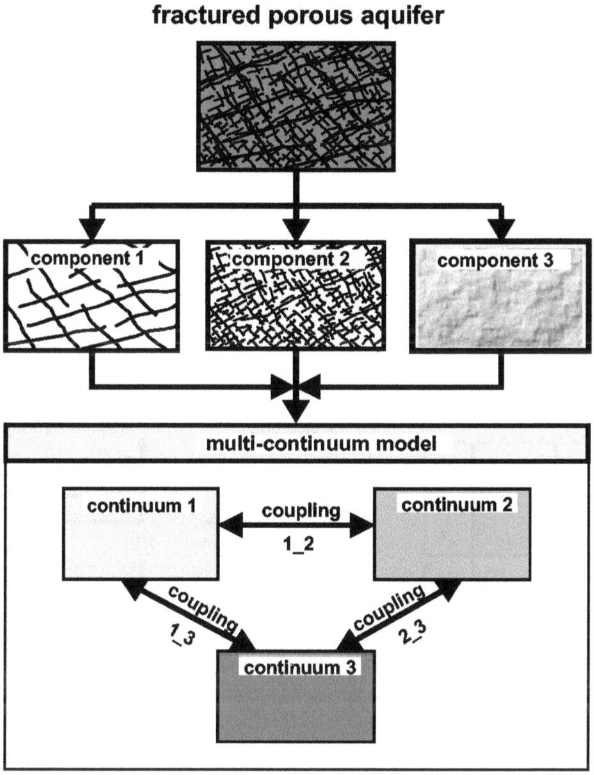

Fig. 2.20. Principle of the multi-continuum approach for fractured permeable formations (Jansen, 1999).

principle of conservation of mass. The multi-continuum approach requires a particular consideration of volume consistency and of exchange terms.

For the denomination of parameters and variables, the number of averaging processes is taken into account. Parameters with a single bar are equivalent continuum parameters and parameters with two bars are averaged over certain areas of the continuum. Equivalent parameters such as porosity are to be averaged once over the REV (cf. Sect. 2.3.1). An example of a second averaging is the averaging of the concentration distribution in a matrix block to obtain one equivalent mean value (cf. Fig. 2.21).

The equations for the continuum correspond to the equation for transient flow and the advection-dispersion equation for conservative tracers. In addition to the familiar terms (storage, flow, source and sink term, advection and dispersion terms), the exchange terms $W_{\alpha\beta}$ and the relative reference volumes Φ_α have to be considered. The latter transform the governing equations for the continuum based on the natural volume of this component. The

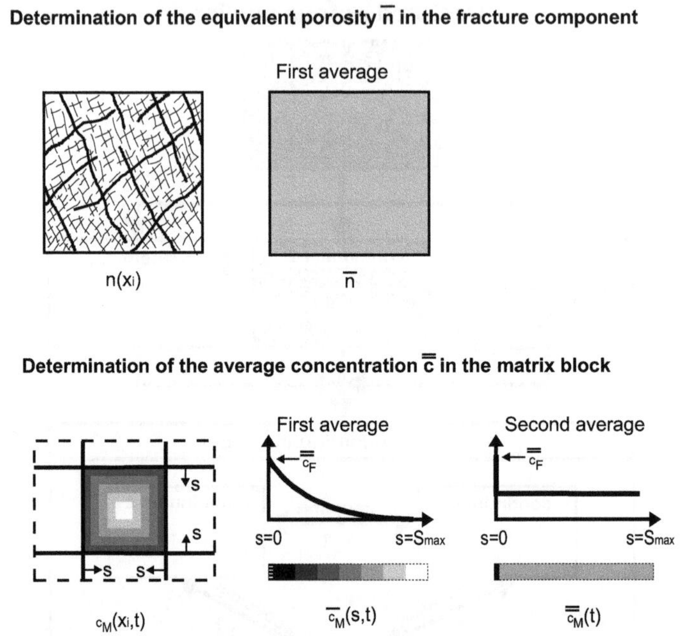

Fig. 2.21. Determination of the first and second average of parameters over certain areas of the continuum (Jansen, 1999).

following equations apply to the continuum α. The index β denotes the continuum coupled with α.

The equation for the flow field, formulated for the hydraulic head as variable, is thus expressed for the permeable component as follows:

$$\Phi_\alpha\left(\bar{S}_{S,\alpha}\frac{\partial\bar{\bar{h}}_\alpha}{\partial t}-\frac{\partial}{\partial x_i}\bar{K}_{ij,\alpha}\frac{\partial\bar{\bar{h}}_\alpha}{\partial x_j}+\bar{Q}_{Q,S,\alpha}\right)+\Phi_\alpha\Phi_\beta W_{Ql,\alpha\beta}=0\,. \tag{2.39}$$

Transport processes within a permeable component are modeled by solving the advection-dispersion equation (2.40):

$$\Phi_\alpha\left(\bar{n}_{e,\alpha}\frac{\partial\bar{\bar{c}}_\alpha}{\partial t}+\bar{\bar{q}}_{i,\alpha}\frac{\partial\bar{\bar{c}}_\alpha}{\partial x_i}-\frac{\partial}{\partial x_i}\bar{\bar{D}}_{ij,\alpha}\frac{\partial\bar{\bar{c}}_\alpha}{\partial x_j}+\bar{Q}_{Q,S,\alpha}(c_R-\bar{\bar{c}}_\alpha)\right) \tag{2.40}$$
$$+\,\Phi_\alpha\Phi_\beta W_{c,\alpha\beta}=0\,.$$

Within this formulation, $\bar{n}_{e,\alpha}$ is the equivalent porosity and $\bar{\bar{q}}_{i,\alpha}$ is the DARCY velocity. $\bar{\bar{D}}_{ij}$ is the tensor of hydrodynamic dispersion (cf. Sect. 2.3.3.2) in this context, written as

$$\bar{\bar{D}}_{ij}=\bar{\alpha}_t\delta_{ij}|q_{ij}|+(\bar{\alpha}_l-\bar{\alpha}_t)\frac{\bar{\bar{q}}_i\bar{\bar{q}}_j}{|\bar{\bar{q}}|}+\bar{n}\bar{D}_{m,ij}\,. \tag{2.41}$$

$\bar{Q}_{Q,S,\alpha}(c_R - \bar{\bar{c}}_\alpha)$ is a source or sink term and W_c is the exchange of mass between the coupled continua α and β. If all exchange processes are taken into account, W_c is given by equation (2.42):

$$W_{c,\alpha\beta} = W_{Dl,\alpha\beta} + W_{Al,\alpha\beta} + W_{Ar,\alpha\beta} \, . \tag{2.42}$$

W_{Dl} is the diffusive solute exchange due to a concentration gradient between the two components. W_{Al} and W_{Ar} are exchange masses due to local and regional advection respectively. The local advection is caused by a fluid exchange resulting from a transient flow, where the tracer is transported advectivly between the coupled continua. If the flow within the subordinate component (e.g. matrix component) is not negligible, a mixing of the flows may occur at the fracture-matrix interface, leading to an exchange of tracer mass. This mixing is represented by the regional advection term (cf. Fig. 2.22).

2.5.2 Types of Coupling

The characteristics of a multi-continuum model are described by the number of identified hydraulic components and the type of coupling and exchange formulation between the continua (cf. Fig. 2.25). The different coupling methods are either parallel, serial or selective (cf. Fig. 2.23). A parallel coupling means that all identified continua are coupled with each other directly, as shown by Gwo *et al.* (1995). A serial coupling (e.g. Lee and Tan, 1987) implies a coupling in the order of hydraulic conductivity (i.e. macro-fracture system, micro-fracture system and matrix). The selective coupling (e.g. Closmann, 1975) would, for example, couple the macro-fracture system to the micro-fracture system and to the matrix, but would not couple the micro-fracture system and the matrix.

2.5.3 Exchange Formulation

Besides the coupling method of the total system, the exchange formulation chosen for the single couplings is important (cf. Fig. 2.25). In the following, double-continuum models are considered, representing two coupled components of a multi-continuum model. The type of continuum model (DPSP or DPDP) defines the appropriate exchange model. Transient approaches (Bertin and Panfiloy (2000), Moyne (1997), Zimmermann *et al.* (1993), Pruess and Narasimhan (1985), among others) and quasi-steady formulations (Quintard and Whitaker (1996), Kazemi (1969),Warren and Root (1963), Barenblatt *et al.* (1960), among others) can be distinguished.

2.5.3.1 DPSP Models with Transient Exchange Formulation

If DPSP models are considered, only molecular diffusion (matrix diffusion), which is of a local nature and does not depend on regional processes, is mod-

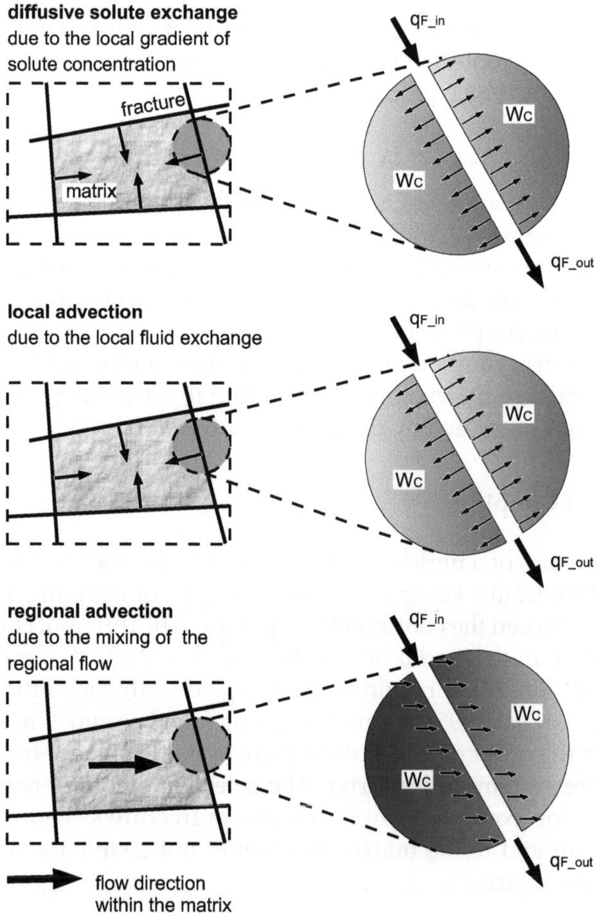

Fig. 2.22. Possible exchange processes for the components (Jansen, 1999).

eled for the matrix. Making use of the local nature of matrix diffusion, transport in the matrix is described by the local one-dimensional diffusion equation (2.43)

$$n_{e,\beta}\frac{\partial c_\beta}{\partial t} - \frac{1}{A_\beta(s)} n_{e,\beta} D_{m,\beta} \frac{\partial}{\partial s}\left(A_\beta(s)\frac{\partial c_\beta}{\partial s}\right) + Q_\beta(c_R - c_\beta) = 0\,, \qquad (2.43)$$

where s is the distance from the exchange interface (the matrix block surface), $A(s)$ is the interface area for diffusive flux at this distance and D_m is the molecular diffusion coefficient. Formulations for $A(s)$ are given by Pruess and Karasaki (1982) and Jansen *et al.* (1996). Within the specific surface to volume ratio of the porous blocks Ω_0, W_D is given by FICK's law:

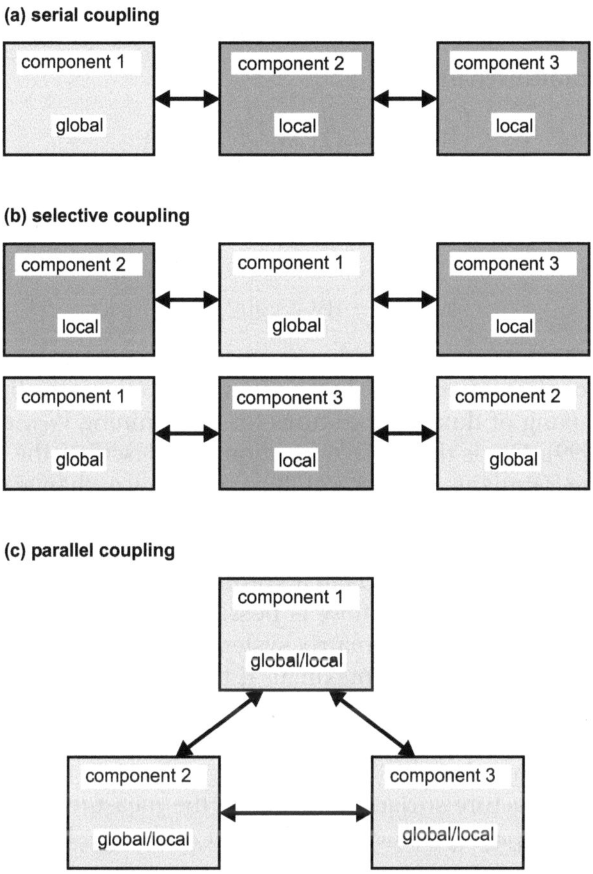

Fig. 2.23. Serial (a), selective (b) and parallel (c) coupling of hydraulic components.

$$W_{\mathrm{D},\alpha\beta} = -\bar{\Omega}_{0,\beta}\bar{n}_{\mathrm{e},\beta}\bar{D}_{\mathrm{m},\beta}\left.\frac{\partial \bar{c}_{\beta}}{\partial s}\right|_{s=0.0} . \tag{2.44}$$

2.5.3.2 DPDP Models with Quasi-Steady Exchange Formulation

If regional transport is considered in both coupled continua, exchange processes are no longer only of a local nature. Therefore, equation (2.40) must be solved for both continua.

The quasi-steady formulation describing the solute exchange W_c can be written for the fluid exchange as follows:

$$W_{\mathrm{Qr},\alpha\beta} = |\bar{\Omega}_{\mathrm{W},ij,\alpha\beta}\,\bar{\bar{q}}_{i,\beta}| \tag{2.45}$$

$$W_{\mathrm{Ql},\alpha\beta} = \bar{\Omega}^2_{0,\beta}\bar{\alpha}_{\mathrm{Q},\beta}(\bar{\bar{h}}_\alpha - \bar{\bar{h}}_\beta) \tag{2.46}$$

and for the solute exchange as

$$W_{\mathrm{Dl},\alpha\beta} = \bar{\Omega}^2_{0,\beta}\bar{\alpha}_{\mathrm{c},\beta}(\bar{\bar{c}}_\alpha - \bar{\bar{c}}_\beta) \tag{2.47}$$

$$W_{\mathrm{Al},\alpha\beta} = |W_{\mathrm{Ql},\alpha\beta}|(\bar{\bar{c}}_\alpha - \bar{\bar{c}}_\beta) \tag{2.48}$$

$$W_{\mathrm{Ar},\alpha\beta} = |W_{\mathrm{Qr},\alpha\beta}|(\bar{\bar{c}}_\alpha - \bar{\bar{c}}_\beta)\ . \tag{2.49}$$

In equation (2.48), W_{Ql} is the fluid exchange due to the local pressure gradient between continuum α and β. In equation (2.49), W_{Qr} is the fluid exchange due to the mixing of fluxes. Procedures for determining W_{Qr} are presented in Jansen (1999). Ω_{W} is the surface function that describes the relevant part of the fracture for mixing. Figure 2.24 illustrates how exchange processes between two components depend on direction. The regional hydraulic gradient induces a fluid flow in the fracture and the matrix system. The direction of flow, e_{qM}, in the matrix system is in accordance with the conductivity characteristics. Inside the fracture, flow is possible only in the direction of the fracture axis. If the flow in the matrix system is perpendicular to a fracture, the regional mass exchange is maximal. If the flow in the matrix system is parallel to the direction of flow in the fracture, the two flows do not mix. Thus, the intensity of mixing of the two flows depends on their relative orientation and magnitude.

The specific fracture surface $\Omega_{\mathrm{W}}(e_{qM})$ is the measure for the surface of interaction participating in the regional fluid exchange. It is defined as the

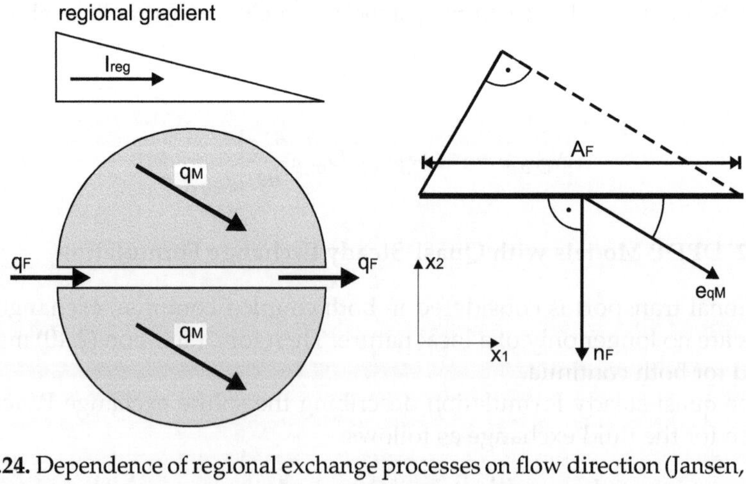

Fig. 2.24. Dependence of regional exchange processes on flow direction (Jansen, 1999).

ratio of the sum of the perpendicular projections of all N fracture surfaces A_F to the direction of flow e_{qM} in the matrix and the volume V_M of the total system under consideration:

$$\Omega_W(e_{qM}) = \frac{1}{V_M} \sum_{k=1}^{N} A_F(k) e_{qM} n_F(k) . \tag{2.50}$$

Ω_W can be obtained by a best-fit analysis. The exchange parameters $\bar{\alpha}_c$ and $\bar{\alpha}_Q$ have to be calibrated.

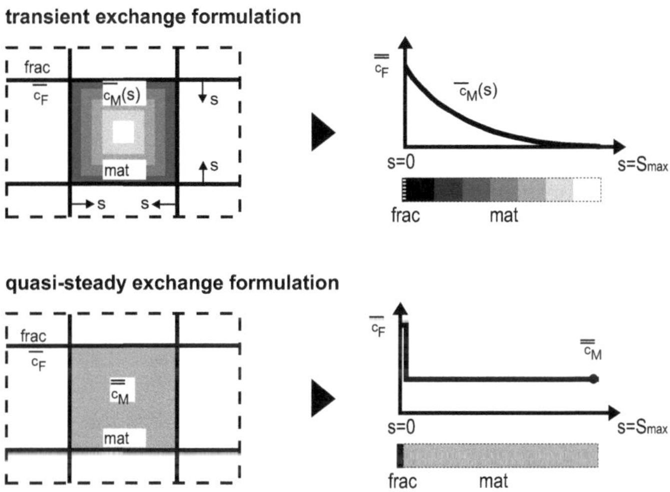

Fig. 2.25. Transient and quasi-steady exchange formulations (Jansen *et al.*, 1998).

2.5.4 Numerical Model

The numerical model STRAFE was developed by the Institute of Hydraulic Engineering and Water Resources Management, Aachen University. The following list provides an overview of the characteristics of the STRAFE multi-continuum model used for the investigations in this book:

- the number of components is theoretically infinite;
- permeable components can only be modeled globally;
- porous components can be modeled locally or globally;
- two permeable components are coupled with the quasi-steady exchange formulation;
- the coupling of a permeable and a porous component can be performed with the quasi-steady exchange formulation or the transient formulation;

- simultaneous coupling of several components can be done in a parallel, serial or selective way;
- the numerical solution is achieved by the standard Galerkin Finite Element method;
- a combination of one-, two- or three-dimensional finite elements is possible.

Birkhölzer (1994b) and Jansen (1999) have presented very detailed descriptions of the numerical model. For the evaluation of flow and transport processes in the unsaturated zone, STRAFE has been further developed to handle these processes in multi-continuum systems by taking the RICHARDS equation into account (Lagendijk (1997) and Thielen (2002)).

2.5.5 Determination of Equivalent Parameters

In the following sections, methods for determining equivalent parameters are presented. These methods are used to obtain parameter sets that are necessary to set up a multi-continuum model as described in the previous section. Applications of these methods are described later in this book (e.g. Chapter 4).

2.5.5.1 Tensor of Equivalent Hydraulic Conductivity

The equivalent hydraulic conductivity tensor describes the flow properties of a fractured and/or porous medium, determined by an averaging process over the corresponding REV. The determination of the hydraulic conductivity tensor is performed according to Long (1983) and Wollrath (1990): a hydraulic gradient is imposed on the system under investigation with varying angles and the corresponding discharge is established (Fig. 2.26, left). The equivalent hydraulic conductivity for each direction is calculated as

$$K_\gamma = \frac{Q_\gamma}{A \cdot \mathrm{grad}(h_\gamma)}, \tag{2.51}$$

where γ is the direction, K is the equivalent hydraulic conductivity, Q is the discharge, A is the area of the outflow boundary and $\mathrm{grad}(h_\gamma)$ is the imposed hydraulic gradient. With the least-square method, an ellipse is fitted to the collection of directional values, representing the equivalent hydraulic conductivity tensor of the system (Fig. 2.26, right).

$$K_\gamma = \underline{e}^T \underline{\underline{K}}\, \underline{e} \qquad \text{with} \quad \underline{e} = \begin{bmatrix} \cos\gamma \\ \sin\gamma \end{bmatrix} \tag{2.52}$$

The better the directional hydraulic conductivities fit the ellipse, the closer the system is to an REV and the better its flow processes can be described by a continuum with the properties of the determined equivalent tensor.

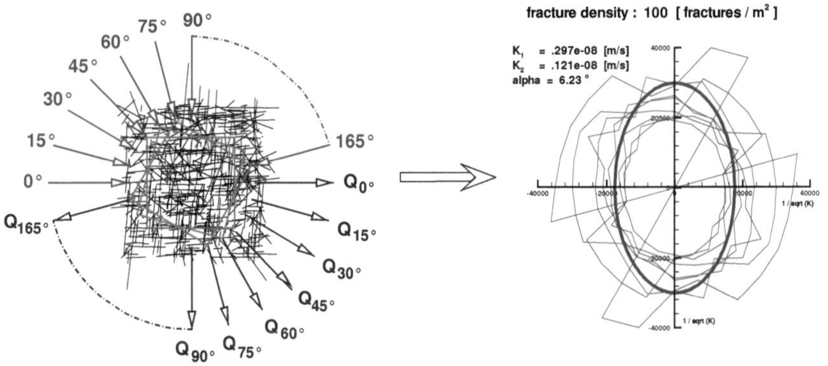

Fig. 2.26. Fracture system and directional discharges (left). Calculated hydraulic conductivities (right) for different fracture system realizations and fitted ellipse (Hemminger *et al.*, 1998).

2.5.5.2 Equivalent Porosity

To describe the mean transport characteristics of a medium, the equivalent porosity $\overline{n}$ is ascertained. The procedure is based on the same principle as for the determination of the tensor of equivalent hydraulic conductivity (cf. Sect. 2.5.5.1).

For each angle γ in which a hydraulic gradient is applied to the system, a transport calculation is performed (Fig. 2.26, left). At the input boundary, a known quantity of an ideal tracer is introduced into the system. At the output boundary, the breakthrough curve is determined and the corresponding cumulative curve is calculated. The equivalent porosity $\overline{n}$ of the system is obtained by:

$$\bar{n}_\gamma = \frac{Q_\gamma t_{\text{median}}}{V} \,. \tag{2.53}$$

The parameter $\bar{n}_\gamma$ is the directional equivalent porosity, Q_γ is the directional discharge, and V is the volume of the total system. t_{median} is the point in time at which 50 % of the tracer mass are detected at the outflow boundary.

Here, only volumes are considered, since the ratio of the volume occupied by flow processes to the total volume is calculated. Initially, the porosity is regarded as being directional, as it is determined on the basis of a breakthrough curve corresponding to a certain direction. It goes without saying that for the calculations, the porosity is considered to be independent of direction and is thus a scalar quantity.

2.5.5.3 Tensor of Equivalent Dispersivity

In order to derive the equivalent dispersion tensor $D_{\mathrm{d},ij}$, first the directional dispersivity $\alpha_{\mathrm{l},\gamma}$ is determined for each of the tracer breakthrough curves.

The determination is based on the one-dimensional transport equation for a DIRAC impulse:

$$c\left(x,t,\alpha_{l,\gamma}\right) = \frac{\Delta\, M}{2\, n_{e,\gamma}\, b\, m\, \sqrt{\pi\, \underline{v}\, t\, \alpha_{l,\gamma}}} \cdot exp\left[-\frac{\left(x - \underline{v}\, t\right)^2}{4\, \underline{v}\, t\, \alpha_{l,\gamma}}\right]. \qquad (2.54)$$

Here, c is the concentration, x is (in this case) the distance between the in- and the outflow boundary, t is the time, $\Delta\ M$ is the injected tracer mass, $n_{e,\gamma}$ is the directional porosity, b the width of the area, m the thickness of the domain, $\underline{v}$ is the mean seepage velocity and $\alpha_{l,\gamma}$ is the directional dispersivity to be determined. With equation (2.54), the $\alpha_{l,\gamma}$ yielding the best reproduction of each of the tracer breakthrough curves is iteratively obtained. For this purpose, the equation is solved with gradually varying $\alpha_{l,\gamma}$ and the deviation from the simulated curve is derived using the least-square method, where N is the number of time sampling points of the tracer breakthrough curve:

$$E = \sum_{i=1}^{N} \left(c_{sim,i} - c\left(x, t_i, \alpha_{l,\gamma}\right)\right)^2. \qquad (2.55)$$

The error is plotted over $\alpha_{l,\gamma}$ and the minimum is established.

From the determined directional dispersion lengths, the directional dispersion coefficient D_γ can be ascertained using the corresponding seepage velocity. The dispersion tensor is established according to the same procedure as for the determination of the hydraulic conductivity (cf. Fig. 2.26). From the principal axes of the tensor, the equivalent longitudinal and transversal dispersion for the angle between the first principal axis and the polar axis can be determined. This concept yields an approximated reference value for the directional dispersivity. The disadvantage of this rather simple approach is that early concentration peaks and strong tailing, both very typical characteristics of the tracer breakthrough curves of fractured porous systems, are not satisfactorily reproduced.

2.5.6 Characteristic Values

Birkhölzer (1994a) and Jansen (1999) develop characteristic values in order to evaluate the importance of exchange processes and to assess the integral transport behavior of the model. A good approximation of characteristic values is necessary to choose an appropriate type of exchange model and to identify the relevant parameters of the aquifer.

As the multi-continuum model presented here allows for both a coupling of fracture-matrix systems and fracture-fracture systems, two conceptual systems are investigated to identify the characteristic values.

Berkowitz *et al.* (1988) present an idealized model area that consists of a regularly fractured fracture network and uniform matrix blocks. The model area and boundary conditions are illustrated in Figure 2.27.

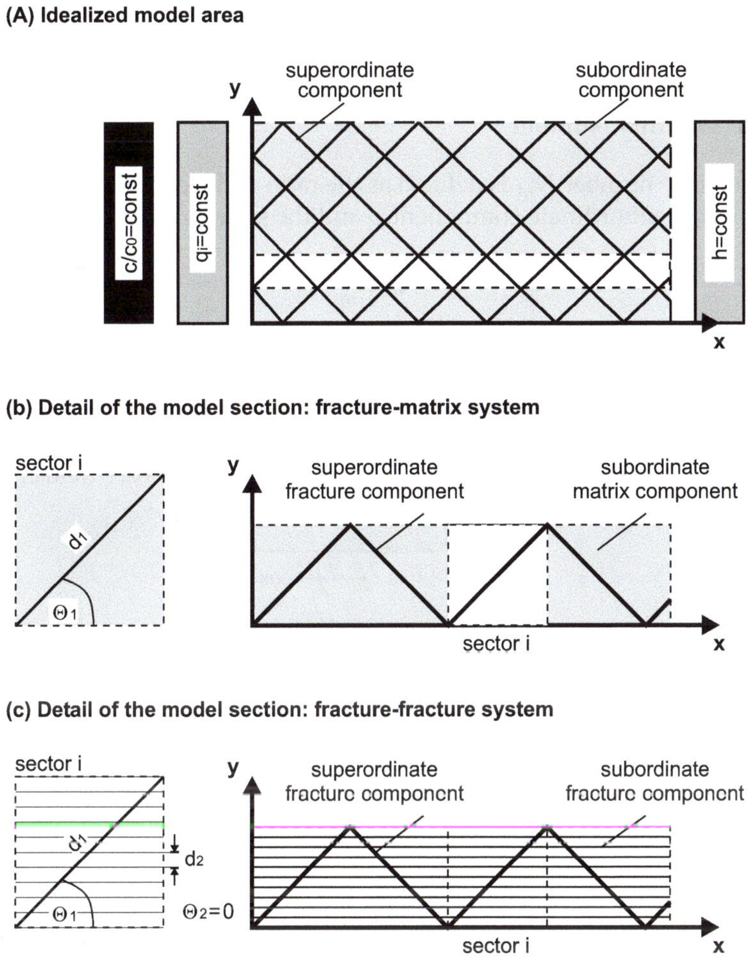

Fig. 2.27. Idealized model area to identify the characteristic values (Jansen, 1999).

The superordinate component consists of two fracture sets that are inclined by θ_1 and $-\theta_1$ respectively. These matrix blocks are quadratic with a side length of d_1 that are supposed to be isotropic and homogeneous. The fracture spacing d_1 and the fracture aperture of $\bar{b}_1$ are assumed to be constant. The subordinate fracture set of the fracture-fracture system consists of horizontal fractures with a spacing of d_2 and an aperture of $\bar{b}_2$.

For the confined model area, steady-state flow conditions are assumed. Continuous tracer injection is performed on the left model boundary.

A regional flow in the direction of the x-axis occurs due to the model area and boundary conditions. This flow field does not cause a dispersive tracer flow at the fracture intersections. Thus, the model area may be reduced to a strip of $d_1 cos\,\theta_1$ in height and a length x.

2.5.6.1 Mobility Number

The mobility number N_M is defined as the ratio of the equivalent DARCY velocity of the subordinate component β and the superordinate component α:

$$N_M = \frac{q_\beta}{q_\alpha}\,. \tag{2.56}$$

It allows for a characterization of the coupled components with regard to the mobility of the fluid in the subordinate component and is therefore a significant criterion on for choosing between a storage or mobility approach.

In the case of isotropic flow conditions and a hydraulic gradient in the x-direction considered here, the mobility number of the fracture-matrix system is defined as

$$N_M = \frac{q_{x,2}}{q_{x,1}} = \frac{K_f}{\frac{g}{6\nu}\frac{(\bar{b}_1)^3}{d_1}cos^2\theta_1}\,, \tag{2.57}$$

whereas, for the fracture-fracture system it is defined as:

$$N_M = \frac{q_{x,2}}{q_{x,1}} = \frac{1}{2}\left(\frac{\bar{b}_2}{\bar{b}_1}\right)^3 \frac{d_1 cos^2\theta_2}{d_2 cos^2\theta_1}\,. \tag{2.58}$$

In order to evaluate the mobility of the fluid in the subordinate component, the relevant parameters for a fracture-matrix system are, according to equation (2.57), the equivalent hydraulic conductivity of the matrix, the fracture width and fracture density of the superordinate component, as well as the orientation of the fractures to the hydraulic gradient.

The flow regime in fracture-fracture systems is determined according to equation (2.58) by the ratio of the fracture width, the ratio of the fracture densities and the ratio of the orientation to the hydraulic gradient.

The reciprocal value of the fracture distances is equal to the fracture density of parallel fractures which is defined as the number of fractures per unit area (Long *et al.*, 1982). Fracture densities are used here because they may also be applied to irregular fracture networks.

2.5.6.2 Diffusion-Advection Number

The diffusion-advection number N_{DA} is defined as the ratio of mass transferred in a diffusive and regional-advective way:

$$N_{DA} = \frac{M_D}{M_{Ar}}\,. \tag{2.59}$$

This parameter allows for the characterization of the importance of exchange processes. It is defined for steady-state flow conditions, where mass transfer due to local advection does not occur.

The diffusive mass transport in the subordinate component results from the spatial and temporal integration of the FICK's second law:

$$\frac{\partial C}{\partial t} = D_{\mathrm{m},2}\frac{\partial^2 C}{\partial s^2} , \tag{2.60}$$

leading to

$$M_D = C_0 n_2 \sqrt{8D_{\mathrm{m},2}T} dA , \tag{2.61}$$

where the equivalent porosity of the subordinate fracture component is defined as:

$$n_2 = \frac{\bar{b}_2}{d_2} . \tag{2.62}$$

Concerning the advective mass transport, only the perpendicular projection $sin\,\theta_1 dA$ in the direction of the flow q_2 is relevant. The mass transferred advectively in the subordinate component during the time T thus results in:

$$M_{Ar} = C_0 q_2 T sin\,\theta_1 dA . \tag{2.63}$$

Taking the flow velocities into account, the diffusion-advection number of the fracture-matrix system is defined as

$$\mathrm{N_{DA}}(T) = \frac{M_D}{M_{Ar}} = \frac{n_2}{K_f I_x sin\,\theta_1}\sqrt{\frac{8D_{\mathrm{m},2}}{T}} \tag{2.64}$$

and, for the fracture-fracture system, one obtains:

$$\mathrm{N_{DA}}(T) = \frac{M_D}{M_{Ar}} = \frac{12\nu}{g\bar{b}_2^2 cos^2\theta_2 I_x sin\,\theta_1}\sqrt{\frac{8D_{\mathrm{m},2}}{T}} . \tag{2.65}$$

Interpreting the importance of mass-exchange processes by the diffusion-advection number leads to the same parameters of the discrete system as for the mobility number (cf. Sect. 2.5.6.1). Additionally, the time scale is taken into account. At early points in time, the gradient of concentration at the block surface is very steep, resulting in a high diffusive mass exchange and a maximum value for N_{DA}. Later, the concentration gradient decreases and the diffusion-advection number exhibits a smaller value. As a characteristic point in time, the loading time T^* is suggested.

2.5.6.3 Loading Time

The loading time T^* is a characteristic time scale describing the period of interaction between the components. In order to determine T^*, both advective and diffusive processes are considered. T^* is defined to account for the time a tracer particle takes to move the distance s_{max} (penetration depth) from the block surface to the middle of the block advectively during the time T_A^* and diffusively during the time T_D^* respectively:

$$T^* = \left(\frac{1}{T_A^*} + \frac{1}{T_D^*}\right)^{-1} . \tag{2.66}$$

The maximum advective transport distance from the block surface to the center of the block is expressed by

$$s_{\max,A} = d1 cos\, \theta_1 . \tag{2.67}$$

The pore velocity within the matrix component is defined as

$$v_{x,2} = \frac{K_f I_x}{n_2} , \tag{2.68}$$

and, for the subordinate fracture component,

$$v_{x,2} = \frac{1}{n_T} \frac{g}{12\nu} \bar{b}_2^2 cos^2\theta_2 I_x , \tag{2.69}$$

where n_T is the porosity of the fracture filling. The advective loading time T_A^* is therefore defined for the matrix component as

$$T_A^* = \frac{n_2 d_1 cos\, \theta_1}{K_f I_x} \tag{2.70}$$

and as

$$T_A^* = \frac{n_t 12 \nu d_1 cos\, \theta_1}{g \bar{b}_2^2 cos^2\theta_2 I_x} \tag{2.71}$$

for the fracture component. The diffusive transport distance perpendicular to the block surface is defined for a matrix component as

$$s_{\max,D} = d_1 sin\, \theta_1 . \tag{2.72}$$

Within the subordinate fracture component, diffusive solute transport may only occur in the direction of the fractures. The diffusive transport distance to the center of the block may thus be determined as

$$s_{\max,D} = d_1 cos\, \theta_1 . \tag{2.73}$$

Therefore, the diffusive loading time of a matrix component is

$$T_D^* = \left(\frac{1}{2} d_1 sin\, 2\theta_1\right)^2 \frac{1}{2D_{\mathrm{m},2}} \tag{2.74}$$

and

$$T_D^* = \left(d_1 cos\, \theta_1\right)^2 \frac{1}{2D_{\mathrm{mol},2}} \tag{2.75}$$

for a subordinate fracture component.

Comparing the loading times for a fracture-matrix system and a fracture-fracture system, special attention has to be given to the equivalent porosity of a matrix and a fracture component because of their different orders of magnitude. The equivalent porosity of a matrix component is, in general, much higher than for a fracture component. Therefore, the relevant time scales for the exchange processes of fracture-matrix and fracture-fracture systems differ.

2.5.6.4 Loss of Identity Length

The loss of identity length L^* is a characterisitic length scale. It is used to estimate the transport distance within which a distinction between the two components is necessary. It is defined by an empirical approach and specifies the transport distance within which the concentration in the superordinate component, introduced at the input boundary, is dissipated to 0.1 % of the initial value, due to exchange processes with the subordinate component. There is a distinction between the advective loss of identity length L_A^* and the diffusive loss of identity length L_D^*, which are calculated as the loss of identity length exclusively for regional advection and diffusive mass exchange respectively. The loss of identity length is defined as:

$$L^* = \left(\frac{1}{L_A^*} + \frac{1}{L_D^*}\right)^{-1} . \tag{2.76}$$

Birkhölzer (1994a) expresses the term for the advective loss of identity length by:

$$L_A^* = -\frac{\ln 0.001}{N_{\mathrm{M}}} . \tag{2.77}$$

According to Tang *et al.* (1981) the diffusive loss of identity length L_D^* is estimated by means of the analytical solution of the diffusion of a tracer from a single fracture into an infinite half space. Making use of the simplification that adsorption and degradation processes can be neglected, the following equation is established:

$$\frac{c}{c_0} = 2\frac{exp(vz)}{\sqrt{\pi}} \int_{l}^{\infty} exp(-\xi^2 - \frac{v^2 z^2}{4\xi^2}) erfc\left(\frac{Y}{2T}\right) d\xi \tag{2.78}$$

with:

$$v = \frac{v_1}{2D_1} \tag{2.79}$$

$$Y = \frac{v^2 \beta^2 z^2}{4A} \xi^{-2} \quad \text{with} \quad \beta^2 = \frac{4D_1}{v^2} \quad \text{and} \quad A = \frac{b}{n_2 \sqrt{D_2}} \tag{2.80}$$

$$T = \sqrt{t - \frac{z^2}{4D_1} \xi^{-2}} \tag{2.81}$$

$$l = \frac{z}{2} \frac{1}{\sqrt{D_1 t}} \tag{2.82}$$

$$D_1 = v_1 \alpha_l + D_{\mathrm{m}} \tag{2.83}$$

$$D_2 = \tau D_{\mathrm{m}} \,. \tag{2.84}$$

z characterizes the transport distance, v_1 is the pore velocity and D_1 the coefficient of dispersion in the fracture. The effective coefficient of molecular diffusion in the matrix is described by D_2 and the tortuosity in the matrix is defined as τ. The period of the observation is t.

By means of equation (2.78), the concentration profiles along the fracture axis are determined for specific points in time (Fig. 2.28). The transport length for which the concentration is equal to 0.1 % of the concentration at the input boundary can be deduced from these profiles. Birkhölzer (1994a) proposes a value of $0.2T_D^*$ as the period of observation. T_D^* is the diffusive loading time of a component. A detailed derivation of this parameter can be found in Birkhölzer (1994a).

2.5.7 Summary

Whereas Sect. 2.4 presents the discrete model concept, this section gives an overview of the alternative multi-continuum model concept. Additionally, necessary equivalent parameters are defined and the determination of these parameters is explained. Characteristic values quantifying the exchange processes in hydraulic systems with more than one component are introduced and their derivation is shown based on a idealized model area. This allows for a better understanding of the numerical investigations presented and discussed within this book.

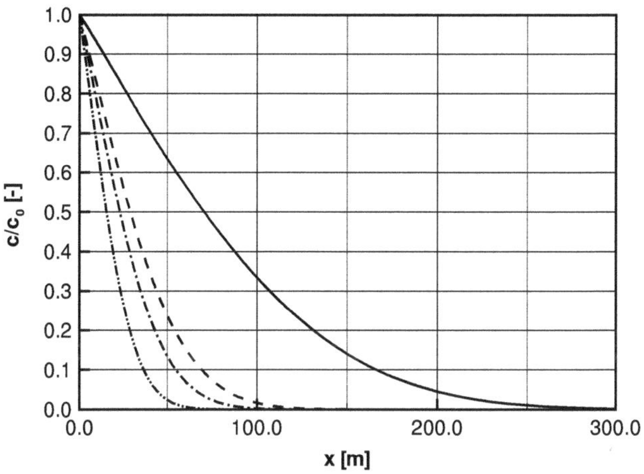

Fig. 2.28. Typical concentration profiles along the fracture for different points in time (Tang *et al.*, 1981).

Multi-continuum models, as a compromise between a detailed discrete approach and a simplified single-continuum approach, substitute the heterogeneous structure of an aquifer by overlaying, coupled homogeneous components that are continuously distributed in the model domain. Special attention has to be paid to the exchange processes and the related coupling terms.

Even though a multi-continuum approach is more general than a discrete model, a basis of solid and reliable experimental data is critical for the success of the model.

2.6 Summary

Following the steps of the transformation from a complex natural system to a simplified numerical model, general characteristics of fractured porous systems, the basic theory and concepts of flow and transport processes in such systems, and two different model approaches are presented.

The discussion has the objective of founding a basis for a better understanding of the research results presented in this book in later chapters.

The general properties of the components of fractured porous systems and in particular the geometrical characterization, often based on statistical approaches, are described.

A basic physical-mathematical description of single-phase flow and transport processes in porous media is given, relevant to the issues discussed in this book.

Depending on the problem scale, the types of heterogeneities (e.g. faults, fractures, fissures) and their frequencies, two different model approaches for the consideration of fractures in porous domains are proposed.

In a discrete model, the matrix and the fractures are locally idealized as continua and the fractures are implemented discretely at their actual location within the domain. Due to the extensive property and geometrical data requirements, as well as the high spatial and temporal resolution necessary to obtain accurate results, discrete models are best suited for principal investigations on a limited scale.

The second approach introduced is the multi-continuum model approach. By transforming, for example, the matrix and the fractures on different scales into separate equivalent continua, the domain is partly homogenized. The exchange between the defined continua is realized by introducing exchange terms for both flow and transport interaction processes. This approach is appropriate for larger problem scales due to the requirement of averaging over a representative elementary volume.

The amount and quality of available data is, for both model approaches, an essential pre-requisite for reliable model results.

Part II

Project Scale Studies

3

Core Scale

S. Baraka-Lokmane, R. Liedl, M. Sauter, G. Teutsch

Activities on the core scale (sample size approx. 0.1 m) focus on determining the permeability and hydraulic conductivity of fractured-porous sandstone samples. Core scale experiments benefit from the fully controlled conditions prevailing in the laboratory. The contribution of individual fractures to overall sample properties and the feasibility of using air as a test fluid instead of water, which represent two major issues in aquifer analog studies, can therefore be analyzed without external perturbations. The development and application of experimental techniques is complemented by simple geometric and hydraulic considerations establishing quantitative expressions for the sample properties mentioned above. In the end, basic knowledge obtained from the experimental and modeling investigations at the core scale may support the characterization of fractured porous media at larger scales.

3.1 Sample Characterization

Eight core samples of 10 cm in diameter and thickness are chosen to demonstrate the feasibility of the experimental methods developed. The samples originate from the middle *Stubensandstein* that belongs to the Middle Keuper of the Southwest German Trias. They are selected according to the following criteria in order to ensure flow governed by fractures (Baraka-Lokmane, 2002a):

- visible and open fractures;
- fractures with the same orientation as the core axis;
- fractures cutting the core from top to bottom.

Mineralogical investigations show that the samples are classified as arkose with well to poorly sorted grains (Baraka-Lokmane, 2002a). Quartz is the dominant framework (70% to 80%), feldspars are present with a proportion of 10% to 25% and micas are rare. Cements are abundant and clay minerals, carbonate (calcite or dolomite) and quartz are most common. Figures 3.1 and 3.2 are scanning electron microscope (SEM) photographs, providing examples of the coexistence of several types of cements.

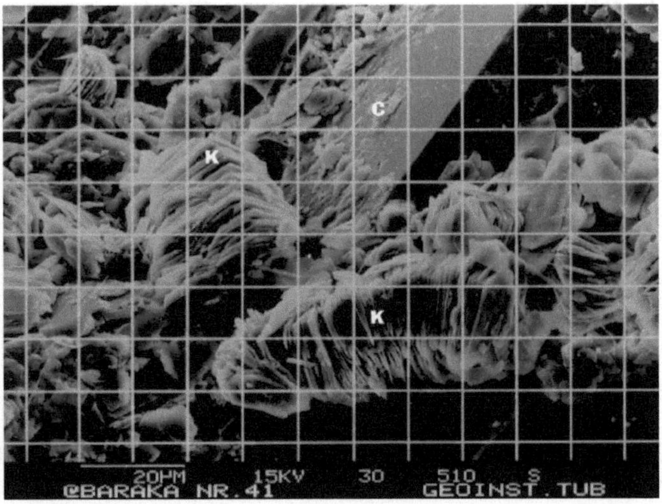

Fig. 3.1. SEM photograph of a cement containing kaolinite (K) and carbonate minerals (C) (Baraka-Lokmane *et al.*, 2003).

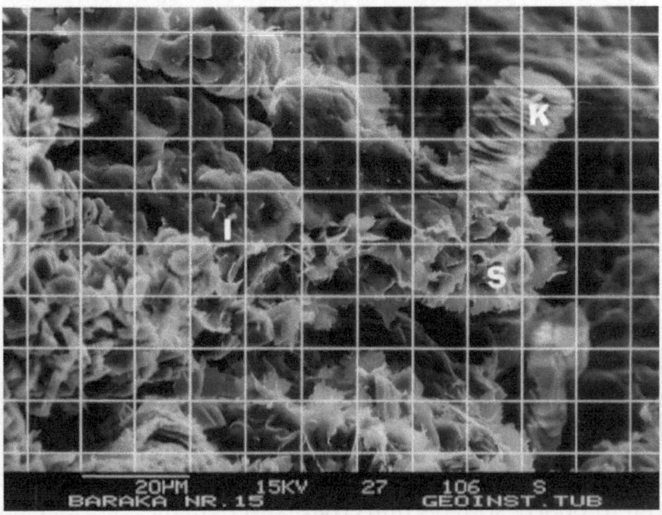

Fig. 3.2. SEM photograph of a cement containing smectite (S), kaolinite (K) and illite minerals (I) (Baraka-Lokmane, 2002a).

According to the type of the dominant cement, the eight samples (Table 3.1) can be divided into three groups:

- samples with carbonate as the dominant cement (samples 4 and 5);
- sample with kaolinite as the dominant cement (sample 8);
- samples with clay minerals (kaolinite or smectite) and carbonate as the dominant cements (samples 1, 2, 3, 6 and 7).

Table 3.1. Characteristics of the eight selected samples (Baraka-Lokmane, 2002a)

Sample Number	Porosity (%)	Saturation (%)	Type of cement (* = dominant)	Fraction characteristics
1	21.54	95.89	kaolinite*, carbonate*, quartz	• more than one fracture • large fracture aperture • presence of clay nodules in the matrix
2	25.69	91.37	kaolinite*, carbonate*	• one fracture • large fracture aperture
3	21.31	91.01	carbonate*, smectite*, illite, kaolinite, quartz	• more than one fracture
4	21.85	93.27	carbonate*, kaolinite	• more than one fracture
5	21.19	93.27	carbonate* kaolinite	• one fracture
6	20.52	98.21	carbonate* quartz, smectite*	• more than one fracture
7	20.17	90.22	carbonate* quartz, smectite*	• more than one fracture
8	17.89	76.69	kaolinite*, quartz carbonate	• one fracture • large fracture aperture

3.2 Experiments

The experiments on the core scale include two approaches. First, detailed investigations of the fracture geometry are performed by resin casting and magnetic resonance imaging. Second, flow tests are conducted using water and air as test fluids (hydraulic and pneumatic experiments, respectively). Both approaches are pursued in order to derive sample permeabilities and conductivities in Sect. 3.3.

3.2.1 Fracture Geometry Investigations

The importance of fracture geometry for fluid flow behavior is widely recognized in the disposal of radioactive waste, petroleum exploitation and underground energy storage. Indeed, the hydrogeological response is governed by the size of the fractures, their connectivity and by the roughness of the fracture surfaces. Various laboratory and field studies have improved the understanding of flow in natural fractures (e.g. Brown and Scholz (1986); Gentier (1986); Pyrak-Nolte *et al.* (1987); Gale (1987); Hakami (1988)). These empirical investigations of fracture void geometry have led to a conceptualization of fractures as two-dimensional heterogeneous porous media (Cox *et al.*, 1990). In the following, a new resin casting technique and magnetic resonance imaging (MRI) are shown to provide relevant information about fracture geometry.

3.2.1.1 Resin Casting

Resin casting is a well known method for obtaining a complete description of the fracture void geometry. Gentier and Billaux (1989) suggest and, for example, Cox *et al.* (1990) and Dollinger *et al.* (1994) actually use a silicone polymer resin as the casting material. During casting, the sample is subjected to a normal stress. The two sides of the fracture are fitted together to act as a mould which is filled with the resin. For calibration purposes, a wedge of known thickness is cast at the same time, using the same batch of silicone material in a machined mould with a flat cover plate. The fracture and wedge moulds are both filled with the resin, and excess polymer is squeezed out by applying pressure on the opposite sides of the fracture. The main problem associated with this technique is that the fracture aperture is no longer natural. Furthermore, the technique cannot be applied to very small fracture apertures (in the order of micrometers).

In order to circumvent these disadvantages, a new resin impregnation method was developed to determine the total and detailed fracture geometry of 10 cm samples without disturbing the fracture (Baraka-Lokmane *et al.* (2001); Baraka-Lokmane (2002b)). In contrast to the studies mentioned above, the focus is not on granitic rocks but on fractured sandstone. The parameters of the fracture geometry to be determined are the fracture aperture, fracture length, fracture width and fracture roughness (Sect. 2.1 and Fig. 3.3).

Using material from the *Stubensandstein* rock described in Sect. 3.1, i.e. the same type of material as samples 1–8, core sections of 1 cm thickness are impregnated under vacuum with the help of the equipment illustrated in Fig. 3.4. The top of the core section is left open and the resin enters through the bottom by capillary forces. In order to make the fracture space more visible for the subsequent digital analysis of the fracture geometry, two dye tracers (Fluoresceine and Methylene blue) and two resins (RTV 141 and EPOFIX) are used for the impregnation. The physical characteristics of these compounds are listed in Table 3.2.

The resin impregnation results are illustrated in Figs. 3.5 and 3.6. For a better visualization of the fractures, the resins RTV 141 and EPOFIX are colored black. The different impregnations show that, due to the improved

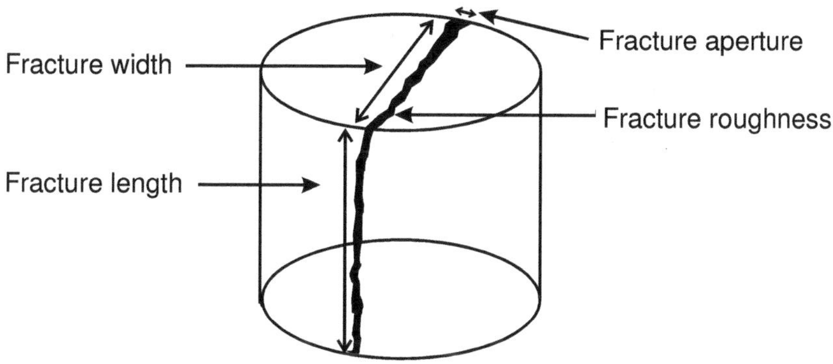

Fig. 3.3. Fracture parameters to be determined by the resin impregnation method (Baraka-Lokmane, 2002a).

Fig. 3.4. Resin impregnation equipment (Baraka-Lokmane, 2002b).

contrast, the results are better with resins than with dye tracers, especially when the resin is colored black.

Figures 3.5 and 3.6 show that Fluoresceine and Methylene blue do not improve the visualization of the fractures. The pixel maps just become paler than the pixel map of the same sandstone section without any impregnation (Fig. 3.7).

Table 3.2. Physical properties of the compounds used for the impregnation (from Baraka-Lokmane (2002a))

Method	Dye tracer		Cast (resin)	
Compound	Fluoresceine	Methylene blue	RTV 141 resin + hardener (Rhône Poulenc)	EPOFIX resin + hardener (Struers)
Experiment	under vacuum			
Dosage	1.5 mg/ 100 ml water	100 mg/ 100 ml ethanol	70 g resin + 7 g hardener	50 g resin + 6 g hardener
Coloration	green	blue	colored black with Araldite coloring paste (Gbrot)	
Viscosity in cps (20°C)	-	-	4000	550 150 (50°C)
Temperature	20°C			
Pot life	-		4 hours	30 minutes
Hardening time	-	-	2 hours at 100°C	8 hours at 20°C

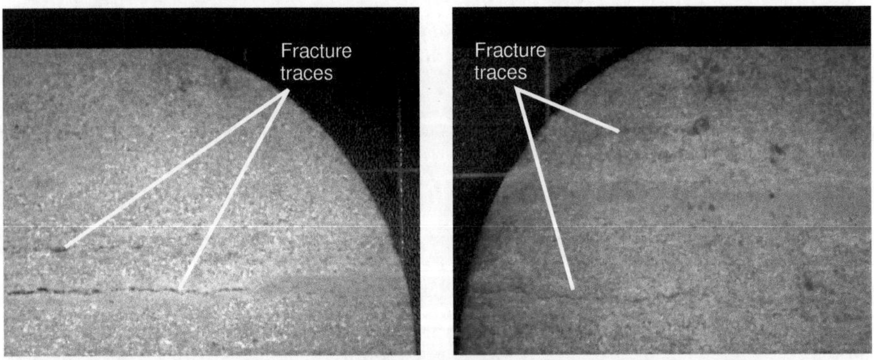

Fig. 3.5. Pixel maps of a sandstone core section impregnated with Fluoresceine (left) and Methylene blue (right) (Baraka-Lokmane, 2002a).

Impregnating core sections of 1 cm thickness under vacuum requires almost five hours under vacuum for each core section. For this reason, Baraka-Lokmane (2002a) developed a resin injection apparatus (Fig. 3.8) to impregnate the complete sample with 10 cm thickness in one experimental run, i.e. within an impregnation time of 1 hour (Baraka-Lokmane, 2002b). In this setup, the sample is wrapped in a latex membrane and then fixed in a core holder. The resin is injected through the sample from the top to the bottom using air pressure. At the same time, the apparatus is connected at its base to a vacuum pump.

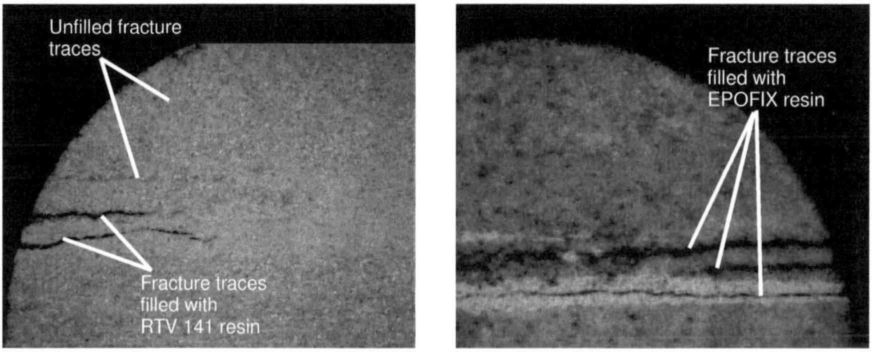

Fig. 3.6. Pixel maps of a sandstone core section impregnated with RTV 141 (left) and EPOFIX (right). Resins are colored black (Baraka-Lokmane, 2002a).

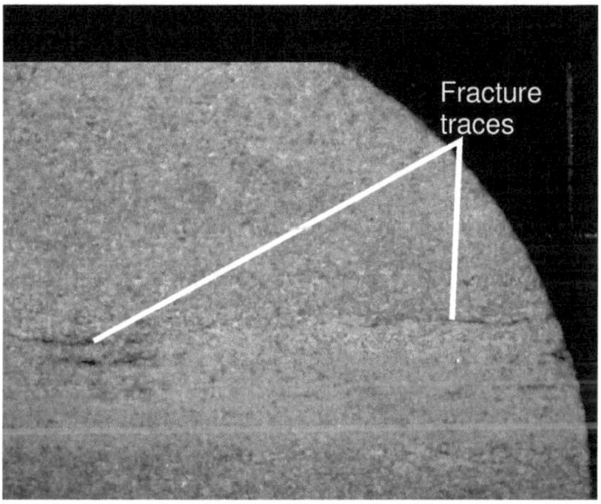

Fig. 3.7. Pixel map of a sandstone core section without any impregnation (Baraka-Lokmane, 2002a).

A series of resin impregnation experiments were performed by Baraka-Lokmane (2002a) using the injection technique. Three resins (Sika Injektion 26, EPOXY and SPURR) were tested to find the most suitable resin to be used for the impregnation. The physical properties of the tested resins are displayed in Table 3.3. The injection pressure was varied, injecting the resin at 6 pressure levels (0.05 bar to 0.1 bar in steps of 0.01 bar) for 10 minutes each.

Observations derived from the experiments can be summarized as follows:

- The sample should be oven-dried and placed under vacuum before injecting the resin.
- The best resin was found to be SPURR because of its low viscosity (60 cps at 20°C) and its long pot life (2 d).

Fig. 3.8. Resin injection equipment (Baraka-Lokmane *et al.*, 2003).

- A latex membrane is superior to silicon or gypsum encapsulation for avoiding leakage of the resin.
- The use of a piston to separate the resin from the air is not necessary.
- 1 h is an appropriate period for resin injection.

Figures 3.9 and 3.10 show two examples of the upper and the lower section of the same sample impregnated with resin using the injection technique described above. The results clearly show the sedimentary layers as well as the location of the fractures.

After impregnation, the cores characterized in Sect. 3.2.1 (Table 3.1) were cut into slices of 5 mm thickness (Figure 3.11). The resulting 128 cross-sections from the eight samples were then photographed and 196 fracture traces were digitized. The image analysis program OPTIMAS (Optimas Corp., Bothell, Washington) was used for the calculation of the parameters of the fracture geometry. Based on this analysis, the rough fracture walls are approximated by a series of rectangles and the length as well as the aperture of each rectangle determined (Fig. 3.12). This approach corresponds to the local parallel plate concept mentioned in Section 2.4.1 (Fig. 2.11). As an example, results for sample 8 are given in Table 3.4.

Various three-dimensional images of the samples were produced by combining a series of slice pictures with the fracture traces (Baraka-Lokmane,

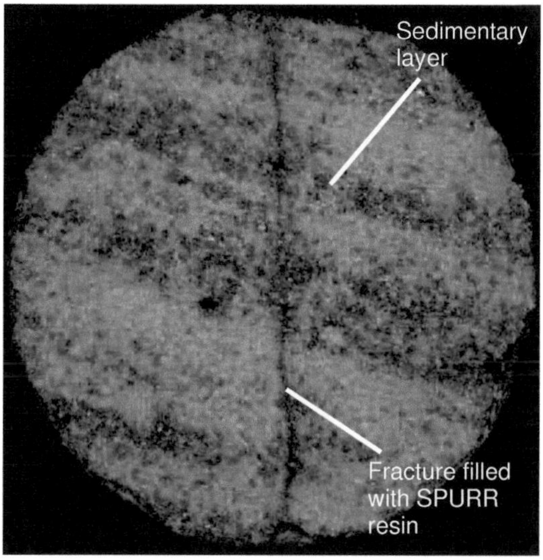

Fig. 3.9. Pixel map of the top of sample 1 impregnated with SPURR resin colored black (from Baraka-Lokmane *et al.* (2003)).

2002a). Figure 3.13 shows sample 8 containing a single fracture, which passes through the core from top to bottom. The study of the slices of 5 mm thickness and the three-dimensional images of the samples allow a detailed description of the fracture system in terms of the fracture aperture, fracture width, fracture length and the connection of the fractures. This can also be

Table 3.3. Physical properties of the tested resins (Baraka-Lokmane, 2002a).

	Resin		
Compounds	Sika Injektion 26 2 compounds (Sika, Nagra, Switzerland)	EPOXY (resin + hardener) (Gbrot, France)	SUPPR 4 compounds (Plano, United Kingdom)
Experiment	injection with pressure		
Dosing	compound A: 1.14 g compound B: 1 g	1.12 g resin + 0.95 g hardener	VCD 10 g DER: 6 g NSA: 26 g DMAE: 1 g
Coloration	colored black with Araldite coloring paste (Gbrot)		
Viscosity in cps (20°C)	100	150	60
Temperature	20°C		
Pot life	2.5 hours	50 minutes	2 days
Hardening time	1-2 days at 20°C	3 hours at 60°C	3 hours at 70°C

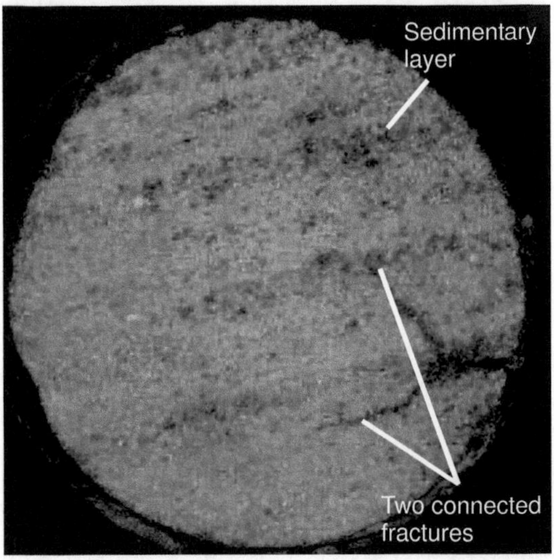

Fig. 3.10. Pixel map of bottom of sample 1 (from Baraka-Lokmane *et al.* (2003)).

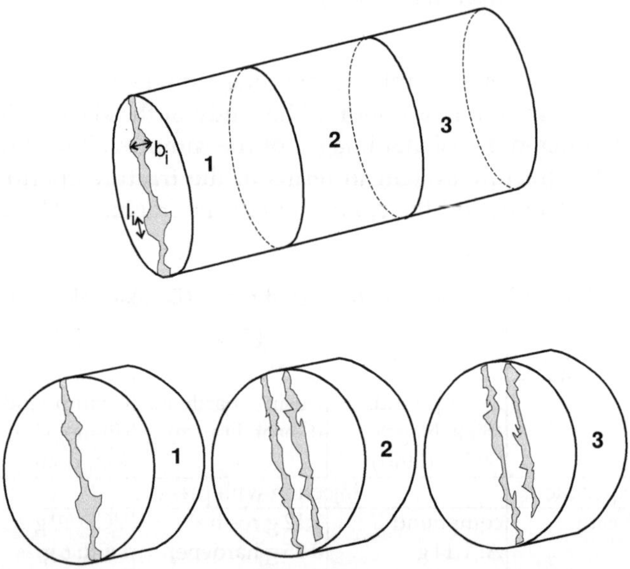

Fig. 3.11. Schematic sketch of a core sample cut in three slices (modified from Baraka-Lokmane *et al.* (2003)).

seen in Fig. 3.14, which shows sample 1 with three fractures, two of which intersect. A fracture which does not connect the top and bottom of the core is present in sample 2 (Fig. 3.15).

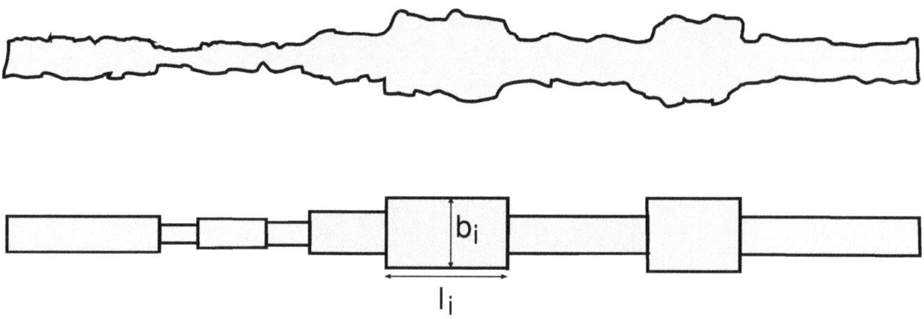

Fig. 3.12. Cross section of a real fracture and the approximation in the digital analysis (modified from Baraka-Lokmane *et al.* (2003)).

Table 3.4. Fracture geometry parameters obtained by OPTIMAS for section 10 of sample 8 (from Baraka-Lokmane (2002a))

Number i of rectangle	Width l_i (mm)	Aperture b_i (mm)
1	2.082	0.061
2	0.796	0.035
3	11.993	0.058
4	0.534	0.033
5	22.353	0.066
6	2.372	0.065
7	0.749	0.055
8	0.551	0.088
9	2.093	0.065

The roughness of each individual fracture can, in principle, be determined by two different methods. In the first method, the surface roughness is measured by profilometry (Barton (1973), Barton and Choubey (1977)), which consists of dragging a sharp stylus in a straight line over the surface to record the surface height in the form of a profile. The fracture roughness in a sample is then approximated by tracing many linear profiles. The second method, which is used here, is the optical method developed by Brown and Scholz (1986).

According to the optical method, square sections (about 1 cm × 1 cm) of the fracture surface are photographed using a SEM (Fig. 3.16). From the photographs, the elevations of the asperities can be calculated to determine the absolute roughness, defined as the mean height of the surface asperities in the fracture with respect to some reference datum.

The fracture surface in Fig. 3.16 is viewed at 45 degrees. Therefore, a correction of each elevation measurement is necessary. For the evaluation of the absolute roughness, ten SEM photographs of a fracture surface were taken for each sample and an arithmetic mean of the real elevations was calculated for each photograph. Arithmetic averaging of the results yields the absolute

Fig. 3.13. Three-dimensional visualization of sample 8 (Baraka-Lokmane *et al.*, 2003).

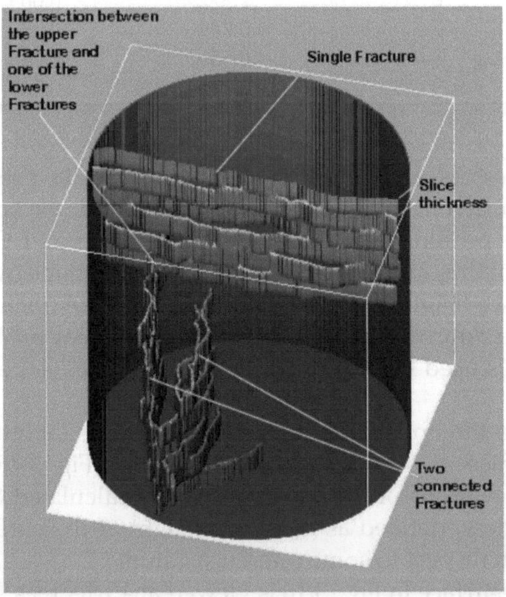

Fig. 3.14. Three-dimensional fracture visualization of sample 1 (Baraka-Lokmane *et al.*, 2003).

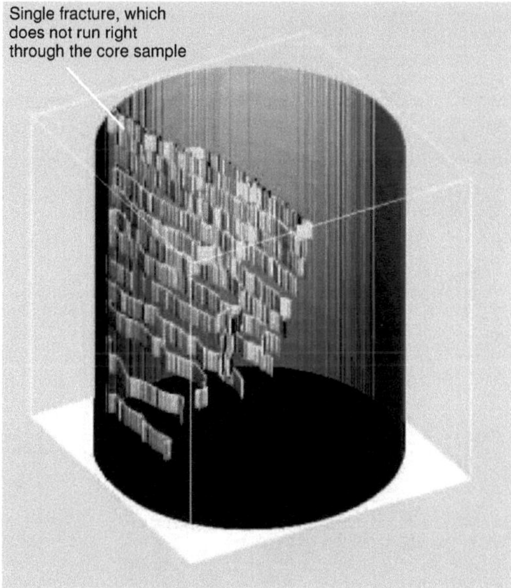

Fig. 3.15. Three-dimensional visualization of sample 2 (Baraka-Lokmane *et al.*, 2003).

fracture roughness values given in Table 3.5. The results are used in Sect. 3.3.1.1 to determine the effective conductivities of the samples.

Table 3.5. Absolute fracture roughness for investigated samples (Baraka-Lokmane, 2002a).

Sample	Absolute roughness (μm)
1	2.77
2	not determined
3	2.31
4	6.93
5	5.08
6	10.17
7	1.85
8	2.31

3.2.1.2 Magnetic Resonance Imaging

Magnetic Resonance Imaging (MRI) is another method which is used here to determine the fracture geometry of one of the eight sandstone cores investigated with the resin casting technique. MRI is a relatively fast method and can be carried out without any perturbation of the fracture aperture. It allows the visualization of mobile fluids and the interaction of these fluids with the

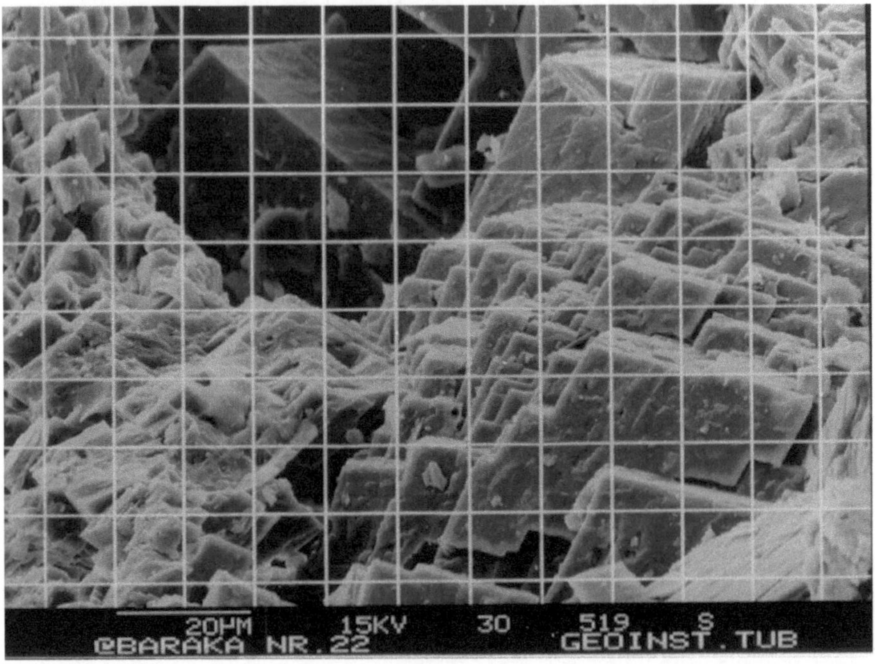

Fig. 3.16. SEM photograph of a fracture surface (Baraka-Lokmane, 2002a).

surface of the pores (Chardaire and Roussel, 1990). Nuclear magnetic resonance imaging adds the extra feature of a three-dimensional visualization of fluids in porous media.

Nuclear Magnetic Resonance (NMR) has long been used in the oil industry as a research tool for the measurement of petrophysical rock properties such as porosity (Timur (1969); Cowgill *et al.* (1981a) Cowgill *et al.* (1981b)), permeability (Headley (1973); Kenyon *et al.* (1986)) and wettability (Brown and Fatt (1956); Kumar *et al.* (1969); Williams and Fung (1982)). It also allows the distribution of pore sizes (Seevers (1966); Kozlov and Ivanchuk (1982); Brown *et al.* (1981); Gallegos *et al.* (1987); Gallegos and Smith (1988); Kenyon *et al.* (1989)) and the saturation in samples (Saraf and Fatt (1966); Saraf and Fatt (1967)) to be determined.

Application of MRI to porous media can be traced back to Gummerson *et al.* (1979), who used NMR to observe the capillary absorption of water as a function of time. They used a magnetic field of 0.7 T. Horsfield *et al.* (1989) and Dechter *et al.* (1989) employed MRI to selectively visualize water or oil by additionally inferring chemical displacement data.

Chardaire and Roussel (1990) applied the MRI technique to study fluid distributions in porous media using a magnetic field of 0.9 T. This work focused on the determination of the local porosity through the intensity of the

NMR signal of a fluid in a porous medium and the visualization of oil and water distribution by chemical-shift selective excitations.

Dijk *et al.* (1999) have used nuclear MRI for measuring the water flow velocities in the fractures and have investigated the effects of wall morphology on flow patterns inside a typical rock fracture. Their study demonstrates that nuclear MRI can effectively and accurately measure even highly heterogeneous flow fields in such systems.

NMR is mainly known as a spectrometric technique. When nuclei with magnetic moments are put in a static homogeneous magnetic field, discrete energy levels are observed. A suitable radio-frequency field is applied to induce transitions between the levels, thus changing the magnetization. At the end of radio-frequency excitation, the nuclei return to equilibrium and the magnetization decays as a function of time. The frequency of this decay depends on the strength of the magnetic field and is characteristic of both the nucleus species and their chemical environment.

To obtain an image of the distribution of a specific nucleus, e.g. hydrogen, in a rock sample, a spatial discrimination is required instead of a chemical one. This discrimination is produced by superposition of three mutually perpendicular, uniform magnetic field gradients on the main magnetic field. In this way, each point of the sample is subjected to a different magnetic field and yields a signal at a different frequency, which is linked to a spatial position. One of the gradients is used to select a slice through the sample. The higher the gradient, the narrower the slice (in conjunction with the length of the radio-frequency pulse). The other two gradients determine the size of the pixels in the slice. The resolution depends on the strength of the three gradients, on the mass of the nucleus and on the transverse relaxation rate. There are two kinds of relaxation: longitudinal relaxation, which governs the recovery of the system towards equilibrium and the frequency of the repetition rate, and transverse relaxation, which characterizes the extinction of the signal and the width of the frequency spectrum, hence the resolution of the method. The time needed to apply and interrupt gradients, especially in echo methods, can be long, and it is necessary to adjust this time with the relaxation rate to obtain a signal.

Relaxation is governed by porosity, permeability, wettability and paramagnetic impurities. In cores, it might be necessary to use sequences of fast nuclear MRI or, in extreme cases, the technique of back projection (Timur (1969); Williams and Fung (1982)). This method enables signals to be received 1 ms or less after excitation instead of 4-6 ms for standard spin-warp sequences. In strong magnetic fields, it is also possible to differentiate species with sufficient chemical resolution (Chemical Shift Selective Sequences) to obtain, for instance, an image of a fluid distribution in core samples (Hinedi *et al.*, 1997). One of the eight fractured porous sandstone samples (sample 1) was chosen to be investigated by the MRI technique, using a magnetic field strength of 1 T. To obtain images of total hydrogen content, a standard spin-warp sequence was applied to five vertical slices of 8 mm thickness, i.e. 40%

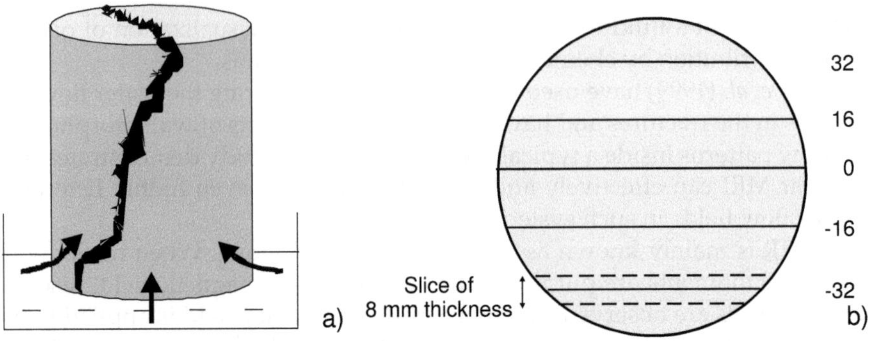

Fig. 3.17. Investigation of sample with MRI (after Baraka-Lokmane *et al.* (2001)): a) sample saturation using a receptacle of water and b) positions where images are taken.

of the sample was imaged (Figure 3.17b). For this type of pulse sequence, the sample was excited by a 90° radio-frequency pulse in a spin-echo sequence. The signal was refocused by a 180° selective pulse which, in conjunction with a magnetic field gradient along the core axis, selected the slice of interest. Discrimination in the slice plane was obtained by two gradients applied as phase encoding and a third one during the acquisition. A two-dimensional Fourier transform produced the corresponding image. The experiments were performed with a spin-echo time of 12 ms. The echo time is the time between the 90° pulse and the middle of the spin-echo production.

The sample was dried and placed in a receptacle of water so that about 1.5 cm of the lower part of the sample was immersed (Fig. 3.17a). This setup was then inserted in the MRI system (Fig. 3.18) in order to observe the development of a wetting front close to a fracture due to capillary action (Fig. 3.19).

During the MRI measurements, the sample was raised a few millimeters from the bottom of the receptacle, allowing water to infiltrate the sample from its base as well. Figure 3.19 shows that the sample is not visible as long as it is not saturated, i.e. only water is observed with MRI measurements and appears black. After two hours, the images show the fracture structure of the saturated sample. It is evident from Fig. 3.19 that sample 1 has two fractures.

The 55 images of sample 1 are combined to yield three-dimensional visualizations of the fracture geometry (Fig. 3.20), showing the two fractures already observed in Fig. 3.19. The first image of Fig. 3.20 also shows the position of the slice illustrated in Fig. 3.19. From 19 minutes after the start of the measurement, only the development of the wetting front in the matrix is observed. This indicates that the two fractures are present only in the lower half of the sample.

According to Baraka-Lokmane (2002a), the results of the MRI measurements are comparable to those obtained from resin casting methods in terms

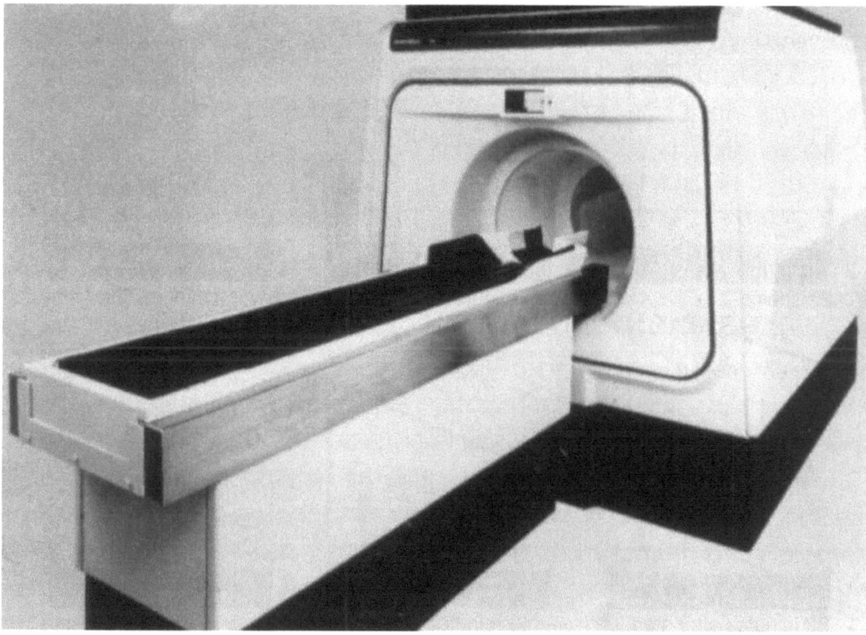

Fig. 3.18. Generic configuration of an MRI system (Baraka-Lokmane, 2002a).

of fracture inventory, orientation, position and connections. The two fractures present in the lower half of sample 1 (Fig. 3.10) are observed with MRI measurements with images taken at the position -16 (Fig. 3.17b). Unfortunately, the selection of imaging positions prohibits the detection of the fracture present in the upper half of the sample (Fig. 3.9).

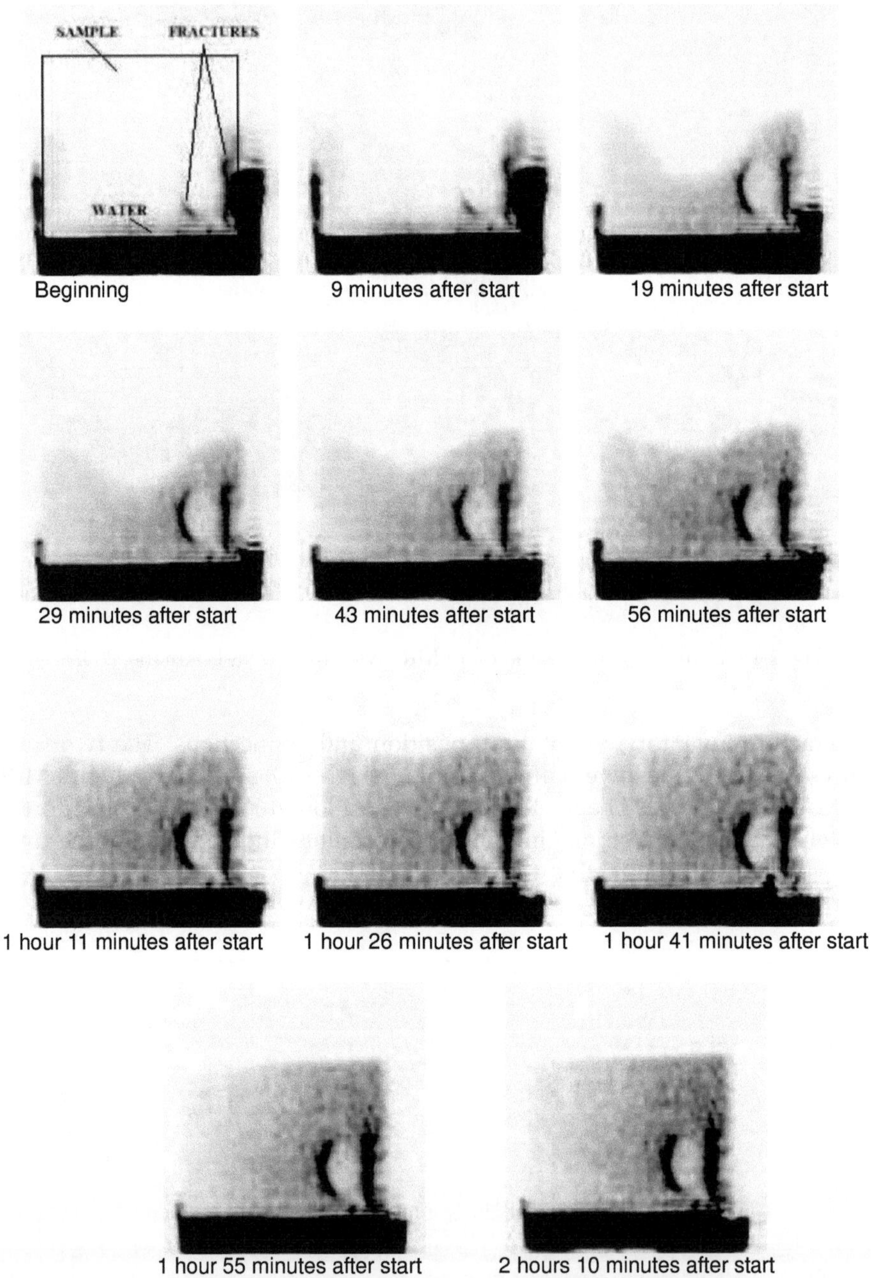

Fig. 3.19. 8 mm thick slice showing the distribution of water inside sample 1 at position -16 (see Figure 3.17b) as a function of time (Baraka-Lokmane *et al.*, 2001).

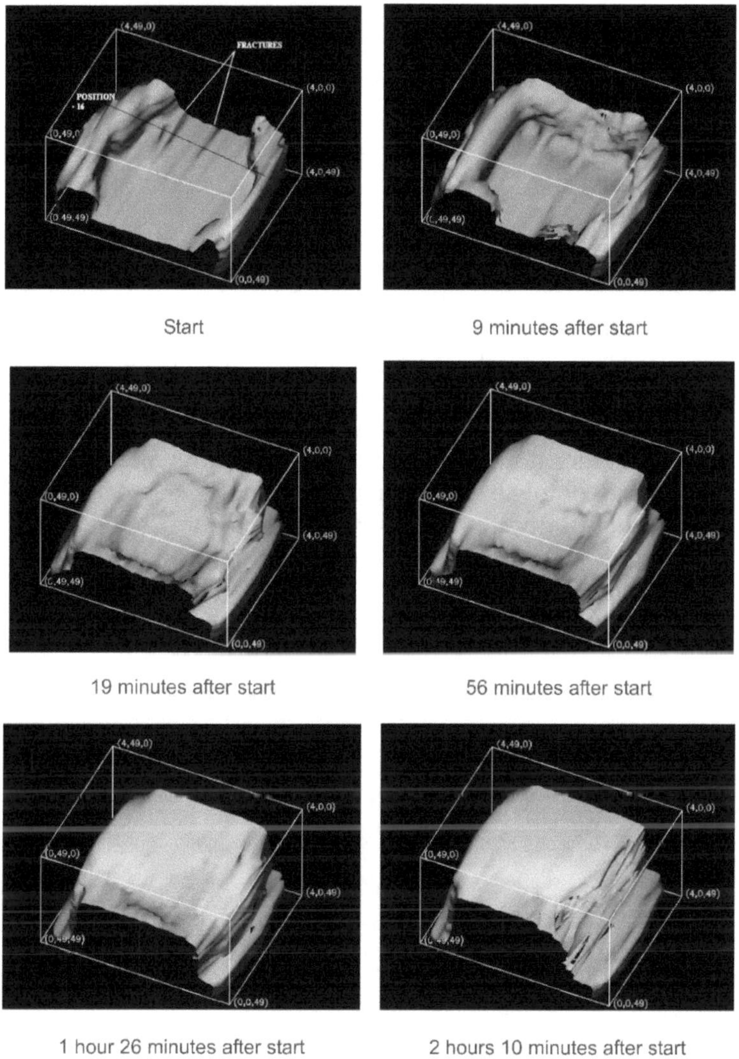

Fig. 3.20. Three-dimensional visualization of the development of the water front in sample 1 as a function of time (from Baraka-Lokmane *et al.* (2001)).

3.2.2 Hydraulic and Pneumatic Experiments

In fractured porous aquifers, average permeability is a complex function of fracture characteristics, e.g. connectivity, aperture and spacing distributions (Bloomfield and Williams, 1995) as well as matrix properties. In the field, it is extremely difficult to characterize fractures and to quantify their contribution to the permeability of the medium. Laboratory experiments can aid in determining the permeability of fractured porous systems. Since gas experiments can be conducted much faster than water flow experiments, gas is a desirable replacement fluid (Jaritz, 1999). This section is devoted to the measurement and comparison of intrinsic permeabilities obtained by using air and water as fluids.

3.2.2.1 Hydraulic Tests

Hydraulic tests rely on water as the test fluid and are carried out under vacuum for the eight fractured porous sandstone samples with the equipment shown in Fig. 3.21. The samples are sealed in a latex membrane to avoid leakage and a confining pressure of at least $3 \cdot 10^5$ Pa is applied (the maximum pressure allowed by the equipment is $7 \cdot 10^5$ Pa). Water is pressed from a storage container through each saturated sample and, with the help of burettes, the volumes of water inflow and outflow are measured. The outflow is routed into another storage container where a saturation pressure is maintained. To ensure complete saturation of the samples, the measurements are carried out at high saturation pressures ranging from $2 \cdot 10^5$ Pa to $5 \cdot 10^5$ Pa. The air residue is compressed and the saturation of the sample increases. Pressure differences across the sample are between $1 \cdot 10^4$ Pa and $7 \cdot 10^4$ Pa and flow rates range from $1.77 \cdot 10^{-8}$ m^3/s to $1.03 \cdot 10^{-7}$ m^3/s.

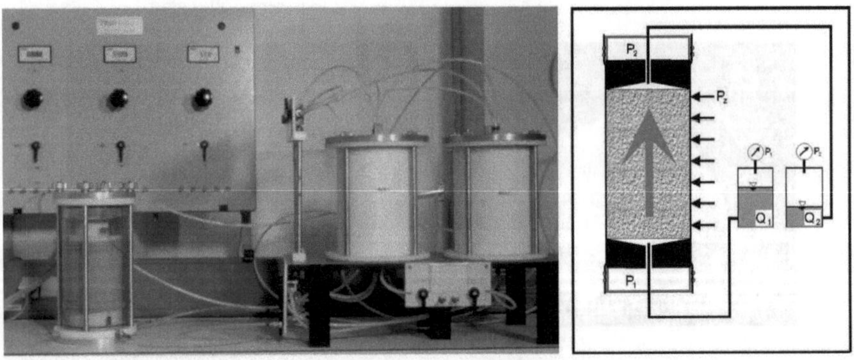

Fig. 3.21. Equipment for hydraulic tests (Baraka-Lokmane, 2002a).

The flow in these experiments is laminar, as can be seen from Fig. 3.22 showing the water flow rate through sample 3 as a function of the hydraulic gradient. Similar results were obtained for the other samples.

3.2.2.2 Pneumatic Tests

The experiments with air as the test fluid were accomplished for the same samples as the hydraulic tests (Sect. 3.2.2.1). Before starting the pneumatic tests, the samples were oven-dried.

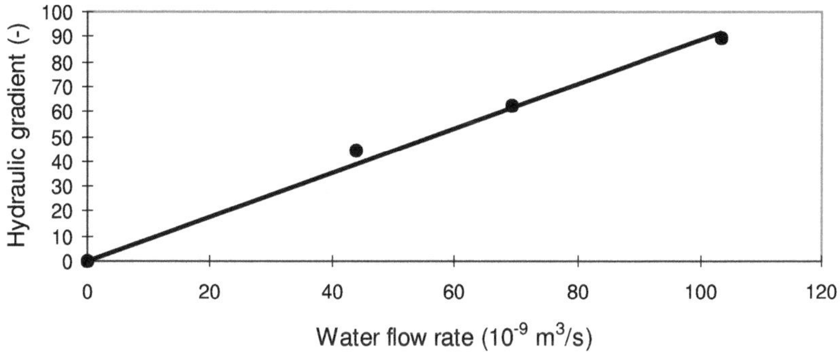

Fig. 3.22. Water flow rate as a function of the hydraulic gradient for sample 3 (Baraka-Lokmane, 2002a).

The equipment for the pneumatic tests is shown in Fig. 3.23. In contrast to the hydraulic experiments, the air pressures are measured directly at the sample input and output with two high-grade steel potential probes placed at either end of the sample. This is done to compensate for pressure losses in the supply. The oven-dried samples are sealed with a flexible latex membrane to avoid leakage. The dry air flows from the bottom to the top of the sample. Because of the compressibility of the gas, flow through the sample is allowed to stabilize before measurements are undertaken. The pressure difference adjuster helps to regulate pressure differences within short periods. For a sample with a permeability higher than 1 md, steady state is achieved within 1 min. The measurement apparatus used allows the mean sample pressure to be increased without increasing the pressure gradients. In this way, the rate of flow stays small and turbulent conditions are avoided.

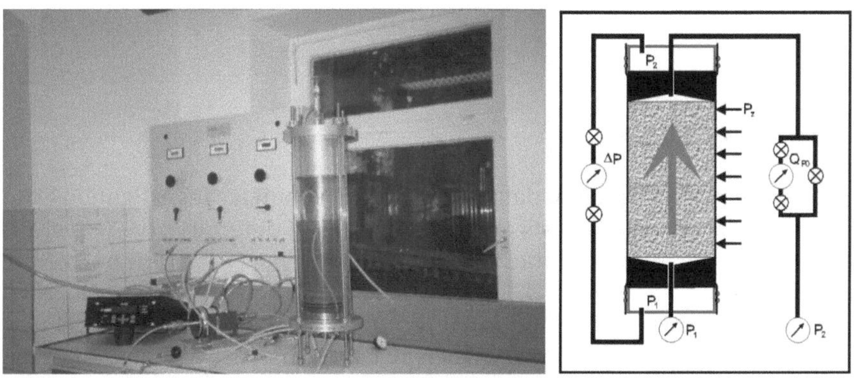

Fig. 3.23. Equipment for pneumatic tests (Baraka-Lokmane, 2002a).

Two parameters are measured. The first is the flow rate of air. It is measured by a flowmeter and ranged from $2.79 \cdot 10^{-7}\,m^3/s$ to $2.95 \cdot 10^{-6}\,m^3/s$ for the eight sandstone samples. The second parameter is the difference in gas pressure across the sample; this ranges between $7.20 \cdot 10^2$ Pa and $8.37 \cdot 10^3$ Pa. The differential pressure is adjusted to be as small as possible to give a suitable rate of air flow.

Turbulent flow conditions can be avoided by applying low gas pressure gradients. To check that this is achieved, the flow rate is measured for each sample for a range of pressure differences. The resulting graph of the gas flow rate plotted against the gas pressure gradient for sample 5 is shown in Fig. 3.24. Measured data are closely approximated by a linear function and it is therefore concluded that the permeability tests were performed under conditions of laminar flow.

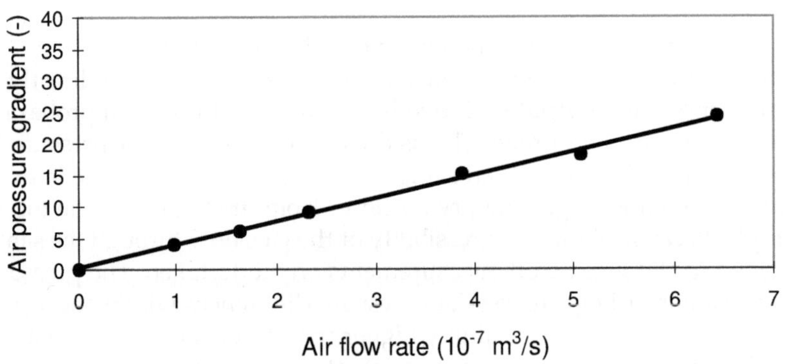

Fig. 3.24. Correlation between normalized pressure gradient and gas flow rate for sample 5 (Baraka-Lokmane, 2002a).

3.3 Interpretation

In this section, experimental data from fracture geometry investigations and hydraulic and pneumatic tests are used in order to determine the (effective) conductivities or permeabilities of each sample. The evaluation of fracture geometry information (Sect. 3.3.1) involves an averaging procedure applied to the conductivities of the individual components of each sample (fractures, matrix). This step is not necessary when permeabilities are obtained from hydraulic or pneumatic experiments 3.3.2, which per se provide pressure differences and flow rates for the entire sample.

3.3.1 Effective Conductivities Obtained from Fracture Geometry Data

Early attempts at modelling fluid flow within fractures were based on the assumption that a fracture could be geometrically represented by a pair of parallel flat plates with a constant aperture between them (Snow, 1968). However, this has been found to be inadequate in describing fluid flow and transport through natural fractures (Abelin *et al.* (1987); Neretnieks (1985)). The parallel plate model can be considered only a rough approximation of flow through real fractures (Sect. 2.4.1).

Real fracture surfaces are not smooth parallel plates, but are rough and may even touch each other at discrete points. Consequently, the fluid is expected to take a tortuous path when moving through a real fracture (de Marsily (1981); Tsang and Tsang (1987); Tsang and Tsang (1989); Brown (1987); Tsang *et al.* (1988); Silliman (1989); Bear and de Marsily (1993)).

Therefore, the determination of effective conductivity accounts for the roughness of individual fractures. However, for the eight fractured-porous samples studied, it is not sufficient to consider flow through the fractures alone. Rather, the contribution of matrix conductivity has to be quantified. To do so, available data are processed in two steps. First, effective fracture apertures are determined for each core slice (Sect. 3.3.1.1). In the second step, these results are used to compute effective conductivities for each sample (Sect. 3.3.1.2). It should be said that this procedure applies to both liquids and gases as test fluids, i.e. effective hydraulic and effective pneumatic conductivities can be determined.

3.3.1.1 Effective Fracture Apertures

Based on to the local parallel plate model (Sect. 2.4.1), the geometrical considerations of this section refer to Figs. 3.11 and 3.12, where fracture cross sections were approximated by an ensemble of rectangles. Considering flow in parallel to the axis of the cylindrical samples, the approximation used here implies that the fracture aperture is assumed to be constant along the flow direction for each rectangle in a given slice, but may vary between rectangles. The total discharge in the nth fracture, $Q_{\mathrm{F}n}$ [L^3/T], can then be represented as

$$Q_{\mathrm{F}n} = \sum_{i=1}^{m_{nz}} Q_{in} \quad , \tag{3.1}$$

where Q_{in} [L^3/T] is the discharge in rectangle i and m_{nz} is the number of rectangles in slice z.

The total cross-sectional area of fracture n in slice z, $A_{\mathrm{F}nz}$ [L^2], is given by

$$A_{\mathrm{F}nz} = \sum_{i=1}^{m_{nz}} A_{inz} \quad , \tag{3.2}$$

where A_{inz} [L^2] is the cross-sectional area of rectangle i. According to Figs. 3.11 and 3.12, A_{inz} is given by

$$A_{inz} = b_{inz} \cdot l_{inz} \quad , \tag{3.3}$$

where b_{inz} [L] is the fracture aperture and l_{inz} [L] is the width of rectangle i in slice z.

The hydraulic or pneumatic conductivity K_F [L/T] of a fracture with constant aperture b [L] is defined as (de Marsily, 1981)

$$K_F = \frac{\rho g b^2}{12\mu\left(1 + 8.8 R_r^{1.5}\right)} \quad , \tag{3.4}$$

where ρ is the density of fluid [M/L^3], μ is the dynamic viscosity of fluid [M/(LT)], g is the acceleration due to gravity [L/T^2] and R_r is the fracture relative roughness [-]. The latter is approximately given by $R_r = \varepsilon/(2b)$ with ε denoting the absolute roughness [L] (Sect. 3.2.1.1). Equation (3.4) can be directly used to express the conductivity of rectangle i of fracture n in slice z as

$$K_{Finz} = \frac{\rho g b_{inz}^2}{12\mu\left(1 + 8.8 R_r^{1.5}\right)} \quad . \tag{3.5}$$

With the fact in mind that the difference in hydraulic head between the faces of slice z is the same for each rectangle, equations (3.1) - (3.5) can be combined with DARCY's Law (2.8) to yield the effective aperture b_{nz} of fracture n in slice z as

$$b_{nz} = \left[\frac{\sum\limits_{i=1}^{m_{nz}} \left(b_{inz}^3 \cdot l_{inz}\right)}{\sum\limits_{i=1}^{m_{nz}} \left(b_{inz} \cdot l_{inz}\right)}\right]^{\frac{1}{2}} \quad . \tag{3.6}$$

Next, the effective aperture b_n [L] of fracture n along the complete sample is computed. To this end, the relationship

$$\Delta h = \sum_{z=1}^{m_z} \Delta h_z \tag{3.7}$$

is employed, which states that the head difference Δh [L] along the sample equals the sum of head differences Δh_z [L] across individual slices. In equation (3.7), m_z denotes the number of slices. An expression for b_n is obtained by combining equations (3.7) and (3.4). This results in

$$b_n = \left(\frac{L}{\sum\limits_{z=1}^{m_z} \frac{L_z}{b_{nz}^3}}\right)^{\frac{1}{3}} \quad , \tag{3.8}$$

where L_z [L] is the thickness of slice z and $L = \sum_{z=1}^{m_z} L_z$ is the length of the sample. According to equation (3.8), a weighted harmonic average of the cubes of b_{nz} has to be used in order to compute the effective fracture aperture b_n.

Effective apertures for the fractured porous core samples under consideration are summarized in Table 3.6. The samples contain up to 5 fractures with effective apertures between 13.5 and 145 μm. Fractures marked with an asterisk (*) were found to be dominant, i.e. traversing the entire sample.

Table 3.6. Effective apertures b_n (in μm) of fractures found in the core samples. Values with an asterisk (*) are related to dominant fractures (from Baraka-Lokmane *et al.* (2003)).

Sample	Fracture 1 (aperture b_1)	Fracture 2 (aperture b_2)	Fracture 3 (aperture b_3)	Fracture 4 (aperture b_4)	Fracture 5 (aperture b_5)
1	43.6*	41.0			
2	54.5				
3	13.5*	41.3	35.8*		
4	15.4*	145	16.3	20.8	115
5	54.9				
6	41.1*	36.9*	35.3	26.3	34.0
7	31.8*	27.7	29.8		
8	49.0				

3.3.1.2 Effective Conductivity

The modelling approach used in the previous subsection is extended to yield the conductivities of the investigated samples. To this end, fracture geometry parameters and matrix conductivities are taken into account by estimating the conductivity of individual slices as a weighted sum of the contribution of the matrix and each fracture. Finally, the conductivity of the bulk sample is obtained by a weighted harmonic average.

First, the conductivity $K_{\mathrm{F}nz}$ of fracture n in slice z [L/T] is calculated by combining equation (3.1) with DARCY's Law (2.8). This results in

$$K_{\mathrm{F}nz} = \frac{\sum_{i=1}^{m_{nz}} (K_{\mathrm{F}inz} \cdot b_{inz} \cdot l_{inz})}{b_{nz} \cdot l_{nz}} \quad , \tag{3.9}$$

where $l_{nz} = \sum_{i=1}^{m_z} l_{inz}$ is the width of fracture n in slice z [L].

Next, the conductivity of slice z is given. It is composed of two terms representing the conductivity of the fracture system and the matrix respectively.

With corresponding ratios of cross-sectional areas used as weighting factors, the conductivity of slice z is given by

$$K_z = \frac{\sum\limits_{n=1}^{m} (K_{Fnz} \cdot b_{nz} \cdot l_{nz})}{\pi R_a^2} + \frac{K_M \cdot \left[\pi R_a^2 - \sum\limits_{n=1}^{m} (b_{nz} \cdot l_{nz})\right]}{\pi R_a^2} , \quad (3.10)$$

where R_a is the core radius [L], m is the number of fractures, and K_M is the conductivity of the matrix [L/T]. From this, the conductivity of the bulk sample, K [L/T], can be obtained as a harmonic average yielding

$$K = \frac{L}{\sum\limits_{z=1}^{m_z} \frac{L_z}{K_z}} . \quad (3.11)$$

Similarly, the conductivity K_F of the fracture system can be computed via

$$K_F = \frac{L}{\sum\limits_{z=1}^{m_z} \frac{L_z}{K_{Fz}}} , \quad (3.12)$$

where K_{Fz} denotes the conductivity of the fracture system in slice z [L/T]. K_{Fz} can be shown to be equal to

$$K_{Fz} = \frac{\sum\limits_{n=1}^{m} (K_{Fnz} \cdot b_{nz} \cdot l_{nz})}{\left[\sum\limits_{n=1}^{m} (b_{nz}^3 \cdot l_{nz})\right]^{\frac{1}{3}} \cdot \left[\sum\limits_{n=1}^{m} l_{nz}\right]^{\frac{2}{3}}} . \quad (3.13)$$

The hydraulic conductivities K and K_F computed with the method outlined above are given in Table 3.7 for the eight samples investigated. In addition, Table 3.7 shows the percentage of discharge flowing through the fracture system.

The results of the values of the effective fracture apertures of all eight core samples (Table 3.6) and the results of the contribution of the fracture to the total hydraulic conductivity of the sample (Table 3.7) are combined in Table 3.8. It can be seen that two different types of sandstone are involved:

- sandstone with fractures conducting less than 80% of the flow and with an effective fracture aperture less than 40 μm (samples 2, 3, 4, 5, 6 and 7);
- sandstone with fractures conducting more than 80% of the flow and with an effective fracture apertures larger than 40 μm (samples 1 and 8).

The results show that, in spite of the large aperture (54.5 μm), the fracture in sample 2 transports only 1% of the flow. The fracture does not dominate the flow because of its limited vertical extent (Fig. 3.15). The partial blocking of the fracture aperture by clay minerals in the case of samples 5 and 6 appears to be the reason for the reduction of the hydraulic conductivity within

Table 3.7. Hydraulic conductivities of the samples and of the respective fracture systems (from Baraka-Lokmane *et al.* (2003)).

Sample	Sample conductivity ($m\ s^{-1}$) after eq. (3.11)	Fracture system conductivity ($m\ s^{-1}$) after eq. (3.12)	$Q_F/(Q_F + Q_M)$ (%)
1	1.41×10^{-6}	1.26×10^{-6}	89
2	1.31×10^{-6}	2.03×10^{-8}	1
3	3.88×10^{-7}	4.42×10^{-8}	11
4	2.14×10^{-6}	1.29×10^{-7}	6
5	1.54×10^{-6}	6.92×10^{-7}	45
6	6.23×10^{-7}	3.74×10^{-7}	60
7	6.31×10^{-7}	4.80×10^{-7}	76
8	9.00×10^{-7}	8.30×10^{-7}	92

Table 3.8. The relationship of the fracture aperture and the amount of the flow through the fracture (from Baraka-Lokmane (2002a).

Fracture aperture (μm)	Share of fracture flow (%)
<20	<30
>30	>50
>40	>80

the fracture. The fractures in samples 5 and 6 transport only 45% and 60% of the flow respectively, in spite of the large apertures (more than 40 μm). The partial sealing of the fractures by clay minerals was verified by two independent methods. The MRI technique was used to determine the geometry of a flood front during a water imbibition experiment for quantifying the distribution of water saturation as the flooding progresses. MRI measurements carried out on sample 5 showed that the fracture shows up as a lighter tone, implying lower water saturation (Baraka-Lokmane *et al.*, 2001). The partial sealing of the fractures in the case of samples 5 and 6 was also verified by resin casting (Baraka-Lokmane, 2002a).

3.3.2 Permeabilities Obtained from Hydraulic and Pneumatic Tests

This section provides permeabilities of the samples derived from hydraulic and pneumatic tests. A comparison of these results with the conductivity values of Table 3.7 will be given in Sect. 3.3.3.

3.3.2.1 Permeabilities Derived from Hydraulic Tests

Hydraulic tests can be evaluated to calculate sample permeability $k\,[L^2]$ by using DARCY's Law (2.8):

$$k_{hydr} = \frac{Q\eta_W L}{A\Delta p} , \qquad (3.14)$$

where $\Delta p = \rho_w g \Delta h$ is the pressure difference [M/(LT)]], Q is the discharge [L^3/T] and A is the cross-sectional area [L^2] of the sample. The subscript hydr is introduced in equation (3.14) to show that intrinsic permeability values are derived from hydraulic tests. Accordingly, the subscript w indicates that η_w and ρ_w are the dynamic viscosity [M/(LT)] and density [M/L^3] of water respectively.

A value for the dynamic viscosity of water can be determined by considering that η_w depends on temperature via

$$\eta_w = \frac{1.002}{1000} \cdot 10^{\frac{1.3272\cdot(20-T)-0.001053\cdot(T-20)^2}{T+105}} . \tag{3.15}$$

Equation (3.15), which has been taken from Schwarzenbach *et al.* (1993), is valid for temperatures $T \geq 20°C$. It is important to note that the temperature T enters equation (3.15) in °C and the resulting values for dynamic viscosity are given in Pa·s.

The permeability values obtained from equations (3.14) and (3.15) for the eight sandstone samples investigated in Sect. 3.2 vary between 10 md and 50 md (Baraka-Lokmane *et al.*, 2003). They are given in Table 3.9.

Table 3.9. Permeabilities derived from hydraulic tests (from Baraka-Lokmane *et al.* (2003)).

Sample number	1	2	3	4	5	6	7	8
Permeability k_{hydr} (md) derived from hydraulic tests by using eq. 3.14	38.6	27.9	17.3	43.2	46.9	12.3	15.1	25.2

3.3.2.2 Permeabilities Derived from Pneumatic Tests

When intrinsic permeability is determined from pneumatic tests, three effects must be taken into account (Bloomfield and Williams, 1995):

- the compressibility of the fluid (e.g. , air),
- the KLINKENBERG effect,
- the turbulent flow regime.

For compressible fluids such as air, it has been shown by, for example, Dullien (1991) that permeability can be calculated from an extended version of DARCY's Law:

$$k_{\text{pneu}} = \frac{2Q\eta_a L p_2}{A(p_1^2 - p_2^2)} , \tag{3.16}$$

where p_1 and p_2 denote the pressure [M/(LT^2)] at the inlet and at the outlet of the sample and η_a is the dynamic viscosity of air [M/(LT)].

The KLINKENBERG effect is responsible for systematic deviations between the permeabilities derived from pneumatic and hydraulic tests. Differences occur due to gas slippage at the pore walls. In order to quantify this process, Klinkenberg (1941) combined the slip theory developed by Kundt and Warburg (1875) with DARCY's Law and thereby established a formula which relates the apparent permeability measured with gas and the true permeability. This relationship, termed the "KLINKENBERG correction", makes results from pneumatic tests usable for predicting water flow behavior. It can be obtained graphically by plotting the permeability derived from pneumatic tests (equation 3.16) against the inverse of the mean absolute gas pressure on the sample, $1/\overline{p}$ (Klinkenberg, 1941). The permeability is evaluated by extrapolation and the KLINKENBERG correction is expressed as follows

$$k^*_{\mathrm{pneu}} = \frac{k_{\mathrm{pneu}}}{1 + \frac{b_{\mathrm{k}}}{\overline{p}}} , \tag{3.17}$$

with b_{k} denoting the KLINKENBERG factor $[\mathrm{M}/(\mathrm{LT}^2)]$ and $\overline{p} = (p_1 + p_2)/2$ the average of inlet and outlet pressure $[\mathrm{M}/(\mathrm{LT}^2)]$.

According to Rasmussen *et al.* (1993), the KLINKENBERG factor b can be written as $b_{\mathrm{k}} = 8C\lambda\overline{p}/d_{\mathrm{p}}$, where d_{p} is the effective pore diameter [L], C is a constant [-] representing a proportionality factor approximately equal to 1, and λ is the mean free path of air molecules [L] at pressure $\overline{p}$, i.e. the average length of free movement of molecules between successive collisions. Gas slippage is significant when the mean free path of the gas molecules is of a similar order of magnitude as the pore size. The average length of this free movement is inversely proportional to the pressure. When the gas mean free path is negligibly small with respect to pore size, the gas slip velocity becomes insignificant (Rasmussen *et al.*, 1993).

On the basis of the KLINKENBERG formula (3.17), the permeability k^*_{pneu} is obtained by recording the flow rates and mean pressures of a gas flow experiment, plotting the apparent permeability k_{pneu} against the inverse of mean pressure (Fig. 3.25), fitting a regression line to the points and locating the intercept which is then taken to be equal to the permeability k^*_{pneu} of the sample.

Generally, the KLINKENBERG corrections vary between approximately 1% and 70% (Mertz, 1991). In this study, the KLINKENBERG corrections vary between 3.70% and 16.67% and the permeabilities obtained from pneumatic tests for the eight samples range from 30 md to 160 md (Table 3.10).

3.3.2.3 Correlating Permeabilities Derived from Hydraulic and Pneumatic Tests

As a next step, the degree of correlation between permeabilities obtained via hydraulic (k_{hydr}) and pneumatic (k^*_{pneu}) tests is determined.

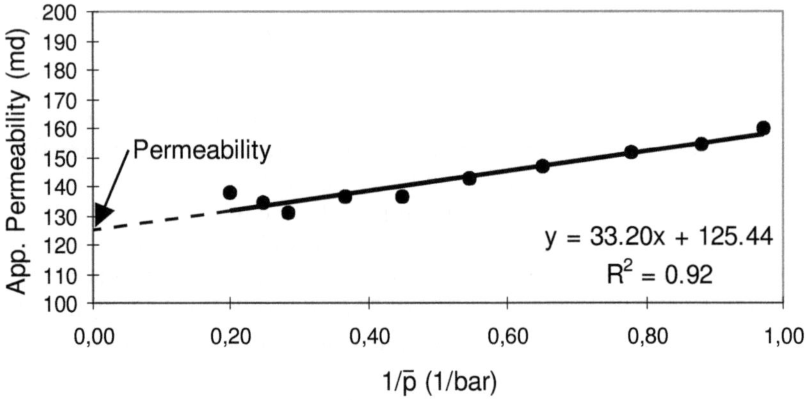

Fig. 3.25. KLINKENBERG correction of the apparent permeability for sample 2 (from Baraka-Lokmane (2002a)).

Table 3.10. Intrinsic permeabilities derived from pneumatic tests including KLINKENBERG correction (from Baraka-Lokmane *et al.* (2003)).

Sample number	1	2	3	4	5	6	7	8
Permeability k^*_{pneu} (md) derived from hydraulic tests by using eqs. (3.16) and (3.17)	159	125	44.2	56.7	55.1	33.4	31.9	130

In Table 3.11, the degree of correlation is analyzed on the basis of the permeability ratios $k^*_{\text{pneu}}/k_{\text{hydr}}$, which are found to be larger than 1 for all samples. Two tendencies in the distribution of this ratio may be observed:

- ratios ranging from 1 to 2 (samples 3, 4, 5, 6, and 7);
- ratios ranging from 4 to 5 (samples 1, 2, and 8).

Table 3.11. Comparison of permeabilities derived from hydraulic and pneumatic tests (from Baraka-Lokmane *et al.* (2003) and Baraka-Lokmane (2002a)).

Sample number	1	2	3	4	5	6	7	8
Permeability k_{hydr} (md) derived from hydraulic tests by using eq. (3.14)	38.6	27.9	17.3	43.2	46.9	12.3	15.1	25.2
Intrinsic permeability k^*_{pneu} (md) derived from hydraulic tests by using eqs. (3.16) and (3.17)	159	125	44.2	56.7	55.1	33.4	31.9	130
Ratios $k^*_{\text{pneu}}/k_{\text{hydr}}$	4.12	4.48	2.55	1.31	1.17	2.72	2.11	5.16

These results may be compared with ratios $k^*_{\mathrm{pneu}}/k_{\mathrm{hydr}}$ reported to be between 1 and 30 in various publications (Table 3.12).

Table 3.12. Permeability ratios $k^*_{\mathrm{pneu}}/k_{\mathrm{hydr}}$ reported in the literature (after Baraka-Lokmane (2002a)).

$k^*_{\mathrm{pneu}}/k_{\mathrm{hydr}}$	Nature of samples	Number of sample	Reference
1→5	Sandstone	1155	Lovelock (1977)
1→10	Sandstone	10	Sampath and Keighin (1982)
2→4	Fractured tuff	105	Rasmussen *et al.* (1993)
1→30	Sandstone	55	Bloomfield and Williams (1995)
1→3	Sandstone	15	Jaritz (1999)

The sandstone used in this study is from the same outcrop as used by Jaritz (1999). The permeability ratio $k^*_{\mathrm{pneu}}/k_{\mathrm{hydr}}$ is in the range 1 to 3 for the sandstone cores studied by Jaritz (1999) and in the range 1 to 5 for the fractured sandstone cores investigated here (Table 3.12).

The factors which may in general affect the observed scatter of permeability ratios are:

- migration of clay particles,
- breakdown of original fabrics,
- undersaturation of samples during hydraulic tests.

Migration of clay particles can be induced by swelling clays which represent strongly hydrophilic minerals known to be important in controlling the water permeability (Sampath and Keighin (1982); Bitton and Gerba (1984); Appelo and Postma (1993); Bloomfield and Williams (1995)). Swelling of clay may be associated with low ionic strength and a large double layer. Accumulation of migrating clay particles in pore throats may lead to clogging and, consequently, to a decrease in permeability. Swelling clays are dispersed, plastic, sticky and have low hydraulic conductivity. During the hydraulic permeability tests, this effect was observed for the samples with a large fracture aperture. This is explained by the presence of the clay minerals as filling materials within the fracture aperture, which may be responsible for the reduction of water flow.

Breakdown of original fabrics may be caused by the passage of wetting fronts across relatively delicate clay mineral complexes (Lovelock (1977); Bloomfield and Williams (1995)). The geometry of some of the flow channels may be physically altered, leading to irreversible changes in the permeability of the medium. This effect is much stronger for water than for air, because water is about 800 times denser and 55 times more viscous than air at 20°C and 1013 hPa.

The undersaturation of samples during hydraulic tests may cause a systematic underestimation of the permeability. It may be practically impossible to fill all the pore space in the sample with water during the hydraulic tests, particularly in finer-grained sandstone (Lovelock, 1977).

In order to understand the reasons for the scatter of water and gas permeability data among the eight samples, physical parameters such as the matrix porosity n and the saturation S_w, as well as matrix permeabilities derived from pneumatic $k^*_{pneu,M}$ and hydraulic tests $k_{hydr,M}$ tests were carried out (Table 3.13).

Table 3.13. The hydraulic, pneumatic and physical parameters of the eight selected samples (Baraka-Lokmane, 2002a).

Sample number	k^*_{pneu} (md)	k_{hydr} (md)	k_r (-)	n (%)	S_w (%)	$k^*_{pneu,M}$ (md)	$k_{hydr,M}$ (md)	$k_{r,M}$ (-)	Characteristics
1	159	38.6	4.12	21.54	95.89	33.26	13.50	2.46	• more than one fracture • large fracture aperture • presence of clay nodules in the matrix
2	125	27.9	4.48	25.69	91.37	123.49	26.60	4.64	• one fracture • large fracture aperture
3	44.2	17.3	2.55	21.31	91.01	31.74	12.45	2.55	• more than one fracture
4	56.7	43.2	1.31	21.85	93.27	34.05	25.99	1.31	• more than one fracture
5	55.1	46.9	1.17	21.19	93.27	31.22	26.68	1.17	• one fracture
6	33.4	12.3	2.72	20.52	98.21	28.34	10.42	2.72	• more than one fracture
7	31.9	15.1	2.11	20.17	90.22	26.84	12.72	2.11	• more than one fracture
8	130	25.2	5.16	17.89	76.69	24.90	7.48	3.33	• one fracture • large fracture aperture

($k_r = k^*_{pneu}/k_{hydr}$, $k_{r,M} = k^*_{pneu,M}/k_{hydr,M}$, n - porosity, S_w - saturation)

From Table 3.13, it is concluded that migration of clay particles and breakdown of original fabrics caused by the passage of wetting fronts across relatively delicate clay mineral complexes affect the observed scatter of permeability data for the eight core samples (Baraka-Lokmane, 2002a). For instance, the presence of clay nodules in the matrix may explain the high permeability ratio of 2.46 for sample 1 (Fig. 3.14).

In addition, the scatter of permeability ratios is particularly high for the samples where a large fracture aperture is observed (samples 1 and 8). In these cases, the two first processes cited above seem to be relevant because of the presence of clay minerals within the large fracture aperture. The mineralogical study shows that kaolinite, which is known as swelling clay, is the dominant cement in these samples (Table 3.1). For sample 8 (Fig. 3.13), the scatter of the permeability ratios is also explained by the undersaturation of this sample during water permeability measurements; indeed, the saturation is lower than 80%. This phenomenon occurs for finer-grained sandstone, as is the case for this sample.

For sample 2, there is no difference between the matrix and the sample permeability ratios (Table 3.13). As mentioned above, the fracture of sample 2 does not run right through the core sample (Fig. 3.15) and, hence, the contribution of fracture permeability to the sample permeability is negligible.

The matrix permeability obtained from pneumatic tests is up to 4.6 times higher than the results derived from hydraulic experiments. Therefore, the clay minerals present as cement in the matrix seem to be responsible for the scatter between groups of permeability values. Pneumatic experiments, however, can be carried out much faster than hydraulic tests and are less prone to experimental problems and errors. The pneumatic technique measures the permeability more closely and it is therefore recommended to use air as a replacement fluid, especially for clay-rich fractured porous rocks.

3.3.3 Comparison of Conductivities

In this section, conductivities obtained by evaluating geometrical information (Sect. 3.2.1) and values derived from hydraulic and pneumatic tests are compared. To this end, permeabilities obtained in the previous section are converted to hydraulic conductivities via

$$K_{\mathrm{w}} = \frac{\rho_{\mathrm{w}} g}{\eta_{\mathrm{w}}} k_{\mathrm{hydr}} \tag{3.18}$$

and to pneumatic conductivities via

$$K_{\mathrm{a}} = \frac{\rho_{\mathrm{a}} g}{\eta_{\mathrm{a}}} k^{*}_{\mathrm{pneu}} \ , \tag{3.19}$$

where subscripts w and a stand for the test fluids "water"and "air".

Figure 3.26 shows the correlation between hydraulic conductivities (K_{w}) based on the geometric approach and results derived from hydraulic experiments. Similarly, pneumatic conductivities (K_{a}) obtained via geometrical considerations are plotted against conductivities resulting from pneumatic tests (Fig. 3.27).

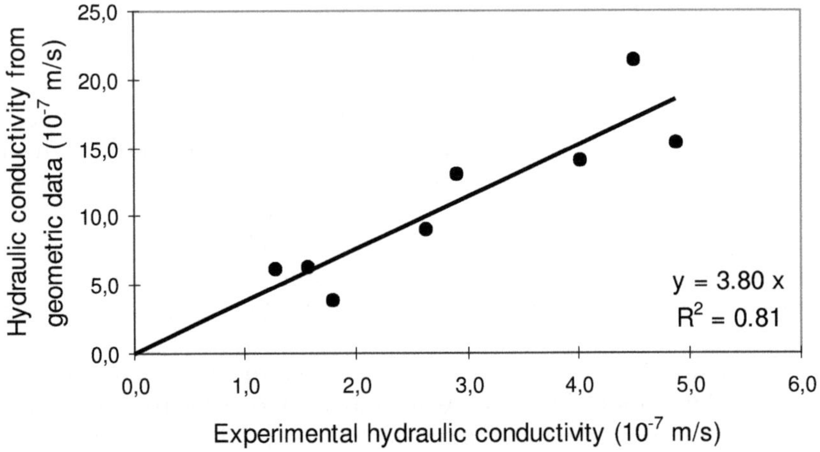

Fig. 3.26. Correlation of hydraulic conductivities obtained via geometrical considerations and those derived from hydraulic experiments (Baraka-Lokmane, 2002a).

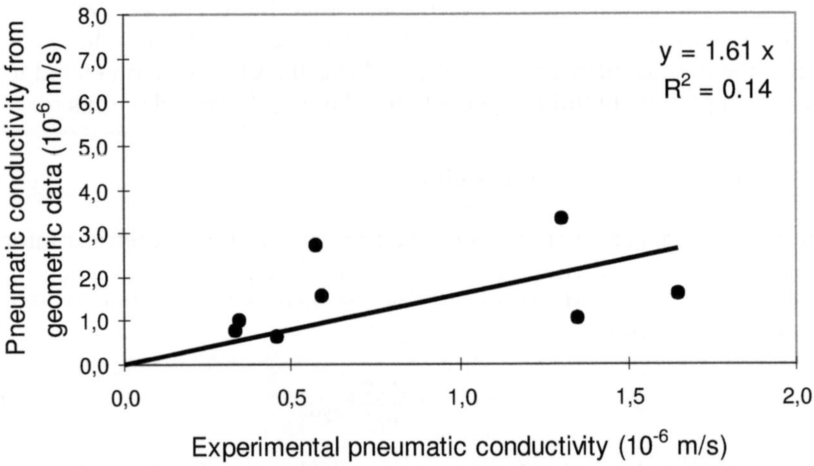

Fig. 3.27. Correlation of pneumatic conductivities obtained via geometrical considerations and those derived from pneumatic experiments (Baraka-Lokmane, 2002a).

The model of Sect. 3.3.1 leads to a good match for hydraulic conductivities with a squared correlation coefficient R^2 of 0.81. However, the model results are about four times higher than conductivities obtained by evaluating hydraulic tests. The results from pneumatic tests seem less reliable than those from the hydraulic experiments, the squared correlation coefficient R^2

being as low as 0.14 and the pneumatic conductivities deduced from geometric investigations twice as high as the values obtained from pneumatic tests.

For samples with fracture flow representing less than 80% of the total flow (Table 3.8), correlations between conductivities calculated according to Section 3.3.1.2 and values derived from experimental results are provided in Figures 3.28 and 3.29. These diagrams show good correlations for hydraulic tests, with a squared correlation coefficient R^2 equal to 0.81, as well as for pneumatic tests, with R^2 equal to 0.71. However, the modelled conductivities are about three to four times higher than the values obtained from pneumatic and hydraulic tests respectively.

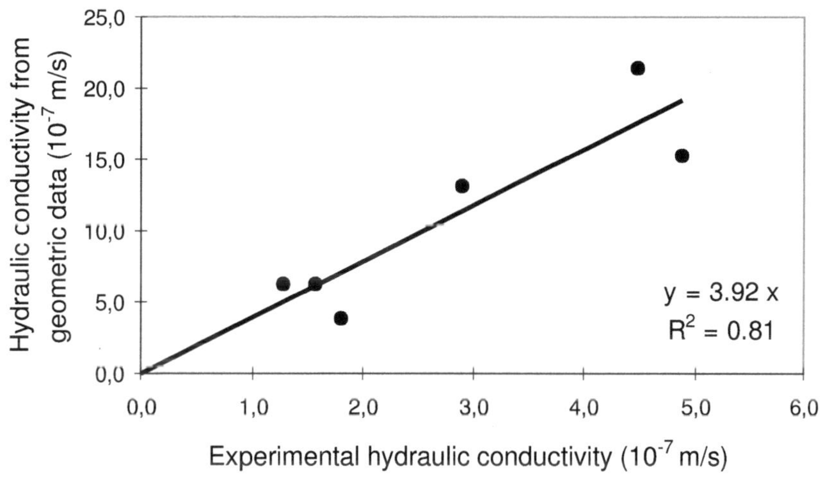

Fig. 3.28. Correlation of hydraulic conductivities obtained via geometrical considerations and those derived from hydraulic experiments for the group where the fracture flow represents less than 80% of the total flow (Baraka-Lokmane, 2002a).

The correlation of hydraulic conductivities based on fracture geometry data with values derived from hydraulic tests show comparable results in the case of the six and all the eight core samples. However, the corresponding correlation of pneumatic conductivities gives better results for the six core samples, where the channelled transport in the fracture comprises less than 80% of the total flow.

Two factors may cause the observed systematic difference between conductivities derived from hydraulic and pneumatic tests on the one hand and values obtained by evaluating fracture geometry data on the other:

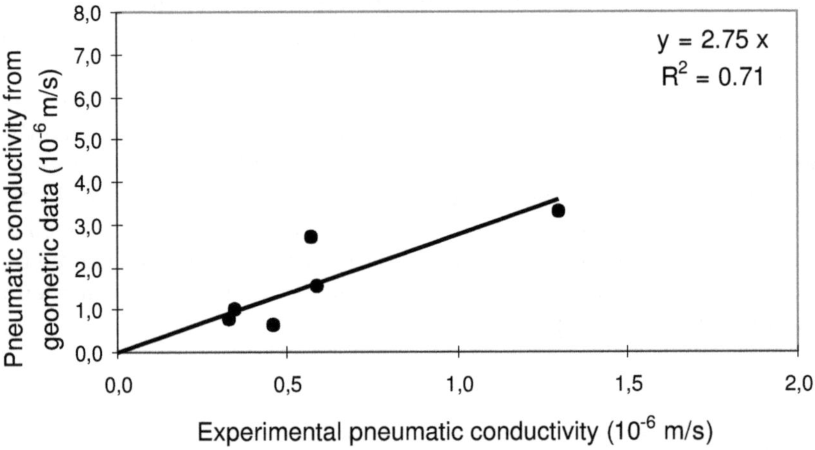

Fig. 3.29. Correlation of pneumatic conductivities obtained via geometrical considerations and those derived from pneumatic experiments for the group where the fracture flow represents less than 80% of the total flow (from Baraka-Lokmane (2002a)).

- the reduction of permeability during hydraulic tests according to the three phenomena already mentioned in Sect. 3.3.2.3,
- a slight overestimation of the fracture aperture.

The second possibility is significant because the fracture conductivity is proportional to the square of fracture aperture (3.4). Therefore, the hydraulic conductivity of each slice K_z (3.10) is proportional, as a first-order estimation, to the cube of the effective fracture aperture. In the case where the conductivities obtained by the geometric approach are about three to four times higher than the measured ones and if we consider only the overestimation of the fracture aperture as a factor causing the difference between these two parameters, the effective fracture aperture determined from the resin casting method would only have to be about 1.6 (hydraulic tests) to 1.2 (pneumatic tests) times higher than the true values. That is, a small systematic error during the derivation of the effective fracture aperture can propagate into a much larger systematic error in the implied hydraulic or pneumatic conductivity.

In the case of the pneumatic experiments, only the second factor (the overestimation of the effective fracture aperture) is applicable. However, when the channelled transport in the fracture comprises more than 80% of the total flow and the effective fracture aperture is more than 40 μm (Table 3.7), the model results are of the same order as the pneumatic conductivity values obtained from pneumatic experiments. The overestimation of the effective fracture aperture from the optical method does not occur for apertures greater than 40 μm.

These results indicate that the determination of the conductivity from fracture geometry parameters such as the effective fracture aperture, fracture width, fracture length and relative roughness measured using an optical method (Sect. 3.2.1) can be valid. This is consistent with the strong geometric constraints imposed on fluid flow in network models of macroscopic scenarios (Zhu and Wong, 1996).

3.4 Summary

The core scale investigations presented in this chapter demonstrate how properties of fractured-porous sandstone samples can be determined by employing different approaches. On the one hand, fracture geometry analysis, which can be based on results from resin casting or MRI, is coupled with a simple but efficient modelling approach yielding hydraulic conductivities. On the other hand, permeabilities are derived from hydraulic and pneumatic experiments using water and air as respective test fluids and applying adequate laws of motion which relate discharges and pressure differences. The comparison of hydraulic and pneumatic test results confirm a correlation between sample permeabilities obtained by both methods. It has to be emphasized, however, that permeabilities from pneumatic tests are found to be up to five times higher than those from hydraulic experiments. This systematic discrepancy seems to be less evident for samples with smaller fracture apertures. A qualitatively similar effect is observed when hydraulic conductivities derived from hydraulic tests are compared with values calculated by the geometry-based modelling approach. Again, a good correlation is obtained for samples without highly dominating fractures although model results are systematically higher than conductivities determined via water or air flow experiments. The reasons for these differences may be explored in future studies based on a larger number of samples, involving a detailed petrographical characterization and an increased resolution of MRI measurements. In addition, the impact of the confining pressure, which is exerted on the samples and could reduce fracture aperture widths during flow experiments, can be regarded as a subject for further analysis.

4

Bench Scale

The investigation of the governing flow and transport processes and the quantification of relevant parameters of fractured porous rocks are often limited by the question as to what extent the parameters measured in the laboratory are applicable to other investigation scales. In order to investigate the effects of scale on pneumatic tomographical investigation techniques for flow and transport in fractured porous rock, cylindrical samples and laboratory block samples are recovered from the field and new experimental techniques are developed.

To bridge the gap between the experimental studies conducted on the core scale as described in Chapt. 3 and on the field block scale (Chapt. 5), on the one hand, cylindrical fractured sandstone cores with a diameter of 0.3 m are collected. To increase the support volume for the estimation of effective parameters without losing the geometrical boundary conditions with respect to the 0.1 m samples. On the other hand, larger sandstone blocks with sample volumes of ca. 0.2 and 0.7 m^3 are selected; these account for the geometrical boundary conditions of the field block scale but still allowing for laboratory experiments under fully controlled boundary conditions.

The description of the experimental studies conducted on the cylindrical samples and on the block samples including their preparartion are described separately for each sample size:

1. cylinders with a diameter of 0.3 m, and
2. laboratory blocks with a size of $90 \times 90 \times 80\,cm^3$ and $60 \times 60 \times 60\,cm^3$.

4.1 Preparation of Fracture Porous Bench Scale Samples for Conducting Flow and Transport Experiments

C.I. McDermott, C. Leven, B. Sinclair, M. Sauter, P. Dietrich

There are no established techniques for recovering either cylindrical or block shaped bench scale samples. This section is divided into two parts, the first covering the recovery and preparation of the cylindrical samples, the second the recovery and preparation of the block samples.

The greatest problem with fractured rocks is that during sampling or preparation the fractures, natural planes of weakness, open up and the sample disintegrates before reaching the laboratory (e.g. Alexander *et al.* (1996); Baraka-Lokmane (2002a)). The problem is by no means new as demonstrated by the work of Wichter and Gudehus (1976) in the recovery of large scale fractured samples. More recently, a number of methods involving resin injection (Freig *et al.*, 1998) have been suggested, but the danger that the resin will impregnate the rock mass and alter the characteristics of the material being sampled remains. A further disadvantage is the sometimes complex preparation of the resin itself under difficult field conditions.

4.1.1 Recovery and Preparation of the Cylindrical Bench Scale Samples

In this section, a method for collecting, preparing and describing very closely spaced fractured samples (¡ 20 mm to 60 mm spacing) is presented where no hardening material is required. This method is used to recover ten relatively undisturbed extremely closely to very closely fractured samples as well as three samples where the fracturing is more discrete. All of the samples are cylindrical with a smooth top and bottom, have a diameter of 30 cm and a length of 30 to 40 cm. From the samples retrieved, five contain clay filled fractures with some open fractures, five contain sand filled fractures, and three samples contain open fracture systems. An example of each of the three types of sample is presented in Fig. 4.1. The samples are ideal for a number of different laboratory applications, including the triaxial investigation of the strength of the fracture network and, in this case, pneumatic tomographical investigation of the flow and transport parameters.

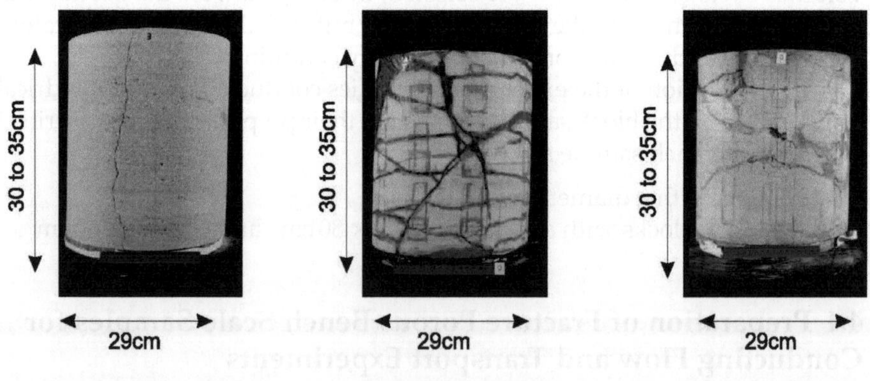

Fig. 4.1. Samples with open (a), sand-filled (b) and clay-filled (c) fracture systems (McDermott *et al.*, 2003b).

4.1.1.1 Sample Size

The selection of the size of the sample is very important. Samples must be large enough so that the fracture network can be investigated, but not too large so that handling in the field or laboratory becomes impossible with conventional techniques. A sample diameter of 30 cm and a length of 30 cm and 40 cm are chosen to fulfil the sampling, field and laboratory requirements. Samples of this size weigh approximately 40 kg and can generally be handled without mechanical assistance. A larger diameter of 40 cm and length of 40 to 50 cm, although technically feasible and containing a much larger selection of the fracture network, would lead to samples in excess of 100 kg, rendering handling difficult.

4.1.1.2 Sampling Location

The sampling locations are chosen on the basis of the experimental requirements and, to some degree, the access to the site. Two different sampling locations, illustrated in Fig. 4.2, are chosen, one in an active quarry (*Pliezhausen*) and one in the vicinity of an abandoned quarry (*Herrenberg*). The criteria for selecting the sampling material are primarily geological, those for the selection of the location are logistic. The quarry faces provide an excellent outcrop and hence an overview of the material to be sampled. In the case of the location *Pliezhausen*, all the samples are taken from material that is freshly exposed. In the case of the sampling location *Herrenberg*, it is necessary to expose a new surface behind the quarry face, as the face itself is a protected area. Samples are taken some five to ten meters behind the face edge of the quarry, and the ground reinstated once work is completed.

All the samples are taken from the upper Triassic *Stubensandstein* formation, a fractured arkose sandstone. In *Pliezhausen* five samples are taken from the forth *Stubensandstein* (Hornung, 1998) containing two types of fracture systems, a dominant closely spaced fracture system predominantly clay filled and a secondary widely spaced open fracture system with a apertures of approximately 1 mm. Three further samples are taken in *Pliezhausen* from the third *Stubensandstein* (Hornung, 1998), containing a number of discrete and interlinked open fractures. Five samples are collected from the second *Stubensandstein* (Hornung, 1998) in the vicinity of *Herrenberg*, containing fractures which are extremely closely spaced and sand filled.

4.1.1.3 Sampling Considerations, Sample Removal and Preparation

The sampling system is a point-sampling system one, and is designed as far as possible to fulfil the following criteria:

1. Random location and hence flexibility regarding the location of sampling,

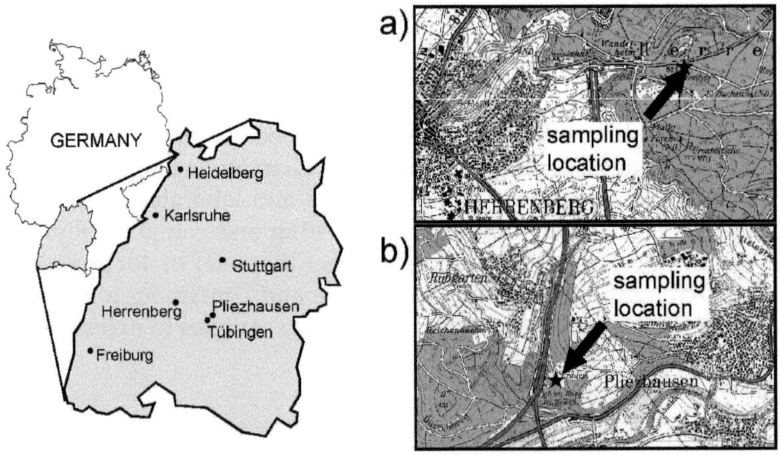

Fig. 4.2. Location of recovery of cylindrical samples: (a) Sampling location *Herrenberg* and (b) *Pliezhausen* (after McDermott *et al.* (2003b)).

2. preservation of in-situ conditions,
3. repeatability,
4. efficiency,
5. fulfilment of later experimental requirements:
   - samples to contain a fracture network, not just discrete individual fractures,
   - samples to be geometrically cylindrical with smooth surfaces, a diameter of 30 cm and a length of 30 to 40 cm.

There are Four main technical problems:

- *drilling of the sample:* the main consideration is how to ensure that there is no washout of the fractures caused by drilling flux;
- *sample recovery:* maintaining the sample and fracture network;
- *preparation of the sample for experimental use:* maintaining the undisturbed nature of the sample and fracture system;
- *sample description:* deriving a three-dimensional impression of the fracture network.

It is clear that some sort of support mechanism for maintaining the stability of the sample is necessary due to the closeness of the fractures to one another. This support mechanism can not be removed from the fractured sample until immediately prior to the experimental investigation. It has to be applied directly after drilling and left in place during the recovery, preparation and description of the sample.

In principle, the support mechanism chosen is a transparent plastic sheath, heated and inserted into the drilling, thereby encompassing the sample. The sheath then cools and contracts around the sample, exerting a limited stabilizing pressure on the sample.

In practice, the sheath comprises 3 mm thick PVC plastic that has to be cut and glued together in such a manner that its internal radius is some 5 mm less than the external radius of the drilled sample. To form the sheath, the ends of the PVC sheet are bevelled and a low-viscosity adhesive with high initial surface tension is used to join the two ends of the plastic (Fig. 4.3). After the ends of the PVC sheet are joined, the properties of the adhesive allows injection of the adhesive into areas of the joint where bonding has not occurred, ensuring an almost perfect joint. In the field, the sheath is warmed in a water bath to a temperature of over 70°C, whereupon it expands to a circumference slightly larger than the drilled sample.

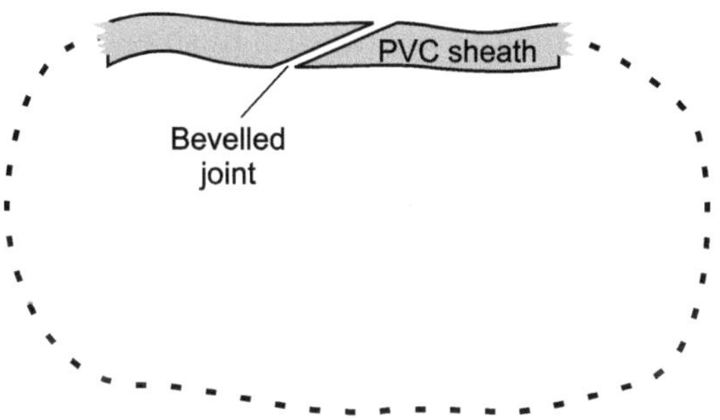

Fig. 4.3. Joining of the PVC sheet to form the protective sheath (McDermott *et al.*, 2003b).

The sampling procedure can be divided into five stages; these are described below and illustrated in Fig. 4.4.

Stage 1: Location The area of porous fractured rock to be sampled is selected and prepared in that any extraneous material is removed until a relatively stable and consistent rock surface is uncovered. Should instauration of the sampling area be necessary once work is completed, as in the case of the *Herrenberg* samples, care is taken to ensure that the extraneous material is removed and stored separately in layers for replacement. Excavated material is stored on a sheet thus protecting the ground surface. The cleared area must have a width of at least 1m to allow for access during sample recovery. The clearance can be undertaken by hand, or by an excavator.

Stage 2: Drilling A portable electric drilling machine, able to drill cores in rock of up to a diameter of 400 mm, is anchored on the prepared surface with the chemical plugs or typical concrete anchoring plugs. The cores are drilled using a water cooled diamond bit with sufficient but not excessive water flux. The pressure of the flush is kept minimal; generally, a head of the order of 1 m is enough to ensure that the drilling barrel is lubricated and that the material that has been loosened as a result of the drill bit action is washed out. The drill flush is applied to the top of the sample and the cuttings (sand washed out) are drawn out on the outside of the drill bit (Fig. 4.5). Given the low hydraulic gradient in the drill flush and the approximate hydrostatic conditions within the drill head during drilling, there is very little outwash of the fractures.

Drilling must be precise and requires the drill bit to be stabilized, particularly for the first cut. In practice, this involves stabilizing the drill head with some heavy weight until the first cut has been made and the drilling barrel rotates smoothly. To recover the bench scale samples, a drill barrel with an external diameter of 300 mm and a cut width of 5 mm is used. Once a sample has been drilled to a depth of approximately 50 cm the drill barrel is removed and the sample, with a diameter of 290 mm, is left in place. During removal of the drill barrel, care is taken to ensure no suction effect is generated inside the barrel. In practice, this means ensuring that the barrel is disconnected from the drill, leaving the opening where the drill flush enters the barrel free, and the drill barrel is removed by hand.

Stage 3: Protection A pre-cut, transparent, 3mm-thick PVC cylindrical membrane is then inserted into the 5mm cut left by the drill (Fig. 4.6). This membrane, prepared before the sampling, has an internal circumference some 5mm less than the external radius of the drilled sample. The membrane is heated in a water bath to a temperature of 70°C to 90°C, causing it to expand, so that it fits over the sample. Once in place the membrane cools and returns to its original size, exerting a stabilizing pressure on the sample and isolating the sample from the surrounding rock.

Stage 4: Removal After the required number of samples has been drilled and protected, the material surrounding the samples is removed with great care by hand or mechanically to at least the depth of the samples. Once the samples are fully exposed, the PVC membrane is pressed radially onto the sample with the use of metal supporting rings. These rings are in a number of positions along the sample, including the top and bottom, thus providing further stability. Care is taken at this stage to avoid applying excess radial pressure to sensitive areas of the sample, such as soft weathered fracture zones. The exposed top of the sample is covered with a thin plastic sheet, followed by a hardening foam which moulds itself to the sample top. Once the foam has obtained some stability, the upper surface is covered with a pre-cut wooden lid, which is fixed in position using tape, and the protective sheath cut flush to the lid. (Fig. 4.7)

After the foam has hardened the sample can be removed from the host rock, using the natural fracture system. This provides a plane of weakness across which the sample can be broken away from the rock. Should the sample be firmly attached to the host rock, a chisel is used to fracture the material artificially at a predetermined position. Finally, the sample is lifted from its position, placed upside down on its lid, and further wrapped with waterproof plastic to ensure that the field moisture content is maintained as far as possible. In addition, if further support and protection are considered necessary, it can easily be done at this stage.

Stage 5: Cutting to fit cell For later use in the experimentation cell the samples must be cylindrical and have smooth bottom and top surfaces (McDermott *et al.*, 1998). To achieve this, the samples are cut with a large circular saw, leaving samples with a length at least as large as the diameter of the sample (Fig. 4.8). Prior to cutting, the samples are further supported using metal rings ensuring that minimum damage occurs during cutting. The extra supporting rings are positioned as close to the proposed cutting surfaces as possible. This exerts a radial pressure inwards at the cutting position and reduces the danger that the blade used for cutting will remove any material from the sample. Once cut, the samples are immediately repacked and protected before transport to the laboratory. Material cut from the samples is retained for other tests, e.g. determination of porosity, specific gravity, etc.

4.1.1.4 Repair

In a few instances, small parts of the samples are lost during the cutting process. As a cylindrical sample is necessary for the later experimental procedure, the missing material is replaced with a moulded plastic element, ensuring that the geometry of the sample is maintained.

4.1.1.5 Description of the Fracture System

The fracture description of the samples is very important for the future experimental investigation and for the development of modeling techniques to analyze the results. To describe the fractures and other important characteristics of the samples, a transparent plastic sheet is wrapped around the samples and significant features on the sample surfaces are drawn onto the plastic. Recording the fracture profile may be carried out whilst the PVC sheath is protecting the cylindrical samples or, if the sample is considered stable enough, without protection. The orientation of the samples is recorded in the field so that the direction north is given. The fracture record of the sample is then digitized, allowing a three-dimensional reconstruction of the fracture network from the surface traces. Where possible, the opening widths of the fractures are recorded for use in discrete modeling approaches. An example of the digitized records is presented in Fig. 4.9. The later experimental investigation techniques are then linked to the location of the fractures within the sample.

4.1.2 Recovery and Preparation of the Block Samples

In the literature,only a few successful attempts to retrieve large scale laboratory samples are reported, including a granite block by Vandergraaf *et al.*

(1997) and a welded tuff block by Tidwell and Wilson (1997). The former was prepared to investigate one fracture plane and latter for examining flow along natural bedding planes. However, as far as the authors are aware, no highly fractured porous sandstone blocks have been recovered because of the risk of sample disintegration due to its weakened nature and logistic problems.

4.1.2.1 Site Location

The two block samples are chosen on the basis of the following criteria:

- $90 \times 90 \times 80\,cm^3$-block: closely fractured with significant fracture apertures,
- $60 \times 60 \times 60\,cm^3$-block: homogeneously distributed fissures with small apertures.

They are taken from the third *Stubensandstein* -formation (Hornung, 1998) adjacent to the test site used for the investigation of a large scale in-situ field block 5 in an active quarry (Fig. 4.10).

4.1.2.2 Sample Removal

Several large scale blocks ($2\,m \times 1.5\,m \times 2\,m$) are removed by a quarrying company from the third *Stubensandstein* formation. These blocks are ideal for experiments and for developing the experimental techniques.

Once the blocks are selected, it is necessary to cut them to get regular sides for easier experimental and modeling work. However, because of the risk of disintegration, it is only possible to cut the $90 \times 90 \times 80\,cm^3$-block with two parallel faces. The instability of the block due to its fractured nature prevented further cuts leaving dimensions of $90 \times 90 \times 80\,cm^3$.

Because the $60 \times 60 \times 60\,cm^3$-block is much more compact, it was possible to cut it to an exact cubical shape with side lengths of 60 cm.

Once cut, the block samples are placed in a heated room for thorough drying. A special extraction system has also been constructed to expedite the drying process (Fig. 4.11).

4.1.2.3 Sample Description

This section describes the physical characteristics of the matrix and the fracture network of the block samples.

Sandstone matrix

Common laboratory experiments (e.g. helium pycnometer and gas permeameter) are used to determine the physical parameters of the porous sandstone matrix of the blocks. Typical values for the third *Stubensandstein* -formation are summarized in Table 4.1.

Table 4.1. Typical parameters of the upper *Stubensandstein* formation ("km4os3").

Specific gravity	ca. 2.7	g/cm^3
Porosity	10 - 25	%
Hydraulic conductivity	$5 \cdot 10^{-8}$ - $1.5 \cdot 10^{-5}$	$m\,s^{-1}$
$CaCO_3$ content	ca. 7	%
C_{org} content	ca. 0.4	%
Saturation from humid air	ca. 12	%
Dominant grain sizes:		
clay	ca. 1 - 5	μm
quartz	ca. 100 - 500	μm
Dominant pore size	ca. 0.1 - 1	μm
Macro pores ($> 1\,\mu m$)	ca. 20	%
Cement of matrix	kaolinit	

The sandstone matrix of both blocks can be described as a hard closely bedded laminated fine to medium sandstone. From a macroscopic point of view, the $90 \times 90 \times 80\,cm^3$-block comprises a more or less homogeneous matrix while the $60 \times 60 \times 60\,cm^3$-block shows several zones of more coarsely and more finely grained sandstone respectively (cf. Figs. 4.12 and 4.13).

Fracture network

To describe the fractures, any significant morphological profile and other important characteristics of the sample, a transparent polyethylene sheet was placed on the block samples and the surface features are drawn onto the sheet. The fracture record of the sample is then digitized, allowing for a three-dimensional reconstruction of the fracture network from the surface traces. The fracture apertures are also gauged using a caliper square and the type of filling is recorded (open, sand, clay-filled). An example of the digitized records of fracturing combined with a surface image of the block is presented in Fig. 4.12 for the $90 \times 90 \times 80\,cm^3$-block and in Fig. 4.13 for the $60 \times 60 \times 60\,cm^3$-block.

The fracture network of the $90 \times 90 \times 80\,cm^3$-block is dominated by a main fracture that intersects sides *I* and *III* at a right angle and has apertures up to 4 mm (in Fig. 4.12 marked with "fracture 1"). Another dominant fracture lies nearly horizontally (marked as "fracture 2") within the $90 \times 90 \times 80\,cm^3$-block.

The $60 \times 60 \times 60\,cm^3$-block comprises only one vertical fracture with an aperture of up to 1.75 mm. As is also evident from Fig. 4.13, further fissures are parallel to the bedding and are open to some extent and mostly clay-filled.

4.1.2.4 Sealing and Further Preparation of the Block Samples

For the two blocks, two different sealing techniques are applied, both based on epoxy resin. The sealing procedure is described for each block individually in the following sections.

Sealing the $90 \times 90 \times 80\,cm^3$-Block

The sealing jacket for the $90 \times 90 \times 80\,cm^3$-block is composed of a resin which had to be poured or painted onto the sample. The top side of the block is prepared and levelled to a flat surface in the field using hand tools. Before the block faces are coated with resin, a thin coating of silicon is applied to all visible fractures, ensuring that an impregnation of the fracture system is prevented. The resin is then poured onto the surface of the block in four or five successive coats until a minimum thickness of 5 mm is reached.

Once the upper surface of the block hardened, the block is rotated and placed upon a specially designed mobile platform (Fig. 4.14). Prior to rotation, the block is reinforced on all sides with metal tension bands, exerting an inward force on the block faces to stabilize any planes of weakness.

The following method is developed for rotating the block without causing any further fracturing: To turn the block, weighing some 1600 kg, and place it on the mobile platform, a front-loading fork-lift truck with tiltable prongs is used. This procedure is illustrated in Fig. 4.15. The block is attached to the prongs, using three material tension bands capable of withstanding 4 tonnes of tensile force. Once attached, the block is lifted and the prongs of the fork lift titled 90° so that the weight of the block is taken by the material bands. The block is then placed on a pallet and the material bands released. The procedure is repeated and the block placed on the mobile platform. Finally, the remaining block faces are coated with four to five coats of epoxy resin until again a minimum thickness of 5 mm is reached.

The technique developed to allow access to the block sample involves drilling holes in the resin coat and a thread cut directly into the resin. However, it is found that drilling dust is forced into the sandstone matrix and fractures. This can not be extracted and leads to the local sealing of the matrix and fractures (Fig. 4.16). In addition, dust coming out of the matrix and fractures over a period of time and entering the airtight connections renders them useless.

To solve this problem, a hole saw is used to extract a core of the same surface area as the required port area from the resin coat. The process of ensuring extraction of the resin coat rather than drilling ensures that the dust sealing described above is prevented. The resulting hole in the resin coat is then sealed with a plastic plug into which a thread is turned so that an airtight connection can be attached to this plug. The plug is inserted into the drilled hole but not allowed to rest on the sandstone, thus ensuring that there is a small air pocket between the sample surface and the plug. This air pocket ensures that later injection through the airtight connection is dispersed over the whole port area and not restricted to specific points. Before the airtight connection is attached, a thread is cut in the underside of the connection to which a filter plug is attached. This ensures that any loose dust generated during the drilling of the port or existing in the matrix of the block sample does not interfere with the operation of the airtight connection. The airtight connection is then mounted in the plastic plug and sealed using a sealing paste in the thread. In Fig. 4.17 the construction of the ports is shown; Fig.Ḟrom Fig. 4.18 the exact location of the ports with respect to their position on the blcok surface can be seen.

However, it is found that the fractures running close and parallel to the surface of the block provide a large plane of action for the force resulting from the internal pressure on the block, attempting to push the resin from the block. During a test phase, a pressure of approximately 1.2 bar was applied to the block and led to failure of the base of the block. The entire base had to be prepared again and loose material removed. In order to counteract this force across the fracture planes and provide the resin with some support, it was necessary to install a number of supporting anchors normal to the fracture planes where this problem could arise.

Sealing thc 60 × 60 × 60 cm³ Block

To circumvent problems resulting from insufficient coating support, a multilayer approach is used for the sealing of the $60 \times 60 \times 60\,cm^3$-block sample, as described in the following. However, because this procedure requires the successive and repeated rotation of the block sample, the procedure is only applicable for sample sizes for which common lifting gear can be used.

To rotate the block sample, the same principle is applied as described for the $90 \times 90 \times 80\,cm^3$-block, except that a small lift is used.

Sealing the block sample comprises the following steps:

1. Attaching ports allowing access to the block faces (Fig. 4.20 (a): In contrast to the preparation of the $90 \times 90 \times 80\,cm^3$-block, ports are attached to the block faces prior to sealing. For this purpose, special cylindrical aluminium plugs are constructed with an air pocket towards the block faces (Fig. 4.19). These plugs are attached with a sealing paste. Attaching the ports prior to sealing has the advantage that
   - the problem of drilling dust can be avoided, and
   - the ports are stabilized by the resin coat applied in the following steps.

2. Sealing fissure traces: as described for the $90 \times 90 \times 80\,cm^3$-block, the traces of the fissures on the block faces are sealed with a thin coat of silicon, preventing impregnation along fissure planes.

3. First resin coating (Fig. 4.20 (b)): in the first resin-coating step, two layers of a thick epoxy resin with low viscosity are painted onto the block faces, giving the surface an airtight seal.

4. Contact coating: in this second resin-coating step, a thin layer of fiberglass scraps suspended in highly viscous epoxy resin is applied. Because of the high specific surface area of the fiberglass scraps, this layer will ensure that the resin coat and the subsequent coats of fiberglass matting will adhere.

5. Coat of fiberglass matting (Fig. 4.20 (c)): In the next steps, the block is covered with two to three layers of fiberglass matting. Each layer has to be soaked in the highly viscous epoxy resin used in the step before. It is important to note that the individual mats should overlap the adjacent block faces by at least one third to ensure a high stability. The layers of fiberglass matting will make the block stable, allowing high pressurization.

6. Fiberglass plasticine coat (Fig. 4.20 (d)): in the final step, all the block faces are coated with a very thick fiberglass plasticine with an extremely low viscosity. This coat can be several centimeters thick and provides additional stability and also acts as a safety coating.

It is important to note that all steps that include the application of epoxy resin have to be carried out before the individual resin coatings harden, which usually takes - depending on the type of resin - approx. 30 to 60 minutes. To apply all the individual coats, the successive and repeated rotation of the block sample is necessary.

Because of this coating, pressures of more than 2 bars could be applied to the $60 \times 60 \times 60\,cm^3$-block sample. Higher pressures are believed to be possible. However, it is also important to consider laboratory safety when working with pressurization above atmospheric pressure.

Controlling the hermetic seal

After the blocks were encased , it was necessary to prove they are hermetically sealed. This is done by pre-pressurizing the blocks to 0.5 to 1.0 bar above atmospheric pressure. Once the system has stabilized the air supply is cut off and the pressure in the blocks monitored using a manometer attached to one or more ports (Fig. 4.21 (a) and (b)).

If the pressure is found to fall with time, it can be assumed that the resin jacket or one of the connections are leaking. At times, it is necessary to submerge the block in water to find the leaks. This is done by rolling it onto a waterproof plastic sheet, then constructing a bath around it using flexible plastic walls (Fig. 4.21 (c)). The bath is then filled with water to cover the block (Fig. 4.21 (d)).

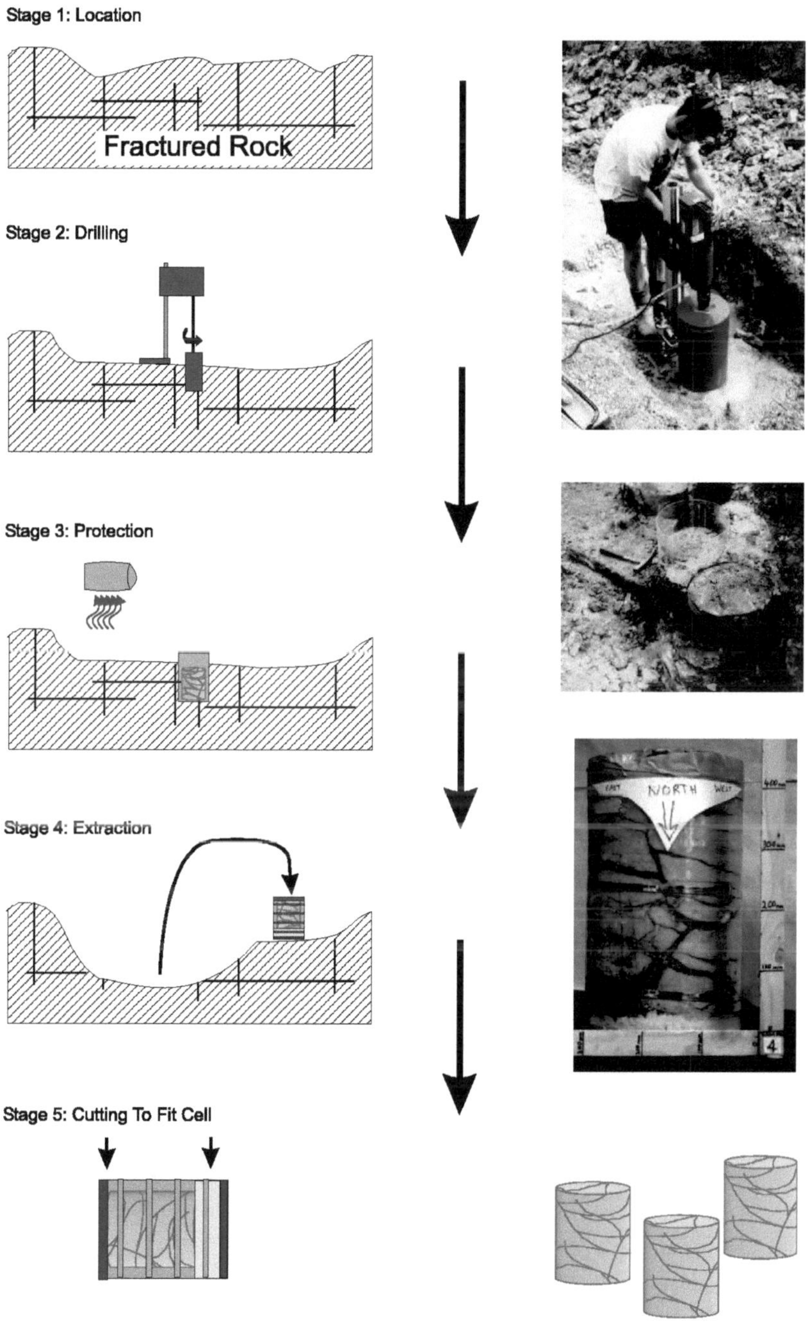

Fig. 4.4. Overview of the steps for sample recovery and preparation (after McDermott *et al.* (2003b)).

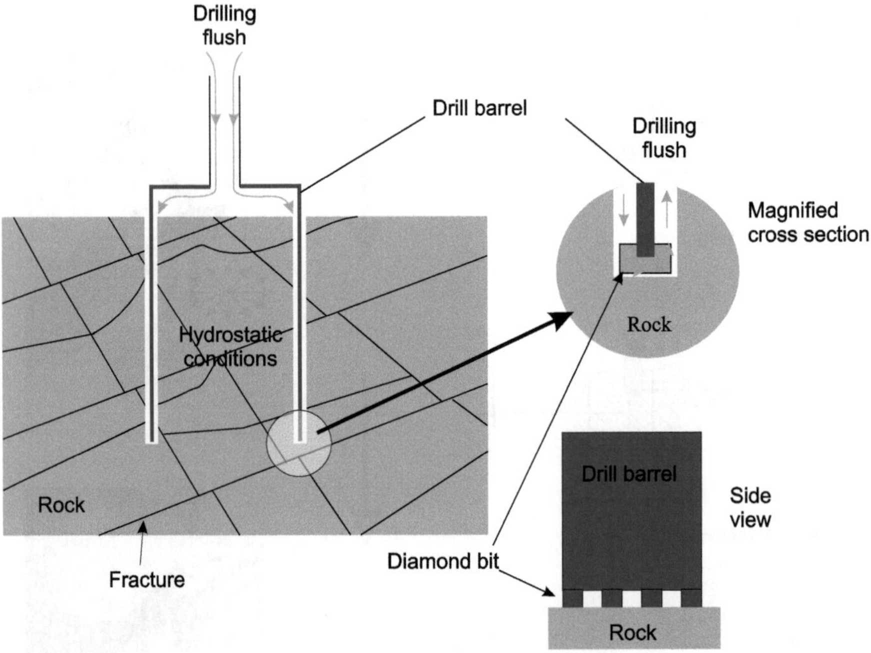

Fig. 4.5. Drilling and flush during the taking of a cylindrical sample (McDermott *et al.*, 2003b).

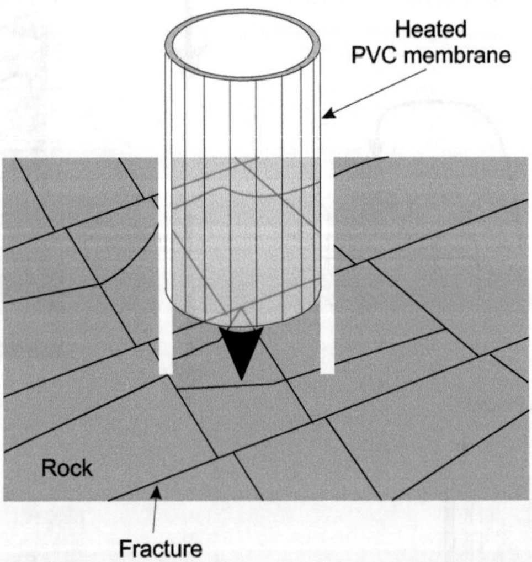

Fig. 4.6. Protection of cylindrical sample with a heated PVC membrane before the removal (McDermott *et al.*, 2003b).

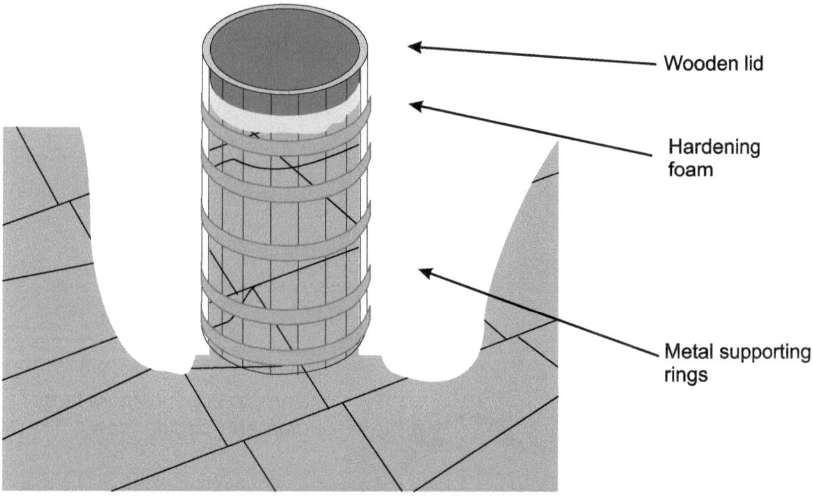

Fig. 4.7. Protection, support and preparation of a cylindrical sample for extraction. The top surface of the sample is protected, then hardening foam is squirted on the top of the sample, a wooden lid is then pressed onto the foam. Once the foam has hardened the sheath is trimmed and taped to the wooden lid. Then the surrounding material is removed and the sample is further supported with metal rings (after McDermott *et al.* (2003b)).

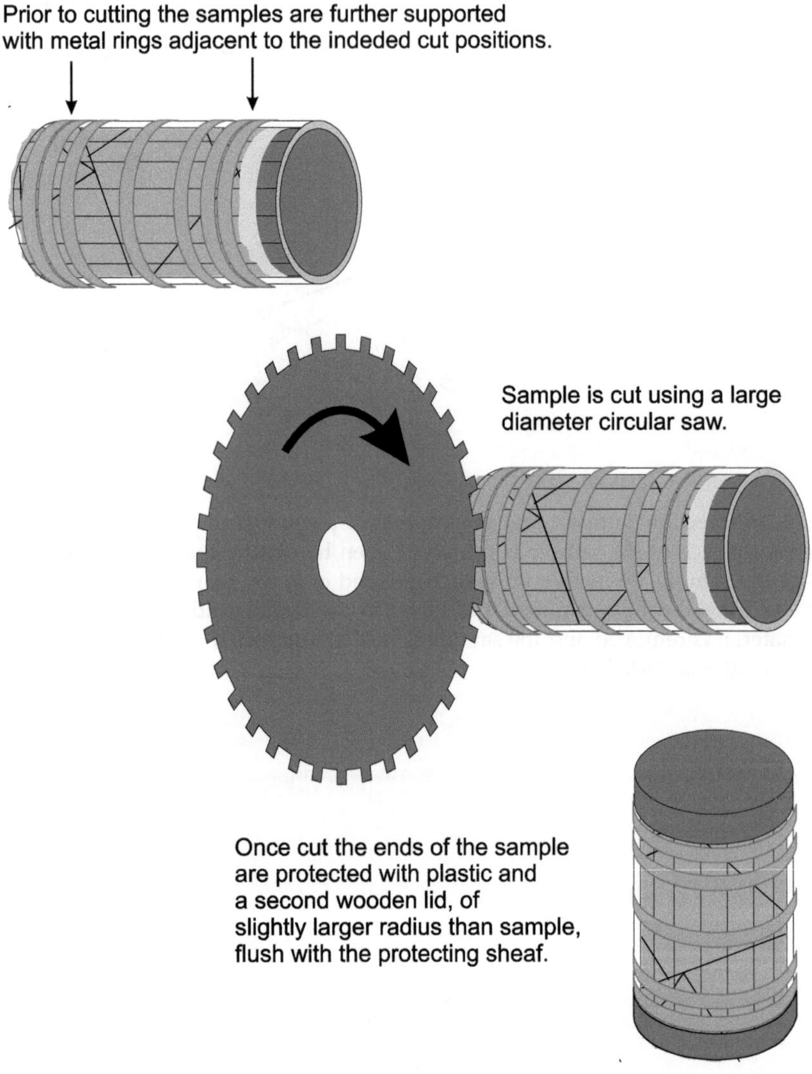

Fig. 4.8. Preparation of the cylindrical sample for laboratory investigation (McDermott *et al.*, 2003b).

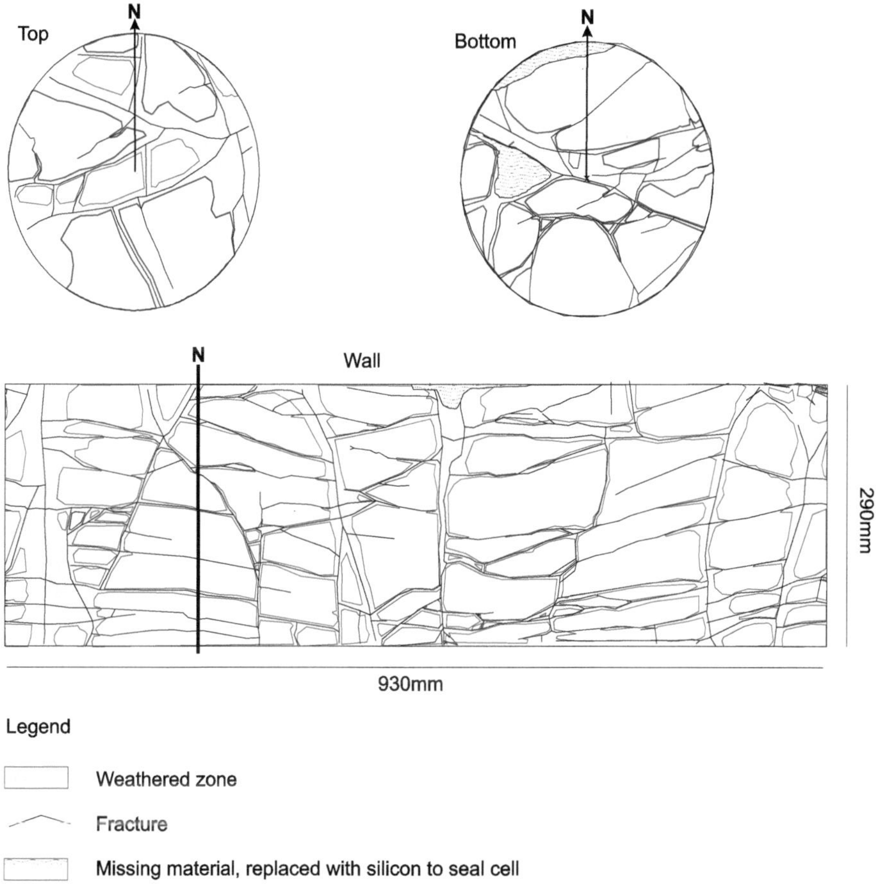

Fig. 4.9. Digitized record of a sample investigated. Example of the nature of fracturing in a sample investigated (height of sample 290 mm, diameter of sample 286 mm) (McDermott *et al.*, 2003b).

Fig. 4.10. Location of recovery of block sample.

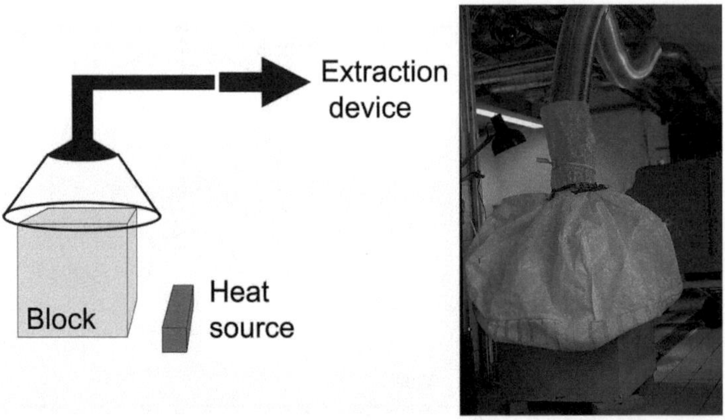

Fig. 4.11. Extraction system for drying the block sample.

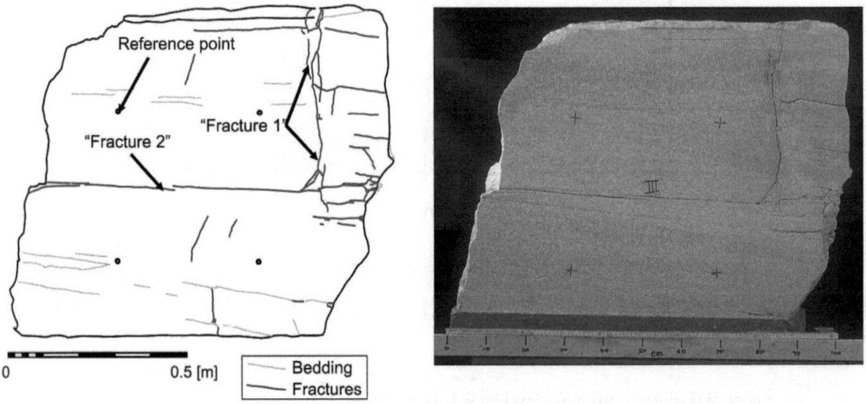

Fig. 4.12. Recording and digitizing the fracture and other important characteristics of the $90 \times 90 \times 80\,\text{cm}^3$-block. "Fracture 1" and "fracture 2" indicate the locations of dominant fractures within the block sample.

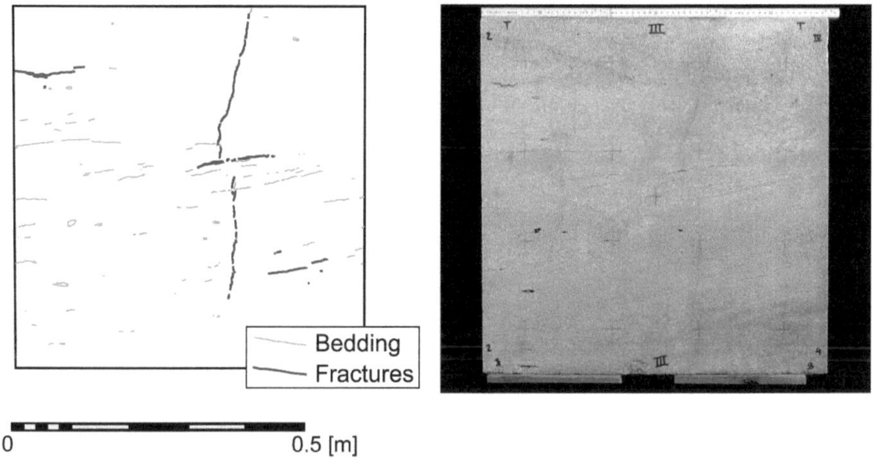

Fig. 4.13. Recording and digitizing the fracture and other important characteristics of the 60 × 60 × 60 cm^3-block.

Fig. 4.14. 90 × 90 × 80 cm^3-block: Block sample on mobile platform.

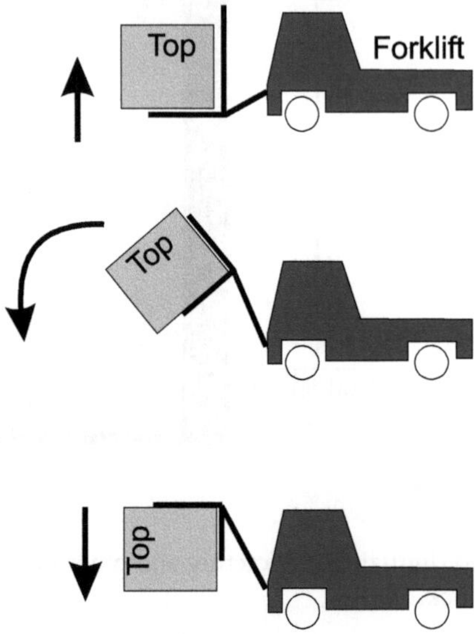

Fig. 4.15. Rotation of block sample.

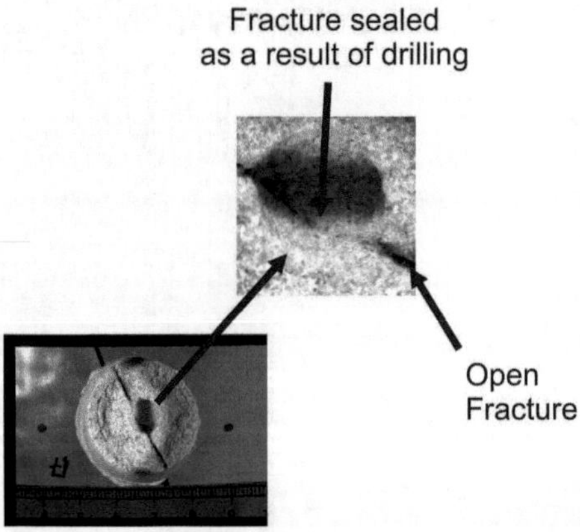

Fig. 4.16. Local sealing of matrix and fractures as a result of drilling.

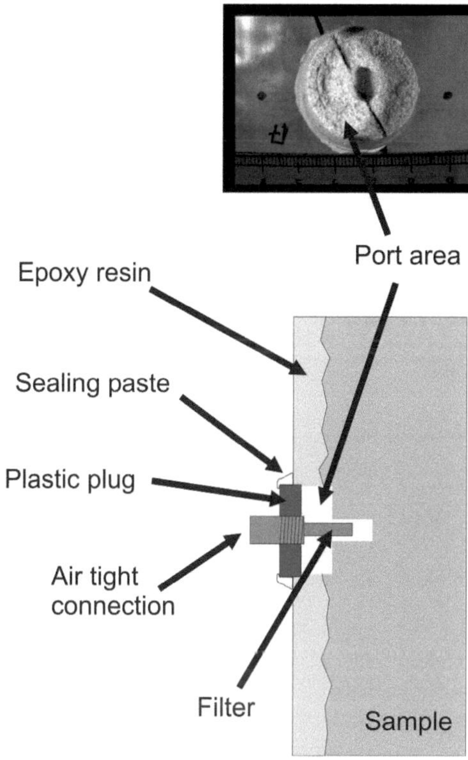

Fig. 4.17. Construction of the ports.

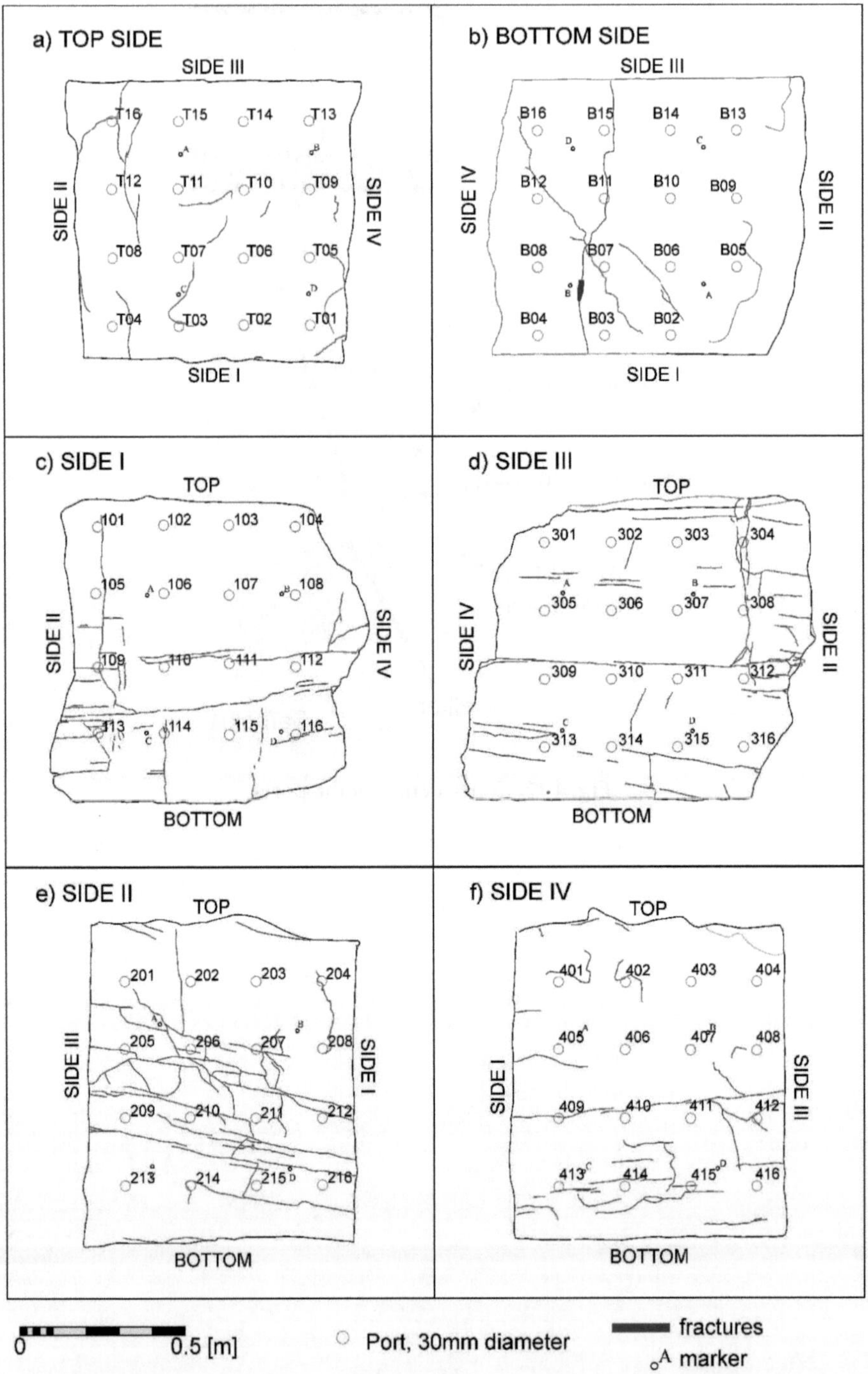

Fig. 4.18. Surface profiles and location of ports at the $90 \times 90 \times 80\,\mathrm{cm}^3$-block.

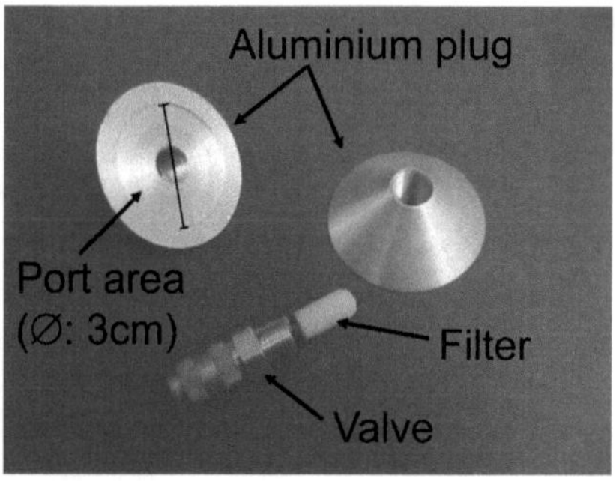

Fig. 4.19. Aluminium plugs used as port connections to the $60 \times 60 \times 60\,\mathrm{cm}^3$-block.

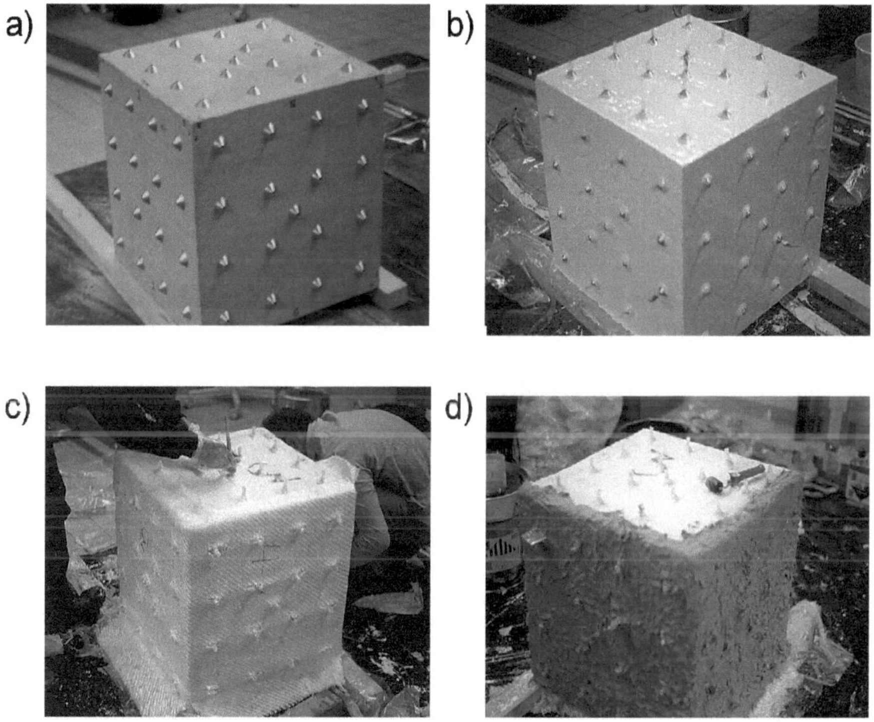

Fig. 4.20. Preparation of the $60 \times 60 \times 60\,\mathrm{cm}^3$-block. (a) Block with attached aluminium plugs. (b) Block with first resin coating. (c) Block with glass fiber mat coating. (d) Block with glass fiber plasticine coat.

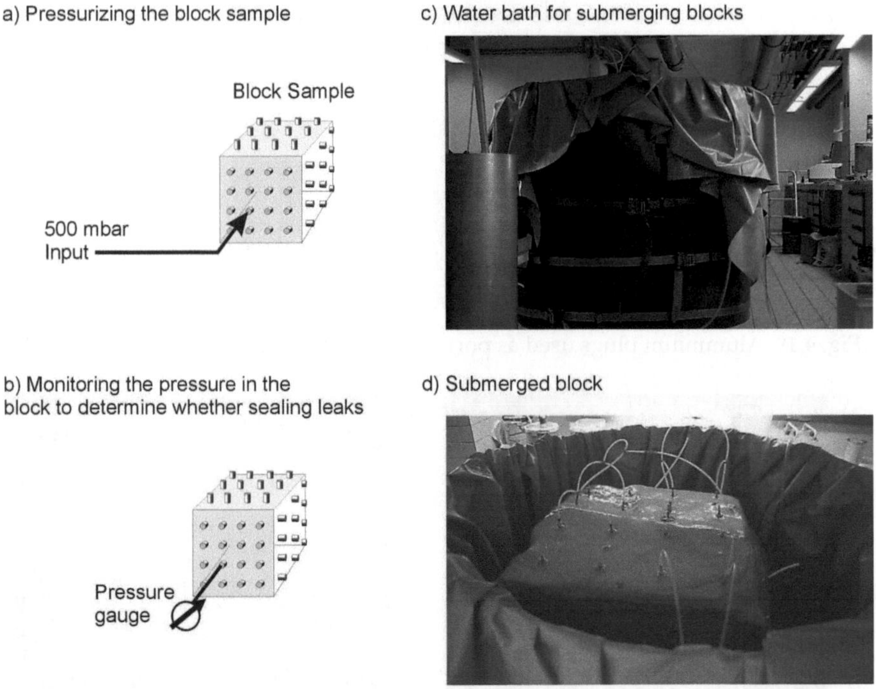

Fig. 4.21. Method for ensuring the hermetic sealing of the block. (a) Pressurizing the block sample. (b) Monitoring the pressure in the block to determine whether leaks are present. (c) Constructed water bath for submerging blocks and checking for leaks. (d) Partly submerged block.

4.2 Flow and Transport Experiments Conducted on Laboratory Cylinders

C.I. M^c^Dermott, M. Sauter, R. Liedl, G. Teutsch

In this section a flexible experimental technique allowing the experimental tomographical investigations of large scale laboratory samples (30 cm diameter x circa 35 cm length) is presented. The samples contain fracture networks and the experimental procedure is designed to allow for the measurement of discrete and integral signals of the constituent elements within the samples, i.e. matrix, individual fractures, fracture matrix interaction, and fracture network response. The signals can easily be evaluated in terms of the geometry of the system and scale of measurement.

4.2.1 Application and Method

In practical experimental terms, when the tomographical investigation concept is applied, it is necessary to make numerous point-to-point, or surface-to-surface measurements of flow and transport across a sample. Several point-to-point measurements are illustrated in 4.22a for a homogeneous isotropic sample and in 4.22b for an anisotropic sample. The results of the measurements are combined, using an appropriate model to produce a parameter field representing the processes causing the signal measured. Here, we concentrate on the experimental procedure and apply a simplified model to present the types of results available from the equipment developed.

To allow such a large number of measurements to be collected, gas flow techniques were applied, enabling the rapid measurement of flow (pressure and flow rate) and transport (tracer concentration) signals. Using gas flow techniques means it is not necessary to saturate the samples prior to the investigation, and very small flow rates could easily be measured rapidly with a high degree of accuracy. In contrast, a direct hydraulic tomographical investigation of such fractured porous samples has proved to be technically very difficult and time consuming (Hagemann, 2001) principally due to the problem of saturation. The saturation of the samples investigated was assumed to remain constant throughout the experimentation, allowing the comparison of the results. The conversion of gas flow parameters to hydraulic parameters has been investigated in detail by Jaritz (1999) for one-dimensional flow systems. In the current experimental cell, flow is anisotropic and multidimensional. The detailed effects of slip flow (Klinkenberg, 1941) and free molecular flow (Carman, 1956) still need to be considered on the basis of a modelled description of the flow system before a direct transfer of gas parameters to hydraulic parameters can be performed.

The difficulties and techniques for collecting fractured samples large enough so that the fracture network could be tomographically investigated, but small enough to remain manageable in the laboratory is discussed in

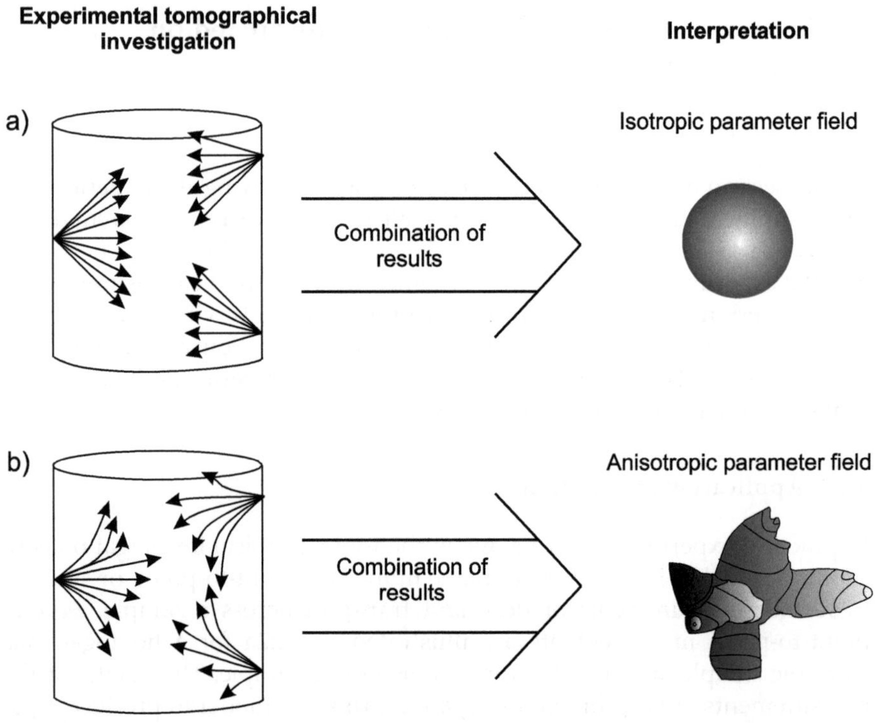

Fig. 4.22. Principles behind the tomographical investigation of a fractured sample with isotropic a) and anisotropic b) parameter distribution (after McDermott *et al.* (2003a)).

McDermott *et al.* (2003b). Following the techniques presented in McDermott *et al.* (2003b) the samples collected for the experimental work have a diameter of 30cm and length of between 30 cm and 40 cm (Fig. 4.23). An example of the fracture record for such a sample is presented in Fig. 4.9.

In accordance with the concepts outlined above and the sample geometry, a special experimental cell was developed which allowed access to preselected positions along the surface of the sample (ports). Spatially orientated point-to-point measurements of the flow and transport characteristics could then be carried out from port-to-port. Additionally, the experimental cell was designed to adjust for samples differing in length (10 cm to 40 cm) and in diameter (290 mm ± circa 15 mm) and was able to maintain stable boundary conditions throughout the period of measurement. The principles and concepts behind the design of the cell to fulfil these criteria for reliable flow and transport measurements are presented in Fig. 4.24.

The experimental cell, named the Multi Input Output Jacket (MIOJ) formed the key element in the experimental set-up, the general arrangement of which is illustrated in Fig. 4.25. To investigate the flow and transport pa-

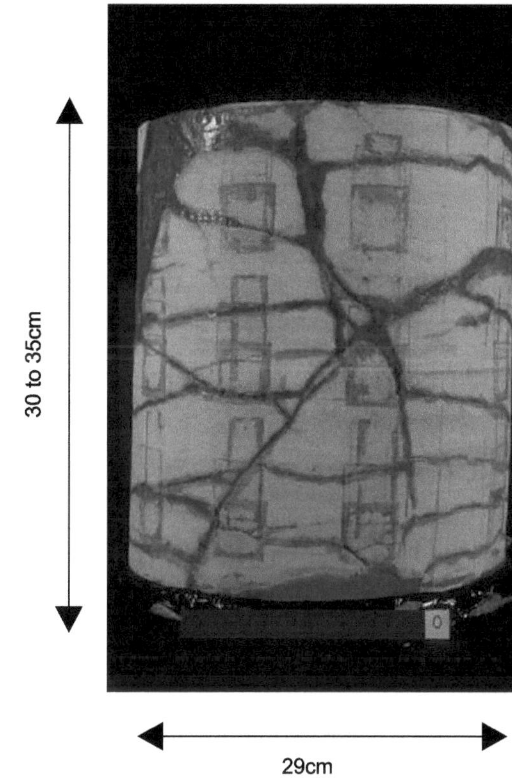

Fig. 4.23. Example of a fractured sample. The visible fractures are filled with sand (McDermott *et al.*, 2003a).

rameters, a stable linear flow field is established across the sample from the input port/ports to the output port/ports. The flow rate across the sample is recorded using a flow meter (a bubble meter is illustrated). Gas tracer is then injected via a flow-through loop (Jaritz, 1999) into the flow field before the input port/ports and the breakthrough of the gas tracer at the output port/ports is recorded using a mass spectrometer. The exact mass of the tracer introduced into the system is known, the location and the time of input, the mass recovered, location and rate of recovery are measured. Further details of the experimental procedure can be found in McDermott (1999).

4.2.2 Technical Details

The Multi Input Output Jacket (MIOJ) consists of two sealing membranes (outer and inner) with predefined open spaces or ports in them, three transparent curved polycarbonate shells (brace shells), two transparent polycarbonate circular plates (top / bottom plates) and a rigid frame. The functionality of the cell is achieved in that the membranes are pressed against the

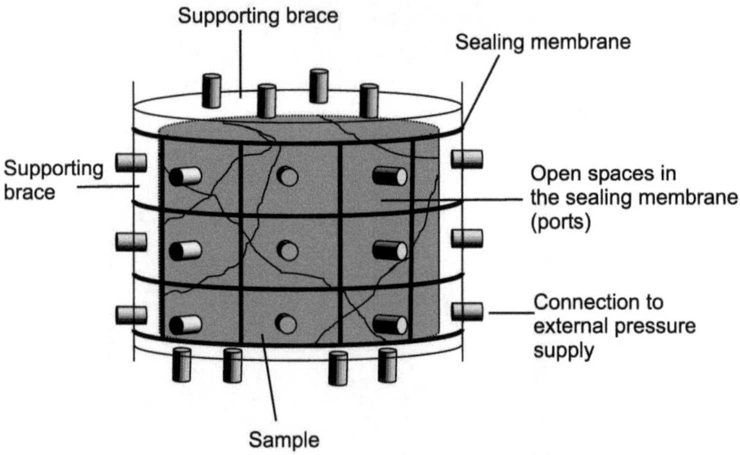

Fig. 4.24. Concepts and principles of the experimental cell enabling tomographic investigation of samples (McDermott *et al.*, 2003a).

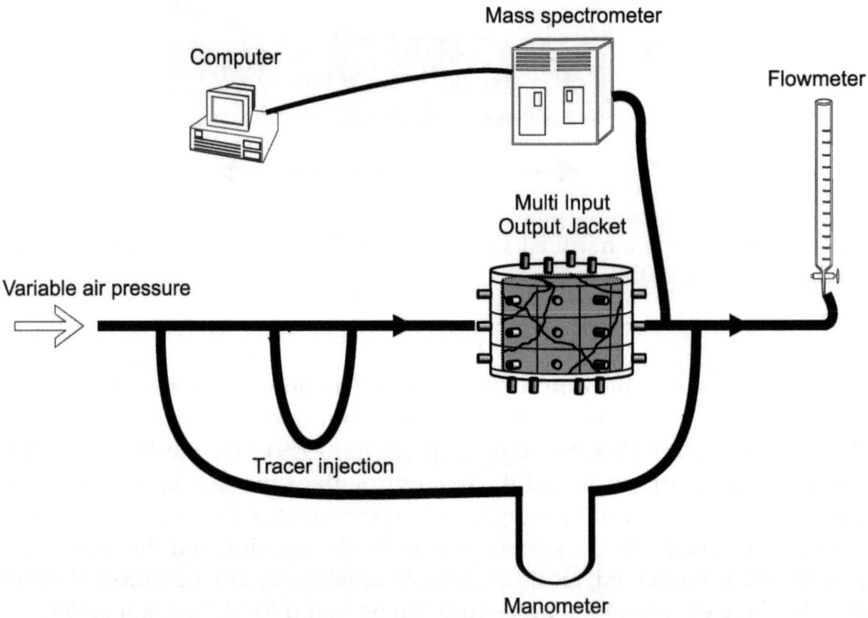

Fig. 4.25. Experimental set-up for investigation of cylindrical samples (McDermott *et al.*, 2003a).

sample with the use of the brace shells and the top and bottom plate, thereby forming a complete cylinder hermetically sealing the sample, Fig. 4.26.

The outer sealing membrane is formed by a soft rubber seal 5mm thick, the inner sealing membrane consists of a self adhesive elastomer 1mm to

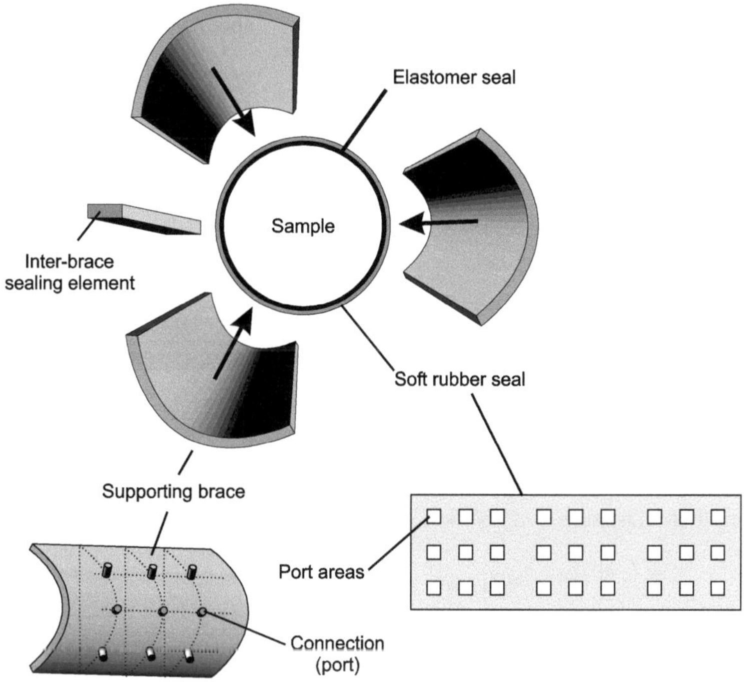

Fig. 4.26. Elements of the Multi Input Output Jacket (MIOJ), (McDermott *et al.*, 2003a).

2 mm thick (Fig. 4.27), stuck to the outer sealing membrane. Under the influence of pressure from the supporting brace, the membrane deforms and adjusts to any unevenness along the surface profile of the sample, thereby sealing any potential leaks. The ports for access to the sample are formed by the spaces in the membranes and each port can be connected to an external pressurized air supply through the connections in the supporting brace (Fig. 4.27).

As the individual brace shells are pressed onto the sealing membrane, the gap between the brace shells is filled with a soft rubber strip of the same material as the outer sealing membrane. The thickness of these inter-brace shell sealing elements can be varied to account for variations in the diameters of the samples.

The supporting brace is held against the sample by four metal bands (Fig. 4.28). These bands are removable and, when screwed together, exert an inward radial force on the brace. The top and bottom braces are pressed against the sample from a rigid frame and a series of screws (Fig. 4.28).

4.2.3 Procedure

Hermetically sealing the sample in the MIOJ and ensuring that flow from port-to-port occurred through the sample and not along the surface of the

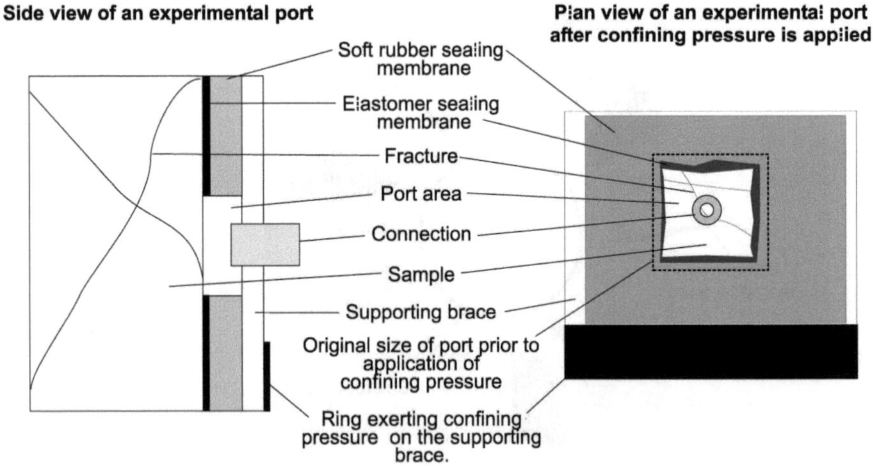

Fig. 4.27. Detailed view of the technique used to isolate port areas on the surface of the sample (McDermott *et al.*, 2003a).

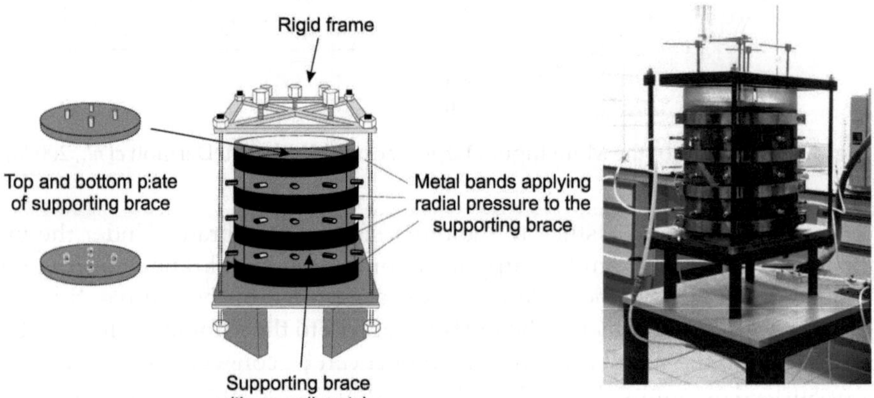

Fig. 4.28. Technical realization of the principle behind the MIOJ (McDermott *et al.*, 2003a).

sample as a result of insufficient contact of the sealing membrane with the surface of the sample (Fig. 4.29) required a lot of care. The techniques employed during the construction of the MIOJ to ensure these conditions included:

- ensuring that the sample was thoroughly cleaned from any fine loose material on the surface;
- any large scale unevenness or missing material from the sample was filled with silicon, thus enabling the sealing membrane of the MIOJ to have a smooth contact with the surface of the sample;

- where the sample had a weak consistency, a thin layer of silicon was placed around the locations of the input and output ports to aid in the sealing of these areas;
- for highly fractured samples, silicon was used to seal all the surface profile of the fractures apart from those areas appearing in ports. The silicon remained only on the surface of the sample and did not enter more than a couple of millimeters into the fracture itself; consequently, the permeability distribution remains practically unaffected. The silicon could be easily removed, allowing re-measurement of the sample using a new geometrical set-up of the ports;
- the outer surface of the sealing membrane was lubricated to ensure an even distribution of the pressure applied by the supporting brace on the sample.

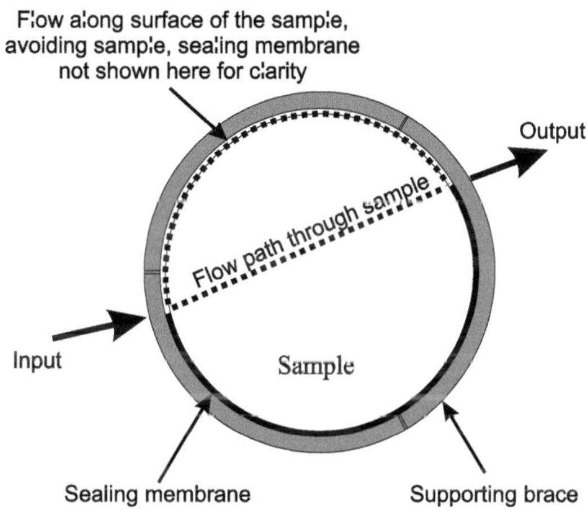

Fig. 4.29. Major flow paths through a cylindrical sample (McDermott *et al.*, 2003a).

Once the sample was sealed in the MIOJ, the apparatus was left for a period of 24 hours to make sure the sealing elastomer (inner sealing membrane) had time to plastically deform and fill any unevenness in the sample profile or block any potential leaks in the system. After this period, the airtightness of the MIOJ (comprising some 30 connections) was tested by ensuring it could maintain a pressure of some 150 mbar. Once it was clear that the system was airtight, stable benchmark flow measurements were made across the sample. After these results had been recorded, the pressure on the supporting brace (both edge and top and bottom) was increased slightly, the benchmark measurements repeated and the results compared. If the rate of flow remained constant, then it was assumed that the flow was through the sample. However, if the flow rate had been reduced by an increase of pressure exerted by the supporting brace, then flow along the surface of the sample and not through the sample was assumed to be occurring. In such cases, the pressure exerted by the supporting brace was increased until no reduction in the flow rate was observed.

The benchmark measurements were repeated throughout the experimental investigation to ensure consistency of results. Both sealing membranes were deformable under light hand pressure; therefore, the effect of the pressure exerted by the supporting brace to ensure complete sealing of the sample on the fracture aperture due to the stiffness of the fractures was assumed to be negligible.

4.2.4 Flexibility of the MIOJ

Depending on the type of investigation required, different surface areas of the sample can be placed under a predefined pressure system, thus allowing a variety of adjustable boundary conditions in an enclosed system. The flow direction across the sample can easily be varied, and the anisotropy of flow and transport parameters can be determined (Fig. 4.30). The MIOJ is not just limited to point-to-point measurements. Any combination of the ports can be used to measure the flow and transport parameters depending on the geometry of the sealing membrane (Fig. 4.31).

As can be seen from Figs. 4.30 and 4.31, the dependency of the flow on scale effects such as the distance between the input and output ports or the size of the port area on the sample at a bench scale can easily be investigated as well as the effect of individual fractures in a fracture system or the fracture network itself.

Measurement of the flow field characteristics. The variable air pressure supply (Fig. 4.25), provided by a regulator valve, enabled a stable pressure difference to be established across the MIOJ. The flow rate across the sample as a result of this pressure difference was recorded using a flow meter. Experimental measurements indicated a linear relationship between flow and pressure difference with the application of a pressure difference of 150mbar in the fractured system and allowed for the assumption that linear flow conditions were present within the samples.

In practice, it was assumed that the stable flow field was established by maintaining stable boundary conditions for a period of five minutes. For the samples investigated, this period of time was determined experimentally by recording the change of the flow in the system against time from the initial conditions of 150 mbar plus atmospheric pressure in the whole sample to the new boundary conditions, i.e. 150 mbar plus atmospheric pressure on the input port to atmospheric pressure on the output port. Fig. 4.32 shows the change in the flow rate with time for such an experimental set-up for the cases where the matrix is connected to the input port. The steeper curve corresponds to the case where a fracture was directly connected to the output port. For the gentler curve, the matrix was connected to the output port. After five minutes, relative stability in the flow measurements can be seen to have

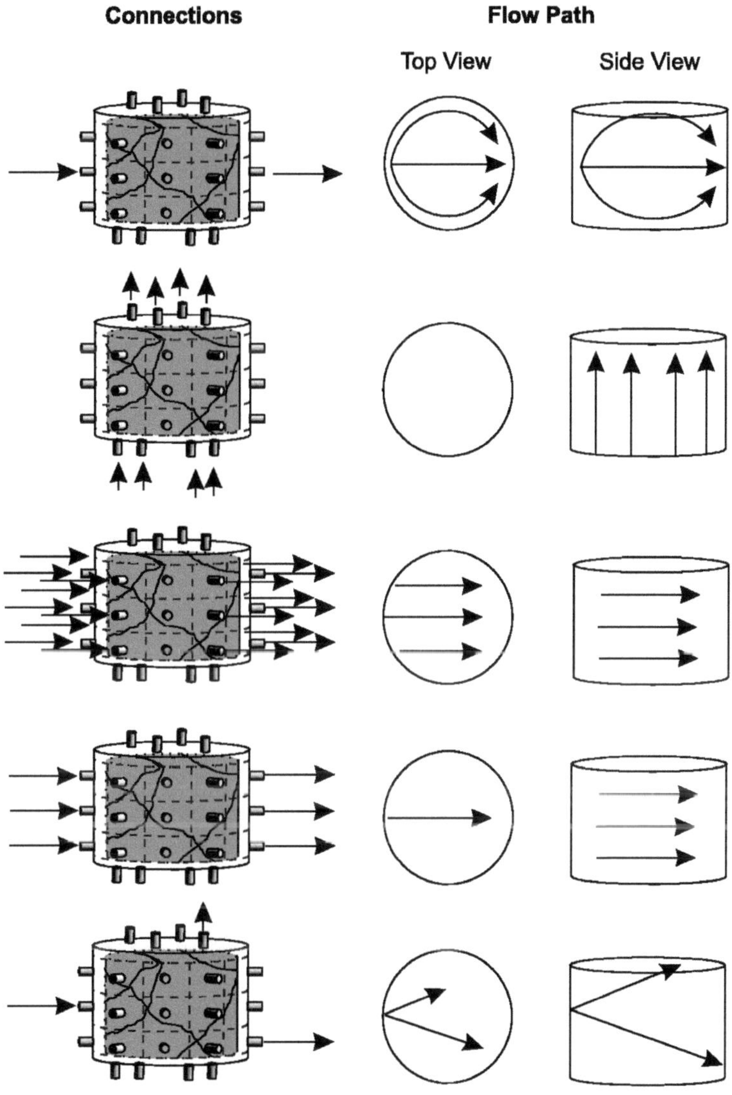

Fig. 4.30. Various configurations of input ports to output ports, allowing the spatial investigation of the flow and transport characteristics (McDermott *et al.*, 2003a).

been achieved. However, true stability in the case of the matrix connection was only achieved after 15 minutes.

Given the large number of measurements necessary to determine the spatially dependent flow and transport parameters, the time of at least 15 minutes per point-to-point experiment required to attain fully stable conditions was not practically possible. For an investigation series comprising several hundred individual point-to-point experiments, the time required

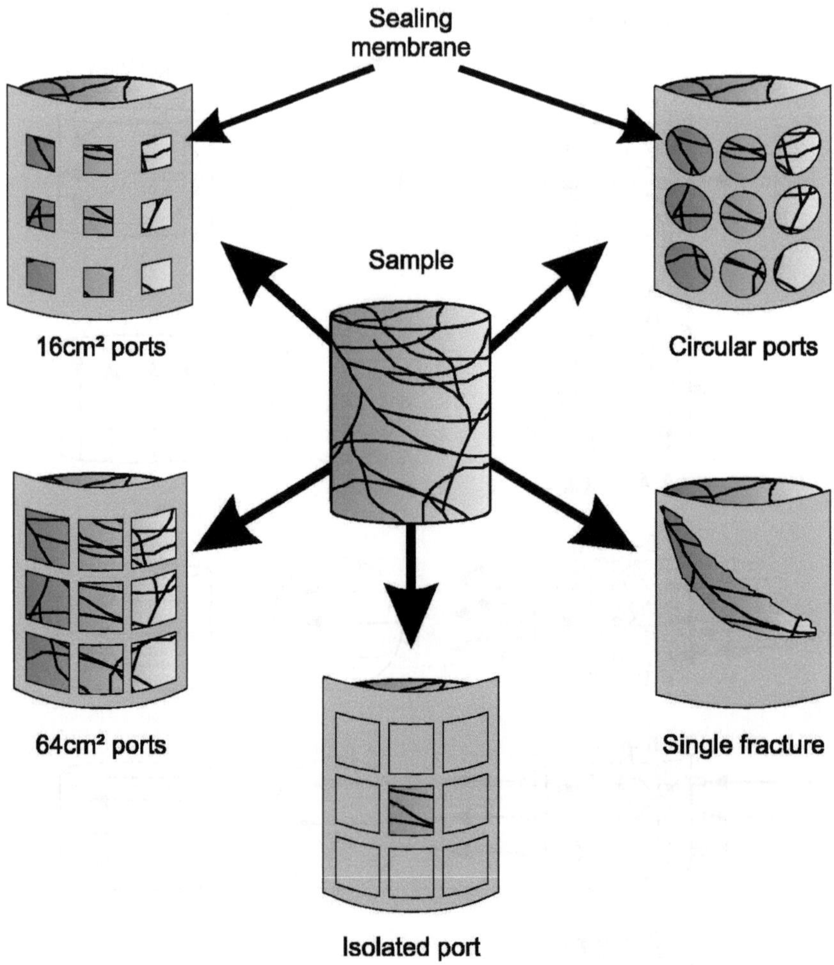

Fig. 4.31. Investigation possibilities of the MIOJ (McDermott *et al.*, 2003a).

for the measurements would have increased by a factor of three, but the accuracy of the results only minimally. Therefore, the error resulting from the possibility of slightly unstable flow conditions existing after five minutes was considered acceptable. Throughout an investigation series the cell was continually tested to prove that it was airtight, i.e. the valves, seals, pipes and any connections were not leaking

Measurement of the transport characteristics. To determine the transport characteristics of the system, it was necessary to undertake tracer tests. Tracer gas (helium) was injected via a flow-through loop into the stable flow field once established across the MIOJ (Fig. 4.25) to provide a Dirac pulse. The breakthrough of the tracer gas was recorded using a mass spectrometer.

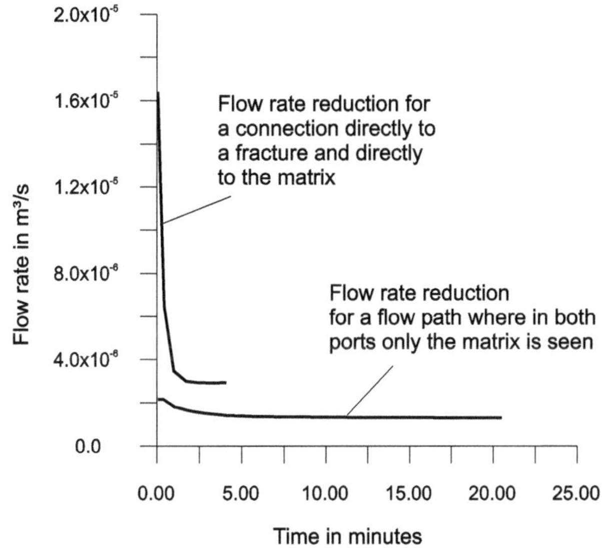

Fig. 4.32. Time taken for stabilization of the flow field in the MIOJ samples (example for sandstone samples with a matrix hydraulic conductivity of approximately $2.7 \cdot 10^{-6}\,\mathrm{m\,s^{-1}}$, where the entire sample was pre-pressurized to 150 mbar, then flow allowed from the output port at 0 mbar), (McDermott *et al.*, 2003a).

This injection technique enables the exact mass of the tracer introduced into the system and the time of input to be determined. If a constant pulse were used, the whole sample would become saturated with the tracer and, prior to further experimentation, the system would have to be flushed for a significant period of time.

To simplify the analysis of the transport information, a non-reactive gas, helium, was chosen as a tracer, thus eliminating the possibility of adsorption, retardation or biological degradation. In addition, the carrier gas used to establish the flow in the samples was air, and therefore the selection of helium was particularly convenient in that practically no helium is found naturally in air.

*Examples of data recorded.*The following data are available after the completion of every individual experiment:

- rate of flow across the sample,
- area of input and output ports,
- pressure at input port and pressure at output port,
- orientation of the input port to output ports and therefore first estimation of the geometry of the flow field,
- type and quantity of the tracer injected into the system,
- the breakthrough curve of the tracer at every extraction point.

After several experiments, the information available for each sample includes:

- differently scaled spatially dependent information about the flow field parameters;
- differently scaled spatially dependent information about the transport parameters.

4.2.5 Flow Experiments

An extract example of the flow rate information recorded during experimentation is presented in Table 4.2. Examples of flow rate tensors derived from the experimental results for a highly fractured sandstone sample are presented in Figs. 4.33 and 4.34. In Fig. 4.33, three different levels of the sample are investigated via point-to-point measurements. In Fig. 4.34, the same sample is investigated along the length of the sides by connecting the input and output ports.

Table 4.2. Extract of data recorded in a pneumatic experimental investigation series. The cross section area of the port used is 0.001225 m^2.

Input port number	Output port number	Pressure diff. (mbar)	Volume flow rate (m^3/s)	Direct distance between ports (m)	Bearing deg. to north	Angle to horizontal
3	18	150	1.25E-05	0.285	166.2	0.0
13	2	150	6.74E-06	0.296	3.6	-19.7
15	2	150	5.39E-06	0.296	3.6	19.7
5	16	150	2.03E-06	0.296	148.8	19.7
1	14	150	4.29E-06	0.296	183.6	-19.7
2	13	150	6.70E-06	0.296	183.6	19.7
5	18	150	2.70E-06	0.296	148.8	-19.7
3	14	150	4.96E-06	0.296	183.6	19.7
1	32	150	8.04E-06	0.318	186.8	-51.8
3	10	150	1.94E-05	0.318	201.0	38.9
3	16	150	7.89E-06	0.348	166.2	35.1
1	18	150	9.44E-06	0.348	166.2	-35.1

The effects of scale on the measurement can immediately be seen from these simple diagrams. With the more discrete point-to-point approach (Fig. 4.30), there is a variation of the flow rate dependent on whether a fracture is directly connected to the port used for investigation. Some port-to-port flow rates are several orders of magnitude greater than others. However, with the more integral surface-to-surface approach, an averaging effect occurs leading to the tensor seen in Fig. 4.34.

This effect of moving from the discrete investigation scale to a more integral investigation according to the size of the ports used for investigation

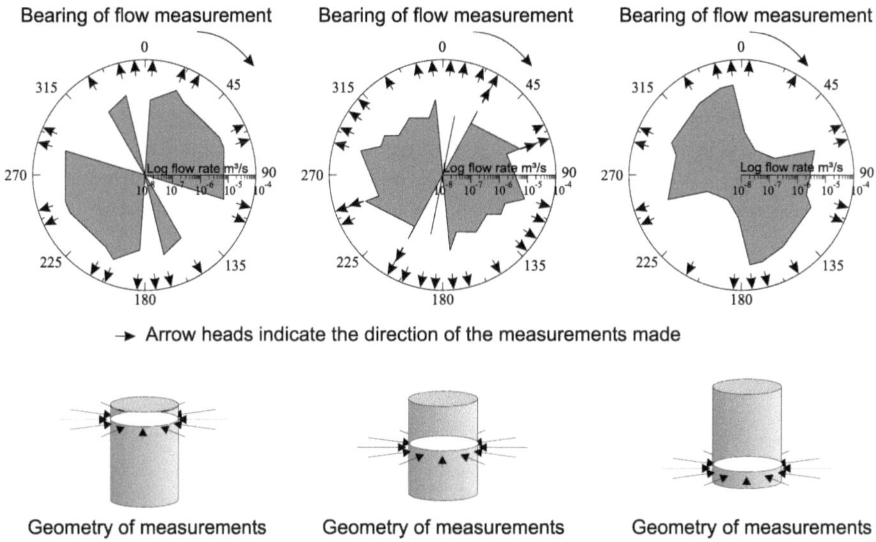

Fig. 4.33. Example of flow rate tensors derived from a highly fractured sample; here, the tomographical investigation proceeds along three separate planes in the sample (McDermott *et al.*, 2003a).

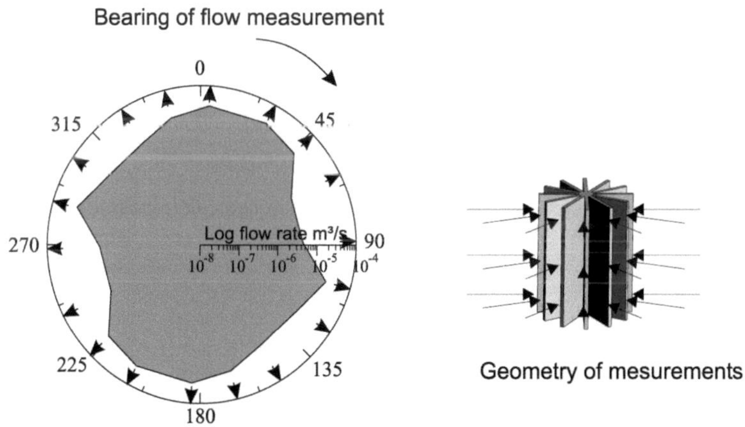

Fig. 4.34. Example of flow rate tensors derived from a highly fractured samples; here, complete side to side measurements are made (McDermott *et al.*, 2003a).

is illustrated in Fig. 4.35. Here, a simple one dimensional DARCYmodel has been used to convert the flow data into apparent hydraulic conductivity to allow the comparison of different scales of measurement. The term "apparent hydraulic conductivity"refers to the fact that this DARCYmodel is a significant simplification of the complex system and provides an apparent hydraulic conductivity. However, the more accurate multiple dimensional analysis of the flow and transport signals to derive actual hydraulic conductivities (e.g. Vesselinov *et al.* (2001), Leven *et al.* (2000), McDermott *et al.* (1998)) is beyond the scope of the current section. In Fig. 4.35, it can be seen that if the investigation areas are smaller then the distribution in apparent hydraulic conductivity is larger. Likewise, if the areas are larger then the variation is smaller. This illustrates the averaging effect of heterogeneities of a more integral signal in the larger areas to the more discrete signals from the smaller areas.

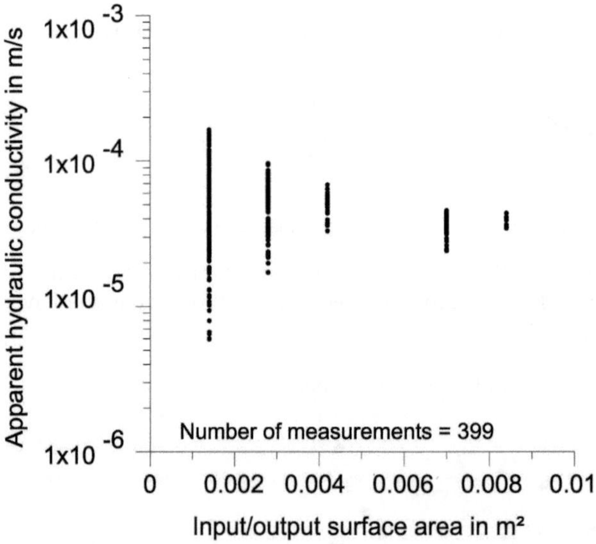

Fig. 4.35. Effect of the surface area on the variation in the apparent hydraulic conductivity (McDermott *et al.*, 2003a).

4.2.6 Transport Experiments

Typical results of transport experiments in the MIOJ are illustrated in Figs. 4.36 and 4.37. In Fig. 4.36, the characteristics are purely dominated by the matrix of the sample, which leads to a signal with a comparatively long tailing and a large arrival time. In contrast, Fig. 4.37 presents a number of different breakthrough curves, illustrating the variation in signal responses found from the MIOJ when a fracture was directly connected either to the input or output ports. Where fracture and matrix are involved in the transport, then interesting double and even triple peaks have been observed. In addition, the fracture network itself may also provide several different channels for flow, leading to different peaks in the breakthrough curve and several different ports may provide a signal from the fracture.

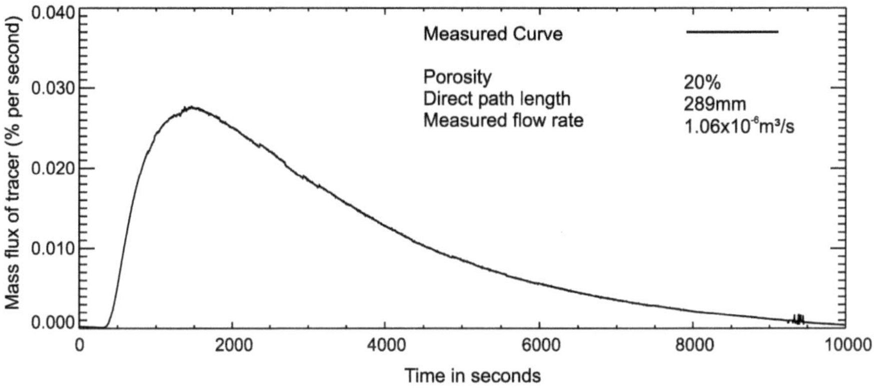

Fig. 4.36. Typical transport result from the MIOJ, dominated by matrix (McDermott *et al.*, 2003a).

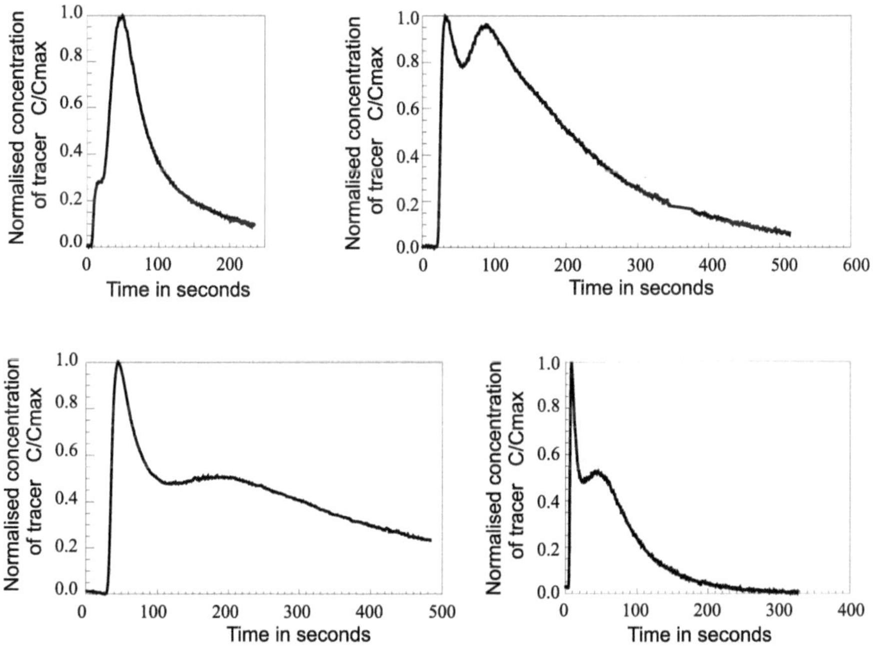

Fig. 4.37. Typical transport results from the MIOJ where fractures are directly connected to at least one of the investigation ports (McDermott *et al.*, 2003a).

4.2.7 Conclusions

The MIOJ provides a flexible innovative approach for the experimental tomographical investigation of fractured porous rock. The experimental method developed provides a wide variety of options for investigating flow and transport in a fracture network. The results gained from experimentation show that it is possible to use the MIOJ to investigate fractured samples using a discrete and integral approach, to define orientated flow and transport tensors, and to investigate upscaling effects.

4.3 Interpretation of Flow and Transport Experiments Conducted on Laboratory Cylinders

By means of numerical model simulation, the significance of parameters and the relevance of occuring flow and transport processes can be investigated. A comparative analysis of measured and simulated results gives information about the properties of the investigated sample on the one hand, and about the possibilities and limitations of the chosen model concept on the other. In this section, investigation results from simulations conducted on one of the cylindrical bench scale samples discussed in Sects. 4.1.1 and 4.2 are presented. A sample with the dimensions 0.290 mm (height) and 0.286 mm (diameter) and a relatively high fracture density is chosen. The digitized record of the sample is shown in Fig. 4.9.

With the discrete model approach (Sect. 2.4), a sensitivity analysis is conducted (Sect. 4.3.1) in order to identify the significance of the flow and transport parameters involved. The results of this analysis are useful for the calibration of the model to the experimentally measured data (Sect. 4.3.2). The possibilities and limitations of using a discrete and a multi-continuum approach (Sect. 2.5) are investigated by determining equivalent parameters for flow and transport, applying the discrete model, and using them as input for a multi-continuum model (Sects. 4.3.3 and 4.3.4). The results from the two approaches are compared, allowing for an assessment of their feasibility on the current scale.

4.3.1 Sensitivity Analysis

L. Neunhäuserer, M. Süß, R. Helmig

When numerical models are used or the simulation results of such models judged, it is always essential to keep in mind that a model is an approximation and not an exact description (Fig. 2.1) of reality. This has a number of reasons:

- the material properties are approximated using model parameters;
- the exact structure of the domain is not known;
- the real problem is spatially and temporally discretized.

A sensitivity analysis is an important tool for determining the influence that these approximations may have. In this section, an analysis of the influence of the parameters permeability, porosity and dispersivity is presented. The aim of the analysis is to found a basis for the later comparison of measured and simulated tracer-breakthrough curves as presented in Sect. 4.3.2.

4.3.1.1 Model Set-Up

The structure information is recorded digitally in the laboratory and is then used to set up the structural model, which serves as a basis for the generation of the finite volume mesh used for the numerical simulation. Even though this example involves a rather small sample, the detection of the spatial distribution of the fractures is not trivial, but requires a certain amount of approximation.

For the purpose of the analyses presented here, a two-dimensional slice from the top of the core sample is selected. In Fig. 4.38, the two-dimensional structural model and the finite volume mesh are shown.

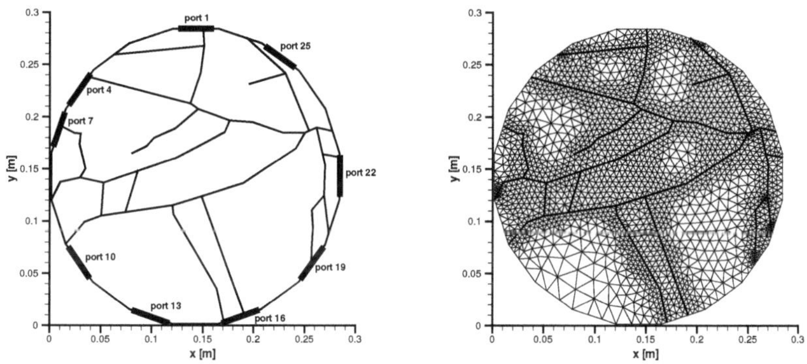

Fig. 4.38. Structural model with location of measurement ports (left) and finite volume mesh (right).

The simulations presented here consist of a steady-state flow calculation for the air-saturated, cylindrical sample and a subsequent transport simulation for a certain number of time steps. The flow field is simulated defining one inflow and one outflow port, where a constant pressure of 116 300 Pa and 101 300 Pa respectively is imposed. All other boundaries are impervious to flow. For the transport simulation, the tracer is injected as a pulse by keeping unit concentration constant until 30 ml of the tracer has entered the domain. The tracer leaves the domain through the outflow port over a free-flow boundary, allowing the concentration to be observed at this point. This set-up is designed in accordance with the actual measurements on the sample.

This set-up is used for sensitivity analysis (Sect. 4.3.1), calibration (Sect. 4.3.2) and determination of equivalent parameters (Sect. 4.3.2).

Measured flow and transport data for several combinations of ports are avaliable. For the comparison of measured and simulated data in this section (Sect. 4.3.1) and also for the calibration of the model as presented in

Sect. 4.3.2, the measurement with input port 1 and output port 10 is chosen (1–10). The tracer-breakthrough curve of this measurement has a double peak, which is considered typical for fractured porous media. Additionally, the port combinations (1–4) and (1–16) are considered in Sect. 4.3.2.

4.3.1.2 Permeability

The influence of the permeability on the flow and transport behavior is presented on the basis of the results of three different simulations. The interesting parameter is the ratio between the fracture and the matrix permeability, and not the absolute values of these parameters. This ratio is a measure of the contrast between the two constituents of the system. Here, the fracture permeability for all fractures is set to $4.1 \cdot 10^{-10}\,\mathrm{m}^2$, which corresponds to a fracture aperture of 70 μm according to the cubic law. Three different matrix permeabilities are applied. Figure 4.39 shows the distribution of the piezometric head for the three cases. The piezometric-head distributions demonstrate the increasing significance of the fractures as the permeability ratio k_F/k_M increases. The discharge is $Q_1 = 7.2 \cdot 10^{-5}\,\mathrm{m}^3\,\mathrm{s}^{-1}$ (left), $Q_2 = 1.3 \cdot 10^{-5}\,\mathrm{m}^3\,\mathrm{s}^{-1}$ (center) and $Q_3 = 0.6 \cdot 10^{-5}\,\mathrm{m}^3\,\mathrm{s}^{-1}$ (right). The decreasing discharge, for a constant discharge area, indicates a decreasing flow velocity in the matrix as its permeability decreases.

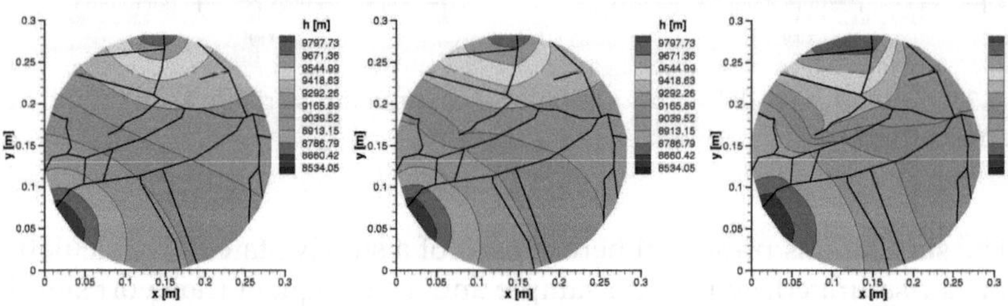

Fig. 4.39. Resulting piezometric-head distributions for $k_M = 3.0 \cdot 10^{-12}\,\mathrm{m}^2$ (left), $k_M = 3.0 \cdot 10^{-13}\,\mathrm{m}^2$ (center) and $k_M = 3.0 \cdot 10^{-14}\,\mathrm{m}^2$ (right) (port configuration 1–10).

For the transport calculation, the matrix and the fracture porosity are fixed at $n_{e,M} = 0.20$ and $n_{e,F} = 0.30$ respectively for all three cases. The dispersivity in the fracture is set to zero as the parallel-plate concept, i.e. the fracture aperture is constant, is applied. For the matrix, a dispersivity of $\alpha_l = \alpha_t = 0.01\,\mathrm{m}$ is assumed. The effective molecular diffusion coefficient is set to $D_m = 1.0 \cdot 10^{-9}\,\mathrm{m}^2\,\mathrm{s}^{-1}$, which is a very low value for air. Figure 4.40 shows the concentration distribution for the three cases at the same point in time. The corresponding tracer-breakthrough curves are presented in Fig. 4.41.

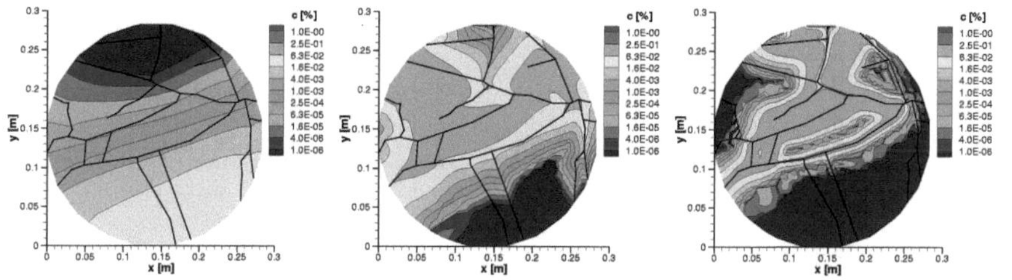

Fig. 4.40. Resulting concentration distributions for $k_\mathrm{M} = 3.0 \cdot 10^{-12}\,\mathrm{m}^2$ (left), $k_\mathrm{M} = 3.0 \cdot 10^{-13}\,\mathrm{m}^2$ (center) and $k_\mathrm{M} = 3.0 \cdot 10^{-14}\,\mathrm{m}^2$ (right) (port configuration 1–10).

For high matrix permeability, the transport process is only slightly influenced by the fractures. The main part of the tracer does not travel significantly more slowly in the matrix than in the fractures towards the output port. This yields an early high peak. The tailing is mainly caused by the asymmetry of the system forcing a portion of the tracer mass to travel a longer distance.

For the intermediate matrix permeability, the transport of the tracer mass in the matrix is significantly slower than for the first case. However, a large part of the mass enters the matrix from the fractures, so that the peak of the tracer-breakthrough curve is substantially lower. The tailing results from the relatively large portion of mass distributed in the total matrix.

Low matrix permeability yields a tracer-breakthrough curve which has an earlier and sharper increase than for the intermediate permeability. The flow and the transport are mainly bundled to the fractures. The double peak of the tracer-breakthrough curve is in this case caused by the two different main flow paths in the fracture system. The tracer mass in the matrix is only removed very slowly, causing a pronounced tailing.

4.3.1.3 Porosity

The influence of the porosity on the flow and transport processes is discussed, using the case with the intermediate matrix permeability. The values of dispersivity and diffusion remain the same as in the previous example. Again, the parameter for the fractures is kept constant, i.e. the fracture porosity is fixed at $n_\mathrm{e,F} = 0.30$, whereas the matrix porosity $n_\mathrm{e,M}$ is set to 0.05, 0.15 or 0.20. Consequently, the right plot in Fig. 4.42 is the same as the center plot in Fig. 4.40.

The change in porosity is reflected in the change of the seepage velocity, i.e. a low porosity yields a high seepage velocity and vice versa. Consequently, the centroid of the tracer-breakthrough curve is shifted to the right for larger porosities as a reaction to the slower advective transport. Addi-

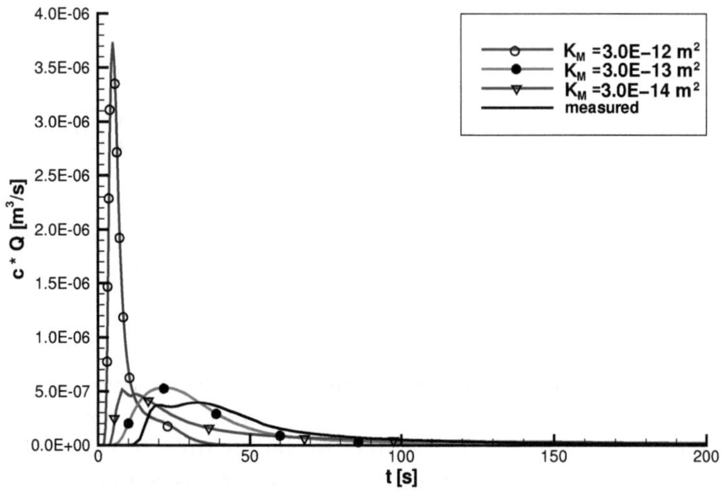

Fig. 4.41. Resulting tracer-breakthrough curves for $k_M = 3.0 \cdot 10^{-12}\,\mathrm{m}^2$, $k_M = 3.0 \cdot 10^{-13}\,\mathrm{m}^2$ and $k_M = 3.0 \cdot 10^{-14}\,\mathrm{m}^2$. The concentration is multiplied by the corresponding discharge in order to achieve comparability.

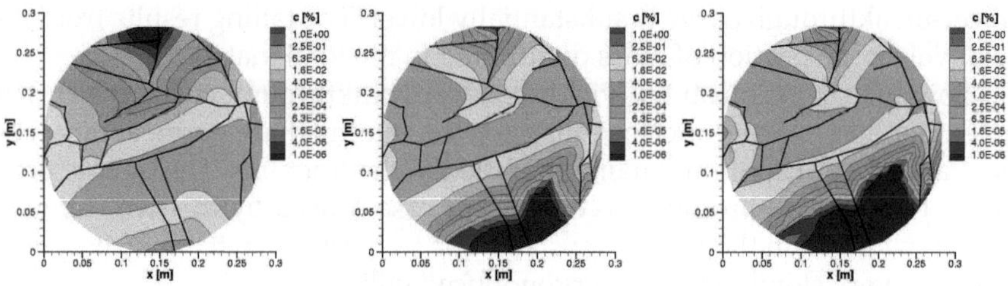

Fig. 4.42. Resulting concentration distributions for $n_{\mathrm{e,F}} = 0.05$ (left), $n_{\mathrm{e,F}} = 0.15$ (center) and $n_{\mathrm{e,F}} = 0.20$ (right) (port configuration 1–10).

tionally, the seepage velocity has an impact on the dispersion term, which increases with decreasing porosity. This yields a smoother concentration front.

Further simulations show that the variation of the fracture porosity, within realistic limits, does not have an essential effect on the processes. This is explained by the small part of the total volume occupied by fractures.

4.3.1.4 Dispersivity

To demonstrate the influence of the dispersivity, again the case with intermediate matrix permeability discussed in Sect. 4.3.1.2 is used as reference. Only the matrix dispersivity is varied. In the first simulation, it is set to

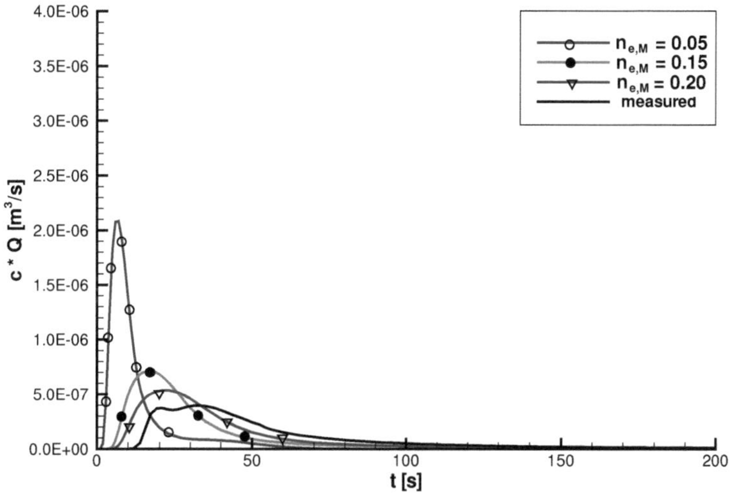

Fig. 4.43. Resulting tracer-breakthrough curves for $n_{e,F} = 0.05$, $n_{e,F} = 0.15$ and $n_{e,F} = 0.20$. The concentration is multiplied by the corresponding discharge in order to achieve comparability.

$\alpha_l = \alpha_t = 0.01$ m, i.e. no change from the reference case. In the second simulation, the value is decreased to $\alpha_l = \alpha_t = 0.001$ m. Figure 4.44 shows the resulting tracer-breakthrough curves.

The decrease of the matrix dispersivity yields a more emphasized advective transport process with a sharper concentration front. The initial arrival occurs earlier and the increase of the tracer-breakthrough curve is faster, due to the smaller portion of the tracer mass entering the matrix from the fractures. Analyzing the concentration distributions for each of the time steps (not presented here) leads to the conclusion that the portion of tracer mass initially entering the matrix and the tracer mass entering the matrix from the fractures is mainly advectively transported through the matrix. This is the main reason for the second peak of the tracer-breakthrough curve. Normally, a portion of tracer mass re-enters the fractures from the matrix because of the steep concentration gradient. Here, however, because of the relatively high permeability of the matrix, the amount that re-enters is negligible.

4.3.1.5 Remarks

In general, it can be stated that a fractured porous system on the considered scale is more sensitive to changes in the matrix parameters (porosity and dispersivity) than to those in the fracture parameters. This is comprehensible, considering the fact that the relative volume of the fractures compared

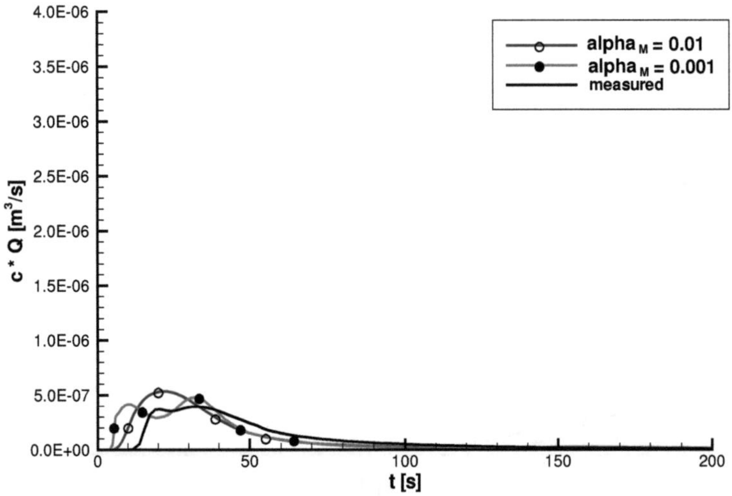

Fig. 4.44. Resulting tracer-breakthrough curves for $\alpha_l = \alpha_t = 0.01\,\mathrm{m}$ and $\alpha_l = \alpha_t = 0.001\,\mathrm{m}$. The concentration is multiplied by the corresponding discharge in order to achieve comparability.

to that of the matrix is very small. Exceptions from this are parameters describing the geometry of the fractures, i.e. fracture density, fracture distance, fracture length and fracture aperture (determining the fracture permeability). The smaller the scale and the stronger the permeability contrast between fractures and matrix, the more important it is weather a fracture is connected to an in- or outflow boundary or not. The same is valid for the connectivity between the fractures.

4.3.2 Comparison of Measured and Simulated Tracer-Breakthrough Curves

L. Neunhäuserer, M. Süß, R. Helmig

In this section, the calibration of the model aiming at the reproduction of the actually measured tracer-breakthrough curves and the corresponding discharges (Sect. 4.2) is discussed. Calibration is in general performed in order to determine a valid set of model parameters that enables the use of a model for the prediction of flow and/or transport behavior for different boundary conditions. Here, the purpose of the calibration is to understand and learn more about the characteristics of the sample investigated rather than actually achieve a calibrated and verified model. The knowledge gained from the sensitivity analysis (Sect. 4.3.1) is a useful support for the calibration process. The model set-up as described in Sect. 4.3.1.1 is applied. The measured data from the port configurations (1–4), (1–10) and (1–16) are selected to represent the system.

4.3.2.1 Calibration

Since the flow field mainly determines the transport process, the first step is to find a parameter set that can reproduce the measured discharges. In order to align the simulated results with the measured values, the fracture apertures and the matrix permeability are varied. For such a task, the conclusions drawn in Sect. 4.3.1 are very helpful. Figure 4.45 presents a possible parameter set together with the measured and the simulated discharges. The corresponding piezometric-head distributions are shown in Fig. 4.46.

The transport simulations are performed for a number of parameter sets (see Table 4.3), all on the basis of specifications based on experimental data

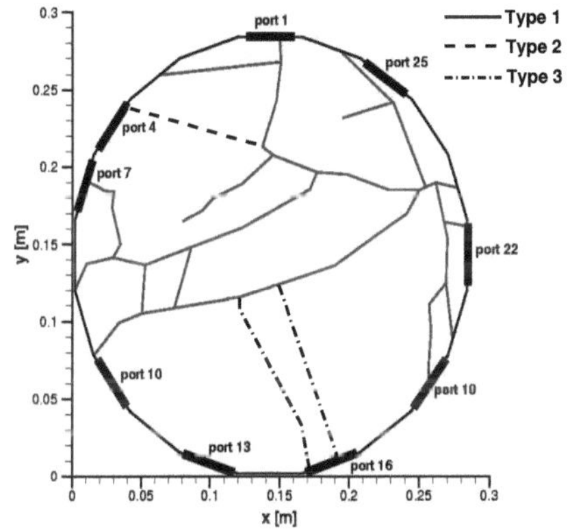

	$k\,(\mathrm{m}^2)$	$b\,(\mu\mathrm{m})$
Matrix	$9.3 \cdot 10^{-14}$	–
Frac. type 1	$6.7 \cdot 10^{-10}$	90.0
Frac. type 2	$9.6 \cdot 10^{-11}$	34.0
Frac. type 3	$1.9 \cdot 10^{-10}$	48.0

Ports	$Q_{\mathrm{meas}}\,(\mathrm{m}^3\,\mathrm{s}^{-1})$	$Q_{\mathrm{sim}}\,(\mathrm{m}^3\,\mathrm{s}^{-1})$
(1-4)	$6.90 \cdot 10^{-6}$	$6.85 \cdot 10^{-6}$
(1-10)	$1.36 \cdot 10^{-5}$	$1.36 \cdot 10^{-5}$
(1-16)	$7.01 \cdot 10^{-6}$	$7.07 \cdot 10^{-6}$

Fig. 4.45. Structural model with ports (left) and parameter set 1 together with the measured and the simulated discharges (right).

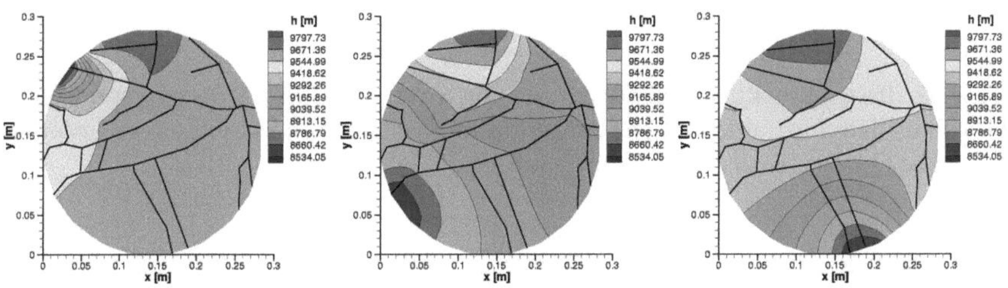

Fig. 4.46. Piezometric-head distribution for parameter set 1 for port configuration 1–4 (left), 1–10 (center) and 1–16 (right).

(Sect. 4.2). The parameters varied are the matrix and the fracture porosities. The corresponding tracer-breakthrough curves for port configuration (1-10) are presented in Fig. 4.47. From this figure, it becomes clear that first, independently of the parameter set, the initial arrival of the tracer occurs too early compared to the measured curve. Second, none of the simulated curves exhibit a double peak. Third, it is pointed out that the curves of parameter set 1-2 and 1-4, only differing in the fracture porosity, can scarcely be distinguished. The lack of sensitivity to this parameter is discussed in Sect. 4.3.1.3.

Table 4.3. Parameter sets for the transport simulation based on flow parameter set 1.

	$n_{e,M}$ (-)	$n_{e,F}$ (-)	$\alpha_{l,M} = \alpha_{t,M}$ (m)	α_F (m)	D_m ($m^2\,s^{-1}$)
Parameter set 1-1	0.20	0.30	0.001	0.0	$6.12 \cdot 10^{-5}$
Parameter set 1-2	0.25	0.30	0.001	0.0	$6.12 \cdot 10^{-5}$
Parameter set 1-3	0.30	0.30	0.001	0.0	$6.12 \cdot 10^{-5}$
Parameter set 1-4	0.25	0.35	0.001	0.0	$6.12 \cdot 10^{-5}$

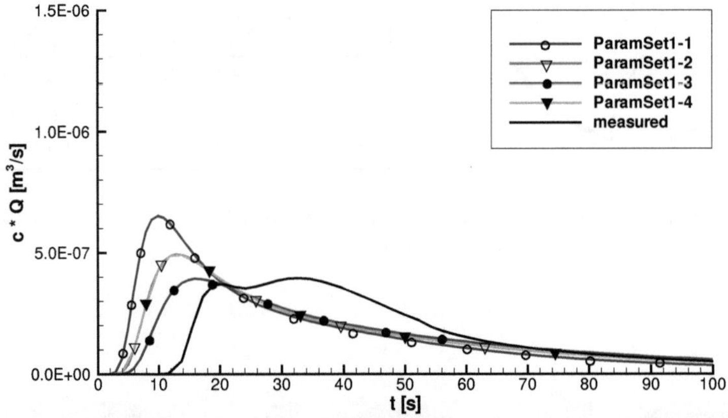

Fig. 4.47. Tracer-breakthrough curves for port configuration (1-10) corresponding to the values listed in Table 4.3. The concentration c is multiplied by the corresponding discharge Q in order to achieve comparability.

In order to fit the initial arrival time of configuration (1-10) better, the fracture aperture of fracture type 1, essential for this port connection, is reduced. As a consequence, the matrix permeability must be increased, in order to maintain the correct discharge. These changes lead to too large a discharge

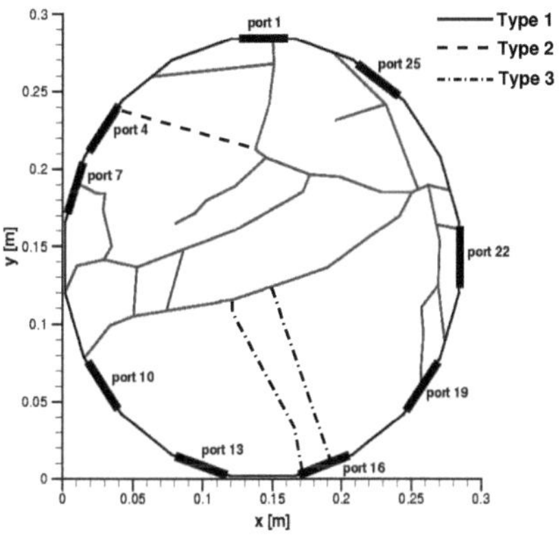

	$k\,(m^2)$	$b\,(\mu m)$
Matrix	$1.5 \cdot 10^{-13}$	–
Frac. type 1	$5.9 \cdot 10^{-10}$	84.5
Frac. type 2	$3.3 \cdot 10^{-11}$	20.0
Frac. type 3	$1.2 \cdot 10^{-10}$	38.0

Ports	$Q_{meas}\,(m^3\,s^{-1})$	$Q_{sim}\,(m^3\,s^{-1})$
(1-4)	$6.90 \cdot 10^{-6}$	$8.03 \cdot 10^{-6}$
(1-10)	$1.36 \cdot 10^{-5}$	$1.36 \cdot 10^{-5}$
(1-16)	$7.01 \cdot 10^{-6}$	$7.05 \cdot 10^{-6}$

Fig. 4.48. Structural model with ports (left) and parameter set 2 together with the measured and the simulated discharges (right).

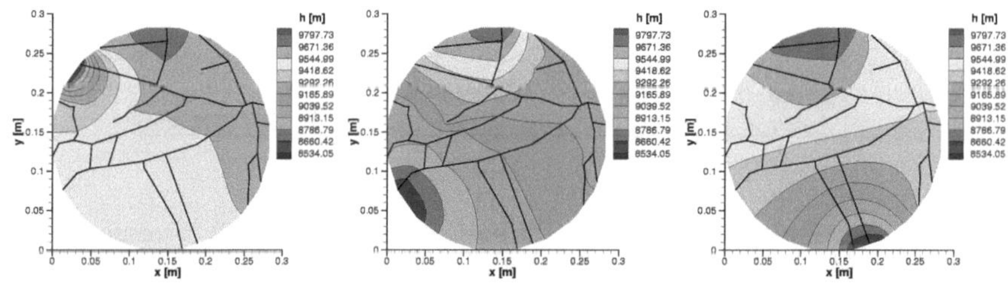

Fig. 4.49. Piezometric-head distribution for parameter set 2 for port configuration 1-4 (left), 1-10 (center) and 1-16 (right).

for port configuration (1-4). To account for this effect, the aperture for fracture type 2 must be set to a diminishingly small value (see Fig. 4.48).

Figure 4.49 shows the piezometric-head distribution for the three port configurations for parameter set 2. Despite the fact that the discharge for port configuration (1-4) could not be satisfactory reproduced, the transport behavior is investigated in order to analyze the influence that parameter changes may have. Table 4.4 summarizes the parameter values used for the transport simulations. The corresponding tracer-breakthrough curves, represented in Fig. 4.50, are divided into two groups according to the values of the matrix dispersivity and the molecular diffusion. These two values affect the shape of the tracer breakthrough curve in the same way.

The first of the two groups (parameter sets 2-1 and 2-2) is characterized by a sharp concentration increase and a high peak. Additionally, the curves in this group have a tendency towards a second local maximum.

Table 4.4. Parameter sets for the transport simulation based on flow parameter set 2.

	$n_{e,M}$ (-)	$n_{e,F}$ (-)	$\alpha_{l,M} = \alpha_{t,M}$ (m)	α_F (m)	D_m ($m^2\,s^{-1}$)
Parameter set 2-1	0.20	0.30	0.005	0.0	$1.00 \cdot 10^{-9}$
Parameter set 2-2	0.25	0.30	0.005	0.0	$1.00 \cdot 10^{-9}$
Parameter set 2-3	0.25	0.30	0.001	0.0	$6.12 \cdot 10^{-5}$
Parameter set 2-4	0.30	0.30	0.001	0.0	$6.12 \cdot 10^{-5}$

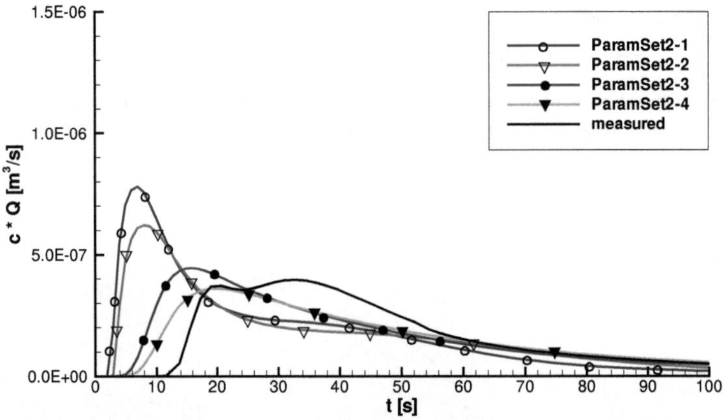

Fig. 4.50. Tracer-breakthrough curves for port configuration (1-10) corresponding to the values listed in Table 4.4. The concentration is multiplied by the corresponding discharge in order to achieve comparability.

The second group (parameter sets 2-3 and 2-4) shows a smoother concentration increase and the peak concentration is significantly lower than the one of the first group. The tailing decreases monotonically without local maxima. Compared to the parameter sets 1-2 and 1-3 (identical transport-parameter values), a later initial arrival time is achieved. However, the peak concentration is lower and the tailing more pronounced than for these sets. This is explained by the changed flow fields, allowing more tracer mass to enter the matrix.

4.3.2.2 Conclusions

The reproduction of the measured discharges proves to be straightforward. However, it must be assumed that multiple parameter sets exist, all satisfying the measured discharges. The assumed permeability distribution must therefore be validated by a transport simulation.

The fracture apertures resulting from the simulations are to be understood as effective model parameters. The fractures are considered to be one-dimensional elements, in which flow conditions according to POISEUILLE are assumed. The transformation from fracture aperture to fracture permeability is realized by applying the cubic law (Sect. 2.4). The fractures of the sample analyzed here show a width of a few millimeters including the weathered zone and, to a large extent, contain a sandy filling material. A possible way of including this in the model is to discretely model not only the fractures but also the complete weathered zone, using two-dimensional elements with a permeability which is independent of the fracture aperture.

The determination of the transport parameters is not trivial. For the chosen example, port configuration (1–10), parameter values are found that reproduce the measured tracer-breakthrough curve relatively well, i.e. in a similar order of magnitude. An essential deficit of the simulated curves is, however, the lack of the characteristic double peak. Therefore, no unique answer can be given to the question as to the reason for this behavior, either the response to two different dominant flow paths or a double-continuum behavior. Additionally, the initial arrival time is not reached. The reasons for these deficits are multifold:

a) Reasons connected with simulation technique:
   - Due to the fractured structure of the core sample, three-dimensional effects may occur, which cannot be reproduced using a two-dimensional model.
   - The description of the fractures using one-dimensional elements may not be suitable, but a realization of the complete weathered zone might be a better solution. Applying a lower permeability of the fractures would allow a better reproduction of the initial arrival time. In this case, the double peak would be the response to two different dominant flow paths.

b) Reasons connected with measurement technique:
   A renewed and detailed inspection of the experimental set-up used to conduct the measurements revealed that the tracer impulse travels through a piece of tubing before entering the core sample as well as after leaving the sample on the way to the concentration measurement. The time recording starts as the pulse is injected into the tubing. Depending on the distance from the point of injection to the inflow port, a delay of the tracer mass must be expected.

From the experience gained with this sample, it can be concluded that, even though the scale is relatively small and the amount of data is large, it is not a trivial task to set up a numerical model that reproduces the measurements correctly. Consequently, the complexity of this task increases as the problem scale is enlarged. This indicates the great challenge that has to be faced when dealing with strongly heterogeneous porous media and especially with fractured porous media.

For a comprehensive and detailed discussion of the influence on structures, such as for example fractures, on flow and transport behaviour and the interpretation of measured data from such systems, we refer to Süß (2004).

4.3.3 Determination of Equivalent Parameters

L. Neunhäuserer, M. Süß, R. Helmig

Flow and transport processes as well as heterogeneities and material properties on local scales affect the system behavior on larger scales. The realization of a large-scale area, based on detailed small-scale information, in a model is generally not possible, first because the required data density is not achievable and, second because the required computational effort is not reasonable. In order to reduce the degree of detail of the system, physically relevant parameters and processes on different scales are identified and transformed into equivalent parameters. In this way, equivalent parameters enable the consideration of small-scale processes on larger scales, achieving a simplified system description.

For the system set-up discussed in Sect. 4.3.1.1, equivalent parameters are determined, applying the discrete model, for the use in a continuum model (Sect. 4.3.4). Methods for the determination of equivalent parameters are explained in Sects. 2.5.5.1 – 2.5.5.3.

For the reasons discussed in Sect. 4.3.2, it is not possible to determine a satisfying parameter set for the cylindrical bench-scale sample under investigation using a discrete model. It is therefore decided to use a fictive parameter set based on measurements and observations given in 4.2 and to compare the results of the discrete and the continuum model results.

Due to the relatively low fracture density of the sample, the current system does not allow the identification of two different continua, representing the fractures and the matrix. Consequently, equivalent parameters for one single continuum are determined.

The parameter values used for the discrete model runs and the notation of the boundaries are presented in Fig. 4.51. For this example, the fluid chosen is water. At the in- and outflow boundaries, constant hydraulic heads of $h = 11.63$ m and $h = 10.13$ m respectively are imposed. Assuming incompressibility, the choice of fluid does not in principle affect the results. Following the conversion of the obtained hydraulic conductivity tensor to a permeability tensor according to equation (2.11), the same result is achieved for both gas and water.

4.3.3.1 Flow Parameters

The determination of the equivalent hydraulic-conductivity tensor is carried out according to the procedure described in Sect. 2.5.5.1. The result is presented in Fig. 4.52. In the graph (left), the calculated values as well as the

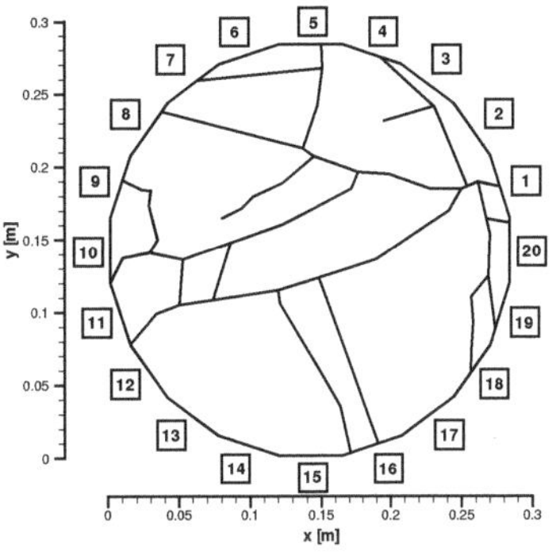

	Fracture	Matrix
Hydr. cond. K (m s^{-1})	$2.5 \cdot 10^{-2}$	$2.5 \cdot 10^{-5}$
Eff. por. n_e (-)	0.30	0.20
Disp. α (m)	0.01	0.00
Mol. diff. D_m (m^2 s^{-1})	$1.0 \cdot 10^{-9}$	$1.0 \cdot 10^{-9}$

Fig. 4.51. Boundary notation (left) and flow and transport parameters (right).

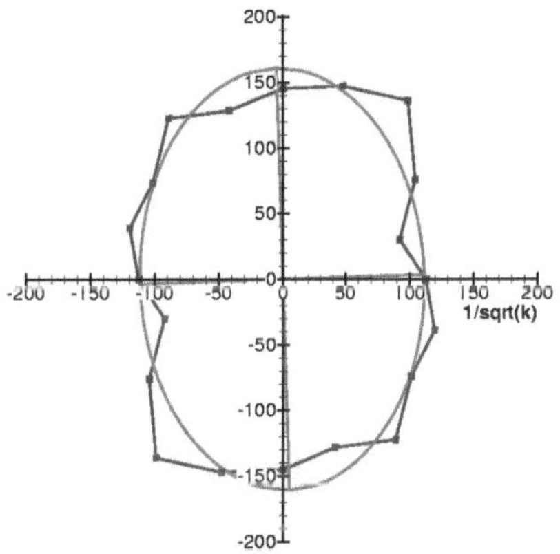

Hydr. cond. $K_{i,j}$, xy-system	
$K_{xx}[ms^{-1}]$	$8.02 \cdot 10^{-5}$
$K_{xy}[ms^{-1}]$	$0.13 \cdot 10^{-5}$
$K_{yy}[ms^{-1}]$	$3.87 \cdot 10^{-5}$
Hydr. cond. $K_{i,j}$, princ. comp. system	
$K_1[ms^{-1}]$	$8.03 \cdot 10^{-5}$
$K_2[ms^{-1}]$	$3.86 \cdot 10^{-5}$
Angle betw. xy- and princ. comp. system	
α_1	1.80382^o
α_2	88.19618^o

Fig. 4.52. Left: Hydraulic conductivity tensor with calculated values and fitted ellipse with principal components. Right: Resulting equivalent parameters.

fitted ellipse are plotted as $1/\sqrt{K}$, i.e. the largest hydraulic conductivity is found in the horizontal direction.

In order to assess the influence of the homogenization on the flow behavior and to check the plausibility of the equivalent hydraulic-conductivity tensor, it is implemented in a homogeneous system without fractures but with the same geometry and boundary conditions as the discrete model. The discharges for the ten port configurations are then determined as listed in Table 4.5 together with the corresponding values for the discrete simulations. The relationship between the discrete and the homogeneous discharge values varies slightly around 100%, except for the configurations where the ports are either well connected by fractures or not connected at all.

Table 4.5. Calculated discharges for the discrete and the equivalent homogeneous system.

	Q_{disc} ($m^3\,s^{-1}$)	Q_{hom} ($m^3\,s^{-1}$)	Q_{hom}/Q_{disc} (%)
10–20	$2.93 \cdot 10^{-6}$	$2.56 \cdot 10^{-6}$	87.5
9–19	$2.35 \cdot 10^{-6}$	$2.39 \cdot 10^{-6}$	101.8
8–18	$2.37 \cdot 10^{-6}$	$2.10 \cdot 10^{-6}$	88.6
7–17	$1.62 \cdot 10^{-6}$	$1.84 \cdot 10^{-6}$	113.1
6–16	$2.04 \cdot 10^{-6}$	$1.67 \cdot 10^{-6}$	81.7
5–15	$1.77 \cdot 10^{-6}$	$1.62 \cdot 10^{-6}$	91.6
4–14	$1.56 \cdot 10^{-6}$	$1.69 \cdot 10^{-6}$	108.8
3–13	$1.32 \cdot 10^{-6}$	$1.88 \cdot 10^{-6}$	143.0
2–12	$2.24 \cdot 10^{-6}$	$2.16 \cdot 10^{-6}$	96.5
1–11	$3.96 \cdot 10^{-6}$	$2.44 \cdot 10^{-6}$	61.7

4.3.3.2 Transport Parameters

To determinie the equivalent transport parameters, porosity and dispersion tensor (i.e. dispersivities) according to Sects. 2.5.5.2 and 2.5.5.3, a transport calculation is performed for all 10 port configurations, analogously to the flow simulation. At the inflow boundary, unit concentration is imposed until 30 ml tracer have entered the domain. The time is dependent on the discharge as shown in Table 4.6. At the opposite boundary, a free-flow boundary is applied, allowing the concentration to be observed.

Table 4.6. Injection times for the tracer.

	Q_{disc} ($m^3\,s^{-1}$)	t (s)		Q_{disc} ($m^3\,s^{-1}$)	t (s)
10-20	$2.93 \cdot 10^{-6}$	10.3	5-15	$1.77 \cdot 10^{-6}$	17.0
9-19	$2.35 \cdot 10^{-6}$	12.8	4-14	$1.56 \cdot 10^{-6}$	19.3
8-18	$2.37 \cdot 10^{-6}$	12.7	3-13	$1.32 \cdot 10^{-6}$	22.8
7-17	$1.62 \cdot 10^{-6}$	18.5	2-12	$2.24 \cdot 10^{-6}$	13.4
6-16	$2.04 \cdot 10^{-6}$	14.7	1-11	$3.96 \cdot 10^{-6}$	7.6

Figure 4.53 presents the determined dispersion tensor together with the corresponding values. The mean equivalent effective porosity of the homogeneous system is $\overline{n}_e = 0.202$, which is slightly higher than the matrix porosity of the discrete system. This confirms the conclusion drawn in Sect. 4.3.1.5 that the system has a low sensitivity to the fracture porosity due to the small ratio between fracture and matrix volume. The maximum deviation from the mean value is ± 0.02.

The deviations of the simulated dispersion values from the determined dispersion tensor show a behavior similar to the hydraulic conductivity tensor. The deviations vary positively and negatively for the same directions, only they are more pronounced for the dispersion. This indicates that the transport behavior is significantly more sensitive to the system properties than the flow behavior.

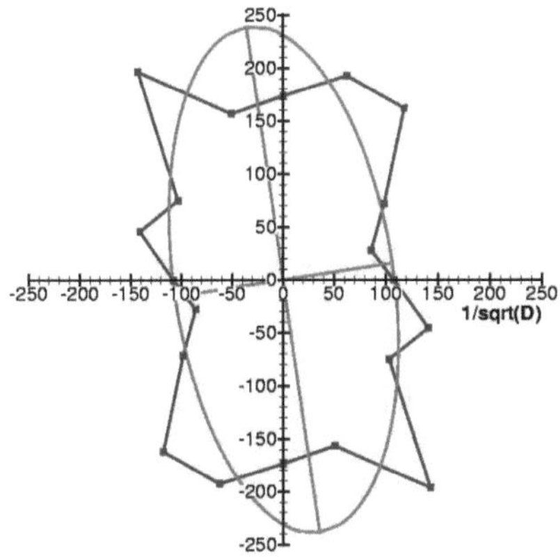

Disp. coeff., xy-system	
D_{xx} ($m^2 s^{-1}$)	$8.52 \cdot 10^{-5}$
D_{xy} ($m^2 s^{-1}$)	$1.02 \cdot 10^{-5}$
D_{yy} ($m^2 s^{-1}$)	$1.87 \cdot 10^{-5}$
Disp. coeff., princ. comp. system	
D_1 ($m^2 s^{-1}$)	$8.67 \cdot 10^{-5}$
D_2 ($m^2 s^{-1}$)	$1.72 \cdot 10^{-5}$
Angle betw. xy- and princ. comp. system	
γ_1	8.6^o
γ_2	-81.4^o
Dispersivities	
α_l (m)	0.067
α_t (m)	0.013
Mean porosity $\overline{n}_e$ (-)	0.202

Fig. 4.53. Left: Dispersion tensor with calculated values and fitted ellipse with principal components. Right: Resulting equivalent parameters.

4.3.3.3 Summary

Equivalent parameters are determined for hydraulic conductivity, dispersion and porosity, using a fictive data set based on actual data and observations of the chosen cylindrical bench-scale sample. In addition to the results presented above, equivalent parameters are determined for the configurations (5–15), (8–18) and (5–12) (notations are shown in Fig. 4.51), where the last one corresponds to configuration (1–10) in Sects. 4.3.1 and 4.3.2. The flow boundary conditions as well as the transport boundary condition at the outflow boundary remain the same whereas, at the inflow boundary, unit concentration is kept constant throughout the entire simulation. The results are discussed in Sect. 4.3.4.

In Section 4.3.4, the determined equivalent parameters are implemented in a continuum model and the suitability of the multi-continuum approach for simulations on this scale is discussed.

4.3.4 Multi-continuum Modeling: Methodology and Approach

T. Vogel, V. Lagendijk, J. Köngeter

Whereas discrete modeling is performed in Sect. 4.3.2 and the effective parameters are determined in Sect. 4.3.3, this section focuses on the development of a multi-continuum model on the cylinder scale. A selection of experimental data and their implementation in the model are presented.

4.3.4.1 Data Preparation

For the various experiments conducted on the cylinder scale (cf. Sect. 4.2), two types of measurement configurations may be distinguished: first, port-to-port measurements that allow the detection of local effects, and second, integral line-to-line measurements, providing information on the averaged characteristics of the sample.

Taking into account the interface distribution inside the specimen and the breakthrough curves resulting from the experiments with both a gas and a tracer, the configurations (port-to-port measurements) situated in the upper level of the cylinder are selected.

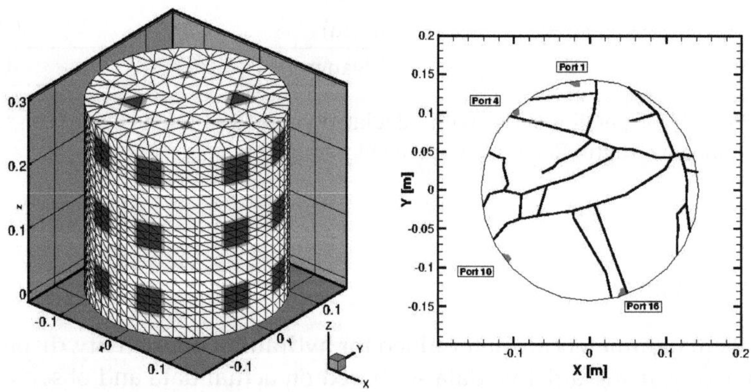

Fig. 4.54. Position of the selected ports; three-dimensional view (left) and view from the top (right).

Figure 4.54 illustrates the location of the measurement ports as well as the selected cross-section with the fractures as found on the cylinder surface. The numerical model is situated at the level of the ports ($z = 0.254$ m). The position of the fractures on the surface of the cylinder shows that, on the small scale, it is not possible to identify a representative elementary volume (REV), by means of which the equivalent parameters could be determined (cf. Sect. 2.5.5).

4.3.4.2 Simulations Concerning the Cylinder

In the following sections, the simulations conducted on the cylinder scale are described and the assumptions are explained.

For the cylinder, the three types of models examined include:

- water with tracer, two-dimensional,
- air with tracer (helium), two-dimensional,
- air with tracer (helium), three-dimensional.

The first series of experiments, *water with tracer*, is an analysis of conceptual character. In the context of this conceptual study, a first approximation of the equivalent parameters for the selected cross-section is determined (cf. Sect. 4.3.3). It has to be pointed out that the equivalent parameters for both flow and transport are identified for the fracture and matrix continua together (not separately). For the analysis of the discrete case, it is not possible to separate the continua as required, since the REV cannot be determined due to the small scale. Therefore, the simulations of the cylinder have a conceptual character. Assuming that the size of the cylinder corresponds to the size of an REV, the multi-continuum simulations are generated for the cylinder. Numerical experiments are conducted with both continuous and temporary infiltration of the tracer.

The method used to obtain the equivalent parameters is presented in Sect. 4.3.3. Figure 4.55 illustrates the discretization and the fractures that can be identified on the top side. It also shows the names of the different sides that are used to identify the different configurations: 5_15 describes a configuration, where side 5 is the input and side 15 the output port. Boundary conditions are applied to the full length of one side (4.5 cm). The cylinder is approximated by a twenty-sided polygon.

The series *Air with tracer (helium), two-dimensional* is performed on the same model as before, but the experimental results are taken into account. A classic single-continuum model, a DPSP model (double-porous, single-permeable) and a DPDP model (double-porous, double-permeable) are set up and different variants are investigated. Boundary conditions are defined according to the experimental design by ports of 35 mm by 35 mm. The location of the ports for certain configurations are presented in Fig. 4.54.

The series *Air with tracer (helium), three-dimensional* is performed by a single-continuum model. The equivalent parameters are determined directly from the experimental results and transferred to the model. The results of the previous two-dimensional studies are taken into account for the parameters of the three-dimensional model. The object of this series is to evaluate the effect of dimensionality. The port configuration is shown in Fig. 4.54 (left).

Figure 4.56 shows the structure of the study. The breakthrough curves presented are used to explain the approach; they do not represent the results. Figure 4.56 suggests that the three different types of model of the cylinder are coupled according to their adaptation. Therefore, an iterative approach is

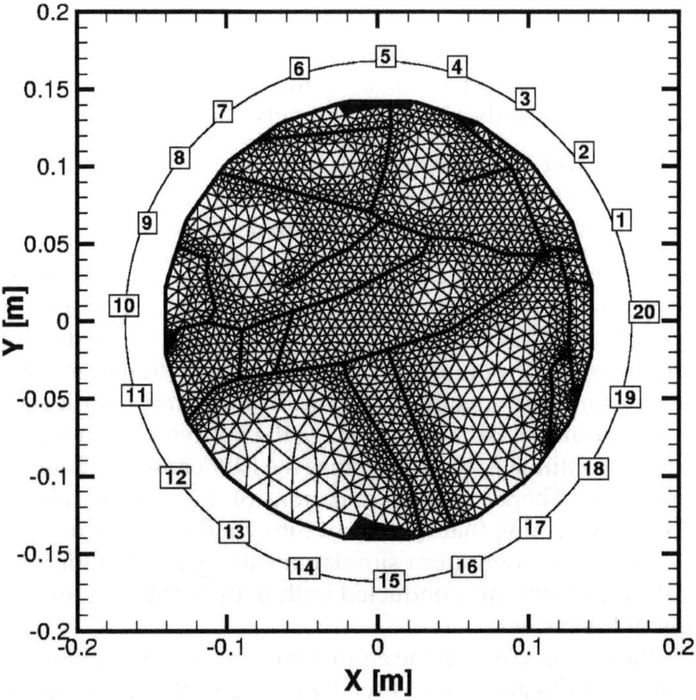

Fig. 4.55. Discretization, side configuration and fracture traces.

chosen. As to the inclusion of the measurements, only data of one configuration (ports 1 and 10) are taken into account at first. The other data are used to evaluate predictional results for the configurations 1_4 and 1_16.

The results of the numerical experiments and the interaction with respect to experimental data and discrete modeling are presented here. Table 4.7 summarizes the characteristic parameters for the double-continuum system of the cylinder. If parameters are varied, the variations are documented in the text.

4.3.4.3 Water with Tracer, Two-Dimensional

The main purpose of this conceptual study is to determine a set of equivalent parameters for multi-continuum modeling. Difficulties arise because neither experimental measurements nor discrete modeling allow the direct separation of the hydraulic components (fracture and matrix). Therefore, equivalent sets of parameters for hydraulic conductivity and dispersivity may only

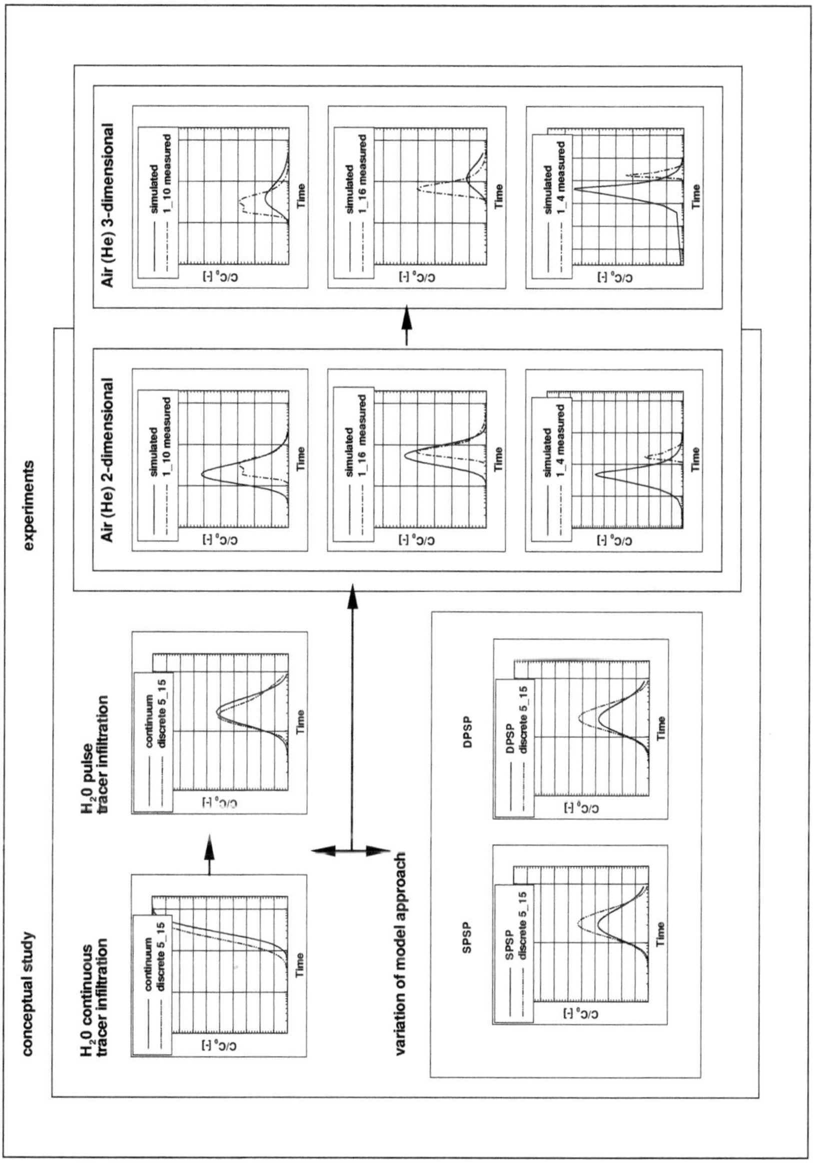

Fig. 4.56. Structure of the cylinder study.

Table 4.7. Characteristic parameters for the double-continuum system of the cylinder.

Equivalent parameters			Fracture	Matrix
Permeability	k_{xx}	(m^2)	$8.02 \cdot 10^{-12}$	$8.02 \cdot 10^{-15}$
	k_{xy}	(m^2)	$1.31 \cdot 10^{-13}$	$1.31 \cdot 10^{-16}$
	k_{yy}	(m^2)	$3.87 \cdot 10^{-12}$	$3.87 \cdot 10^{-15}$
Porosity	n_e	(-)	0.195	0.195
Thickness of the sclice	d_z	(m)	0.05	0.05
Dispersivity				
-longitudinal	α_l	(m)	$4.00 \cdot 10^{-2}$	$4.00 \cdot 10^{-2}$
-transversal horizontal	α_{th}	(m)	$4.00 \cdot 10^{-3}$	$4.00 \cdot 10^{-3}$
Molecular				
Diffusivity coefficient	D_m			
-water		(m^2s^{-1})	$4.00 \cdot 10^{-9}$	$4.00 \cdot 10^{-9}$
-gas		(m^2s^{-1})	$4.00 \cdot 10^{-6}$	$4.00 \cdot 10^{-6}$
exchange parameters				
Max. penetration depth	s_{max}	(m)	0.08	
Specific surface	Ω_0	(m^{-1})	52.47	
Specific fracture surface	$\Omega_{W,xx}$	(m^{-1})	17.58	
	$\Omega_{W,xy}$	(m^{-1})	0.59	
	$\Omega_{W,yy}$	(m^{-1})	15.82	
Shape factor	ϵ	(-)	2.00	
Exchange coefficient	α_c	(m^2/s)	$1.76 \cdot 10^{-6}$	

be determined for the complete model. As described in Sect. 4.3.3, the directional hydraulic conductivities are determined using a discrete model.

Continuous Tracer Infiltration

In order to identify the type of multi-continuum model, transport simulations with continuous tracer injection are performed. The mobility number N_M (cf. Sect. 2.5.6.1) is defined as the ratio of the equivalent flow velocities of the subordinate component, β, to the superordinate component, α. The mobility number allows for a characterization of the coupled components with respect to the mobility of the fluid in the subordinate component, β, and is therefore decisive for storage or mobility models:

$$N_M = \frac{q_\beta}{q_\alpha} . \tag{4.1}$$

To determine the tensor of equivalent hydraulic conductivities, a ratio of $q_\beta / q_\alpha = 0.001$ is assumed. The loss of identity length, L^*, in Table 4.8 represents a characteristic length scale for estimating the transport distance for which a differentiation between the two components is necessary. A detailed discussion of the characteristic values can be found in Sect. 2.5.5. The importance of the dependency of the model type and the scale is shown, for example, by Teutsch and Sauter (1992).

Table 4.8. Characteristic tracer breakthrough curves for double continuum models (Jansen, 1999).

<table>
<tr><th colspan="2">Shape of the breakthrough curve</th><th>Characteristic values</th><th>Exchange model</th></tr>
<tr><td>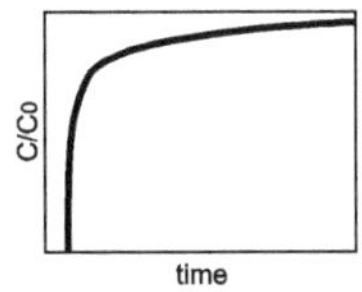
</td><td>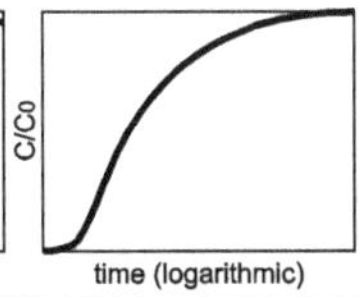
</td><td>$N_M \leq 0.001$ and $L^* > 1$</td><td>DPSP</td></tr>
<tr><td>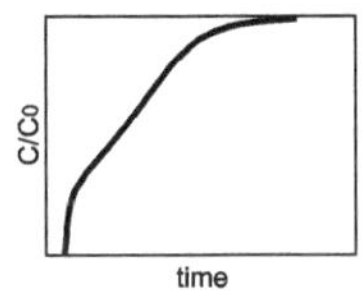
</td><td>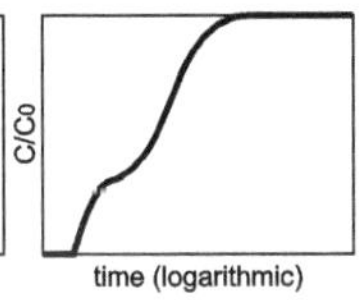
</td><td>$0.001 \leq N_M \leq 0.5$ and $L^* > 1$</td><td>DPDP</td></tr>
<tr><td>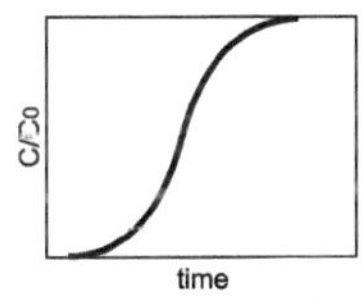
</td><td>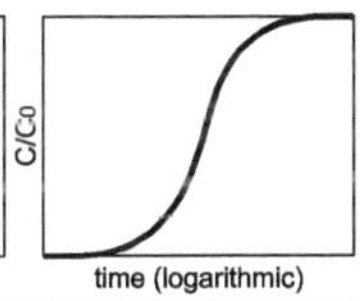
</td><td>$N_M \geq 0.5$ and $L^* \leq 1$</td><td>SPSP</td></tr>
</table>

A rough calculation of the loss of identity length shows that the transport distance in the cylinder (maximum of 0.286 m) is less than the loss of the determined identity length. Therefore, a distinction of the continua is not necessary. Despite this calculation, the identification is performed by means of describing the breakthrough curve. Allowing a characterization of the breakthrough curves of configuration 5_15 (Fig. 4.57) by means of Table 4.8, a DPSP model may be determined for the linear projection, whereas a SPSP model may be identified when the logarithmic figure is regarded. However, the identification should be independent of the choice of diagram. It is preferable to place the logarithmic scale on the time axis.

The investigations related to the SPSP model are presented, and the results of the simulations with continuous injection are shown in Figs. 4.57 and 4.58 for different porosities. The other parameters are identical to those summarized in Table 4.7. A good approximation concerning the discrete simulations is obtained for a porosity of $\bar{n}_e = 0.195$. The calibrated porosity is almost equal to the porosity obtained using the discrete model ($\bar{n}_e = 0.202$).

In the following sections, the breakthrough curves are presented on a linear scale. If, however, a logarithmic scale yields for a better visualization, it is chosen and mentioned explicitly.

Figure 4.59 shows the results for a variation of exchange parameters if a double-porous single-permeable model (DPSP) is used. A transient exchange

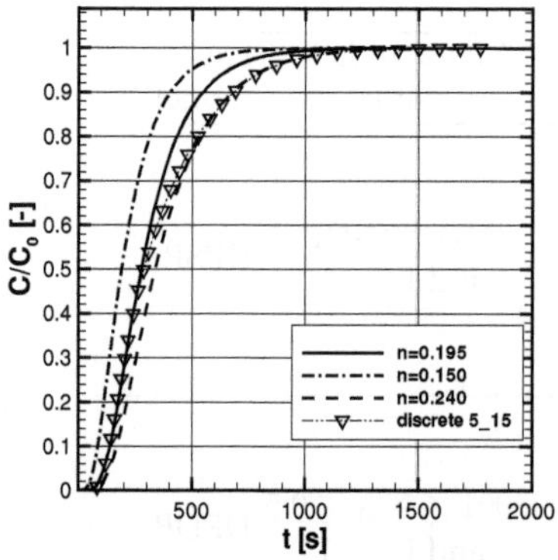

Fig. 4.57. Tracer breakthrough curves for continuous tracer injection; port-to-port connection 5_15; linear time scale (left), logarithmic time scale (right).

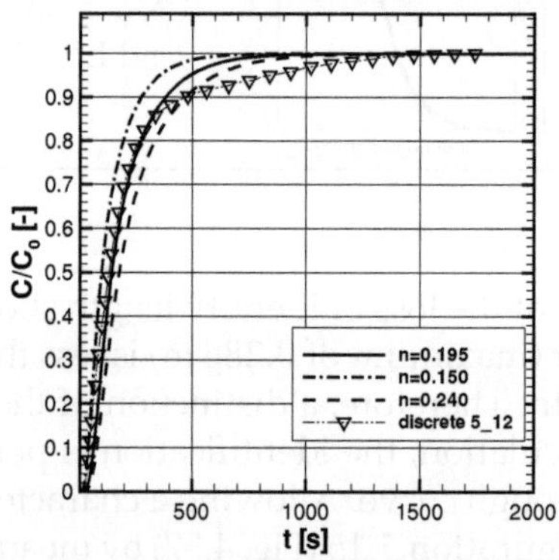

Fig. 4.58. Tracer breakthrough curves for continuous tracer injection; port-to-port connection 5

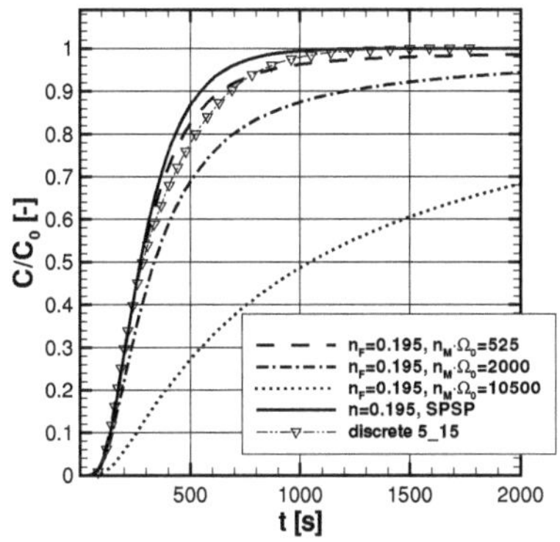

Fig. 4.59. Variation of exchange parameters for the DPSP model.

formulation is applied to the multi-continuum model (cf. Sect. 2.5.3). As a steady-state flow field is assumed for this series of experiments, there is no local fluid exchange between the components. Therefore, only diffusive exchange processes and related parameters must be considered to understand their influence on the exchange between the continua. The initial value of the specific surface Ω_0 is obtained from geometric considerations and is determined to be equal to 52.47 m^{-1}. The estimated matrix porosity is set to 20 %. The influence of the storage medium can clearly be seen in Fig. 4.59. If the specific surface or the matrix porosity is reduced, the intensity of mass transfer between the components is reduced and the results show a good approximation of the discrete simulation for a value of $n_M \cdot \Omega_0 = 525$. This would be comparable to a matrix porosity of 20 % and a specific surface of 10 m^{-1}. The comparison with the single-porous single-permeable model shows that deviations occur only after 200 s. In contrast to the SPSP solution, the DPSP approach falls short of the discrete solution for late time steps. The storage effect of the matrix component is overestimated by the DPSP model. The studies with continuous tracer injection indicate that the SPSP model is the continuum model which is more suitable for the examined cylinder.

Limited Tracer Infiltration

In analogy to the experiments conducted on the cylinder scale, transport analyses with limited tracer injection are made. For these studies, there is a pressure at the input of 116 300 Pa (Dirichlet boundary condition) and, at the outlet, the discharge is set as specified by the discrete model (Neumann boundary condition). The duration of the tracer injection is set in such a way that the total volume injected equals 30 ml.

With the results of the studies with continuous tracer injection in mind, for the following analyses with limited tracer injection the SPSP model is considered first.

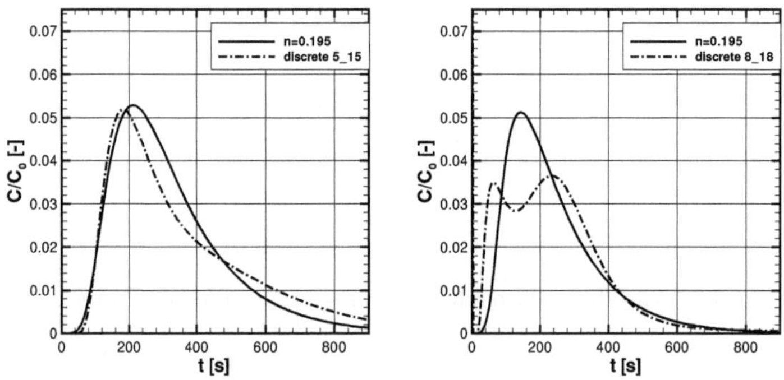

Fig. 4.60. Tracer-breakthrough curves for limited tracer injection; port-to-port connections 5_15 (left) and 8_18 (right).

In Fig. 4.60, the results of the continuum simulation for the configurations 5_15 and 8_18 are compared to the results of the discrete calculations by means of breakthrough curves. For both configurations, the time of the first increase in concentration calculated by the model mimics the one found in the discrete calculations. The maximum of configuration 5_15 nearly coincides with the one of the discrete calculations. The greatest deviations occur in the tailing. In configuration 8_18, the double peak cannot be modeled as determined in the discrete calculations. However, the average of both curves shows a good match.

Configuration 8_18 is presented here because it demonstrates the difficulties encountered in homogenizing the cylinder specimen. Figure 4.61 presents the distribution of the relative concentration calculated by means of the discrete model for three points in time. The first time, $t = 60$ s, is the time at which the first maximum occurs in the breakthrough curve (Fig. 4.60). The point in time $t = 120$ s corresponds to the local minimum, and the next time, $t = 200$ s, lies just before the second maximum of the breakthrough curve. This illustration makes it obvious that the double peak of configuration 8_18 originates from discrete local effects rather than from the interaction of two heterogenous components. Figure 4.62 illustrates the breakthrough curves of configuration 10_20, which contains a continuous fracture connecting the two ports, for both discrete and continuum modeling. As already observed for configuration 5_15, the two curves match very well. Only the time of maximum concentration appears to lag a little in the continuum simulation.

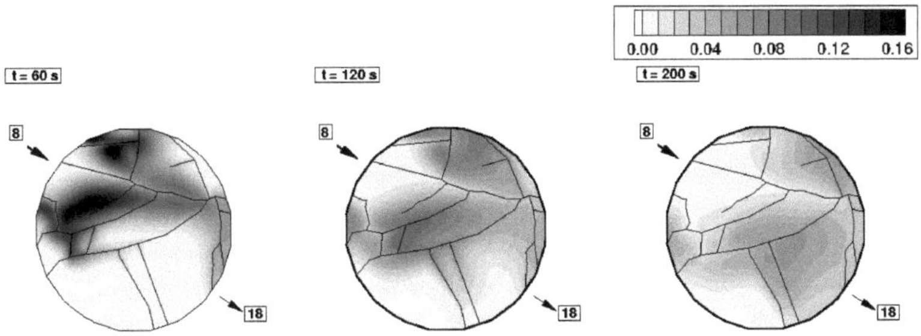

Fig. 4.61. Distribution of relative concentration for limited tracer infiltration; side 8_18.

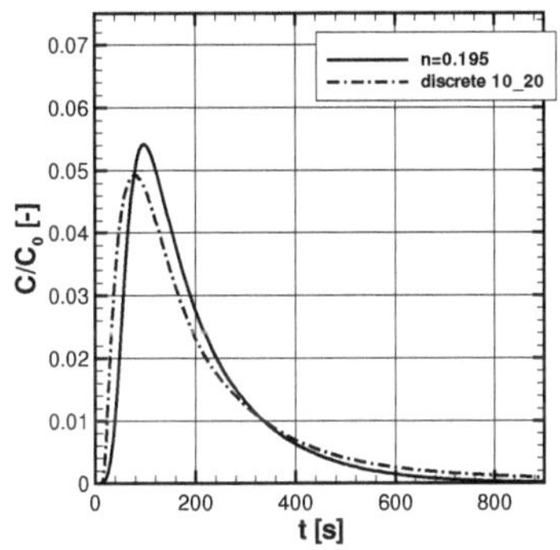

Fig. 4.62. Breakthrough curves for limited tracer infiltration; side: 10_20.

Finally, a DPSP approach with limited tracer injection is compared to the SPSP results presented. The comparison is made exemplarily for configuration 5_15, as illustrated in Fig. 4.63. The parameters required to formulate the component interaction have been chosen according to the results of the studies with continuous tracer injection, i.e. a matrix porosity of 20 % and a specific surface of $10\,m^{-1}$. The storage effect of the matrix is clearly overestimated in this approach, which results in a reduced maximum of the relative concentration. Due to the high storage effect of the rock matrix, the tailing determined by means of the DPSP approach is longer and goes beyond the range displayed in Fig. 4.63.

Remarks

The studies on *water with tracer* on the cylinder scale demonstrate that homogenization is only possible in some cases. The numerical analysis shows that the SPSP approach gives the best results for the kind of problem with a

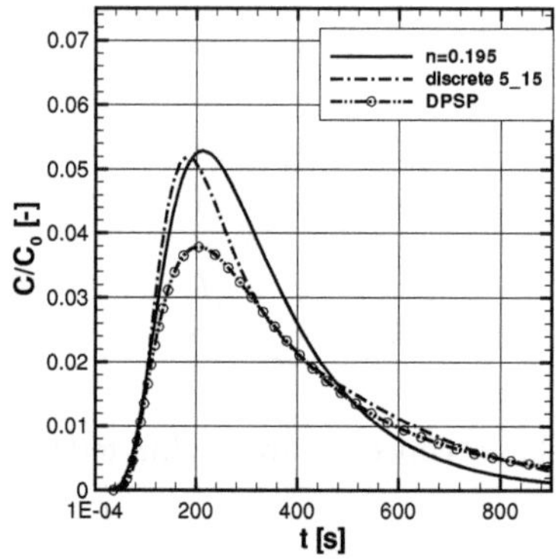

Fig. 4.63. Tracer-breakthrough curves for limited tracer injection; comparison of SPSP and DPSP model; side: 5_15.

two-dimensional model chosen here. It has to be considered, however, that for the superordinate continuum, equivalent parameters taking into account the flow and transport characteristics of the rock matrix are specified. On the present scale, it is impossible to undertake a distinct typing. Further investigations transferring the fracture patterns of the cylinder to a larger area could help determine separate equivalent parameters and thus lead to better continuum modeling.

4.3.4.4 Air with Tracer (Helium), Two-Dimensional

In the following section, the results of the experiments *air with tracer (Helium), two-dimensional* are presented. The model is developed on the basis of the results of the previous section, and the analysis is made by means of a single continuum model. The equivalent parameters are listed in Table 4.7. The molecular diffusion for air as the transport medium is orders of magnitude higher than for water (cf. Table 4.7). The density is assumed to be 1.0 kg m^{-3}, and the dynamic viscosity is $1.814 \cdot 10^{-5}\,m^2s^{-1}$.

The results are analyzed by means of breakthrough curves. The tracer breakthrough curves determined experimentally all show a recovery rate of less than 90 % after 800 s of measurement (cf. Table 4.9). Therefore, the comparison of the numerical analysis and the experimental results is made by means of an unmodified breakthrough curve and one scaled to 100 % respectively. The amount of tracer that has passed the outlet at $t = 800$ s in the numerical simulations can be retrieved from Table 4.9. For sufficiently long observation periods ($t > 2\,000$ s), the numerical model gives recovery rates of 100 %.

Table 4.9. Recovery rates for the selected configurations after 800 s.

Configuration	Experiment (%)	Model (%)
1_10	73.60	90.99
1_16	80.70	93.44
1_4	66.20	89.60

The selected configurations (cf. Fig. 4.54) are all located on the same level. In analogy to the studies of *water with tracer*, the two-dimensional model is taken as a slice with a thickness of 5 cm situated on the level of the port centers.

The results of the two-dimensional modeling are presented in Figs. 4.64 and 4.65. The single-continuum simulations are carried out with a porosity of 19.5 % and 24.0 %. Qualitatively, the breakthrough curves for the port connections 1_10 and 1_16 correspond well with the curves determined experimentally. As previously observed for configuration 8_18 (cf. Sect. 4.3.4.3), the double peak determined in the experiments for configuration 1_10 cannot be modeled with a single-continuum model. Here as well, the double peak results from local effects (splitting of the transport paths along the fractures) rather than from the interaction of the sub- and superordinate component. For all configurations, the transport signal is detected earlier in the numerical analyses than in the experiments. The largest deviations occur for configuration 1_4. However, it seems to be unrealistic for the distance between ports 1 and 4 that the signal occurs after only 11.5 s. If the curve is displaced, a good qualitative correspondence is reached for this configuration as well.

Remarks

The comparison of the results of the single continuum model for *air with tracer (helium)* with the experiments shows that the transport behavior is qualitatively well reproduced. A comparison of the two-dimensional model with the three-dimensional continuum model (cf. Sect. 4.3.4.5) indicates that the dimensionality has a considerable impact on the results for the geometry studied here. The assumption of the thickness available for the two-dimensional model has a crucial influence on the behavior of the flow and transport of the modeled system.

4.3.4.5 Air with Tracer (Helium), Three-Dimensional

To evaluate the influence of the dimensionality, a three-dimensional model is developed for the studies on *air with tracer (helium)*. The characteristic parameters are listed in Table 4.7. The permeability in the vertical direction k_{zz}

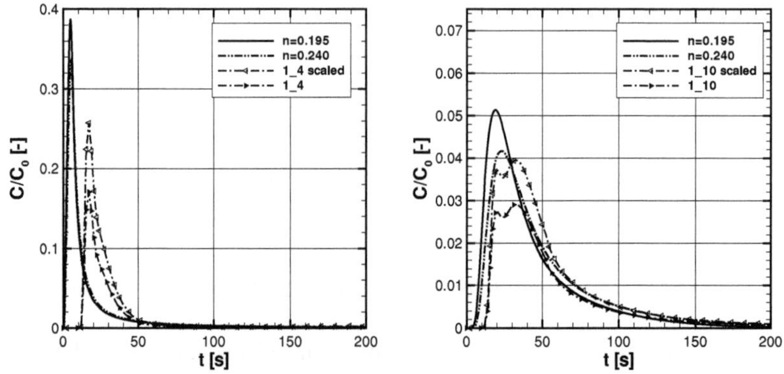

Fig. 4.64. Tracer-breakthrough curves for port-to-port connections 1_4 (left) and 1_10 (right), measured curves scaled and unchanged.

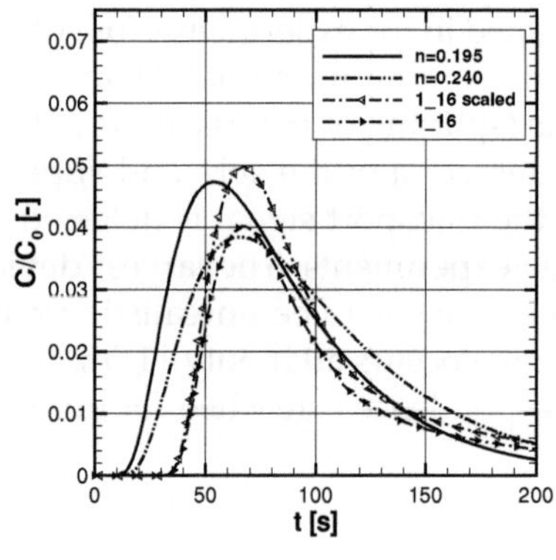

Fig. 4.65. Tracer-breakthrough curves for port-to-port connection 1_16, measured curves scaled and unchanged.

is assumed to be equal to k_{yy}. Furthermore, it is presumed that the z-axis corresponds to a principal axis of the permeability tensor, so that the additional entries on the secondary diagonals are equal to zero.

For the simulation of the flow, Fig. 4.66 contrasts the pressure distribution of the three-dimensional model with the results for two configurations of the two-dimensional model. For the given discharge and inlet pressure, a minimal pressure of 113 086 Pa is reached at the outlet of the two-dimensional model, for which a thickness of $dz = 5$ cm has been assumed. For the three-dimensional model, a pressure of 115 177 Pa is determined.

Thus, in the three-dimensional model the thickness of the region used by the flow is bigger than assumed for the two-dimensional model. To verify this, a second two-dimensional calculation is conducted, and the thickness is assumed to be $dz = 15\,\text{cm}$. For this second configuration, a minimum pressure of 115229 Pa is determined for the outlet of the two-dimensional model (cf. Figs. 4.66 b and d). In addition to the pressure distributions, Figs. 4.66 a-d present streamlines to illustrate the flow paths. The differences between the three-dimensional model and the two-dimensional model are due to the illustration, as Fig. 4.66 b is a top view of the three-dimensional diagram.

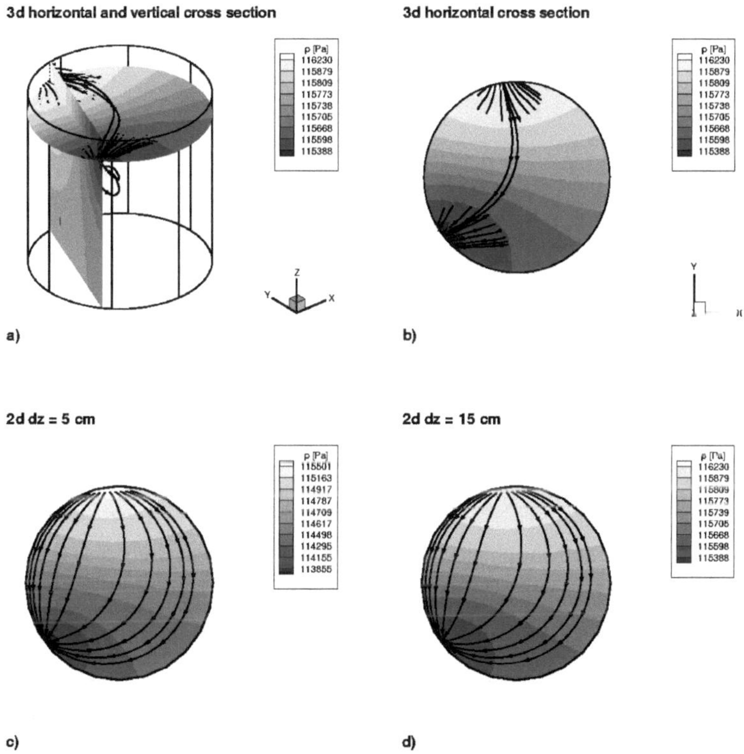

Fig. 4.66. Pressure distributions and stream lines for the 3d-model, compared to the 2d-model.

The results of the three-dimensional transport simulations for the cylinder are compared to those of the two-dimensional model with a thickness of $dz = 5\,\text{cm}$. Increasing the thickness for the two-dimensional model would, in conjunction with a present amount of injected tracer, imply an injection distributed over the height, which would lead to an unrealistic test arrangement. This model set-up would correspond to the line-to-line measurements (cf. Sect. 4.2).

The results of the transport calculations for port connections 1_4 and 1_10 conducted with the three-dimensional model are illustrated in Fig. 4.67. In contrast to the previous breakthrough curves, a logarithmic scale is chosen for the time axis of these figures in order to enable a definite differentiation. The influence of dimensionality is less evident for the short distance between ports 1 and 4 than for the longer connection 1_10. For configuration 1_10, the point in time in the experiments at which the maximum occurs is reproduced better by the three-dimensional model. However, the intermediate storage of tracer is overestimated, so that the maximum is too small and the following tailing is longer than observed in the experiment.

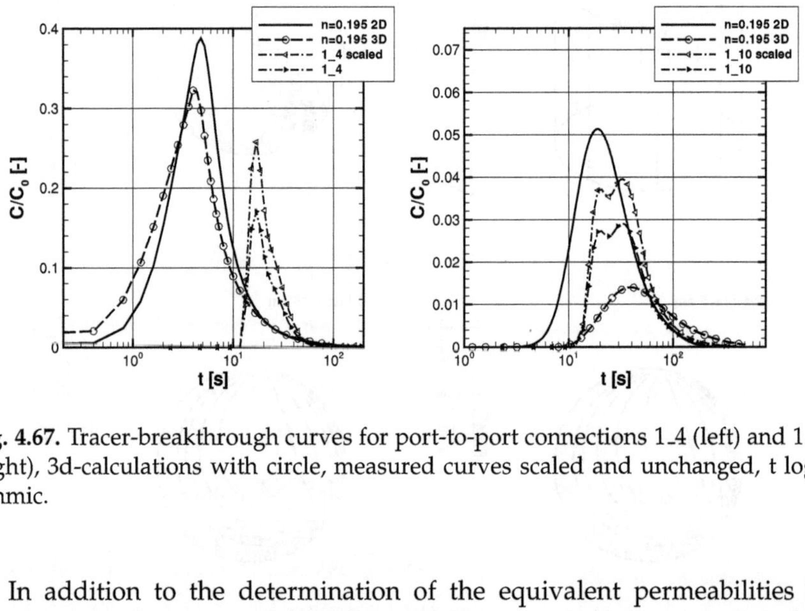

Fig. 4.67. Tracer-breakthrough curves for port-to-port connections 1_4 (left) and 1_10 (right), 3d-calculations with circle, measured curves scaled and unchanged, t logarithmic.

In addition to the determination of the equivalent permeabilities by means of the discrete model (cf. Sect. 4.3.3), one set of parameters is calculated directly from the series of permeability measurements in McDermott (1999) and is referred to as He2MS44. The values of the tensors employed are presented in Table 4.10, and the resulting ellipsoids are illustrated in Fig. 4.68.

The order of magnitude of the permeabilities determined using the two-dimensional discrete model corresponds well with the directional permeabilities obtained from the experiments. The assumption that the vertical axis is a principal axis of the permeability tensor is not confirmed by the experiments. However, a comparison of the results of the transport calculations (cf. Fig. 4.69) shows that the "incorrect" assumption has only a minor influence on the experimental set-up orientated horizontally.

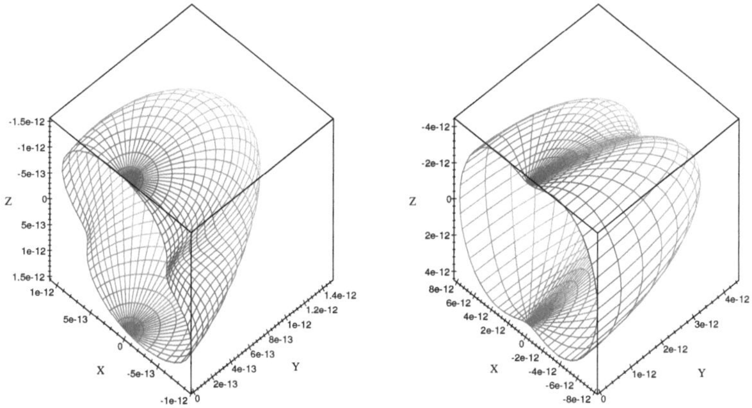

Fig. 4.68. Optimized K-tensors from the experiments (left) and from the discrete modeling with assumptions for the vertical direction (right).

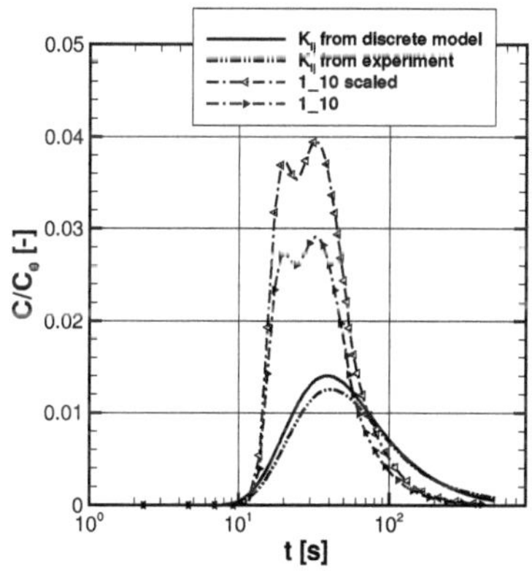

Fig. 4.69. Breakthrough curves for port-to-port connection 1_10; measured curves scaled and unchanged.

Table 4.10. Comparison of the equivalent permeabilities.

	Series He2MS44	Discrete model
k_{xx} (m^2)	$6.98 \cdot 10^{-13}$	$8.02 \cdot 10^{-12}$
k_{xy} (m^2)	$-2.95 \cdot 10^{-13}$	$1.31 \cdot 10^{-13}$
k_{xz} (m^2)	$-2.41 \cdot 10^{-13}$	0.00
k_{yy} (m^2)	$1.42 \cdot 10^{-12}$	$3.87 \cdot 10^{-12}$
k_{yz} (m^2)	$1.50 \cdot 10^{-14}$	0.00
k_{zz} (m^2)	$1.46 \cdot 10^{-12}$	$3.87 \cdot 10^{-12}$

4.3.4.6 Summary

The selection of an appropriate multi-continuum model is complicated by the fact that the experimental data does not allow for a separation of possible continua (e.g. macro-fractures, micro-fractures and host matrix). However, in conjunction with discrete modeling, parameter sets may be determined that allow the development of a multi-continuum model. This model is based on experimental data, discrete modeling and conceptual studies. Thus, a qualitatively good model of the cylindrical core is obtained. The investigations show that local effects may not be reproduced, as the scale of investigation is too small for a multi-continuum model.

Effects of dimensionality are discussed, emphasizing the importance of determining an adequate model thickness when using only a two-dimensional approach.

4.4 Flow and Transport Experiments Conducted on Laboratory Blocks

C. Leven, R. Brauchler, M. Sauter, G. Teutsch, P. Dietrich

4.4.1 Integral Measuring Configuration

The *integral* measuring configuration is set up to obtain information on the integral system response, i.e., flow and transport parameters are averaged over the matrix and the fracture network. Figure 4.70 shows the measuring set-up used in the laboratory to conduct flow and transport experiments in a one-dimensional flow field. For this set-up all ports on one side are used for injection and all ports on the opposite side for extraction.

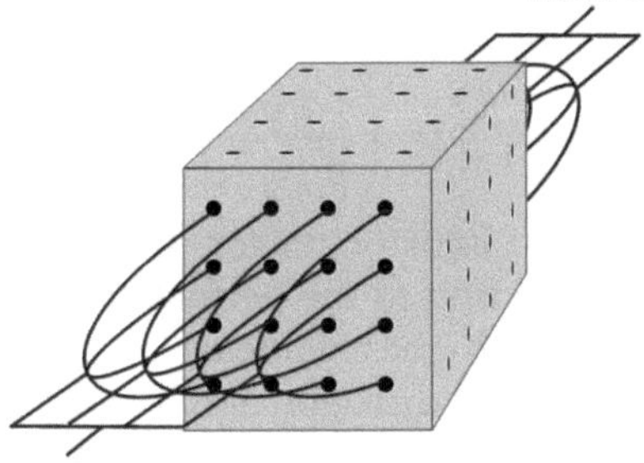

Fig. 4.70. Measuring set-up for the integral flow and transport measurements.

4.4.1.1 The 90 × 90 × 80 cm^3-Block

Flow Experiments

Within the laboratory block a steady state flow field is established by applying a constant injection (p_{in}) and extraction pressure (p_{out} = atmospheric pressure) and using the experimental setup shown in Fig. 4.25. The pressure difference between the injection and the extraction side and the flow rate at the extraction side are measured for each experiment. The data recorded for this experimental series can be seen in Table 4.11.

On the basis of DARCY's law, a ratio of flow rate Q to pressure difference Δp scaled by the measuring distance r is used to define a measure for the conductivity of the block with respect to gas flow. With this ratio, low values of $Qr/\Delta p$ will indicate low permeabilities since the potential gradient will be high and the resulting flow rate low for low permeabilities. In turn, for high permeabilities, like for a single fracture, low pressure gradients and high flow rates can be expected. In Table 4.11, the ratios are listed for the three *integral* flow experiments. It is evident that the connection from *bottom* to *top* has the highest pressure difference and the lowest $Qr/\Delta p$ ratio. In turn, the lowest pressure difference and highest ratio is found for the connection from side *III* to *I*.

On the assumption of compressible flow of an ideal gas and the cross-sectional area of the block as an effective flow-through area A, the perme-

Table 4.11. 90 × 90 × 80 cm^3-block: Data from the *integral* pneumatic flow experiments (r - distance, Δp - pressure difference, Q - flow rate).

from	to	r (m)	Δp (mbar)	Q (m^3/s)	$Qr/\Delta p$
III	*I*	0.8	63.8	$8.22 \cdot 10^{-5}$	$1.03 \cdot 10^{-6}$
II	*IV*	0.9	105.9	$6.74 \cdot 10^{-5}$	$5.73 \cdot 10^{-7}$
bottom	*top*	0.9	225.1	$8.57 \cdot 10^{-5}$	$3.43 \cdot 10^{-7}$

ability k for the *integral* measurements can be calculated following Carman (1956):

$$k = \frac{Q}{A}\,\mu_g\left(\frac{2\,p_{out}\,r}{p_{in}^2 - p_{out}^2}\right) \tag{4.2}$$

with μ_g for the dynamic viscosity, p_{in} for the pressure at inlet, and p_{out} for the pressure at outlet, leading to the permeability values listed in Table 4.12. Due to cross bedding, the highest values of permeability and conductivity can be expected in horizontal direction and the highest resistivity to flow in vertical direction. This is indicated by the highest values of permeability for the connections in the horizontal direction (*III-I* and *II-IV*) and the lowest values in the vertical direction (*bottom-top*). Because of the orientation of the main fractures (Sect. 4.1.2.3) the connection *III-I* indicates the best conductance.

Table 4.12. $90 \times 90 \times 80\,cm^3$-block: Permeability and conductivity values derived from the one-dimensional flow experiments (K_g - pneumatic conductivity, K_f - hydraulic conductivity).

from	to	k (m^2)	K_g (m/s)	K_f (m/s)
III	*I*	$2.25 \cdot 10^{-13}$	$1.57 \cdot 10^{-7}$	$2.21 \cdot 10^{-6}$
II	*IV*	$1.38 \cdot 10^{-13}$	$9.67 \cdot 10^{-8}$	$1.36 \cdot 10^{-6}$
bottom	*top*	$7.92 \cdot 10^{-14}$	$5.54 \cdot 10^{-8}$	$7.78 \cdot 10^{-7}$

Transport Experiments

After the flow field reaches steady state, a volume of 105 ml helium is injected into the system at the injection side via a bypass loop and the tracer breakthrough is recorded at the extraction side as described in Sect. (4.2.1). In Fig. 4.71 the recorded breakthrough curves are shown. The first initial breakthrough is detected for connections *III-I* and *II-IV* after the same time with a slightly stronger concentration increase for connection *II-IV*. The initial breakthrough for the *bottom-top* connection occurs significantly later. Although the highest hydraulic conductivity can be observed for connection *III-I*, the dominant breakthrough, i.e. the time t_{peak} when the maximum concentration is reached, occurs first for connection *II-IV* even though the transport distance is longer. The lowest breakthrough concentrations and velocities are observed for the connection *bottom-top*. The seepage velocities (Fig. 4.71) derived from the dominant breakthrough (t_{peak}) times show corresponding estimates.

Discussion

The *integral* flow and transport investigations described indicate that, for integral measurements, the contribution of the matrix tends to become negligi-

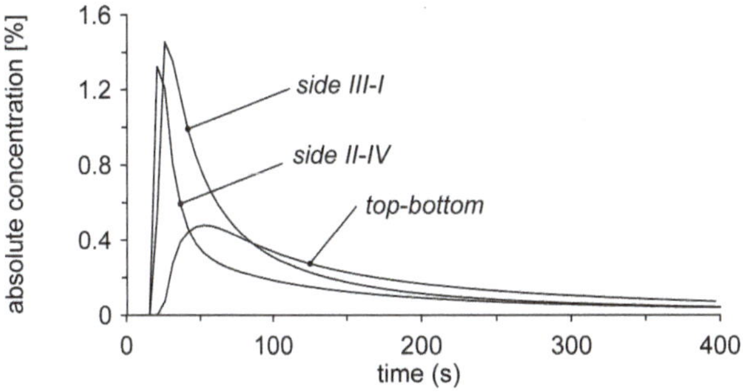

Seepage velocities		
from	to	v [m/s]
III	*I*	0.031
II	*IV*	0.043
bottom	*top*	0.017

Fig. 4.71. $90 \times 90 \times 80\,\text{cm}^3$-block: Breakthrough curves recorded for the *integral* gas tracer experiments (left). Seepage velocities v derived from the *integral* transport experiments (right).

ble if the integral connections include fractures with high apertures (connections *III-I* and *II-IV*). As described in Sect. 3 (Table 3.8) similar results were found for integral flow experiments with fractured sandstone cores where the contribution of the matrix to the overall conductivity is less than 20% for fracture apertures of less than 40 μm.

The results of the *integral* transport experiments indicate fracture-dominated transport for the connections that include fractures with high apertures. However, a significant influence of the matrix can be observed as indicated by the tailing of the breakthrough curves. The direct injection of tracer into the matrix as well as matrix diffusion processes within the fracture network may be the cause for these observations with integral measuring configurations.

4.4.1.2 The $60 \times 60 \times 60\,\text{cm}^3$-Block

Flow Experiments

As in the case of the experiments described in the previous section (4.4.1.1: 90 x 90 x 80 cm^3-block), the *integral* measuring configuration is applied to characterize the $60 \times 60 \times 60\,\text{cm}^3$-block in terms of an integral system response. In Table 4.13, the data recorded for this experimental series are given. The ratio $Qr/\Delta p$ listed in Table 4.13 for all three connections reveals the least resistance to gas flow in the direction *I-III* and the highest resistance in the vertical direction *bottom-top* which is due to the sedimentological cross bedding (cf. Sect. 4.1). Following eq. (4.2), permeability and conductivity values are given for the *integral* measuring configuration in Table 4.14.

Transport Experiments

For this experimental series of transport experiments, the same set-up is used as described above: 105 ml helium is injected via a by-pass loop into the system and the helium breakthrough is recorded at the outlet. The resulting

Table 4.13. 60 × 60 × 60 cm^3-block: Data from the *integral* pneumatic flow experiments (r - distance, $\Delta\ p$ - pressure difference, Q - flow rate).

from	to	r (m)	$\Delta\ p$ (mbar)	Q (m^3/s)	$Qr/\ \Delta p$
I	*III*	0.6	490.0	$4.00 \cdot 10^{-5}$	$4.90 \cdot 10^{-8}$
II	*IV*	0.6	490.5	$3.23 \cdot 10^{-5}$	$3.94 \cdot 10^{-8}$
bottom	*top*	0.6	489.5	$1.05 \cdot 10^{-5}$	$1.29 \cdot 10^{-8}$

Table 4.14. 60 × 60 × 60 cm^3-block: Permeability and hydraulic conductivity values derived from the one-dimensional flow experiments.

from side	to side	intrinsic permeability k (m^2)	pneumatic conductivity K_g (m/s)	hydraulic conductivity K_f (m/s)
III	*I*	$1.99 \cdot 10^{-14}$	$1.10 \cdot 10^{-9}$	$1.95 \cdot 10^{-7}$
II	*IV*	$1.60 \cdot 10^{-14}$	$8.82 \cdot 10^{-10}$	$1.57 \cdot 10^{-7}$
bottom	*top*	$5.21 \cdot 10^{-15}$	$2.87 \cdot 10^{-10}$	$5.12 \cdot 10^{-8}$

breakthrough curves are shown in Fig. 4.72. Except for the breakthrough curve of connection *bottom-top*, the breakthrough curves show double peaks and strong tailing. This behavior is assumed to reflect the response of particular parts of the system (e.g. coarse-grain matrix vs. fine-grain matrix, cf. Fig. 4.13). In terms of seepage velocities (Fig. 4.72), the transport behavior is consistent with the derived flow parameters (Table 4.13 and 4.14) with the highest values in horizontal directions (connections *I-III* and *II-IV*).

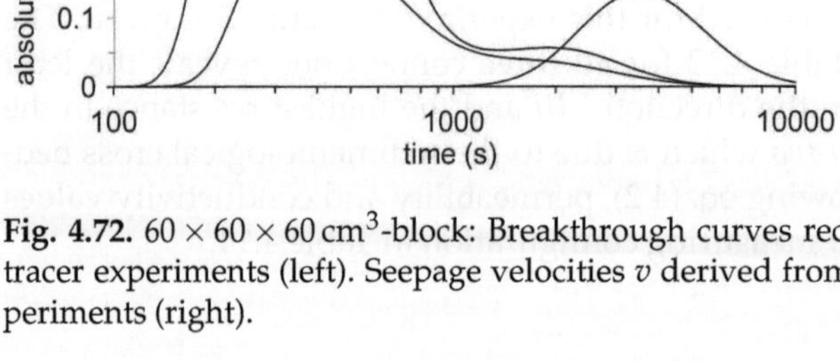

Seepage velocities

from	to	v_i (m/s)
I	*III*	$2.45 \cdot 10^{-3}$
II	*IV*	$1.15 \cdot 10^{-3}$
top	*bottom*	$1.71 \cdot 10^{-4}$

Fig. 4.72. 60 × 60 × 60 cm^3-block: Breakthrough curves recorded for the *integral* gas tracer experiments (left). Seepage velocities v derived from the *integral* transport experiments (right).

Discussion

The *integral* flow and transport investigations described clearly show the anisotropic flow and transport behavior within the $60 \times 60 \times 60\,\text{cm}^3$-block. The *integral* flow experiments clearly indicate the relation between the direction of the experiment with respect to the main axis of the permeability tensor (Fig. 4.73). For the $60 \times 60 \times 60\,\text{cm}^3$-block, the permeability vector K_B from Fig. 4.73 corresponds to the direction of connection *I-III* which results in the highest permeability (cf. Table 4.13). In contrast, the lowest permeability is detected by the flow experiments for connection *top-bottom*. This connection would correspond to a direction with an inclination to the permeability vector K_C in Fig. 4.73.

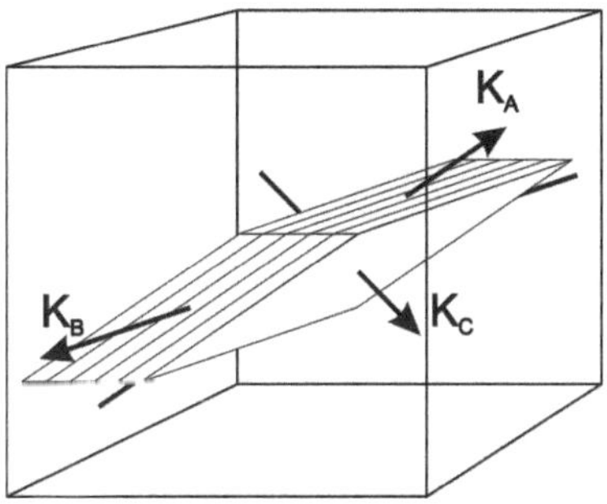

Fig. 4.73. Principle sketch of the $60 \times 60 \times 60\,\text{cm}^3$-block oriented with respect to the cross bedding foresets and approximate definition of the three mutually perpendicular permeability components K_A, K_B, and K_C (after Fondeur (1964)).

The results from the *integral* transport experiments yield a more detailed picture of the hydraulic properties. As indicated by the seepage velocities, the same situation as that described for the flow experiments can be concluded if it is assumed that the three determined seepage velocities v are parallel to the principal permeability components. Furthermore, the breakthrough curves with the observed double peaks represent the response of different parts of the systems. A fast response can be interpreted as resulting from flow and transport parallel to the inclination of the cross bedding foresets (connection *I-III* corresponding to the direction of K_B in Fig. 4.73) and the coarse grained matrix located at the top of the $60 \times 60 \times 60\,\text{cm}^3$-block (connection *II-IV*). Secondary peaks in the breakthrough curves are interpreted as due to the finer matrix components within the block.

4.4.2 Port-Port Measuring Configuration

For the measuring configuration *port-port*, gas and tracer is injected and extracted at single ports (cf. Sect. 4.1) positioned on opposite block faces. This configuration allows for a permutation of different port connections and a change in flow-field orientation, thus enabling a tomographical hydraulic investigation of the fractured sandstone block with a very high resolution. By these means, the anisotropy and heterogeneity of parameters relevant to flow and transport relevant parameters can be estimated. The *port-port* configuration is employed to generate three-dimensional flow fields. Figure 4.74 shows an example of one possible port-to-port connection.

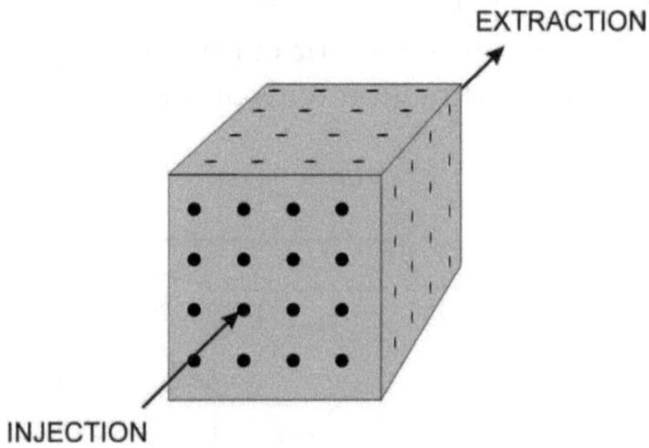

Fig. 4.74. Measuring set-up for the *port-port* measurements.

4.4.2.1 The 90 × 90 × 80 cm^3-Block

Flow Experiments

For this type of experiment, the same set-up is used as that shown in Fig. 4.25. From the 47 conducted flow experiments, 26 measurements are connected to matrix-dominated ports while the injection and extraction ports are connected to the matrix on one side and directly to a fracture on the opposite side during a further 8 measurements (Table 4.15). For the remaining measurements, at least one of the ports is connected to a matrix port in the direct vicinity of a fracture (at a distance of less than 20 mm).

To establish a steady state flow field, a constant injection (p_{in}) and extraction pressure (p_{out} = atmospheric pressure) are applied and resulting the flow rate and the pressure difference between injection and extraction port are recorded. The recorded data for these experimental series can be gathered from Table 4.15. According to Sect. 4.4.1.1, a ratio of $Q/\Delta p$ is calculated, providing a measure of the conductivity of the system with respect to gas flow. Low values of $Q/\Delta p$ will indicate low conductivities and vice versa. In Table 4.15, the ratios are given corrected for the measuring distance.

A comparison of the ratios to the type of connection (matrix (*m*), fracture (*f*), or matrix connection in the direct vicinity to a fracture (*fm*)) shows that the lowest ratios arise from *m* connections and highest ratios from *f* or *fm* connections. However, in some cases, unexpected observations are made:

connections that include a direct fracture connection or a connection in the direct vicinity of a fracture reveal small ratios of $Q/\Delta p$ (e.g. connections *316-113*, *t16-b16*, *t13-b13*, *t11-b11*, *210-411*, and *211-410*; cf. Fig. 4.18). This reduction in $Q/\Delta p$ is mainly due to limited fracture lengths. Since the probability of fracture intersections decreases with decreasing fracture length (Long and Witherspoon, 1985), the influence of the fracture system on the resulting flow field will also decrease. In particular, for connections from ports on the *top* to ports on the *bottom* side, the limited length in a vertical direction of the main fracture (Sect. 4.1.2.3) and the restraining function of the horizontal bedding planes come into play.

Generally, as stated by Streltsova (1988), "it is fracture continuity and inter-connectivity, however, not the characteristics of individual fractures that are responsible for the particulars of a fractured reservoir". In Figs. 4.75 and 4.76 the spatial distribution of the ratio $Q/\Delta p$ corrected for the measuring distance is given for sides *I* and *II*. It is evident that the conductivity with respect to gas flow increases in regions in the vicinity of the main fractures (e.g. ports on the far right of side *I* in Fig. 4.75). On the other hand, matrix-dominated regions within the block are indicated by lower values of the ratio $Qr/\Delta p$ (e.g. central region on side *I*).

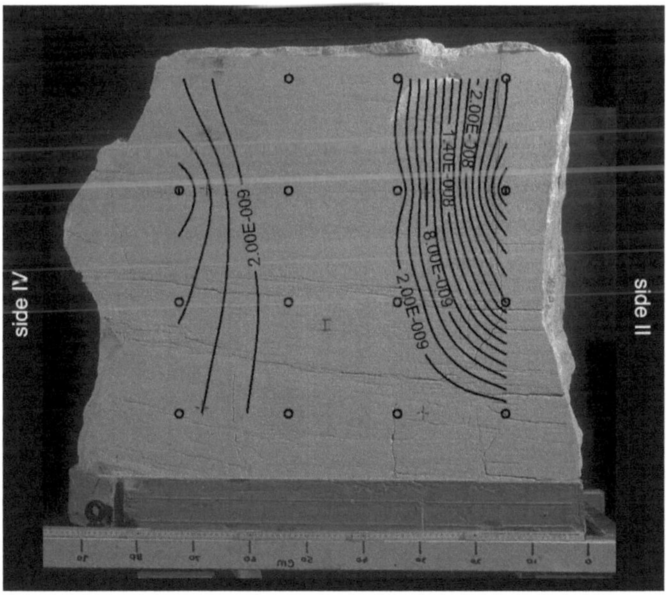

Fig. 4.75. $90 \times 90 \times 80\,\text{cm}^3$-block: Example of spatial distribution of ratio $Qr/\Delta p$ for side *I*.

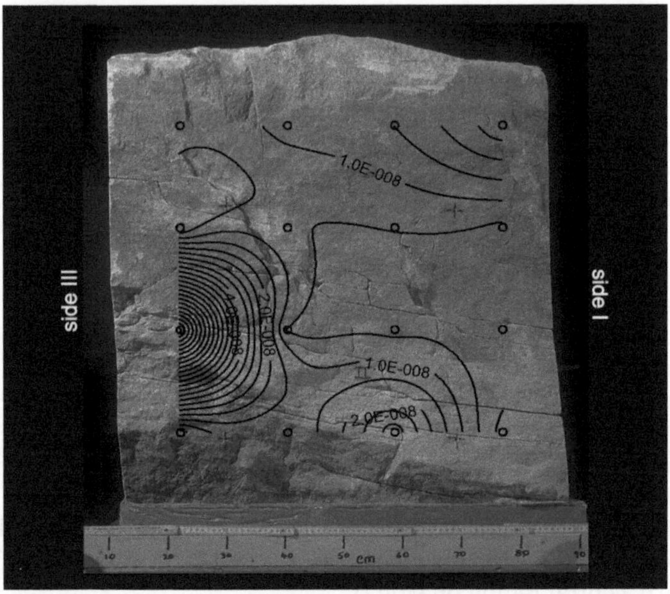

Fig. 4.76. $90 \times 90 \times 80\,\text{cm}^3$-block: Example of spatial distribution of ratio $Qr/\Delta p$ for side *II*.

Table 4.15. $90 \times 90 \times 80\,\text{cm}^3$-block: Data from *port-port* flow experiments (m: matrix, f: fracture, fm: fracture in direct vicinity to matrix port). The exact position of the ports can be gathered from Fig. 4.18.

injection - extraction port number	port type (f, m, fm)	pressure difference Δp (mbar)	flow rate Q (m^3/s)	distance r (m)	ratio $Qr/\Delta p$
201-404	m-m	302	$3.3 \cdot 10^{-6}$	0.9	$9.8 \cdot 10^{-9}$
202-403	m-m	302	$3.4 \cdot 10^{-6}$	0.9	$1.0 \cdot 10^{-8}$
203-402	m-fm	299	$4.0 \cdot 10^{-6}$	0.9	$1.2 \cdot 10^{-8}$
204-401	f-fm	295	$9.7 \cdot 10^{-6}$	0.9	$3.0 \cdot 10^{-8}$
205-408	m-m	301	$7.1 \cdot 10^{-7}$	0.9	$2.1 \cdot 10^{-9}$
206-407	f-fm	313	$2.3 \cdot 10^{-6}$	0.9	$6.6 \cdot 10^{-9}$
207-406	f-m	303	$1.2 \cdot 10^{-6}$	0.9	$3.6 \cdot 10^{-9}$
208-405	m-m	306	$1.7 \cdot 10^{-6}$	0.9	$4.9 \cdot 10^{-9}$
209-412	fm-fm	300	$3.8 \cdot 10^{-5}$	0.9	$1.1 \cdot 10^{-7}$
210-411	fm-m	300	$6.0 \cdot 10^{-7}$	0.9	$1.8 \cdot 10^{-9}$
211-410	f-m	516	$1.1 \cdot 10^{-6}$	0.9	$2.0 \cdot 10^{-9}$
212-409	m-fm	301	$9.9 \cdot 10^{-7}$	0.9	$3.0 \cdot 10^{-9}$
213-416	m-m	302	$8.1 \cdot 10^{-7}$	0.9	$2.4 \cdot 10^{-9}$
214-415	fm-m	298	$3.2 \cdot 10^{-6}$	0.9	$9.6 \cdot 10^{-9}$
215-414	f-m	303	$1.1 \cdot 10^{-5}$	0.9	$3.1 \cdot 10^{-8}$
216-413	m-m	301	$3.0 \cdot 10^{-7}$	0.9	$9.0 \cdot 10^{-10}$
301-104	m-m	304	$1.4 \cdot 10^{-6}$	0.8	$3.7 \cdot 10^{-9}$
302-103	m-m	300	$5.0 \cdot 10^{-7}$	0.8	$1.3 \cdot 10^{-9}$
303-102	m-m	300	$1.1 \cdot 10^{-6}$	0.8	$3.0 \cdot 10^{-9}$
304-101	f-fm	300	$7.8 \cdot 10^{-6}$	0.8	$2.1 \cdot 10^{-8}$
305-108	m-m	297	$3.4 \cdot 10^{-6}$	0.8	$9.1 \cdot 10^{-9}$
306-107	m-m	306	$4.5 \cdot 10^{-7}$	0.8	$1.2 \cdot 10^{-9}$
307-106	m-m	300	$3.3 \cdot 10^{-8}$	0.8	$8.9 \cdot 10^{-11}$
308-105	f-fm	300	$1.1 \cdot 10^{-5}$	0.8	$3.0 \cdot 10^{-8}$
309-112	m-m	306	$1.6 \cdot 10^{-7}$	0.8	$4.2 \cdot 10^{-10}$
310-111	fm-f	300	$1.1 \cdot 10^{-6}$	0.8	$3.0 \cdot 10^{-9}$
311-110	m-fm	300	$4.8 \cdot 10^{-7}$	0.8	$1.3 \cdot 10^{-9}$
312-109	fm-f	300	$7.4 \cdot 10^{-6}$	0.8	$2.0 \cdot 10^{-8}$
313-116	m-m	306	$8.0 \cdot 10^{-8}$	0.8	$2.1 \cdot 10^{-10}$
314-115	fm-m	307	$2.8 \cdot 10^{-7}$	0.8	$7.3 \cdot 10^{-10}$
315-114	m-f	300	$1.7 \cdot 10^{-6}$	0.8	$4.4 \cdot 10^{-9}$
316-113	m-f	300	$5.0 \cdot 10^{-8}$	0.8	$1.3 \cdot 10^{-10}$

Table 4.15. continued.

port number	port type	Δp	Q	r	$Qr/\Delta p$
t02-b02	m-m	299	$9.3 \cdot 10^{-7}$	0.9	$2.8 \cdot 10^{-9}$
t03-b03	f-m	296	$4.4 \cdot 10^{-6}$	0.9	$1.4 \cdot 10^{-8}$
t04-b04	m-m	299	$1.6 \cdot 10^{-6}$	0.9	$4.8 \cdot 10^{-9}$
t05-b05	m-m	298	$3.7 \cdot 10^{-6}$	0.9	$1.1 \cdot 10^{-8}$
t06-b06	m-m	300	$2.5 \cdot 10^{-7}$	0.9	$7.4 \cdot 10^{-10}$
t07-b07	m-m	300	$5.8 \cdot 10^{-7}$	0.9	$1.7 \cdot 10^{-9}$
t08-b08	m-m	300	$3.4 \cdot 10^{-7}$	0.9	$1.0 \cdot 10^{-9}$
t09-b09	m-m	295	$1.8 \cdot 10^{-6}$	0.9	$5.5 \cdot 10^{-9}$
t10-b10	m-m	299	$7.9 \cdot 10^{-7}$	0.9	$2.4 \cdot 10^{-9}$
t11-b11	m-f	300	$3.9 \cdot 10^{-7}$	0.9	$1.2 \cdot 10^{-9}$
t12-b12	m-m	301	$2.2 \cdot 10^{-7}$	0.9	$6.5 \cdot 10^{-10}$
t13-b13	f-m	323	$4.0 \cdot 10^{-7}$	0.9	$1.1 \cdot 10^{-9}$
t14-b14	m-m	297	$1.6 \cdot 10^{-7}$	0.9	$4.9 \cdot 10^{-10}$
t15-b15	m-m	299	$4.3 \cdot 10^{-8}$	0.9	$1.3 \cdot 10^{-10}$
t16-b16	f-m	301	$1.4 \cdot 10^{-7}$	0.9	$4.3 \cdot 10^{-10}$

Transport Experiments

As described in Sect. 4.2.4, a certain volume of helium (in this case 105 ml) is injected into the system at the injection port via a bypass loop after the flow field has reached steady state. The tracer breakthrough is recorded at the outlet of the extraction port (Fig. 4.77) using a mass spectrometer. In Fig. 4.77a, the recorded breakthrough curves of all 28 transport experiments are shown. In Fig. 4.77b, the cumulative concentration is given for the same set of breakthrough curves.

Breakthrough curves detected at matrix-dominated port connections are indicated by mainly broad and flat curves in contrast to breakthrough curves recorded at the outlet of direct fracture connections. In this latter case, the initial and dominant breakthrough passes at significantly earlier times with a sharp concentration increase up to the highest maxima. Figure 4.78a gives two examples of tracer breakthrough curves for a fracture- and a matrix-dominated measuring connection. It is obvious that transport experiments with fracture-dominated connections result in much sharper and steeper concentration increases and higher peak concentrations than experiments with matrix-dominated connections. From the absolute values of tracer recovery (Fig. 4.78b) for the given examples, it becomes evident that the tracer recovery of the fracture-dominated connection occurs much faster and with higher absolute values than that of the matrix-dominated connection (in this case, 68% vs. 28%).

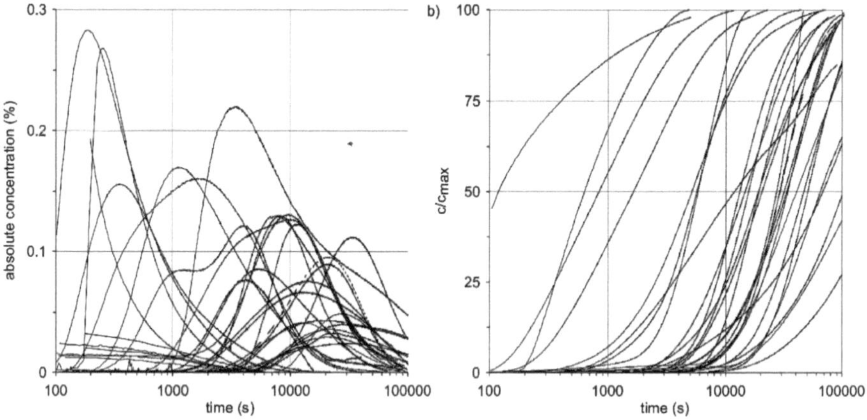

Fig. 4.77. 90 × 90 × 80 cm^3-block: Breakthrough curves of gas tracer experiments for all *port-port* configurations. a) Breakthrough curves [%]. b) Cumulative concentrations.

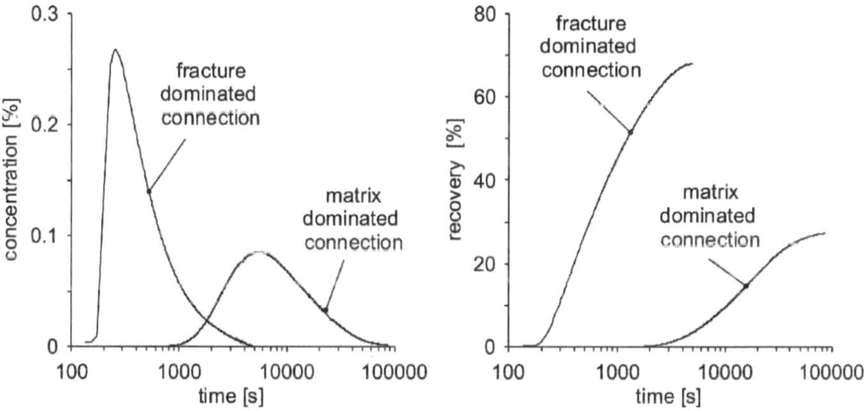

Fig. 4.78. 90 × 90 × 80 cm^3-block: Comparison of fracture- (connection *308-105*) and matrix-dominated transport (connection *208-405*). a) Tracer breakthrough curves. b) Tracer recovery curves.

However, as Fig. 4.79 shows, there is no clear correlation between the type of connection (e.g. matrix, fracture) and the recovery rate. For mainly matrix-dominated connections, a wide range of recovery rates can be detected (Fig. 4.79b, columns '*m*' and '*f-m*'). It should also be mentioned that the recovery was greater than 100% for 20% of the transport experiments conducted (Fig. 4.79a). This is mainly because of tracer remaining from previous experiments. Therefore, for comparable tracer experiments, the flushing period between consecutive experiments must be long enough to ensure that the amount of remaining tracer is negligible and the initial background concentration has been reached.

a)

Tracer recovery	Relative number [%]
< 25 %	13
25 - 50 %	17
50 - 75 %	23
75 - 100 %	27
> 100 %	20

b)

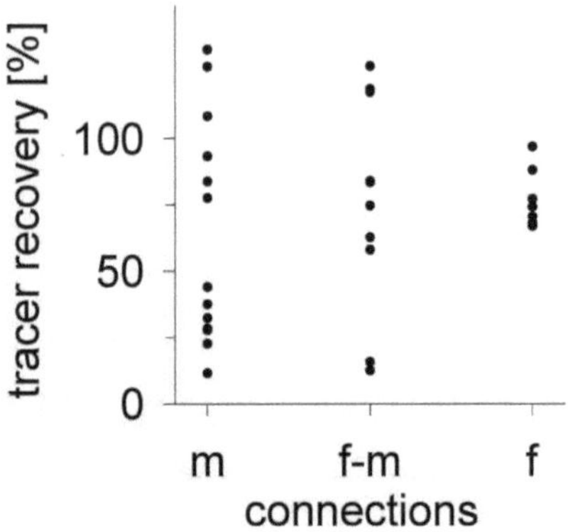

Fig. 4.79. 90 × 90 × 80 cm^3-block: Tracer recovery: a) Relative number of the *port-port* transport experiments with corresponding tracer recovery rates. b) Type of connection versus recovery rates (matrix (m), fracture and matrix (f-m), fracture (f) dominated connections.

An inspection of the breakthrough curves in Fig. 4.77 reveals some other transport phenomena:

- some curves with dominant breakthrough times between $t_{peak} = 1000$ and 10,000 seconds show double peaks with very broad curves.
- some curves show a smooth concentration decrease which can be described as "tailing".
- the plot in Fig. 4.77b indicates an "accumulation" of breakthrough curves at later times.

Table 4.16 lists different breakthrough times ($t_{initial}$, t_{peak}, and t_{median}) and seepage velocities calculated on the basis of the dominant breakthrough time (t_{peak}) for all transport experiments conducted. However, because of limitations of the mass spectrometer, only *port-port* connections are used for the transport experiments with flow rates greater than 50 ml/min.

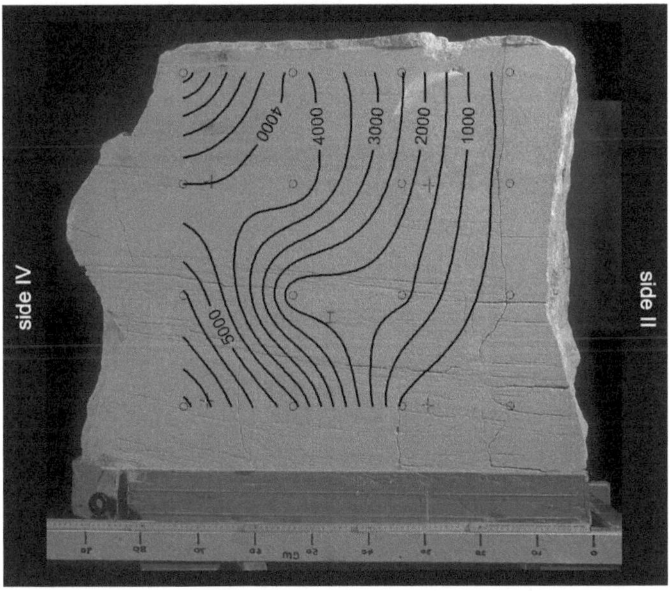

Fig. 4.80. $90 \times 90 \times 80\,cm^3$-block: Plot of initial tracer breakthrough times recorded for *port-port* connections from side *I* to *III*. The plot shows side *I*.

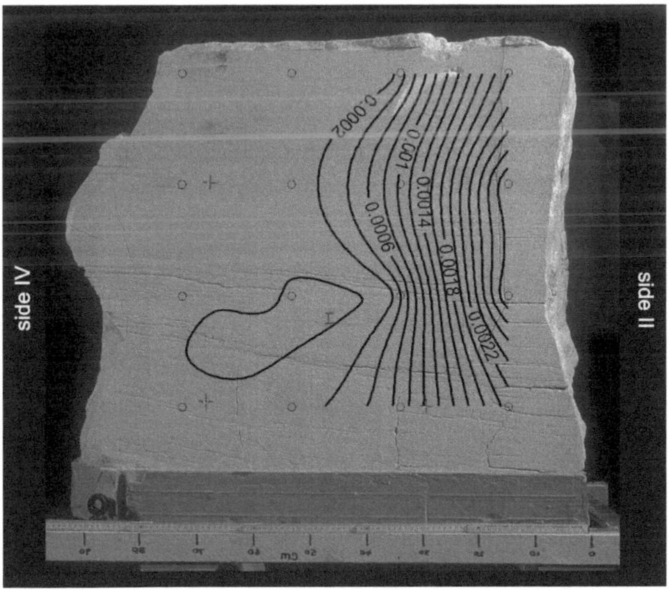

Fig. 4.81. $90 \times 90 \times 80\,cm^3$-block: Plot of the seepage velocity v recorded for *port-port* connections from side *I* to *III*. The plot shows side *I*.

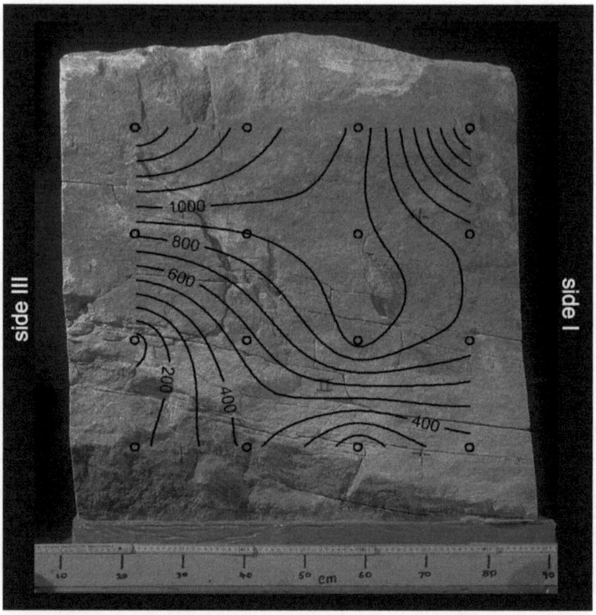

Fig. 4.82. 90 × 90 × 80 cm^3-block: Plot of initial tracer breakthrough times recorded for *port-port* connections from side *II* to *IV*. The plot shows side *II*.

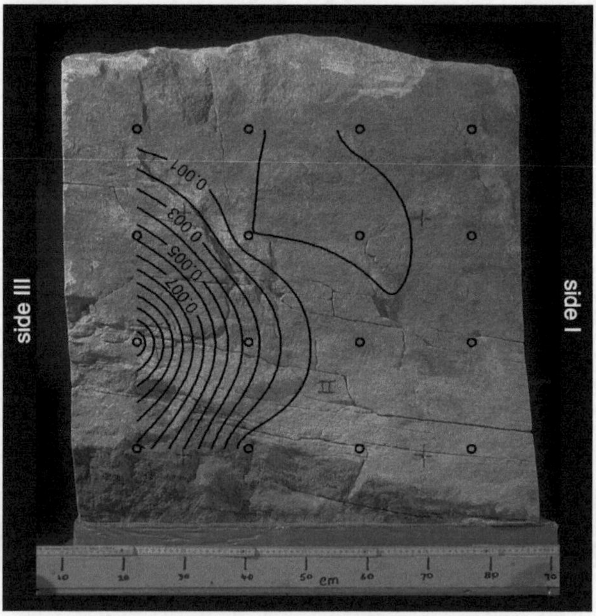

Fig. 4.83. 90 × 90 × 80 cm^3-block: Plot of the seepage velocity v recorded for *port-port* connections from side *II* to *IV*. The plot shows side *II*.

Table 4.16. $90 \times 90 \times 80\,cm^3$-block: Data from *port-port* transport experiments (m: matrix, f: fracture, fm: fracture in direct vicinity to matrix port). The exact position of the ports can be gathered from Fig. 4.18.

injection and extraction port number	port type (f, m, fm)	$t_{initial}$ (s)	t_{peak} (s)	seepage velocity v (m/s)	t_{median} (s)
201-404	m-m	1521	7702	$1.2 \cdot 10^{-4}$	20411
202-403	m-m	1150	9592	$9.4 \cdot 10^{-5}$	15615
203-402	m-fm	975	8207	$1.1 \cdot 10^{-4}$	13084
204-401	f-fm	156	3907	$2.3 \cdot 10^{-4}$	5661
205-408	m-m	1)	1)	1)	1)
206-407	f-fm	900	14387	$6.3 \cdot 10^{-5}$	31620
207-406	f-m	1)	1)	1)	1)
208-405	m-m	751	5550	$1.6 \cdot 10^{-4}$	68221
209-412	fm-fm	52	63	$1.4 \cdot 10^{-2}$	117
210-411	fm-m	1)	1)	1)	1)
211-410	f-m	950	12819	$7.0 \cdot 10^{-5}$	33293
212-409	m-fm	752	3406	$2.6 \cdot 10^{-4}$	68221
213-416	m-m	1)	1)	1)	1)
214-415	fm-m	428	9675	$9.3 \cdot 10^{-5}$	19021
215-414	f-m	104	1608	$5.6 \cdot 10^{-4}$	5145
216-413	m-m	1)	1)	1)	1)
301-104	m-m	1168	4135	$1.9 \cdot 10^{-4}$	5801
302-103	m-m	4184	21871	$3.7 \cdot 10^{-5}$	130614
303-102	m-m	2520	14567	$5.5 \cdot 10^{-5}$	40665
304-101	f-fm	84	362	$2.2 \cdot 10^{-3}$	4040
305-108	m-m	4005	13455	$5.9 \cdot 10^{-5}$	40632
306-107	m-m	4502	22318	$3.6 \cdot 10^{-5}$	46767
307-106	m-m	1)	1)	1)	1)
308-105	f-fm	154	253	$3.2 \cdot 10^{-3}$	635
309-112	m-m	5608	20651	$3.9 \cdot 10^{-5}$	2)
310-111	fm-f	1594	33416	$2.4 \cdot 10^{-5}$	27240
311-110	m-fm	1641	18619	$4.3 \cdot 10^{-5}$	101984
312-109	fm-f	125	247	$3.2 \cdot 10^{-3}$	839
313-116	m-m	7200	26098	$3.1 \cdot 10^{-5}$	64885
314-115	fm-m	4719	18753	$4.3 \cdot 10^{-5}$	33927
315-114	m-f	360	1499	$5.3 \cdot 10^{-4}$	9531
316-113	m-f	1)	1)	1)	1)

Table 4.16. continued.

port number	port type	$t_{initial}$	v	t_{peak}	t_{median}
t02-b02	m-m	1)	1)	1)	1)
t03-b03	f-m	3324	11254	$8.0 \cdot 10^{-5}$	19177
t04-b04	m-m	1)	1)	1)	1)
t05-b05	m-m	1312	20416	$4.4 \cdot 10^{-5}$	29212
t06-b06	m-m	1)	1)	1)	1)
t07-b07	m-m	1)	1)	1)	1)
t08-b08	m-m	1)	1)	1)	1)
t09-b09	m-m	3927	34255	$2.6 \cdot 10^{-5}$	50599
t10-b10	m-m	1)	1)	1)	1)
t11-b11	m-f	1)	1)	1)	1)
t12-b12	m-m	1)	1)	1)	1)
t13-b13	f-m	1)	1)	1)	1)
t14-b14	m-m	1)	1)	1)	1)
t15-b15	m-m	1)	1)	1)	1)
t16-b16	f-m	1)	1)	1)	1)

1) - no breakthrough curve measured
2) - estimation not possible.

Discussion

The phenomena observed in the *port-port* experiments can be interpreted as follows:

- Relatively short breakthrough times with high concentration maxima correspond to a fast transport of the tracer through the fracture system with less pronounced interaction with the porous sandstone matrix.
- The broad and flat breakthrough of tracer reflects transport mainly through the matrix with dominant diffusive and/or dispersive transport mechanisms.
- Several reasons can be given for the tailing in the breakthrough curves:
  - Matrix diffusion: Due to the injection of tracer directly into the fracture network, a concentration gradient between fracture network (higher concentration) and matrix (lower concentration) is established. Following FICK's law, a tracer flux from regions of higher concentration to regions of lower concentration occurs, i.e. from the fracture network to the matrix. After the main tracer mass passes through the fracture network, the concentration within the fractures decreases again. After a certain time, a gradient in the opposite direction is established, i.e. from the matrix to the fracture, and the tracer is released into the

fractures. This process leads to a retarded tracer breakthrough at the observation point, and is typically observed as tailing in the breakthrough curve.
  - Differential advection: The dimensionality of the flow field which is assumed to be three-dimensional for the *port-port* connections. As described by McDermott (1999), such tailing can be due to pathways of varying length which are established through the dimensionality of the flow field.
- The occurrence of the double peaks can be due to transport through two or more fractures or combined transport through matrix and fracture network.

4.4.2.2 The $60 \times 60 \times 60\,cm^3$-Block

Flow Experiments

With the experimental set-up shown in Fig. 4.25, flow experiments are conducted using *port-port* connections as described in Sect. 4.4.2.1. In Table 4.17, the flow rate information recorded during the experiments is presented for all *port-port* connections as well as the ratio $Qr/\Delta p$ which provides a measure for the conductivity of the system with respect to gas flow. In Fig. 4.84, the ratio is plotted for side *II*.

In comparison to the $90 \times 90 \times 80\,cm^3$-block, the spatial distribution of the ratio reflects the bedding and the matrix composition of the sandstone block in particular. Beyond this, the differences of the absolute values of the ratio are much smaller. Both, indicates that the influence of the fractures is not dominating the system but the structural composition of the matrix plays an important role. The matrix of the sandstone block can be separated in two parts. The upper part consists of a coarser grained matrix while the lower part consists of a finer grained matrix. This classification agrees with the conducted flow experiments. From the 51 flow experiments, 8 measurements were conducted in the upper part of the sandstone block. The recorded flow rates in this region of the sandstone block are up to four times higher than in the finer grained matrix. An influence of the fractures on the flow experiments can not be determined.

Table 4.17. $60 \times 60 \times 60\,cm^3$-block: Data from *port-port* flow experiments (m: matrix, f′: fissure).

injection extraction port number	port type (f′, m)	pressure difference Δp (mbar)	flow rate Q (m^3/s)	distance r (m)	ratio $Qr/\Delta p$
101-304	m-m	505	$1.40 \cdot 10^{-5}$	0.6	$1.66 \cdot 10^{-8}$

Table 4.17. continued.

port number	port type	Δp	Q	r	$Qr/\Delta p$
102-303	m-m	500	$1.14 \cdot 10^{-5}$	0.6	$1.37 \cdot 10^{-8}$
103-302	m-f′	502	$7.92 \cdot 10^{-6}$	0.6	$9.47 \cdot 10^{-9}$
104-301	m-m	505	$3.65 \cdot 10^{-6}$	0.6	$4.34 \cdot 10^{-9}$
105-308	m-m	502	$7.48 \cdot 10^{-7}$	0.6	$8.94 \cdot 10^{-10}$
106-307	m-m	502	$1.07 \cdot 10^{-6}$	0.6	$1.28 \cdot 10^{-9}$
107-306	m-m	502	$6.77 \cdot 10^{-7}$	0.6	$8.09 \cdot 10^{-10}$
108-305	f′-f′	502	$5.33 \cdot 10^{-7}$	0.6	$6.38 \cdot 10^{-10}$
109-312	m-m	501	$2.83 \cdot 10^{-6}$	0.6	$3.39 \cdot 10^{-9}$
110-311	m-f′	501	$2.17 \cdot 10^{-6}$	0.6	$2.60 \cdot 10^{-9}$
111-310	f′-f′	500	$1.03 \cdot 10^{-6}$	0.6	$1.23 \cdot 10^{-9}$
112-309	f′-m	499	$1.14 \cdot 10^{-6}$	0.6	$1.37 \cdot 10^{-9}$
113-316	m-m	500	$1.92 \cdot 10^{-6}$	0.6	$2.30 \cdot 10^{-9}$
114-315	m-m	499	$2.12 \cdot 10^{-6}$	0.6	$2.54 \cdot 10^{-9}$
115-314	m-m	501	$1.30 \cdot 10^{-6}$	0.6	$1.55 \cdot 10^{-9}$
116-313	m-m	502	$1.03 \cdot 10^{-6}$	0.6	$1.23 \cdot 10^{-9}$
117-317	m-m	490	$1.75 \cdot 10^{-7}$	0.6	$2.14 \cdot 10^{-10}$
201-404	f′-m	506	$5.83 \cdot 10^{-6}$	0.6	$6.92 \cdot 10^{-9}$
202-403	m-m	499	$7.95 \cdot 10^{-6}$	0.6	$9.57 \cdot 10^{-9}$
203-402	m-m	500	$8.75 \cdot 10^{-6}$	0.6	$1.05 \cdot 10^{-8}$
204-401	m-m	500	$6.37 \cdot 10^{-6}$	0.6	$7.65 \cdot 10^{-9}$
205-408	m-m	501	$4.27 \cdot 10^{-7}$	0.6	$5.11 \cdot 10^{-10}$
206-407	m-m	498	$1.11 \cdot 10^{-6}$	0.6	$1.33 \cdot 10^{-9}$
207-406	f′-m	508	$1.10 \cdot 10^{-6}$	0.6	$1.29 \cdot 10^{-9}$
208-405	f′-m	508	$1.17 \cdot 10^{-6}$	0.6	$1.38 \cdot 10^{-9}$
209-412	m-m	504	$1.05 \cdot 10^{-6}$	0.6	$1.25 \cdot 10^{-9}$
210-411	m-m	504	$1.75 \cdot 10^{-6}$	0.6	$2.08 \cdot 10^{-9}$
211-410	m-m	504	$1.75 \cdot 10^{-6}$	0.6	$2.08 \cdot 10^{-9}$
212-409	f′-m	502	$1.06 \cdot 10^{-6}$	0.6	$1.26 \cdot 10^{-9}$
213-416	m-m	500	$1.92 \cdot 10^{-6}$	0.6	$2.30 \cdot 10^{-9}$
214-415	f′-m	500	$1.20 \cdot 10^{-6}$	0.6	$1.44 \cdot 10^{-9}$
215-414	m-m	500	$1.12 \cdot 10^{-6}$	0.6	$1.34 \cdot 10^{-9}$
216-413	m-m	500	$6.83 \cdot 10^{-7}$	0.6	$8.21 \cdot 10^{-10}$
217-417	m-m	503	$4.02 \cdot 10^{-7}$	0.6	$4.79 \cdot 10^{-10}$
b01-t01	m-m	503	$3.42 \cdot 10^{-6}$	0.6	$4.07 \cdot 10^{-9}$
b02-t02	m-f′	502	$2.83 \cdot 10^{-6}$	0.6	$3.39 \cdot 10^{-9}$
b03-t03	m-m	501	$2.13 \cdot 10^{-6}$	0.6	$2.55 \cdot 10^{-9}$
b04-t04	m-m	501	$2.33 \cdot 10^{-6}$	0.6	$2.79 \cdot 10^{-9}$
b05-t05	m-m	500	$2.83 \cdot 10^{-6}$	0.6	$3.40 \cdot 10^{-9}$

Table 4.17. continued.

port number	port type	Δp	Q	r	$Qr/\Delta p$
b06-t06	m-f′	502	$2.08 \cdot 10^{-6}$	0.6	$2.49 \cdot 10^{-9}$
b07-t07	m-m	501	$1.33 \cdot 10^{-6}$	0.6	$1.60 \cdot 10^{-9}$
b08-t08	m-m	502	$1.68 \cdot 10^{-6}$	0.6	$2.01 \cdot 10^{-9}$
b09-t09	m-m	500	$1.57 \cdot 10^{-6}$	0.6	$1.88 \cdot 10^{-9}$
b10-t10	m-m	498	$2.17 \cdot 10^{-6}$	0.6	$2.61 \cdot 10^{-9}$
b11-t11	m-m	497	$2.25 \cdot 10^{-6}$	0.6	$2.71 \cdot 10^{-9}$
b12-t12	m-m	496	$2.92 \cdot 10^{-6}$	0.6	$3.53 \cdot 10^{-9}$
b13-t13	m-m	498	$1.48 \cdot 10^{-6}$	0.6	$1.78 \cdot 10^{-9}$
b14-t14	m-m	497	$1.48 \cdot 10^{-6}$	0.6	$1.78 \cdot 10^{-9}$
b15-t15	m-m	497	$1.40 \cdot 10^{-6}$	0.6	$1.69 \cdot 10^{-9}$
b16-t16	m-m	496	$1.27 \cdot 10^{-6}$	0.6	$1.54 \cdot 10^{-9}$
b17-t17	m-m	495	$1.87 \cdot 10^{-6}$	0.6	$2.26 \cdot 10^{-9}$

Transport Experiments

For the experimental series of transport experiments conducted on the $60 \times 60 \times 60\,cm^3$-block, 105 ml helium is injected into the system and the tracer breakthrough is recorded as described in Sect. 4.2.4. Fig. 4.85a shows breakthrough curves recorded for all 28 transport experiments. In Fig. 4.85b, the cumulative concentration is given for the same set of breakthrough curves. Table 4.18 lists the different breakthrough times ($t_{initial}$, t_{peak}, and t_{median}) and seepage velocities calculated on the basis of the dominant breakthrough time (t_{peak}) for all conducted transport experiments. In Figs. 4.86 and 4.87 plots of the initial breakthrough times and the seepage velocities respectively are given for side *II*.

Breakthrough curves recorded in the coarse grained matrix of the sandstone block are characterized by an initial and dominant breakthrough occurring at significantly earlier times with a sharp concentration increase up to the peak, while the remaining curves are characterized by a more or less broad and flat shape and later initial and dominant breakthrough times. A more detailed investigation of the remaining breakthrough curves indicates that it is not possible alone by means of the shape of the breakthrough curves and the different breakthrough times ($t_{initial}$, t_{peak}, and t_{median}) to determine if the breakthrough is recorded parallel or orthogonal to the bedding. One reason for this is the influence of the boundary condition on the transport behavior. In Sect. 10 a multivariate statistical approach is proposed that allows to distinguish breakthrough curves recorded parallel or orthogonal to the bedding.

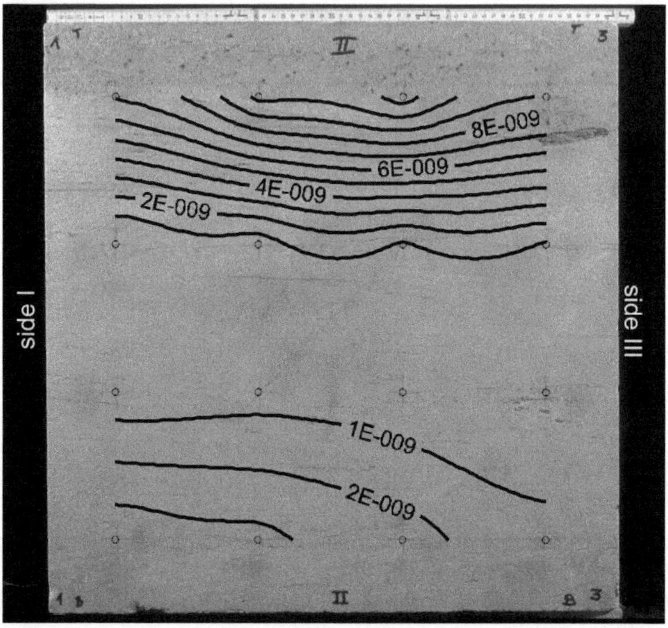

Fig. 4.84. $60 \times 60 \times 60\,\text{cm}^3$-block: Spatial distribution of ratio $Qr/\Delta p$ for side *II*.

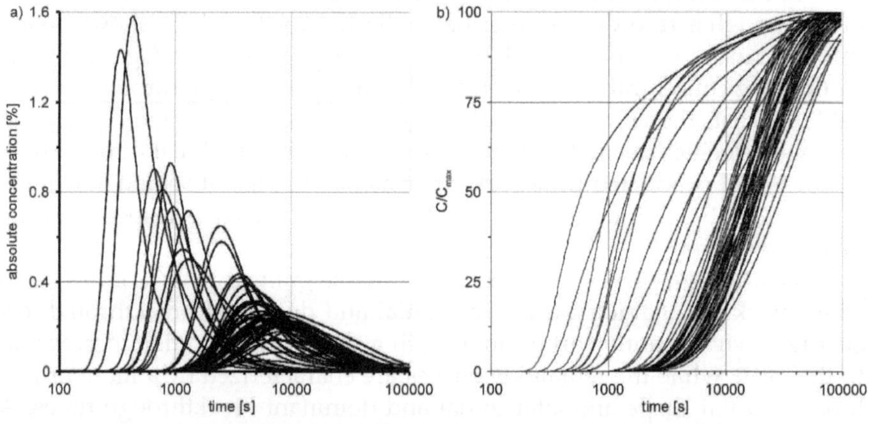

Fig. 4.85. $60 \times 60 \times 60\,\text{cm}^3$-block: Breakthrough curves of gas tracer experiments for all *port-port* configurations. a) Breakthrough curves [%]. b) Cumulative concentrations.

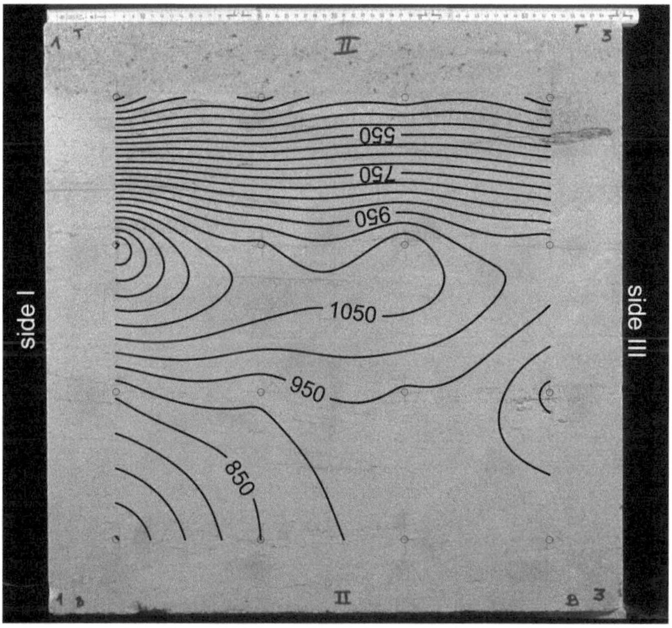

Fig. 4.86. 60 × 60 × 60 cm^3-block: Plot of initial tracer breakthrough times recorded for *port-port* connections from side *II* to *IV*. The plot shows side *II*.

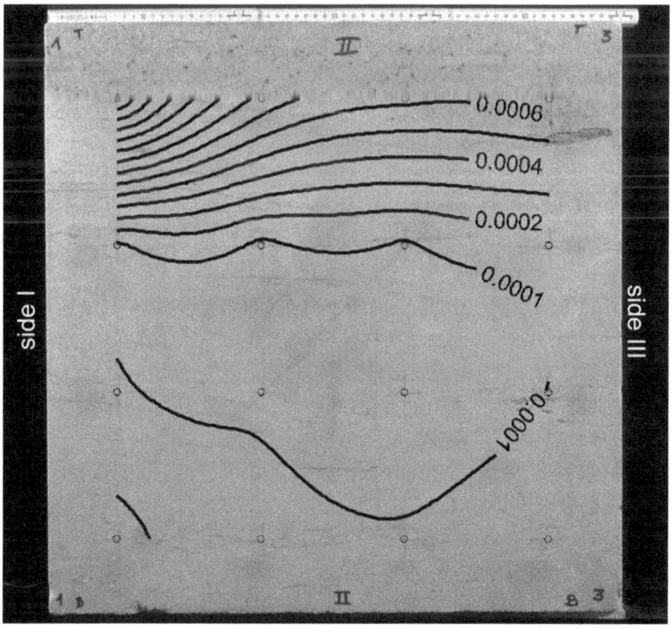

Fig. 4.87. 60 × 60 × 60 cm^3-block: Plot of the seepage velocity v recorded for *port-port* connections from side *II* to *IV*. The plot shows side *II*.

Table 4.18. $60 \times 60 \times 60\,\text{cm}^3$-block: Data from *port-port* transport experiments (m: matrix, f′: fissure).

injection and extraction port number	port type (f, m, fm)	$t_{initial}$ (s)	t_{peak} (s)	seepage velocity v (m/s)	t_{median} (s)
101-304	m-m	146	335	$1.79 \cdot 10^{-3}$	594
102-303	m-m	228	654	$9.18 \cdot 10^{-4}$	1007
103-302	m-k	463	1140	$5.26 \cdot 10^{-4}$	1918
104-301	m-m	401	1296	$4.63 \cdot 10^{-4}$	3599
105-308	m-m	872	4373	$1.37 \cdot 10^{-4}$	15195
106-307	m-m	1018	6535	$9.18 \cdot 10^{-5}$	16981
107-306	m-m	991	9279	$6.47 \cdot 10^{-5}$	22825
108-305	k-k	1100	6833	$8.78 \cdot 10^{-5}$	27183
109-312	m-m	400	1316	$4.56 \cdot 10^{-4}$	5176
110-311	m-k	704	4598	$1.30 \cdot 10^{-4}$	10194
111-310	k-k	1188	8783	$6.83 \cdot 10^{-5}$	19320
112-309	k-m	957	5462	$1.10 \cdot 10^{-4}$	15864
113-316	m-m	645	2446	$2.45 \cdot 10^{-4}$	5314
114-315	m-m	834	3573	$1.68 \cdot 10^{-4}$	7183
115-314	m-m	929	4953	$1.21 \cdot 10^{-4}$	12904
116-313	m-m	913	3657	$1.64 \cdot 10^{-4}$	11148
201-404	k-m	180	428	$1.40 \cdot 10^{-3}$	1168
202-404	m-m	280	778	$7.71 \cdot 10^{-4}$	1481
203-402	m-m	365	961	$6.24 \cdot 10^{-4}$	1531
204-401	m-m	315	904	$6.64 \cdot 10^{-4}$	1993
205-408b	m-m	1361	7728	$7.76 \cdot 10^{-5}$	30524
206-407	m-m	1065	8118	$7.39 \cdot 10^{-5}$	18862
207-406	k-m	1100	7037	$8.53 \cdot 10^{-5}$	18705
208-405	k-m	990	4750	$1.26 \cdot 10^{-4}$	17954
209-412	m-m	860	4969	$1.21 \cdot 10^{-4}$	19286
210-411	m-m	910	6438	$9.32 \cdot 10^{-5}$	14067
211-410	m-m	941	7273	$8.25 \cdot 10^{-5}$	15104
212-409	k-m	833	5894	$1.02 \cdot 10^{-4}$	15324
213-416	m-m	646	2561	$2.34 \cdot 10^{-4}$	6818
214-415	k-m	850	4775	$1.26 \cdot 10^{-4}$	14947
215-414	k-k	930	5781	$1.04 \cdot 10^{-4}$	15485
216-413	m-m	950	4264	$1.41 \cdot 10^{-4}$	19576
b01-t01	m-m	840	4430	$1.35 \cdot 10^{-4}$	14100
b02-t02	m-k	840	4974	$1.21 \cdot 10^{-4}$	11071
b03-t03	m-m	910	5915	$1.01 \cdot 10^{-4}$	13756
b04-t04	m-m	940	4336	$1.38 \cdot 10^{-4}$	13066

Table 4.18. continued.

port number	port type	$t_{initial}$	v	t_{peak}	t_{median}
b05-t05	m-m	910	4703	$1.28 \cdot 10^{-4}$	10705
b06-t06	m-k	1200	7910	$7.58 \cdot 10^{-5}$	14488
b07-t07	m-m	1360	10993	$5.46 \cdot 10^{-5}$	20699
b08-t08	m-m	973	6939	$8.65 \cdot 10^{-5}$	17586
b09-t09	m-m	1220	8672	$6.92 \cdot 10^{-5}$	18854
b10-t10	m-m	1150	9185	$6.53 \cdot 10^{-5}$	14415
b11-t11	m-m	1170	8756	$6.85 \cdot 10^{-5}$	13787
b12-t12	m-m	1140	5737	$1.05 \cdot 10^{-4}$	11468
b13-t13	m-m	1143	6572	$9.13 \cdot 10^{-5}$	16347
b14-t14	m-m	1263	8506	$7.05 \cdot 10^{-5}$	16729
b15-t15	m-m	1420	8401	$7.14 \cdot 10^{-5}$	18114
b16-t16	m-m	1252	6503	$9.23 \cdot 10^{-5}$	20538

Discussion

From the *port-port* experiments conducted on the $60 \times 60 \times 60\,cm^3$-block, it becomes evident that the flow and transport behavior is mainly dominated by the particulars of the matrix. The plots shown in Figs. 4.84, 4.86 and 4.87 make it obvious that the flow and transport behavior is dominated by a layer of coarse-grained matrix at the top of the $60 \times 60 \times 60\,cm^3$-block and that the fissures have no evident influence on the flow and transport. Further information can be gathered by performing a multivariate statistical evaluation of the experiments (Sect. 10).

4.5 Interpretation of Flow and Transport Experiments Conducted on Laboratory Block

The following section of this chapter focuses on the interpretation of experimental data of the laboratory block by means of numerical modeling and apparent parameters.

4.5.1 Interpretation of Flow and Transport Experiments Based on Apparent Parameters

C. Leven,T. Vogel,V. Lagendijk

To interpret the experiments conducted on the $90 \times 90 \times 80\,cm^3$-block and to quantify the relevant parameters, a numerical finite element model was employed (Sect. 2.5.4). Table 4.19 lists the input parameters of the model.

With the numerical model, a standard response for a homogeneous, isotropic block was generated, taking into account the geometry of the block and the spatial position of the ports connected during the experiments.

With a procedure described in McDermott (1999) and Chap. 6, the variation in the hydraulic conductivity of the block is given by a comparison of the flow rates from the experiments and the numerical model. In this interpretation it is assumed that any difference in measured and modeled flow rates can be assigned to the heterogeneous nature of the block since the model solely assumes homogeneity. However, this linear relationship is only valid for DARCY-type flow. Therefore, we will refer to the hydraulic conductivity of the fractured sandstone block in the following as *apparent* hydraulic conductivity K', which is derived from

$$K'_{measure} = K_{model} \frac{Q_{measure}}{Q_{model}} \tag{4.3}$$

with $K'_{measure}$ resulting apparent hydraulic conductivity,
K_{model} hydraulic conductivity used as model input parameter,
$Q_{measure}$ flow rate estimated for particular flow experiment,
Q_{model} flow rate resulting from numerical model.

Table 4.19. Model input parameters to provide a standard flow and transport signal with which the experimental data are compared.

Intrinsic permeability [m^2]	$4.0 \cdot 10^{-14}$
Total porosity [-]	0.15
Longitudinal dispersivity [m]	$6.7 \cdot 10^{-2}$
Transverse dispersivity [m]	$1.3 \cdot 10^{-2}$
Molecular diffusion coefficient [m^2 s^{-1}]	$4.4 \cdot 10^{-6}$

Figure 4.88 shows the distribution of the hydraulic conductivity derived from (4.3) projected on the block faces of sides *II* and *III*. The heterogeneity of the sandstone block can easily be seen, with the highest values near the dominating fractures (e.g. right part of *III* in Fig. 4.88 b). The analysis of the hydraulic conductivity of Figure 4.88 indicates that fracture dominated flow occurs only in cases of direct fracture connections. At connections to the matrix, the influence of the fracture network decreases strongly.

In a quantitative interpretation of the transport experiments, the variation in the transport signal is presented as a percentage of the modelled standard signal.

Figure 4.89 illustrates the comparison of the simulated and measured transport signals for one particular port-to-port connection. The breakthrough curve represented by a dashed line in Figure 4.89 (a) shows the

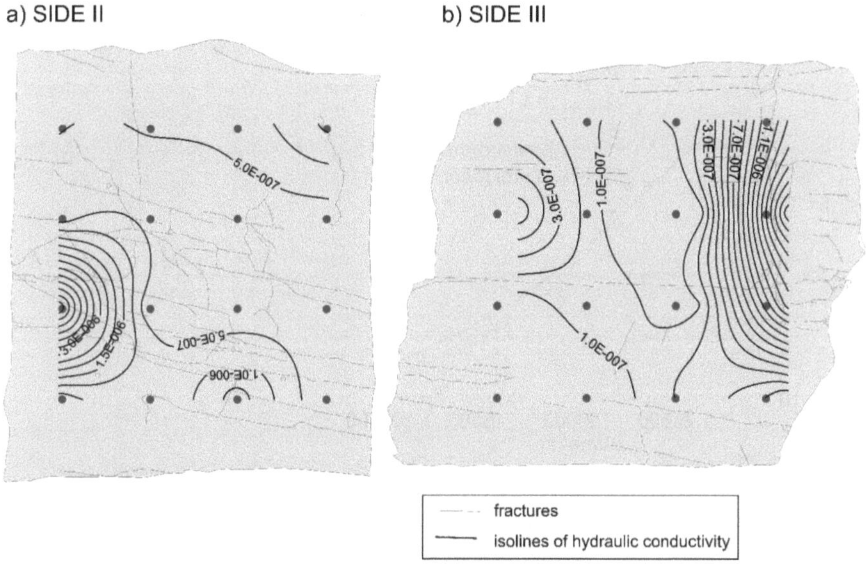

Fig. 4.88. $90 \times 90 \times 80\,\text{cm}^3$-block: Isolines of apparent hydraulic conductivity in (m/s) for side *II* (a) and side *III* (b). The circles indicate the spatial position of the ports.

transport signal which would be recorded for port connection *211-410* if the sandstone block were a homogeneous, isotropic medium with the parameters listed in Table 4.19. The tracer breakthrough in the experiment (solid curve in Fig. 4.89) occurs slightly earlier than in the numerical simulation. This can be due to preferential flow ("channeling"), as is also indicated by the slightly earlier dominant breakthrough time in the experiment. The tailing indicated by the observed transport signal is most likely due to the tracer spreadig more slowly in the sandstone matrix than along the pathways of the fracture network.

Figure 4.90 gives the results of the comparison between the transport signals gained from the simulations of the homogeneous, isotropic medium and the gas-transport experiments. It should be noted that effects due to different boundary conditions are eliminated by comparing each experimental result with the corresponding numerical simulation of the particular *port-port* connection.

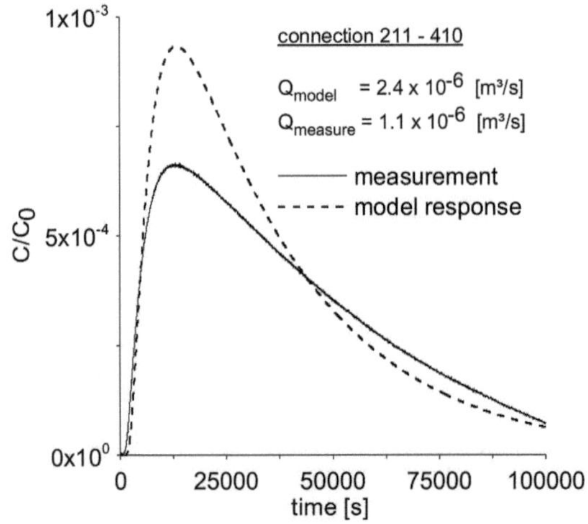

Fig. 4.89. $90 \times 90 \times 80\,\text{cm}^3$-block: Comparison of the breakthrough curves gained from the numerical simulation and from the tracer experiments for the connection between ports 211 and 410. The initial breakthrough time of the numerical simulation occurred after 1380 s, in the experiments it occurred after 950 s.

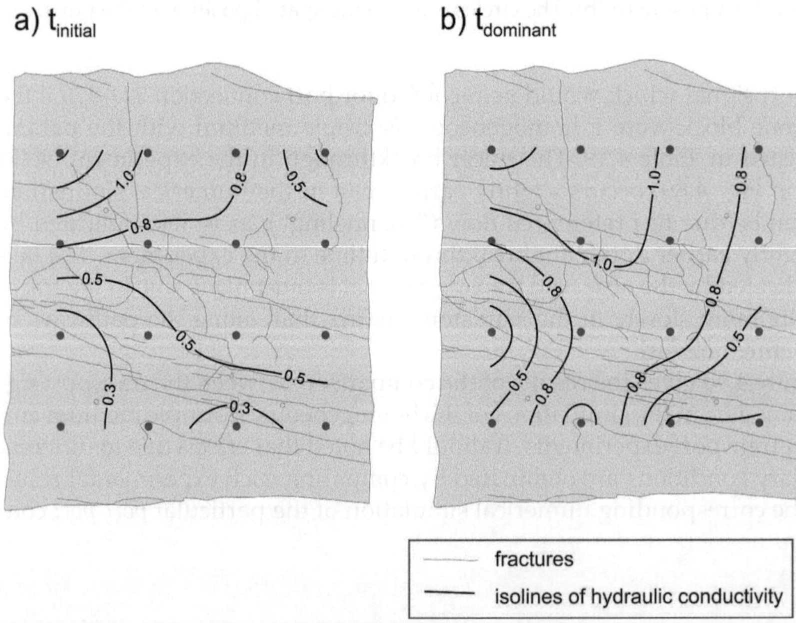

Fig. 4.90. $90 \times 90 \times 80\,\text{cm}^3$-block: Quotient of initial (a) and dominant (b) breakthrough times derived from the experiment and the numerical simulation for a homogeneous, isotropic medium. Values less than one indicate earlier breakthrough times for the experiment than those calculated with the numerical model.

4.5.2 Multi-continuum Modeling: Methodology and Approach

T. Vogel, V. Lagendijk, J. Köngeter

In the following discussion, a brief outline of the approach and methodology used for developing a multi-continuum model is presented, along with the implementation of experimental data in the model.

The following studies carried out on the scale of the $90 \times 90 \times 80\,\text{cm}^3$-block, were similar to those conducted on the scale of the cylinder. First, numerical investigations were performed. The parameters were obtained directly from the experiments (Sect. 4.4). For this configuration, the measured concentrations were very low, which caused problems concerning the recovery rate within the experiments.

Figures 4.91 and 4.92 illustrate the simplified block for the FE calculations and the open and filled fractures detected on the block surface. A more detailed description of the different sides of the block can be found in Sect. 4.4. The grid illustrated in Figs. 4.91 and 4.92 corresponds to the measurement port grid. The dimensions of the model block are 0.9 m in the direction of the x- and the z-axes and 0.8 m in the direction of the y-axis, based on the coordinate system as shown.

4.5.2.1 Laboratory-Block Simulations

In the following discussion, the simulations conducted on the bench scale of the $90 \times 90 \times 80\,\text{cm}^3$-block are described and the assumptions are explained. The bench-scale block approach differs from the one performed on the cylinder scale as the simulations are based on different sets of data and no discrete model is available. For the numerical investigations, only the data of the tracer experiments between sides II and IV are taken into account as these are the most reliable.

The array of pictures illustrated in Fig. 4.93 within a 4x4 matrix corresponds to side II of the laboratory block. Each picture illustrates the connection between the ports by the intersection of the planes. Tracer is injected at the positions indicated by the black colour. The empty positions within the 4x4 matrix indicate configurations without experimental data.

Figure 4.94 shows the original breakthrough curves and the modified breakthrough curves for two configurations studied. The modified curves have been shifted by the initial value of concentration at the beginning of the measurements (background concentration). The absolute value of the background concentration varies between the experimental configurations and must be taken into account in the interpretation of the data. All of the breakthrough curves in this section have been shifted according to their value of background concentration.

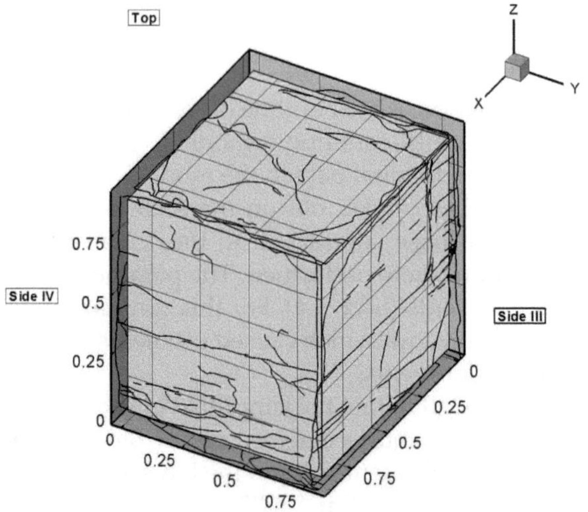

Fig. 4.91. View of the laboratory block from above with a grid of the ports.

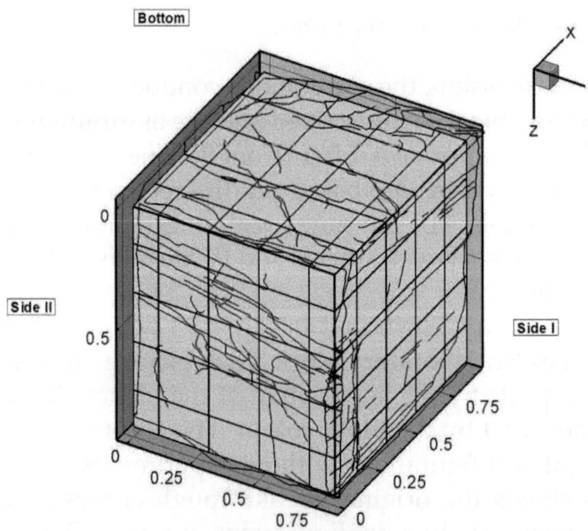

Fig. 4.92. View of the laboratory block from below with a grid of the ports.

4.5.2.2 Determination of Characteristic Model Parameters

The equivalent parameters for the laboratory block can only be determined by analyzing the experimental data because a discrete model of the block

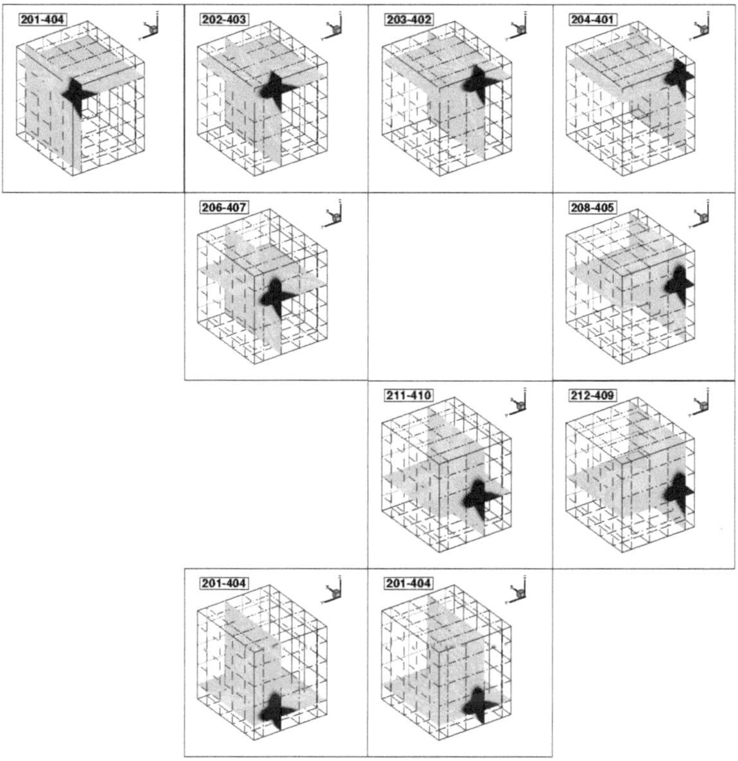

Fig. 4.93. Position of selected ports of sides II and IV.

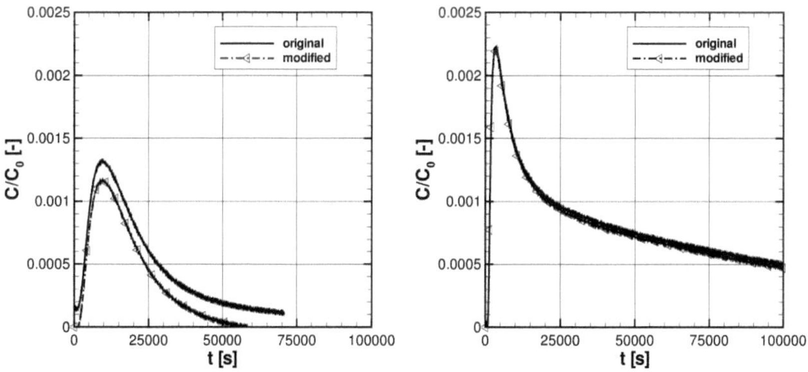

Fig. 4.94. Background concentrations of the experimental data of the laboratory block: Ports 202_403 (left side), Ports 212_409 (right side).

Table 4.20. Characteristic parameters for the single-continuum model.

Equivalent parameters			Continuum
Permeability	k_{xx}	(m^2)	$2.08 \cdot 10^{-12}$
	k_{xy}		0.00
	k_{xz}	(m^2)	0.00
	k_{yy}	(m^2)	$4.37 \cdot 10^{-13}$
	k_{yz}	(m^2)	0.00
	k_{zz}	(m^2)	$7.15 \cdot 10^{-13}$
Porosity	n_e	(-)	0.25
Dispersivity			
-longitudinal	α_l	(m)	$6.67 \cdot 10^{-2}$
-transversal horizontal	α_{th}	(m)	$1.32 \cdot 10^{-2}$
-transversal vertical	α_{tv}	(m)	$1.32 \cdot 10^{-2}$
Molecular diffusivity coefficient	D_m	(m^2s^{-1})	$4.40 \cdot 10^{-6}$

does not exist. Based on the investigations on the cylinder, a single-continuum model is used.

The equivalent permeabilities are determined by averaging the permeability measurements in the direction of the coordinate axes. The values of porosity, dispersivity and molecular diffusion are taken from McDermott (1999). The parameters are summarized in Table 4.20.

4.5.2.3 Boundary Conditions

A total of 105 ml of tracer is injected into the system. The boundary conditions at the input and output ports are chosen to ensure a constant pressure of 101 300 Pa at the input port, and the experimental flow rate is taken at the output port.

4.5.2.4 Transport Simulation

The results for a representative selection of simulations are illustrated in Fig. 4.95 and Table 4.21. Input and output ports are connected to a fracture, to the matrix or to both. The recovery rates ascertained in experiments and numerical simulations after certain periods of time are also presented in Table 4.21. Some of the experimental rates - time-integrated concentration at the output over infiltrated amount of tracer - are above 100 %. A possible explanation could be the low concentrations detected at the output port, which are of the

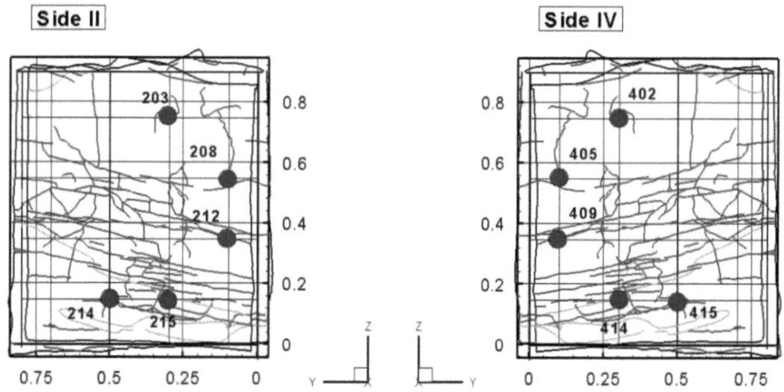

Fig. 4.95. Position of ports for a representative selection of results.

Table 4.21. Ports and recovery rates of representative configurations.

Configuration	Input port	Output port	Rate experiment (%)	Rate model (%)	Time (s)
203_402	Matrix	Fracture	83.53	95.01	$1.50 \cdot 10^{+5}$
208_405	Fracture	Matrix	26.96	79.84	$1.50 \cdot 10^{+5}$
212_409	Fracture	Fracture	113.36	66.56	$1.50 \cdot 10^{+5}$
215_414	Matrix	Matrix	132.58	99.81	$1.50 \cdot 10^{+5}$
214_415	Fracture/Matrix	Fracture/Matrix	111.49	76.93	$7.50 \cdot 10^{+4}$

same order of magnitude as the background concentration (cf. Fig. 4.94). Additionally, it is possible that this effect is due to tracer remaining in the block from a previous experiment.

Figs. 4.96 to 4.98 show that the transport behaviour of the laboratory block may be reproduced qualitatively with the determined equivalent parameters and the chosen model concept. As regards detecting the maximum concentration, the best results are obtained by simulations with matrix ports or mixed ports (215 414 and 214 415). For the model configuration 203 402, where the input port is connected to the matrix and the output is attached to a fracture, the arrival of the maximum relative concentration is too early. On the other hand, for an inverse configuration (208 405), the peak of concentration is predicted too late and the tailing of the breakthrough curve is more distinct. The model underestimates the maximum for configurations with

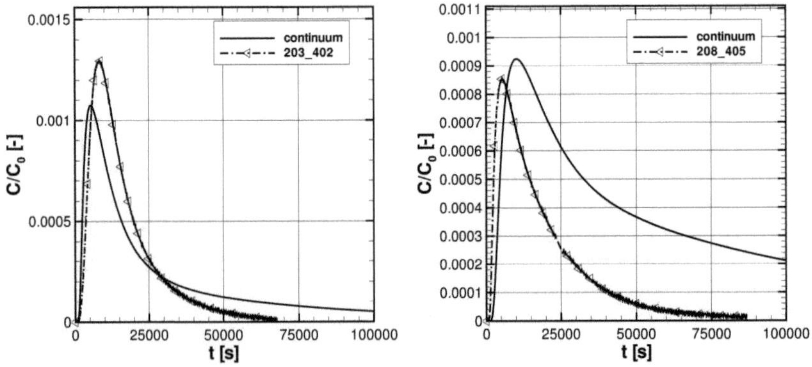

Fig. 4.96. Breakthrough curves for a pulse injection; ports 203_402 and 208_405.

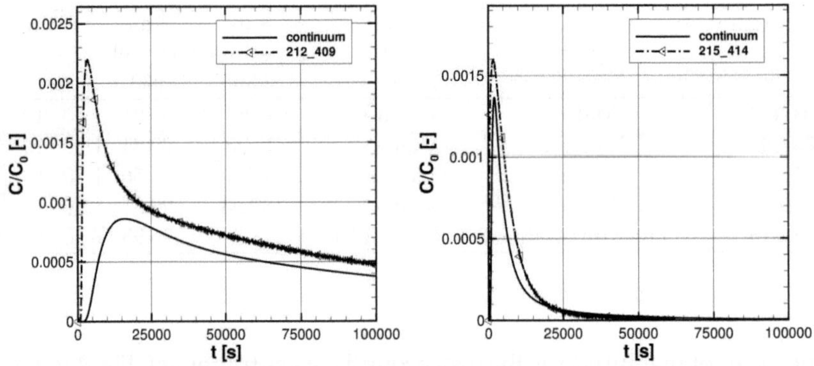

Fig. 4.97. Breakthrough curves for a pulse injection; ports 212_409 and 215_414.

fracture-fracture connections (212_409) and the recovery rate after 15 000 seconds is low compared to the experimental results.

Thus, it can be shown that the single-continuum model of the laboratory block, with the equivalent parameters determined directly from the experiments, reproduces the transport behaviour well for configurations with significant matrix involvement. For configurations with direct fracture connections, the results underline that there are local effects within the experiment that cannot be represented by a single-continuum model.

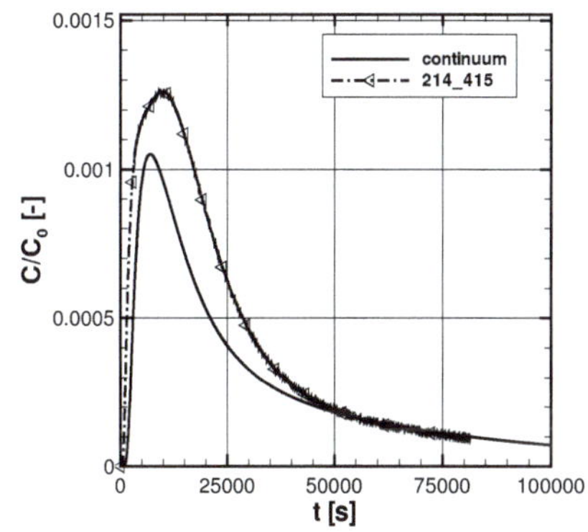

Fig. 4.98. Breakthrough curves for a pulse injection; Ports 214_415.

4.5.2.5 Summary

A continuum model developed for the $90 \times 90 \times 80\,cm^3$-block on the basis of the experimental data emphasizes the importance of interpreting experimental data in order to ensure the quality of the numerical model.

Configurations that are mainly influenced by the matrix may be modeled qualitatively by applying the equivalent parameters determined directly from experimental data. Simulations of configurations with a direct fracture connection indicate that local effects exist that may not be reproduced by a single-continuum model.

5

Field-Block Scale

The field-block scale represents the largest scale which is investigated experimentally in this context. The installation of an in-situ test site allows all experimental approaches to be employed on the field-block scale in a certain area of the real aquifer. In this manner, it becomes possible to check the scale-dependence and assignability of core-scale or bench-scale results to almost natural aquifer conditions. Thus, the idea behind investigations on the field-block scale is to create a link between the experimental results on the core- or bench-scale and the real aquifer environment. Depending on the geological characteristics of the system under investigation and on the kind of experiment, the size of the chosen area has significant influence on the transferability of the evaluated data to the real aquifer. To make this upscaling process possible, the investigated area should ideally describe at least one representative elemental volume (Luckner and Schestakow (1986)) of the whole heterogeneous aquifer system. The size of this REV has to be chosen individually in each case and depends primarily on the characteristics of the fracture system such as orientation, density and length of fractures.

Usually, fractured porous aquifers are not readily accessible without major expenses incurred by drilling and sampling. Therefore, an approach involving an equivalent system above ground was chosen for the tests under controlled boundary conditions. This hard-rock aquifer-analogue approach is based on the investigation of accessible outcrops of the aquifer material in the unsaturated zone such as can be found in a quarry. Their sedimentological and lithological properties can be considered to be analogous to the less accessible groundwater system at greater depth.

5.1 Choice of the Field Block Location

C. Thüringer, M. Weede, H. Hötzl

The test site has to fulfil two conditions: on the one hand, the rock matrix porosity should be exceptionally high and, on the other, the material should be fractured densely and as homogeneously as possible. The local infrastructure had to be kept in mind as well. The search thus soon concentrated on quarries of Middle- or Upper Triassic sandstones in the central and southern parts of Germany.

5.1.1 Regional Positioning

The search for a suitable location of the test site was subject to the following fundamental criteria. For both geological and technical reasons, the best place to set up a hard-rock test site is generally an active quarry. There, it is likely that the chosen cutout of the hard-rock aquifer has been exposed to weathering processes for only a short period of time. In addition, the infrastructure necessary to install and operate a hydrogeological test site is usually available at an operating quarry, with water and power supply and sufficient access roads.

The rock material to be investigated should preferably have both a high matrix porosity and a homogeneously distributed and dense fracture system. Thus, the Triassic sandstones of *Buntsandstein* and *Keuper* in Baden-Württemberg were shortlisted. Table 5.1 gives an overview of the physical properties of the various sandstones initially selected.

A comparison of the data shows that the *Buntsandstein* of the Southern Black Forest (Lahr District) and the *Schilfsandstein* of the *Kraichgau* seem to have the highest matrix porosities of 18 % to 21 %. In contrast, sandstones of the *Odenwald* and the greater part of the *Stubensandstein* of the Pliezhausen area show comparably low porosities of only 13 %.

In order to choose a suitable location for the test site, an inquiry concerning active sandstone quarries in Baden-Württemberg was undertaken with the following criteria:

- densely and homogeneously fractured hard rock with as high a matrix porosity as possible;
- absence of single fractures significantly dominating flow and transport;
- availability of the location and its infrastructure for the duration of the project;
- possibility of excavating a fresh cut from the rock formation.

Initially, all the active sandstone quarries in Baden-Württemberg were registered. Böttger (1989) collected the data of 22 active quarries in all: *Buntsandstein* (7 quarries), *Schilfsandstein* (11 quarries), *Stubensandstein* and

Table 5.1. Physical properties of selected sandstone quarries in Baden-Württemberg (Grimm (1990), Kulke (1967), Ufrecht (1987)).

Formation	Name	Stratigraphic Horizon	Density [g/cm^3]	Porosity [Vol.-%]	Max. Water Absorption [Vol.-%]
Bunt-sandstein	Lahrer Sandstein	su	2.16	18.29	18.29
	Neckartäler, red	su	2.25	15.23	15.23
	Neckartäler, red-white	su	2.33	11.83	11.83
Schilf-sandstein	Maulbronner Sandstein	km2	2.19	15.3	-
	Pfaffenhofener Sandstein	km2	2.00	21.1	-
	Niederhofener Sandstein	km2	2.06	17.9	-
	Weiler Sandstein	km2	2.15	18.1	-
	Mühlbacher Sandstein	km2	2.15	15.9	-
Stuben-sandstein	Pliezhausener Sandstein (Lower Bed)	km4	2.31	13.77	13.77
	White Variety	km4	-	-	13.5
	Yellow Variety	km4	-	-	17.2
Stuben-sandstein	Pliezhausener Sandstein (Upper Bed)	km4	2.24	21.5	20.6

Rötsandstein (together 4 quarries). The resources database of the Geological Survey of Baden-Württemberg (LGRB) listed 21 companies with 26 active sandstone quarries in early 1995. Eleven of these quarries were working *Buntsandstein*, 12 *Schilfsandstein* (Middle Keuper, km2) and one each *Stubensandstein* (Middle Keuper, km4), *Schilfsandstein* (Upper Keuper, ko), and *Lettenkohlensandstein* (Lower Keuper, ku). In addition, some information could be gained from geologists, with experience of the geology of South-Western Germany. In this way, the number of inspected *Stubensandstein* quarries increased, although most of these were either used for the extraction of sand or were abandoned and recultivated.

The inquiry revealed that the highly porous *Schilfsandstein* (Kraichgau) and the *Buntsandstein* (Black Forest) are inconsistent with the aims of the project due to their fracture systems. They are either dominated by very high fracture distances of up to 2 m or by widely gaping fractures. The criterion of a homogeneous and dense fracture system cannot be fulfilled for the field block. This also applies to the *Stubensandstein* layer which is extracted from the quarry near Pliezhausen and which is called 'Lower Bed'. The overlying

'Upper Bed', however, is characterized by quite a dense fracture system with fracture distances of an estimated 0.25 m to 0.5 m (Figs. 5.1 and 5.4). Later on, these values are confirmed by a statistical fracture-system analysis (Section 5.3.1).

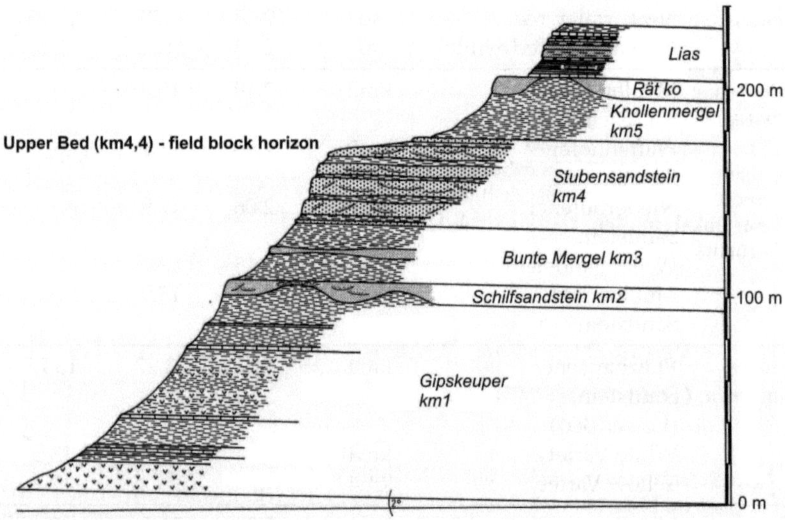

Fig. 5.1. Stratigraphic overview from Middle Keuper to Lower Lias (Einsele and Agster (1986), mod.).

The intensive inquiry into *Stubensandstein* quarries showed quarry *Fauser*, at Pliezhausen-Rübgarten near Reutlingen, to be the only *Stubensandstein* quarry in Baden-Württemberg which is still in use. At other quarries, the sandstone disintegrated during cutting, so that it was used for the extraction of sand (region of Löwensteiner Berge and Mainharder Wald) or limestone production (e.g. quarry *Bayer*, Kernen/Esslingen). Numerous other quarries had already been filled in the course of recultivation processes.

The *Stubensandstein* of the Upper Bed turned out to have more favorable rock characteristics (smaller mean fracture distance) and better rock-matrix properties (expected porosity of approximately 20 %) than the *Schilfsandstein* of the Kraichgau area. Thus, following an inspection of the potential areas, the quarry *Fauser* in Pliezhausen-Rübgarten (Fig. 5.2) was selected to be the location of the hard-rock test site.

Fig. 5.2. Position of the test site near Pliezhausen/Baden-Württemberg.

5.1.2 Local Positioning

After quarry *Fauser* near Pliezhausen has been chosen as the best location for setting up the test site, the exact field-block boundaries had to be fixed. In this context, the following limiting aspects had to be kept in mind:

- limited thickness and lateral extension of the sandstone facies of the Upper Bed due to numerous quick lithofacial changes within the *Stubensandstein* horizon,
- limited amount of space for the test site due to the nearby slope of a neighboring landfill site, north of the quarry area,
- intended and approved enlargement of the quarry to the east with an access road from the south-east.

At quarry Fauser, there are outcrops of the *3rd* and the *4th* *Stubensandstein* (Lower Bed and Upper Bed with field-block horizon). Above these sandstone layers follows the silty *Knollenmergel* (km5) with a thickness of more than 3 m. Only the Upper Bed of the *Stubensandstein* features the high matrix porosity and small fracture distances, so that it was finally chosen as field-block horizon (Fig. 5.1).

During the search for a suitable test site within the quarry area, a spur next to a small biotope pond in the north-western part of the quarry was initially considered (Fig. 5.3). This spur presented an isolated sandstone block with only a rather thin coverage of loose material of 1.7 m to 2.9 m at that time. However, based on a facial analysis (Bengelsdorf, 1997), this area was identified as part of an older slip-off-slope sedimentation. Its petrographic characteristics are the same as those of the Lower Bed (*3rd Stubensandstein*) with fracture distances of up to 2 m. Thus, it turned out to be unsuitable as a location for the test site.

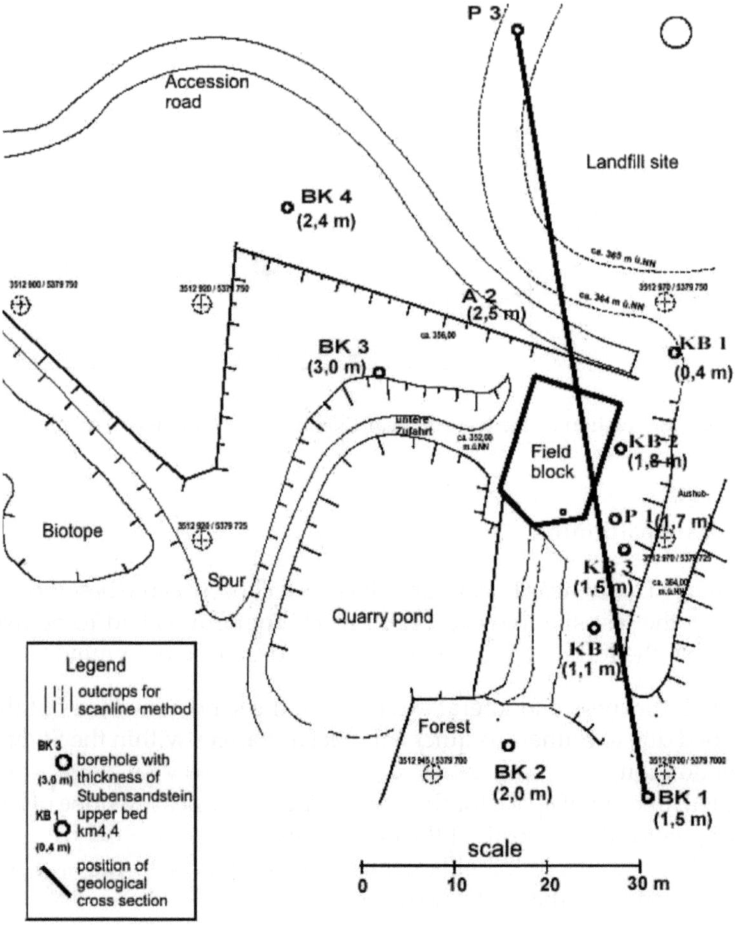

Fig. 5.3. Map of the quarry area with position of the test site when the project started in 1996 and position of geological cross-section (cf. Fig. 5.4).

While the exact test site boundaries were being established, information gained from former quarry-investigation boreholes could be used. Boreholes in the north-eastern part of the quarry indicated a thickness of the 4th *Stubensandstein* of 2.2 m to 2.5 m. An analysis of the variations in thickness of the Upper Bed at the different boreholes and a correlation of these data with the current outcrops led to the assumption that the maximum thickness of this sandstone layer was in the north-eastern part of the quarry with an average of to 2 m to 3 m. This part of the quarry is bordered to the north by a landfill site for building rubble and ground excavation material (Fig. 5.3).

At the test site, the technique of ground-penetrating radar was applied in order to gain additional information on the sandstone thickness. However,

3 m-thick layers of silty and clayey *Knollenmergel* (km5) significantly reduced the penetration depth of the electromagnetic waves. As a result, it was not possible to determine the thickness of the sandstone layer on the basis of the measured reflectograms.

To investigate depth and thickness of the Upper Bed (km4,4 - sandstone horizon) in the north-eastern part of the quarry in more detail, four core drillings were set up at the projected field-block location. The drillings indicated a sandstone thickness of 1.8 m to 2.1 m, quickly decreasing further to the north-east (Fig. 5.4). In spite of a covering of more than 3 m of *Knollenmergel*, this area was finally selected as the best location for setting up the test site.

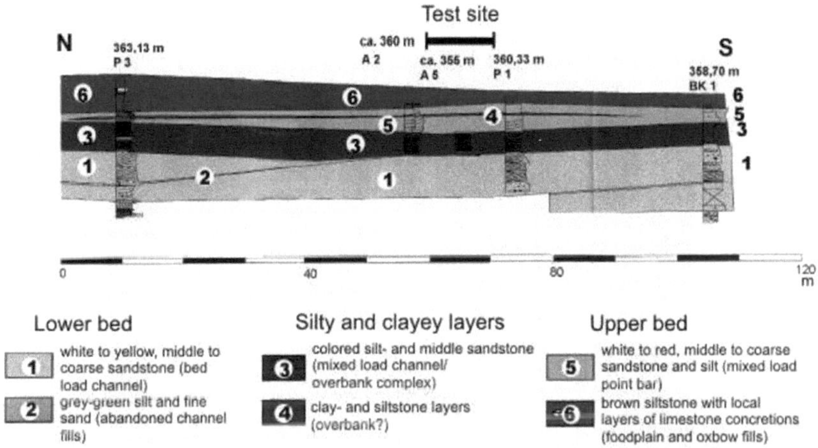

Fig. 5.4. Geological cross-section of the test-site quarry (Bengelsdorf (1997), mod.).

5.2 Preparing a Test Site on the Field-Block Scale

C. Thüringer, M. Weede, H. Hötzl

The preparation of the test site is very extensive and time-consuming and can be divided into two main phases. During the first phase, the rock formation is excavated and shaped for complete encapsulation during the second phase. This sealing is one fundamental requirement for setting up controlled experimental boundary conditions.

5.2.1 Excavating and Cutting

The field block was prepared in several steps. First, the raw field block was carefully excavated step by step during the expansion of the quarry area to the east. For this, excavators were used. At that time, digging in the lower sandstone bed in the south-eastern part of the quarry was scheduled as part of the expansion plan and this fixed the southern boundary of the field block. To the north, the field block is bordered by the foot of the neighboring landfill *Schindhau*. To the west, the raw shape of the field block was delimited by the old quarry digging line and, to the east, the decreasing thickness of the sandstone layer limited the size of the test site.

Finally, the raw field block was shaped like an irregular polygon, excavated on three sides (Fig. 5.5).

Fig. 5.5. Raw field block from the south (length of scale rod in the center of the figure: 2 m).

South of the designated test-site area, a step-by-step removal of the Upper Bed (km4,4) (Figs. 5.1 and 5.4) was carried out. In this way, three more or less parallel outcrop walls were generated. These three walls, approximately 2 m high and 15 m to 20 m long, were used together with one orthogonally oriented wall for a statistical determination of the fracture-system parameters as described in Sec. 5.3.1. Figure 5.3 shows the position of these walls.

The raw, irregular field block was cut to its final shape of an axis-symmetrical oblong pentagon by using a diving saw. Alternative methods, such as a wire saw or a sword saw, had been tested in the run-up to the project and found to be impracticable. When the diving saw technique is applied, the contours of the later field block initially have to be redrawn by vertical boreholes (diameter of boreholes: 210 mm) at a distance of 60 cm to 80 cm to each other. Then the guide rail of a diving-saw system can be lowered into each borehole. The blade of the diving saw cuts into the sandstone from each hole to the right and to the left. By inserting the saw into the next hole, cutting down and repeating this procedure for every single borehole, the final shape of the field block is achieved. Figure 5.6 illustrates the hydraulically powered diving-saw system used at test site Pliezhausen.

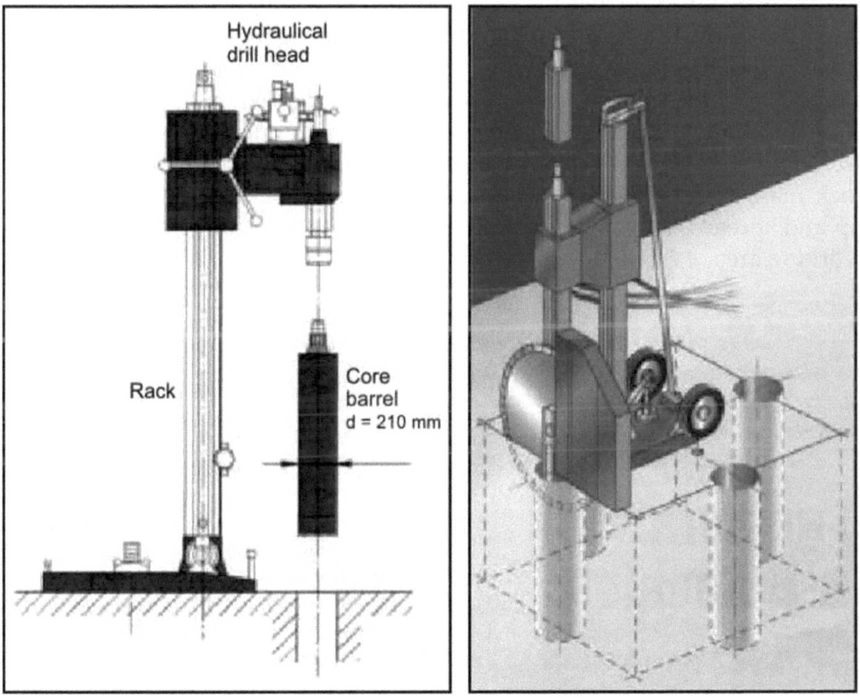

Fig. 5.6. Hydraulically driven diving-saw system used to shape the field block (Hilti Corporation, mod.).

In order to create field block side walls that are as plain as possible, a lateral overlapping of the cuts had to be achieved. To this end, the distance between two neighboring boreholes had to be chosen in accordance with the saw blade diameter. With saw blades with a maximum diameter of 100 cm, the borehole distance was chosen to vary between 60 cm and 80 cm. The core-drilling technique was used in connection with a rinsing system, with water supplied by the small quarry pond.

The cuttings must follow the same cutting plane to avoid a blocking of the diving-saw system and material damage. When the saw is moved to the next borehole, the orientation of the saw blade must not be changed horizontally or vertically. To ensure this in spite of the rough and bumpy sandstone surface, the diving-saw system was mounted on a platform, which could be moved along two parallel rails. This construction was fixed on the sandstone surface using rock dowels, with the platform oriented horizontally. In this way, it was possible to shift the drilling and diving-saw systems a distance of approximately 3 m without lifting them. Afterwards, the whole rail construction was moved and fixed again.

Despite this significant improvement of the saw system, meeting a fracture at an acute angle turned out to be a problem. Sandstone fragments very often broke off the field block and wedged between saw blade and sandstone. This caused a slowdown of the cutting progress and sometimes even a blocking of the saw blade because its cutting plane shifted.

After the raw field block had been cut into a pentagon the cutting slots were widened by hydro-pressure cushions (Fig. 5.7). In this way, separated rock mass could be lifted off the field block. The sandstone was chopped up and spread around the test site by an excavator, creating a level and dry working area.

Fig. 5.7. Widening the cutting slots with hydro-pressure cushions (left) while shaping the field block (right); direction of view: south-east.

At this point, the field block was shaped. However, in order to create controlled flow fields under controlled boundary conditions inside the sandstone, the whole block had to be sealed.

5.2.2 Sealing Process and Installations

One of the main advantages of experiments on the laboratory scale is the possibility to control the boundary conditions completely when hydrogeological tests are carried out. Within the hard-rock aquifer-analogue project, this advantage is transferred to the field-block scale by sealing the whole natural aquifer cut-out.

As already described in Sec. 5.1.1, the *Stubensandstein* formation is characterized by a changing sequence of sandstone beds and silty to clayey layers. Several core drillings from the test-site location investigation phase show that also the field block bases on a clay layer of 3 to 4 m thickness. This clay is nearly impermeable, as revealed by laboratory tests. Thus, it may be regarded as a natural basal sealing of the field block. Consequently, in order to encapsulate the block, only the five side walls (approx. 85 m^2 altogether) and the surface (approx. 90 m^2) have to be covered artificially. In terms of the provided investigation program, the coating should be both water- and gas-proof. During the starting phase of the project, a two-component epoxy-resin system was used. Previous adhesion tests with fragments of the *Stubensandstein* had shown this to be the most suitable system. One fundamental prerequisite for the successful application of the epoxy resin is the complete dryness of the sandstone surface. Thus, the coating could only be applied during completely rainless days. This was done between July and September 1998 in four steps.

At first, loose rock material was removed and every surface of the field block was sand-blasted thoroughly. In a second step, the walls were primed with an epoxy primer in order to improve the adhesion of the resin. Afterwards, the main fracture traces and some points of disruption were filled with cement. Smaller fractures were closed by a grout of resin and silica sand. In the last step, the prepared surfaces of the field block were covered by at least two layers of two-component epoxy-resin coating. The mean thickness of this coating was approximately 5 mm.

Although the coating was applied with great care and despite the promising results already gained with resin coatings in the laboratory, this method proved in time to be completely unsuitable for field conditions. The great differences in temperature, which may be up to 50 °C at the test site Pliezhausen in the course of one year, induced stresses between resin and sandstone. Cracks appeared in the coating after two years and the sealing started to flake. The now unusable resin coating had to be ablated and removed in the following weeks.

The failure of the resin coating illustrates the enormous operational demands on the coating material over the years. Temperature changes may create significant stresses between materials of different coefficients of heat expansion. From then on, a multi-layer coating of special water- and gas-proof concrete with an expansion comparable to that of sandstone was used. The application of this coating can be divided into three main steps:

1. building a basal ring of concrete around the field block as a foundation for the casing walls;
2. erecting casing walls of limestone bricks, enclosing the field block and setting up a core wire frame as a reinforcement,
3. filling in the gap between wall and field block with concrete and sealing the block surface.

During the first construction phase, a 40 cm-thick ring of concrete is cast around the base of the field block. It has to perform two main tasks. On the one hand, it represents the foundation of the side walls of the encapsulation. Without this ring, the construction of the 2.20 m-high supporting walls would have been impossible on the soft and plastic clay layers. On the other hand, a drainage pipe is installed inside the concrete foundation in contact with the sandstone base. This pipe allows a continuous irrigation of the sandy to silty basement of the field block. This is to prevent these layers from drying out and leaking during gas-flow and gas-tracer experiments. When a suitable hydraulic head is connected to this inlet pipe, the whole sandstone block may be saturated from its basement step by step. In addition, drainage pipes are installed inside the concrete ring, through which the seepage water from the outcrop walls behind the field block can be discharged. Figure 5.8 shows the unsealed field block after construction phase 1 with the cast concrete ring as the foundation of the supporting walls. The white tarpaulin acts as a temporary provisional weathering protection during the construction.

What cannot be seen in this photo are lengths of reinforcing wire, which are fixed in the concrete foundation at a lateral distance to the field block

Fig. 5.8. Unsealed field block with concrete ring foundation; direction of view: northeast.

of approximately 20 cm. Reaching once around the whole field block and protruding 15 cm vertically from the concrete strip around it, they provide fixing points for a construction of reinforcing wire, installed in construction phase 2.

During this second phase, retaining walls of limestone brick (40 cm × 10 cm × 20 cm) are erected and the whole field block is enclosed in a cage of reinforcing steel wire. Steel mats (2.50 m × 5.00 m, diameter of wire: 5 mm) are placed and fixed onto the protruding wire strips installed during construction phase 1 so that they overlap (Fig. 5.9).

Fig. 5.9. Reinforcing wire at the north-western edge of the field block and overlapping steel mats, tied by twisted wires.

If the steel mats are higher than the field block, they are bent at right angles and fixed to the steel mats lying on the field-block surface. Reinforcing bar spacers ensure that the steel mats are not in direct contact with the sandstone. In this way, a completely closed cage of reinforcing wire is built, in order to compensate for tensile stresses caused by frost-thew-changes. It should be mentioned that all ground-penetrating radar techniques (Sec. 5.5.2) had been applied before the reinforcing wire was used. The rather close-meshed steel mats make any measurements based on the penetration of electromagnetic waves impossible.

Afterwards, a 2.50 m-high wall of limestone bricks is built, around the whole field block and the steel cage at a distance of approximately 30 to 40 cm (20 cm distance between steel and sandstone). The wall can be adapted by shaping the individual bricks and installations (see below) can be integrated.

During the third and final construction phase, the gap between the retaining walls and the sandstone block is filled with special concrete using a concrete pump. During this process, the wall has to be supported from the outside because of the immense pressure exerted by the concrete when the whole 2.50 m gap is filled up. Ideally, the gap is filled in two to three layers and the concrete should be mixed in such a way that it can flow, but will set quickly. Every concrete layer has to be consolidated accurately in order to avoid significant inclusions of air between the reinforcement wire and the field-block side wall. The block surface should also be covered with concrete layer by layer. Applying this method avoids major inhomogeneities inside the encapsulation. In hot sunshine, the setting process of the concrete on the block surface should be delayed to avoid cracking as a result of over-rapid hardening. This can be done with a wet tarpaulin, for example.

In all, approximately 54 m^3 of special concrete were used to seal the test site. Figure 5.10 shows the completely sealed field block with the retaining walls and the block surface completely covered with concrete.

Fig. 5.10. Completely encapsulated field block; direction of view: south-east.

The completely sealed field block is accessible for hydrogeological experimentation by means of different installations. First, there are several boreholes, with which different investigations can be carried out. The boreholes are the results of core drilling with a diameter of 50 mm, carried out before the field block was sealed. Each borehole is between 1.80 and 2.20 m deep and thus reaches down through the sandstone layer into the underlying impermeable clay. It is necessary to distinguish between the marginal boreholes at the edges of the field block, made at an early stage of the project (Sec. 5.6.2), and the ring of boreholes in the central part of the test site (Sec. 5.6.3). Furthermore, measuring ports are fixed on both the side walls and the field-block surface, forming a regular grid with some additional ports on the traces of some of the main fractures (Fig. 5.11).

Each port consists of a PVC tube with a diameter of 40 mm at the block and 25 mm at the outer end. Without changing the natural aquifer properties, the tubes are fixed to the sandstone using a circular cut of 1 cm depth and

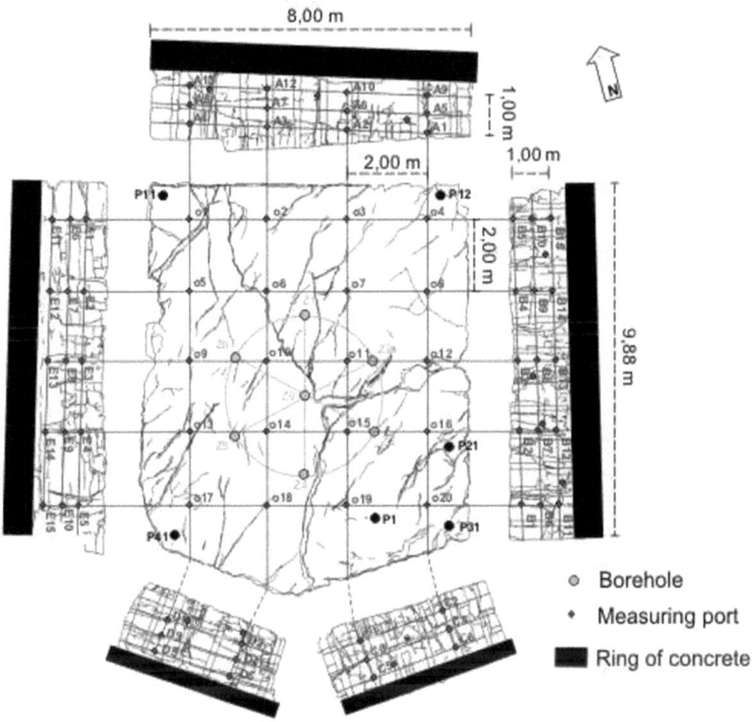

Fig. 5.11. Boreholes and measuring ports used at the field block.

caulking grout. By means of these tubes, the pressure distribution and tracer concentration can be logged, independently of the boreholes. The tubes have to be attached to the sandstone, fixed to the reinforcing steel and be inserted through holes in the retaining wall before the cement is poured. Figure 5.11 displays all the installations available for hydrogeological experimentation at the test site.

5.3 Characterization of the Rock-Matrix and Fracture-System

K. Witthüser, M. Weede, C. Thüringer, H. Hötzl

In order to assess the influence of the structural properties of the porous fractured aquifer on flow and solute transport processes, several structural parameters have to be estimated. Parameters of the fracture system and those of the hard rock matrix can be distinguished. In the following, the methods applied to the *Stubensandstein* field block and the associated evaluated parameters are described.

5.3.1 Statistical Evaluation of Fracture Parameters

In order to gain reliable and meaningful data on fracture allocation and orientation, as many fissures as possible have to be registered. The scanline method (Priest, 1993) is a feasible way of achieving this aim. In this method, the investigation is carried out at easily accessible outcrop walls, where a preferably representative part of the fracture system is visible. Depending on the fracture system, these walls should cross the fractured aquifer in several different orientations. In the case of the field block, three parallel outcrop walls and one orthogonally oriented wall can be analyzed (Fig. 5.3). First, the outcrop walls are covered with a grid of parallel marking lines. Then, every line is systematically scanned from one end to the other and analyzed for fractures which cross the line. For every single one of these fractures, parameters such as fracture orientation, fracture length (fracture-trace length at the outcrop wall), fracture spacing (distance between two neighboring fractures on the scanline), aperture and fracture characteristics (e.g. roughness, coating) are systematically registered.

With this method, about 300 fractures are recorded at the test site. After favored orientations among this large amount of fractures have been decided on, elements with an orientation within a certain limited deviation are assembled in clusters. Each cluster represents one main fracture orientation, to which properties such as aperture or fracture length can be statistically assigned. In this manner, a statistical model of the field block is obtained.

In addition to the scanline technique, another, less labor-intensive fracture registration method is used, namely stereophotogrammetric shooting. The method requires a stereometric camera and an analytical plotter. Initially, photographs are taken from all sides of the field block, using a stereometric camera. Fracture traces, visible in the stereo pictures, are then analyzed by an analytical plotter. At the test site, in all five walls undergo stereophotogrammetric fracture-orientation evaluation. Among these are two side walls of the raw field block and four exposed walls of the quarry, partly orthogonally oriented. The stereo pictures are analyzed by an analytical plotter at the Department of Photogrammetry and Remote Sensing at Karlsruhe University. Figure 5.12 shows a 3D view of the unsealed field block with all the fracture traces determined by the stereophotogrammetric shooting.

With the cosine-exponent-weighting method, the intersection points of the normal vectors of the fracture traces with the lower hemisphere are calculated. The result is visualized in a two-dimensional contour plot, as shown in Fig. 5.13. The main fracture-trace orientations are determined by identifying the peak discontinuity densities and choosing an acceptable angle of deviation for each cluster. The identified clusters are assumed to satisfy a Fisher distribution.

The measurement of fracture-trace lengths along an outcrop results in several sampling errors (Kulatilake *et al.* (1993), Pointe and Hudson (1985), Baecher *et al.* (1977)). These errors include the proportional length bias, the

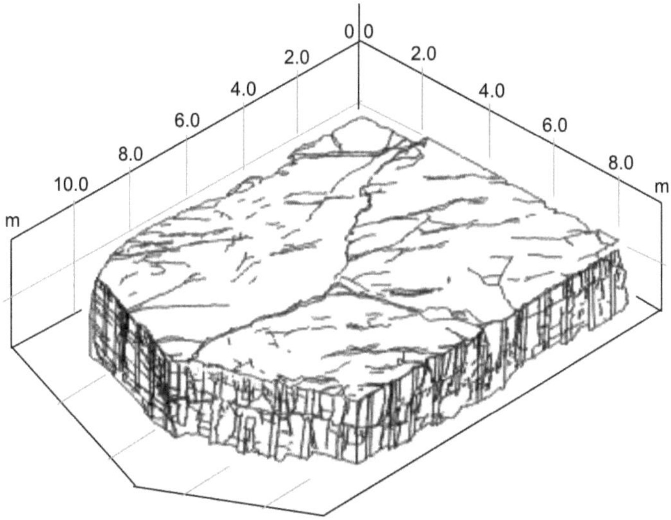

Fig. 5.12. 3D view of the field block, with stereophotogrammetrically registered fracture traces and boreholes; direction of view: north-west.

censoring bias and the truncation bias. While intensive field surveys have shown that the truncation bias is negligible for a short cut-off length, all other sampling errors can be treated and corrected mathematically. Since the trace lengths at the outcrop are assumed to follow an exponential distribution, the Erlang-2-distribution is fitted to the empirical scanline data for each cluster, always yielding high levels of statistical significance. The quality of this adaption can be tested by applying the Kolmogoroff-Smirnov test (Hartung, 1995).

The fracture spacings evaluated by the scanline technique are adapted by an exponential distribution, which can be derived for a random allocation of fracture elements (Priest and Hudson, 1976). For a further examination of a possible areal correlation, fracture-density variograms (Chilès and de Marsily, 1993) are calculated. The mean three-dimensional fracture density, i.e. the mean fracture plain [m^2] per rock mass volume [m^3] is calculated according to Chilès and de Marsily (1993).

Altogether, 947 fracture elements are measured at the test site, 300 by the scanline method and 647 by the stereophotogrammetric method. Fig. 5.13 shows the cluster orientation of all registered fractures. The associated statistical parameters are listed in Tab. 5.2. In the diagram, three main directions can be identified. There are two nearly vertical fracture orientations (clusters II and III), striking 56°and 139°, and one almost horizontal bedding plane (cluster I).

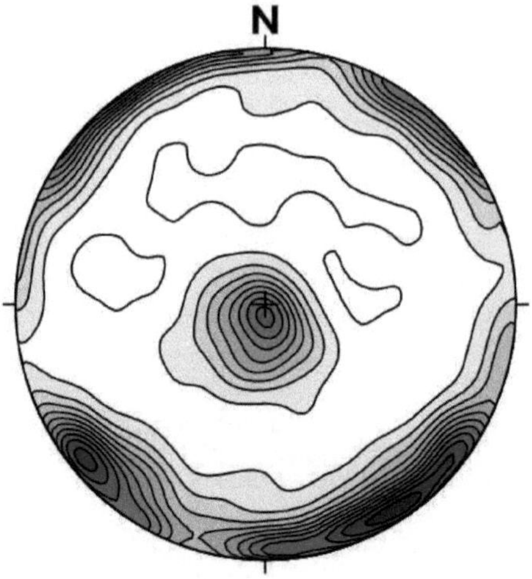

Fig. 5.13. Fracture-orientation-distribution diagram of Pliezhausen test site.

Table 5.2. Statistical parameters of the three main clusters.

Cluster	Dip Direction	Dip	Spherical Aperture
I	201°	85°	11.22°
II	146°	7°	12.05°
III	229°	8°	10.20°

Looking at the fracture-orientation-distribution diagram while distinguishing between scanline- respectively shooting-direction, the blind zones (Terzaghi, 1965) of the particular measurements can be clarified. Evaluating the orientation of the bedding plane is not possible by this means because it is parallel to the horizontal stereophotogrammetric shots. Thus, a correction of the geometric error is not practicable. Only by evaluating the data in three directions in space using the scanline technique or the stereophotogrammetric method does a realistic statistical description of fracture orientation become possible.

The empirical fracture-distance distribution, measured by the scanline method, can be adapted by an exponential distribution very well. The maximum deviation between the two distribution functions is only 0.0446. The mean fracture distance according to all scanline measurements is 0.231 m. The semivariograms of fracture density do not indicate a spatial correlation between the fracture intersections and the scanlines (Fig. 5.15).

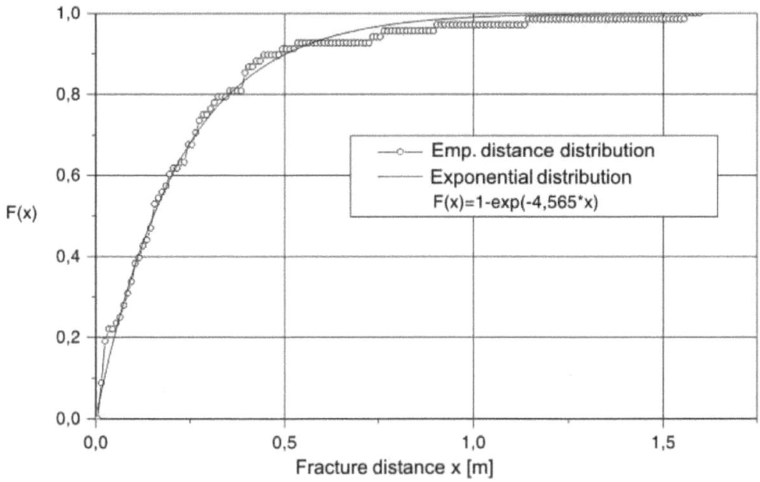

Fig. 5.14. Fracture-distance distribution at the test site (scanline technique).

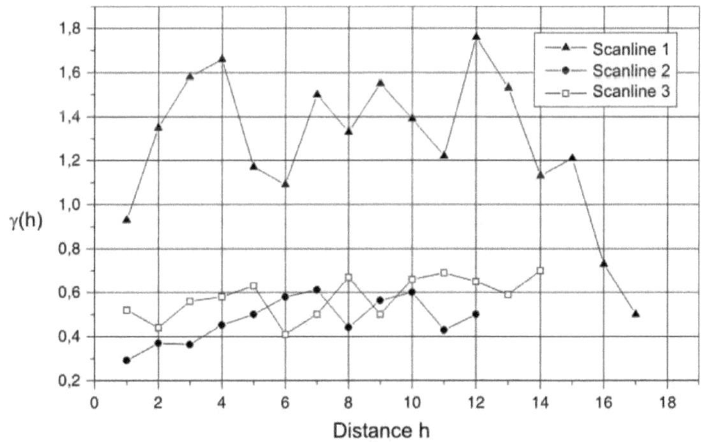

Fig. 5.15. Semivariograms of fracture density according to different scanlines.

A random fracture distribution in space is confirmed by a highly significant adaption of the expected exponential distribution for random fractures and the random variograms of fracture density. The mean fracture density is $11.337\,m^2/m^3$.

Fitting the horizontal and vertical fracture-trace lengths to Erlang-2-distributions shows high statistical significance. Figure 5.16 additionally describes the exponential distributions of the whole outcrop, corrected with regard to the proportional length bias.

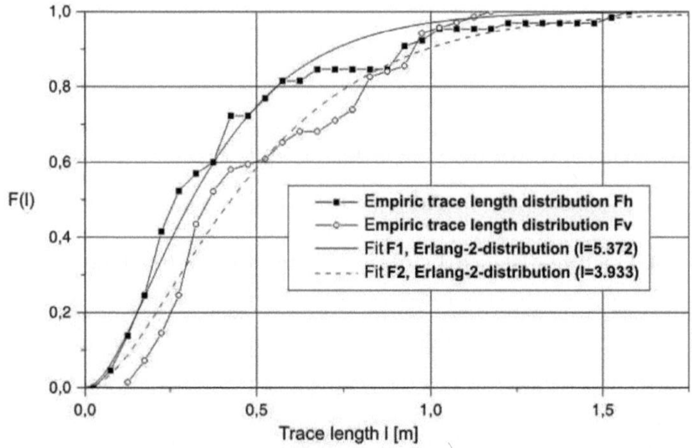

Fig. 5.16. Fracture trace lengths of the scanline measurements.

Describing the fracture-trace lengths according to cluster affiliation does not yield any significant difference in the distribution parameters. Fractures preferentially end at sedimentary boundaries, so that a common distribution function for two almost vertical fracture clusters and the horizontal bedding can be assumed.

For both the side walls and the surface of the field block, visible fracture traces have been recorded (Fig. 5.17). A 3D illustration of the test block can be created with the available information (Fig. 5.12).

5.3.2 Determination of Rock-Matrix Properties

Not only the fracture system, but also the properties of the hard-rock matrix have great influence on flow and transport processes. Thus, hydraulic and physical parameters such as density, porosity or permeability are determined in accompanying laboratory tests. Besides, X-ray diffractometry is an effective tool for estimating the retarding influence of fracture coatings and fracture fillings.

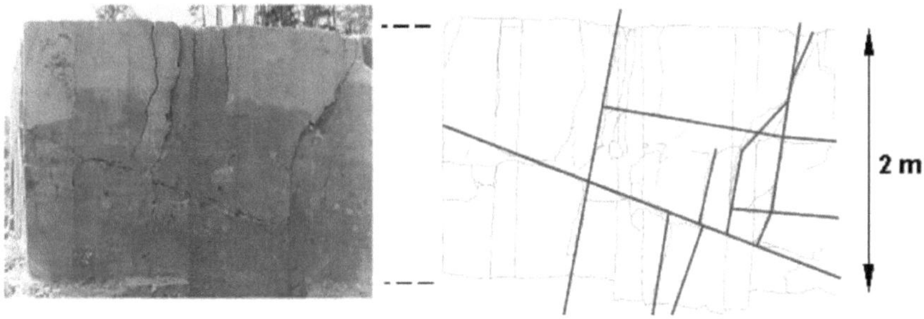

Fig. 5.17. Fracture traces on photo and digital plot (east face of the block).

5.3.2.1 Mercury Porosimetry

Mercury porosimetry is a popular method for determining the porosity of any solid material. Apart from the total porosity, the mean pore diameter and the pore-diameter distribution can also be evaluated. The mode of operation is based on the high surface tension (0.480 N/m) and the large contact angle of mercury. According to Washburn (1941), mercury, as a non-wetting liquid, only intrudes into pores if pushed by an external force. If the pores are pictured as a bunch of capillaries, the pressure which is necessary to push the mercury into a pore is inversely proportional to the pore diameter. This dependence is described by the WASHBURN equation:

$$d = -\frac{4\gamma \cdot \cos\theta_C}{p} \tag{5.1}$$

During the measuring process, the sample is surrounded by a gravimetrically determined volume of mercury. Then the whole sample chamber is pressurized step by step. With continuously increasing pressure, the volume of the mercury which is forced to intrude into the sample pores is recorded at every pressure step. According to the WASHBURN equation, every pressure can be associated with a certain diameter of the pores which get filled. Thus, it is possible to make a statement on porosity and pore-diameter distribution by measuring the volume of intruded mercury (Fig. 5.18).

The porosity of the sandstone matrix is determined, using mercury porosimetry on samples from different parts of the field block. The results range from about 19 % in the upper part of the block up to 23 % near the finely grained basement. In this context, it should be noticed that these results are not necessarily representative of the effective porosity, the volume accessible for groundwater movement. According to Busch *et al.* (1993), a flow of water due to gravity cannot take place through pore channels smaller than 0.008 mm (Fig. 5.18). Hydrostatic forces make all water inside these channels adhere to the walls, so that any flow is averted. The distribution of the

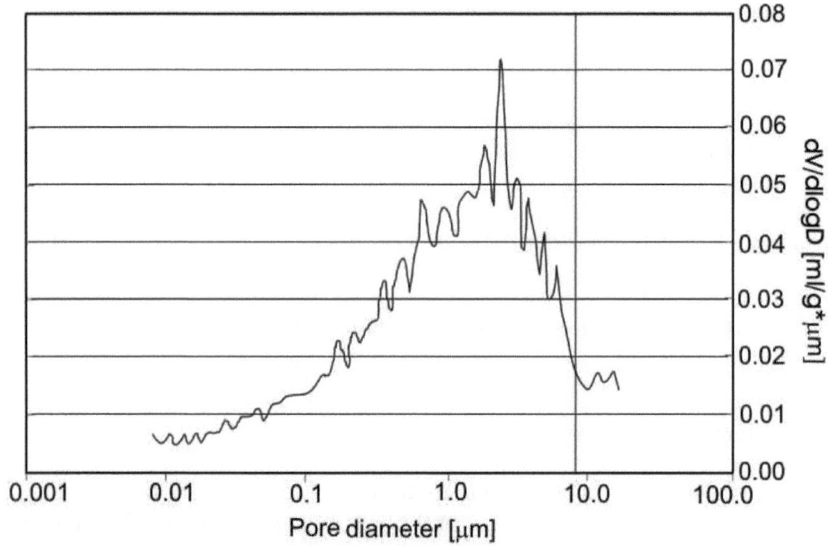

Fig. 5.18. Pore diameter distribution of Stubensandstein sampled at test site Pliezhausen, evaluated by mercury porosimetry; total porosity of this sample: 21.2 %.

sandstone pore diameters shows that the pore diameters in the field-block matrix are mostly between 0.001 and 0.005 mm (Fig. 5.18). These pores may not be added to the effective porosity. Besides, there are a certain number of dead-end pores which are not available for groundwater flow.

In summary, it should be noticed that the measured rock-matrix porosity can be seen as a guiding value. Even if these results are feasible for the assessment of diffusive transport processes between fracture and matrix on the one hand, they must on the other hand not be equated blindly with the hydraulically effective porosity of the rock matrix.

5.3.2.2 Permeability

In order to estimate the influence and the role of the rock matrix during flow and transport processes in fractured aquifers, the permeability of a compact, unfractured sandstone core sample should be evaluated in the laboratory by DARCY experiments. This approach simplifies the rock matrix to a homogeneous medium. The influence of dead-end pores, caused by the cementation of the hard rock is, together with the grain size distribution of the rock, the most significantly limiting factor for the hydraulic permeability.

The method is applied to completely saturated cylinder samples (∅ = 10 cm) installed in pressure chambers. It is very time-consuming to achieve as complete a saturation of the cylinder samples as possible. The best results are obtained rather quickly, by at first drying the samples in a drying chamber, until a constant weight is reached. Then, only the base plane of the cylindrical sample is irrigated with the aid of filter plates. Thus, water is pulled into the samples from the bottom by capillary forces, displacing trapped air upwards. When the water content inside the samples is stable, the water level is raised slightly and a new balanced water content is awaited. During the entire saturation process, the increasing water content of the sample should be monitored gravimetrically. In this way, the sandstone samples of 5 cm and 10 cm length can be saturated to approximately 90 %, during a period of only 3 to 4 days.

When a satisfactory saturation is reached, a constant hydraulic gradient is created, which causes a constant flow through the samples (Fig. 5.19). On the assumption of a homogeneous porous matrix throughout the sample, the hydraulic conductivity can be determined on the basis of the DARCY'S law by measuring the flow rate depending on the hydraulic gradient and the sample dimensions.

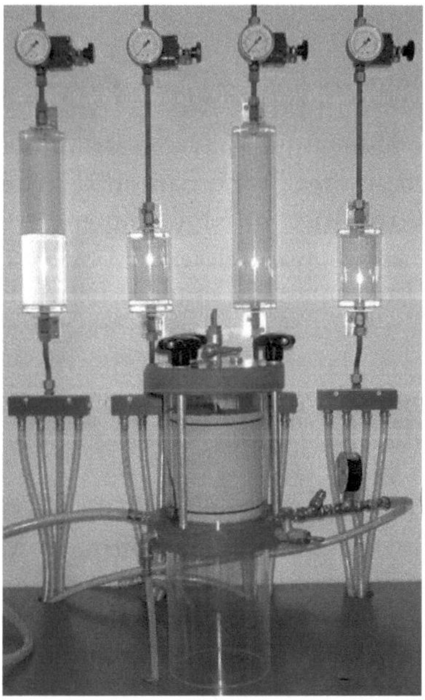

Fig. 5.19. Experimental set-up of the DARCY experiment, applied to samples of the field-block rock matrix.

The results of applying this experiment to core samples of the *Pliezhausen* test site were hydraulic conductivities varying from about $1.0 \cdot 10^{-7}$ m/s to $1.0 \cdot 10^{-9}$ m/s. Table 5.3 exemplifies the measured pressure gradients, flow rates and sample dimensions.

Table 5.3. Results of the DARCY experiment, applied to cylindrical samples of *Stubensandstein* (d = diameter, l = length).

Sample dimension	Cross-sectional area [m^2]	Flow distance [m]	Hydraulic gradient [-]	Flow rate [m^3/s]	Permeability coefficient [m/s]
cylinder d = 9.8 cm l = 10.3 cm	$7.54 \cdot 10^{-3}$	0.103	13.235	$4.98 \cdot 10^{-10}$	$4.9 \cdot 10^{-9}$
cylinder d = 9.8 cm l = 10.1 cm	$7.54 \cdot 10^{-3}$	0.101	14.265	$6.99 \cdot 10^{-9}$	$6.5 \cdot 10^{-8}$
cylinder d = 9.8 cm l = 5.2 cm	$7.54 \cdot 10^{-3}$	0.052	18.854	$6.00 \cdot 10^{-9}$	$4.22 \cdot 10^{-8}$
cylinder d = 9.8 cm l = 5.1 cm	$7.54 \cdot 10^{-3}$	0.051	19.608	$3.15 \cdot 10^{-8}$	$2.13 \cdot 10^{-7}$

5.3.2.3 X-Ray Diffractometry

Fracture coatings and fillings may have great influence on the retardation potential of a fractured aquifer. To prepare the clay analysis, samples taken at the field test block are dried at 60 °C and milled in an agate mill. The qualitative mineral inventory was determined by X-ray diffractometry, using the JCPDS files of the International Center of Diffraction Data (ICDD, 1980). This method is based on the diffraction of monochromatic X-rays hitting the sample at the crystal grid layers. With the BRAGG equation, the grid-layer distance can be designated by means of the angle of maximum diffraction. These different layer distances of the crystal grid are characteristic for each mineral.

By comparison with the JCPDS files (ICDD, 1980), the mineral spectrum can be analyzed and identified. X-ray diffractometrical investigations of powder samples (particle size < 125 mm) at an angle range between 0° and 63° provides a review of the mineral inventory. A further identification of clay minerals occurs with air-dried texture compounds, so-called LuBOs. 1 g of the powder sample (grain size < 125 mm) is charged with 100 ml of distilled water, sodium pyrophosphate is added and this is then dispersed in an ultrasonic bath for 10 min. After separation of the clay fraction by centrifugation, the residual sample suspension is pipetted onto microscope slides. During the slow drying procedure, the clay mineral layers adjust in a horizontal position and thus increase the base reflections significantly. Further investigations occur in connection with the following usual clay-mineralogical procedures (Jasmund and Lagaly (1993), Wilson (1987)). To identify swellable clay minerals, the texture samples are vaporized with ethylene glycol in the drying chamber at 60 °C for 96 hours after the first test series. The glycol saturation displaces the base reflections of swellable clay minerals so they can be distinguished from non-swellable species.

In order to differentiate between kaolinite and chlorite, the texture samples are cauterized in a muffle furnace at 550 °C and then reinvestigated with X-ray diffractometry. During the cauterization of the kaoline, the grid is destroyed completely while a diffraction peak caused by chlorite persists. The measurements are carried out with a SIEMENS D 500 X-ray-diffractometer with copper-potassium-tube.

To interpret the texture sample diffractograms, the benchmarks of Thorez (1975) and Thorez (1976) are used. The fraction of swellable clay minerals is determined by methylene blue adsorption (Hofmann *et al.*, 1976). In this method, the fraction of swellable clay minerals can be evaluated by the absorption of methylene blue, using a calibration value (Wyoming bentonite with 84 % by weight smectite).

The fracture fillings taken from a side wall at the raw test site and from borehole SB 46 at about 1.0 m depth are compared. X-ray diffractograms of both powder samples show a significant content of swellable clay minerals at an angle range from about 1° to 9°. These minerals are characterized by a grid-layer distance of 10 Å to 18 Å (illites, chlorites, smectites, vermiculites). The fraction of swellable clay minerals is about 30 % to 40 %, determined by the consumption of methylene.

The fraction of kaolinite in the side wall sample is estimated to be lower than 5 % on the basis of its peak face at 12.4°. By the cauterization of the normal LuBO sample, the 124°peak (basal distance 7 Å) disappeared completely. Thus, a content of chlorite in this sample can be ruled out. Halloysite is also apparently not present as a corresponding peak in the powder-sample diffractogram could not be found at 8.8°.

Small quantities of quartz could be verified at 20.9° (basal distance 4.26 Å) and 26.7° (3.34 Å). The fraction of kaolinite and feldspar is assumed to be lower than 5 % each, that of quartz should be lower than 10 %. The loss by combustion reaches about 4 %. The calcium carbonate content is very low (0.7 %).

Illite with its diffraction angle of 8.8°(basal distance 10 Å) can be verified in higher concentrations in the LuBO samples. It shows a second peak at 17.7°(5 Å) together with small amounts of muscovite at 19.95°(4.47 Å).

The fracture material taken from borehole SB 46 shows a higher fraction of quartz of about 30 % (Fig. 5.21) than the side wall sample (Fig. 5.20), which is caused by the higher content of sand in this material. This sample contains mainly illite and mixed-layer clay minerals.

The powder sample peak measured at 8.8° (basal distance 10 Å) is not only caused by illite but also by hydrated halloysite. This is confirmed by the 12.4°peak of the normal, air-dried sample, which cannot be caused by kaolinite because of the missing 7 Å-signal of the powder sample. In fact, this identifies the dehydrated halloysite, whose basal distance is reduced from

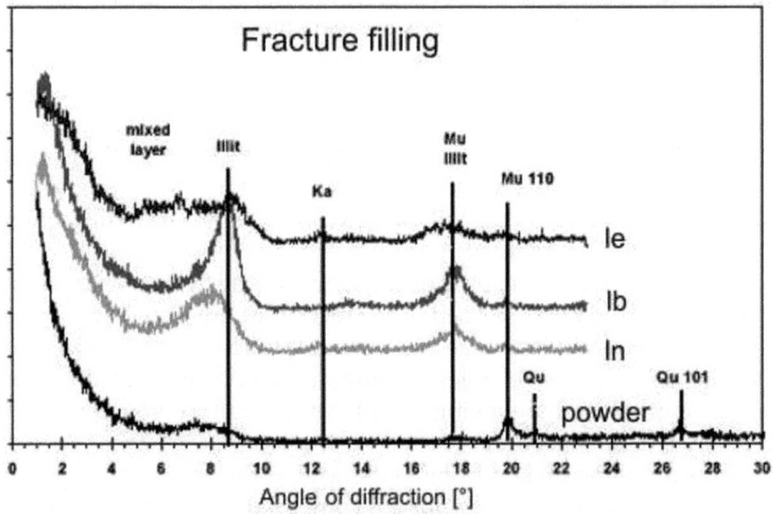

Fig. 5.20. X-ray diffractogram of a fracture coating at the field block side wall (le = LuBO ethylene glycol vaporized, lb = LuBO cauterized, ln = normal LuBO).

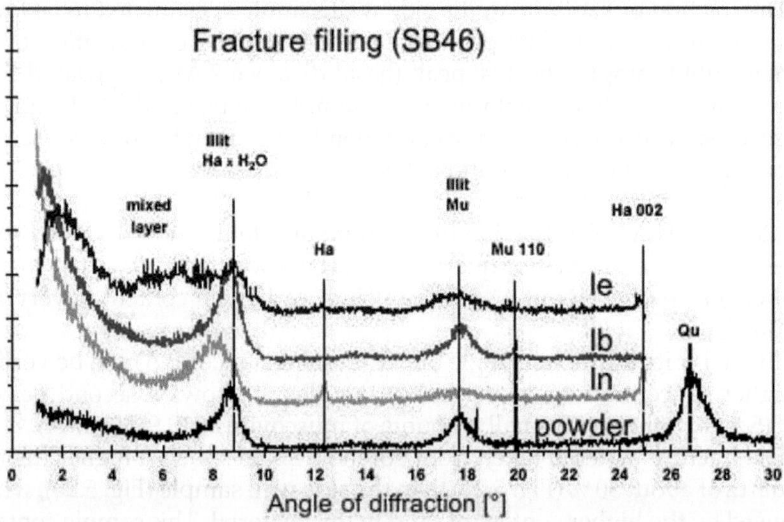

Fig. 5.21. X-ray diffractogram of fracture coating inside investigation borehole SB 46 (le = LuBO ethylene glycol vaporized, lb = LuBO cauterized, ln = normal LuBO).

10 Å (diffraction at 8.8°) to 7 Å (diffraction at 12.4°) by the loss of water during the drying process (Schachtschabel *et al.*, 1992).

The appearance of montmorillonite is indicated by a diffuse increase of concentration at about 5° within the ethylene glycol vaporized samples of fracture fillings. The absence of one well-defined single peak shows that there is presumably no pure montmorillonite but a mix of different smectites and mixed-layer minerals. The diffractograms suggest that the swellable clay minerals make up about 15 %.

5.4 Geostatistical Analysis of the Fracture Lengths and Fracture Distances

A. Silberhorn-Hemminger, Y. Rubin, R. Helmig

In Sect. 5.3.1, the results of a univariate evaluation of fracture-trace parameters such as orientation, length, aperture, distance, and fracture density are presented. Here, a geostatistical analysis, which includes the spatial distribution of the investigated parameters, is conducted. This leads to an improved understanding of the spatial variability of discontinuities in fractured media. In a detailed geostatistical analysis of the fracture density and the fracture orientation, La Pointe and Hudson (1985) show that fractured media can follow a systematic spatial pattern. Further geostatistical investigations in fractured media are done by Lunn and Mackay (1996). They analyze the vertical and horizontal correlation structure of the permeability of core sample data in fractured media. Desbarats and Bachu (1994) investigate the heterogeneity of the hydraulic conductivity of a sandstone aquifer using a geostatistical approach. The investigation considers the core scale up to the regional scale.

In the following sections, the spatial variability of the fractured media is analyzed by considering fracture-trace lengths and fracture distances. The fractures are recorded on a fracture-trace map of the surface of the field block shown in Fig. 5.22. First, the analysis concept of the spatial variability of the fractures using a modified scanline method is introduced in Sect. 5.4.1. The geostatistical analysis of the side walls is presented in Sect. 5.4.2. Finally, the results of the analysis are discussed in Sect. 5.4.3.

5.4.1 Strategy

The field block is characterized by high matrix porosity and is densely fractured. Three main fracture clusters are identified. Figure 5.22 clearly shows these three clusters on the side walls and on the top of the block: a strong horizontally oriented and two strong vertically oriented clusters. The data of the fracture clusters can be seen in Table 5.2. A detailed presentation of the field block including the geological characterization and the description of the material and hydraulical properties can be found in Sect. 5.3.

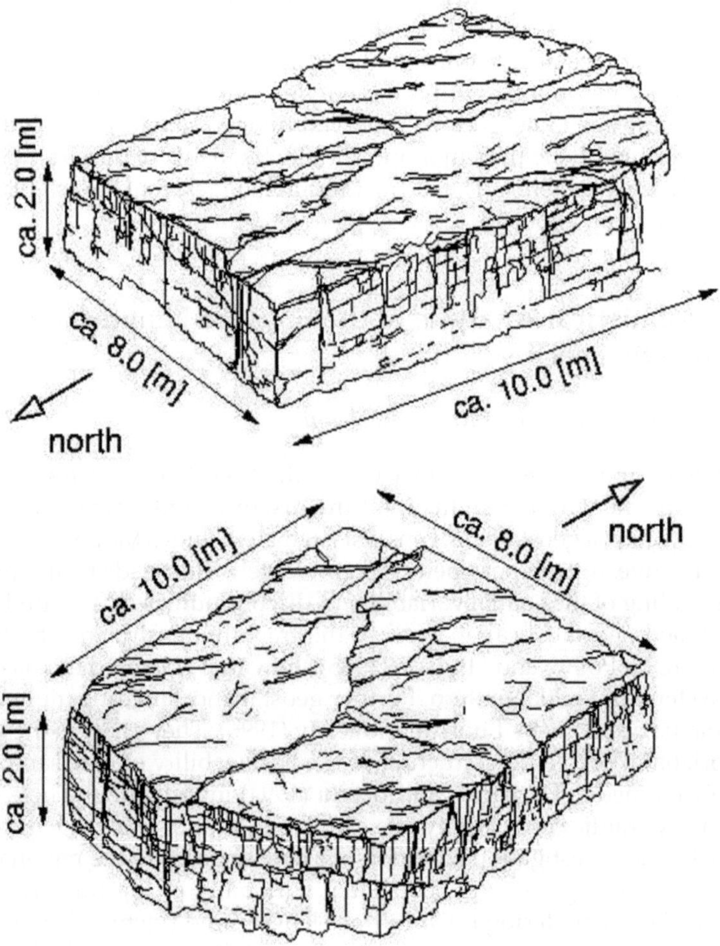

Fig. 5.22. View of the field block from the north-west and the south-east.

The spatial analysis of the fracture system is conducted using a modified scanline method. A set of parallel scanlines is distributed over the investigation domain. The scanlines themselves are divided into segments of equal length l. The scanlines are arranged perpendicular to one main fracture cluster. Furthermore, only fractures which are almost perpendicular to the scanlines are investigated with respect to intersection points with the scanlines. This allows a separate variogram investigation for each fracture cluster. Along the single scanline segment, the number of intersection points between fracture traces and the scanline segment itself is recorded.

In order to investigate the spatial variability of fractures (fracture arrangement and extension), the fracture frequency $FF(x_i)$ and the indicator variable $I(x_i)$ are analyzed. Here, the fracture frequency is indicated as the number of intersection points *(fracture - scanline segment)* which is recorded along the scanline segment with center x_i. The indicator variable $I(x_i)$ represents the existence or non-existence of one or more intersection points *fracture - scanline segment* which are recorded along the scanline segment with center x_i

$$I(x_i) = \begin{cases} 1 & \text{intersection "fracture - scanline" exists} \\ 0 & \text{no intersection "fracture - scanline" exists.} \end{cases} \quad (5.2)$$

The property of *fracture* is not continuously present in space. When a certain point in space is observed, a fracture either exists or not. Therefore, the indicator variable seems to be the right choice in order to investigate the spatial variability of fractures in space. Additionally, observations of the field block show that single fractures often do not appear as a single trace. They appear as two traces laying very close together. The fracture maps in Figs. 5.23 and 5.24 of the south-east, south-west and west wall show this very clearly. The advantage of the indicator variable is that two fracture traces which actually represent one single fracture are interpreted as one single fracture. However, the investigation of the variable fracture frequency interprets the two traces as two single fractures. The disadvantage of the indicator variable becomes clear when larger scanline segments are used. The indicator variable means a loss of information: two or more real fracture traces cannot be interpreted correctly, because the indicator variable just records the existence or non-existence of intersection points, but not the number of intersection points. Therefore, the investigation of the variable fracture frequency is preferable for larger scanline segments.

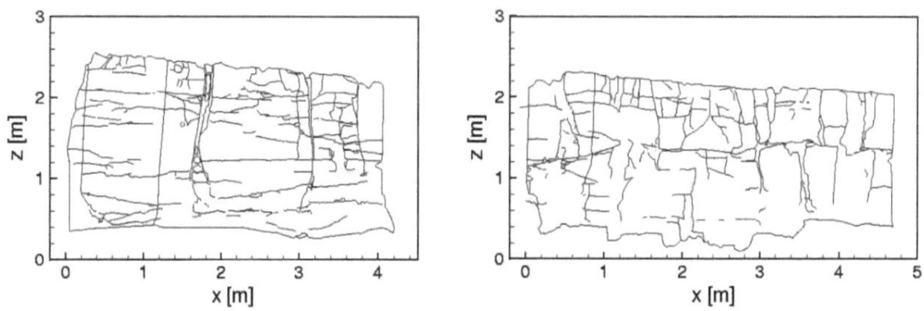

Fig. 5.23. Fracture map of the south-west and the south-east wall.

Within this study, different lengths l of scanline segments are investigated ($l = 0.04\,\text{m}, 0.10\,\text{m}, 0.20\,\text{m}, 0.30\,\text{m}$). With smaller scanline segments, the natural fracture structure and the small-scale behavior can be recorded in detail. The use of larger scanline segments leads to a higher smearing of the small-scale behavior and to the indication of spatial patterns which can be found

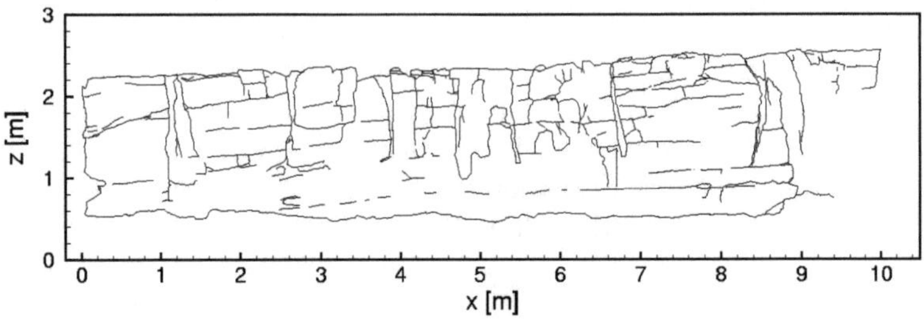

Fig. 5.24. Fracture map of the west wall.

on larger scales. The connection between fracture system, set of parallel scanlines, scanline segments, fracture frequency, and indicator variable is shown in Fig. 5.25.

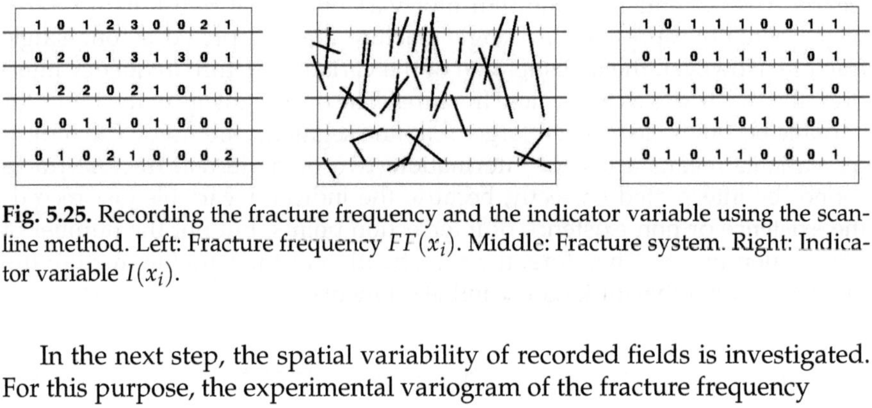

Fig. 5.25. Recording the fracture frequency and the indicator variable using the scanline method. Left: Fracture frequency $FF(x_i)$. Middle: Fracture system. Right: Indicator variable $I(x_i)$.

In the next step, the spatial variability of recorded fields is investigated. For this purpose, the experimental variogram of the fracture frequency

$$\gamma_{\mathrm{FF}}(h_i) = \frac{1}{2}\frac{1}{n(|h_i|)}\sum_{\alpha=0}^{n}\left[FF(x_{i_\alpha}+h_i) - FF(x_{i_\alpha})\right]^2 \tag{5.3}$$

and the experimental variogram of the indicator variable

$$\gamma_{\mathrm{I}}(h_i) = \frac{1}{2}\frac{1}{n(|h_i|)}\sum_{\alpha=0}^{n}\left[I(x_{i_\alpha}+h_i) - I(x_{i_\alpha})\right]^2 \tag{5.4}$$

are determined. Here, h_i is the separation vector, n is the number of pairs of distance $|h_i|$, and α is the index of the sample number.

The experimental variograms are determined in the directions parallel and perpendicular to the scanlines. The variograms are calculated using *GSLIB – Geostatistical Software Library* (Deutsch and Journel, 1992).

5.4.2 Geostatistical Analysis of the Side Walls

The fracture traces which can be detected on the side walls and the top of the field block are analyzed as described in Sect. 5.4.1. The fracture-trace map of the five side walls (north, east, south-east, south-west and west), which have uneven surfaces, is taken in three-dimensional spaces and then it is projrcted into two-dimensional planes. The error between the real and the projected surface is negligible.

The analysis of the spatial variability of the indicator variable and the fracture frequency is done by a variation of the parameters shown in Table 5.4.

Table 5.4. Experimental variogram investigation: variation of the parameter.

Aim of investigation	Scanline direction	Distance d between scanlines	Length l of scanline segments	Direction of separation vector h_i
Vertical fractures	Horizontal	0.10 m	0.04, 0.10, 0.20 m	h_i = (x,0), $\parallel$ to scanlines h_i = (0,z), $\perp$ to scanlines
Horizontal fractures	Vertical	0.20 m	0.04, 0.10, 0.20 m	h_i = (0,z), $\parallel$ to scanlines h_i = (x,0), $\perp$ to scanlines

The detailed analysis of the variogram investigation is shown later for the segment length $l = 0.10$ m. The results of the variogram investigation for the segment lengths $l = 0.04$ m, 0.10 m, 0.20 m are discussed in Sect. 5.4.3.

5.4.2.1 Horizontal Scanlines: Segment Length $l = 0.10$ m

The indicator and fracture-frequency fields of the south-west and the west walls can be seen in Figs. 5.26 and 5.27. The black cells of the indicator field represent the indicator value $I(x_i) = 0$ and the white cells represent the indicator value $I(x_i) = 1$. Comparing the variable fields of the south-west and the west wall with the real fracture traces of the two walls (Fig. 5.23, left, and Fig. 5.24) shows the basic structure of the fracture traces within the variable fields. A similar result is obtained using a scanline-segment length of $l = 0.04$ m (not shown here); however, for the shorter segment length, smaller scale variations are captured.

The indicator fields no longer show the small scale information. On the one hand, the fields only show the *no existence / existence* of one or more intersection points along a scanline segment in accordance with equation (5.2). On the other hand, the fracture-frequency fields represent the small scale differences indirectly by a higher fracture-frequency value. The fracture-frequency field of the south-west wall shows this clearly. The number of segments with a fracture frequency of $FF(x_i) \geq 2$ is quite high compared to the total amount of segments with a frequency of $FF(x_i) > 0$. One can even find segments with a frequency of $FF(x_i) = 5$. The fracture-frequency field of the west wall includes a high number of segments with a fracture frequency $FF(x_i) = 1$ and $FF(x_i) = 2$. The number of segments with a frequency $FF(x_i) > 2$ is relatively small compared with the total number of segments with a frequency of $FF(x_i) > 0$. The number of segments without intersection points (fracture frequency $FF(x_i) = 0$) is equal for both types of variable field. The trend of smearing the smaller scale differences with larger segment length becomes more obvious for the variable fields of segment length $l = 0.20$ m.

Figures 5.28 and 5.29 show the experimental variograms of the indicator variable and the fracture frequency. Some of the variograms of the investigation with vector h_i=(x,0) parallel to the scanlines (Fig. 5.28) show a slightly oscillating and periodic behavior. In order to infer a periodic behavior from the variogram, the shape of the graph should be more distinct. Hence, the behavior represented by the variograms is interpreted as a nugget-effect behavior. This means that the appearance of one or more fractures and their spatial location perpendicular to their extension (fracture trace, fracture plane) is independent of the neighboring fractures. There is no correlation. The appearance of the fractures is based on a random process.

Figure 5.29 shows the experimental variograms with vector h_i=(0,z) perpendicular to the scanlines. The variograms focus on the spatial fracture extension, the length of the fracture traces. The graphs of the two investigation variables show very similar characteristics. The curves increase strongly for small separation distances $|h_i|$. For larger separation distances $|h_i|$, the rate of increase is reduced until it reaches a horizontal plateau for a separation distance of about $|h_i| = 0.70$ m. The variograms show a correlated behavior approximately up to the distance $|h_i| = 0.70$ m. Within this range, the spatial extension of the fractures is based on a correlated process. This means that, within this range, there is a spatial dependency of the fracture extension (fracture length) between the neighboring points in the direction of the fracture trace. The feature *no existence/existence of one or more fractures* at point x_i depends on the features of the neighboring points which are located within a distance smaller than the range $|h_i| = 0.70$ m in the direction of vector h_i. The features of the points which are located at a distance greater than the range $|h_i| = 0.70$ m have no influence of the feature at point x_i.

The similarity between the indicator and the fracture-frequency fields can be observed in the variograms as well. A comparsion of the average indicator variogram with the average fracture-frequency variogram (e.g. Fig. 5.29) shows very similar characteristics. However, the $\gamma_{\mathrm{FF}}(|h_i|)$ values of the fracture-frequency variogram are higher than the $\gamma_{\mathrm{I}}(|h_i|)$ values of the indicator variogram. This behavior can be observed in all graphs. The use of larger scanline-segment lengths (e.g. $l = 0.10$ m) even increases this difference. This behavior is explained by the variogram equations. On the basis of

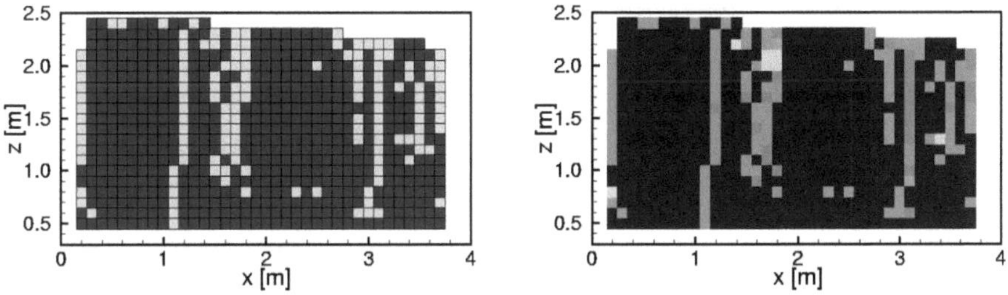

Fig. 5.26. South-west wall: indicator field and fracture-frequency field, scanlines in x-direction, segment length $l = 0.10\,\text{m}$, distance between the scanlines $d = 0.10\,\text{m}$.

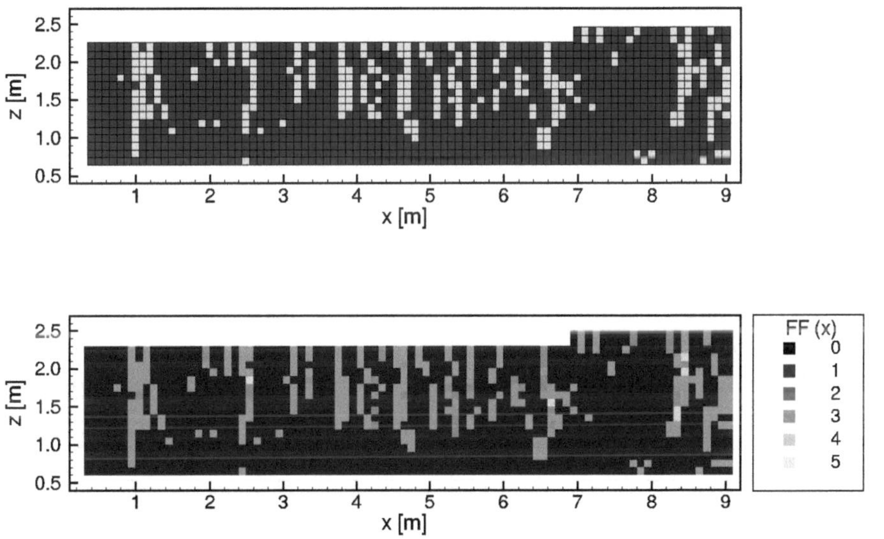

Fig. 5.27. West wall: indicator field and fracture-frequency field, scanlines in x-direction, segment length $l = 0.10\,\text{m}$, distance between the scanlines $d = 0.10\,\text{m}$.

the variable fields, the indicator variogram is calculated using equation (5.4) and the fracture-frequency field is calculated using equation (5.3)

The increment $[I(x_{i_\alpha} + h_i) - I(x_{i_\alpha})]$ of the indicator variogram is exclusively 0 or 1. In contrast, the increment $[FF(x_{i_\alpha} + h_i) - FF(x_{i_\alpha})]$ of the fracture-frequency variogram can have values higher than 1. The total number of point pairs n and the number of the point pairs with increment 0 are

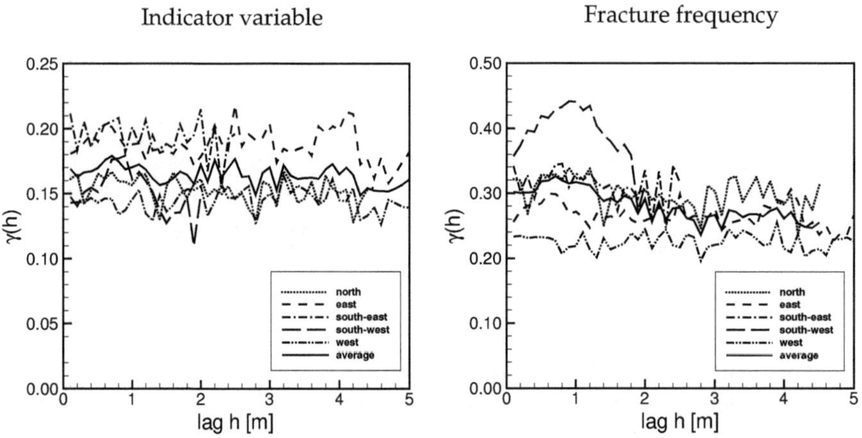

Fig. 5.28. Experimental variograms of the five walls: vector h_i parallel to the scanlines, segment length $l = 0.10\,\mathrm{m}$.

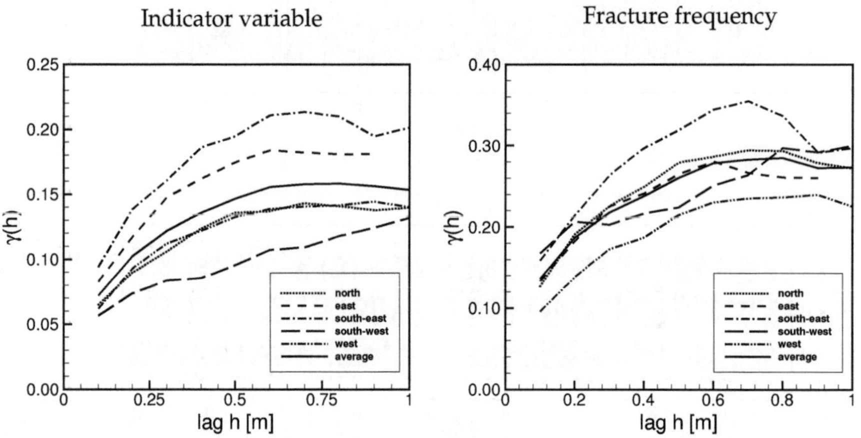

Fig. 5.29. Experimental variograms of the five walls: vector h_i perpendicular to the scanlines, segment length $l = 0.10\,\mathrm{m}$.

equal for both variogram types. Resulting from this, there is a higher variability in the fracture-frequency variograms, which can be observed in a higher mean and a higher variance. The values of the univariate statistic (mean, variance) and the number of segments for the indicator variable and the fracture frequency are shown in Fig. 5.34 for the five side walls.

5.4.2.2 Vertical Scanlines: Segment Length $l = 0.10\,\text{m}$

Figures 5.30 and 5.31 show the variable fields of the south-west and the west wall as exemples of the five walls of the field block. The indicator and the fracture-frequency field of the single walls are rather similar. The number of segments with a fracture frequency $FF(x_i) > 1$ is not very high compared to the total number of segments. The horizontal fracture distance is generally $\geq 0.10\,\text{m}$, and the natural fractures are indicated clearly by single fracture traces. The opposite case can be observed in the fracture-frequency field of the south-west wall shown in Fig. 5.26. The two variable fields of the south-west wall (Fig. 5.30) show many fracture traces which are longer than 1.0 m. On the other hand, the number of long fracture traces is not very high in the variable fields of the west wall. Here, the short fracture traces dominate the field.

Figures 5.32 and 5.33 show the experimental variograms. As can be seen in the four plots, the variogram values of the south-west wall are clearly higher than the values of the other four walls. The higher proportion of the number of segments with $I(x_i) = 0$ to the number of segments with $I(x_i) = 1$ leads to the higher variogram values. This fact can be observed in the mean and in the variance as well. Figure 5.35, left, shows the values of the univariate statistic of the five different walls.

The variogram analysis with separation vector h_i parallel to the vertical scanlines indicates a nugget effect. This means that the location of the fractures in space is based on a random process. However, the variograms differ in their variability: the variogram values are significantly higher and the expected range of the fracture extension (see Fig. 5.33) is about $2.0 - 2.5\,\text{m}$. The higher variability is based on the larger segment length, as described in Sect. 5.4.2.1 in the case of the horizontal scanlines. The larger segment length leads to a reduced total number of segments. Additionally, the number of segments with $I(x_i) = 1$ within the indicator fields increases, and the number of segments with $I(x_i) = 0$ decreases. This fact leads to higher variogram values.

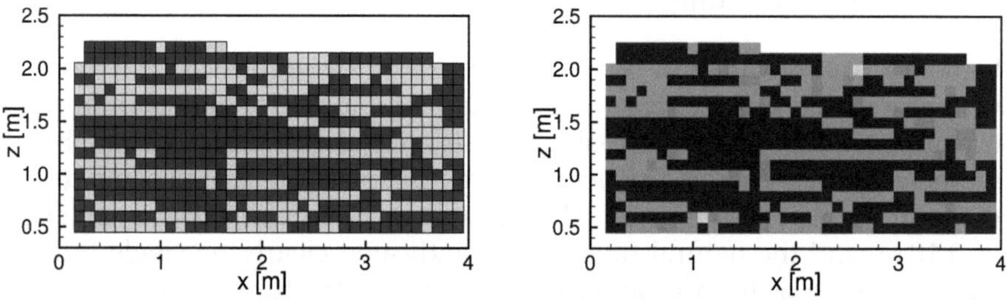

Fig. 5.30. South-west wall: indicator field and fracture-frequency field, scanlines in z-direction, segment length $l = 0.10\,\mathrm{m}$, distance between scanlines $d = 0.20\,\mathrm{m}$.

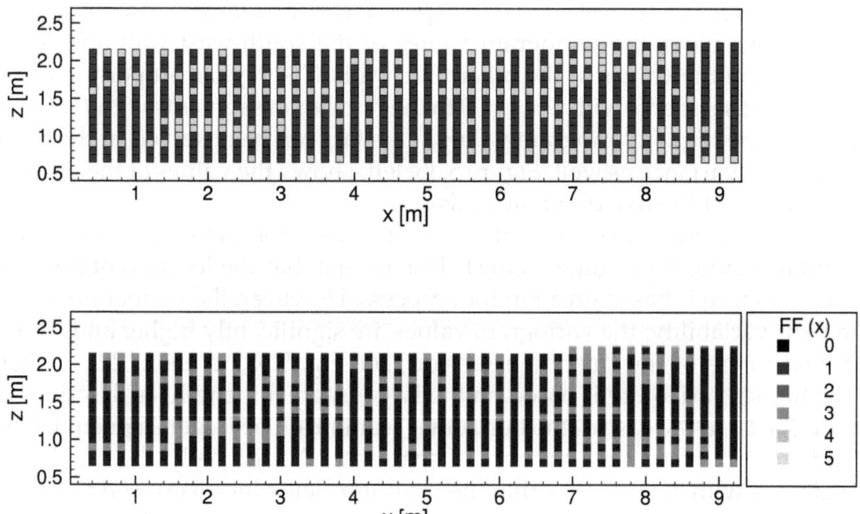

Fig. 5.31. West wall: indicator field and fracture-frequency field, scanlines in z-direction, segment length $l = 0.10\,\mathrm{m}$, distance between scanlines $d = 0.20\,\mathrm{m}$.

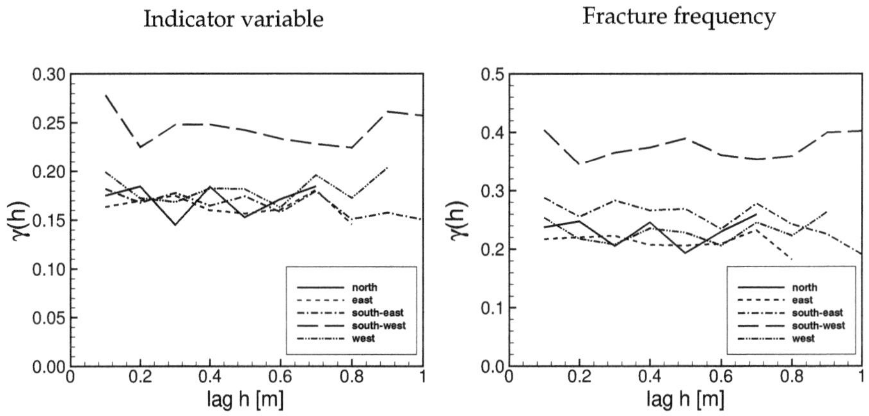

Fig. 5.32. Experimental variogram of the five walls: vector h_i parallel to the scanlines, segment length $l = 0.10\,\text{m}$.

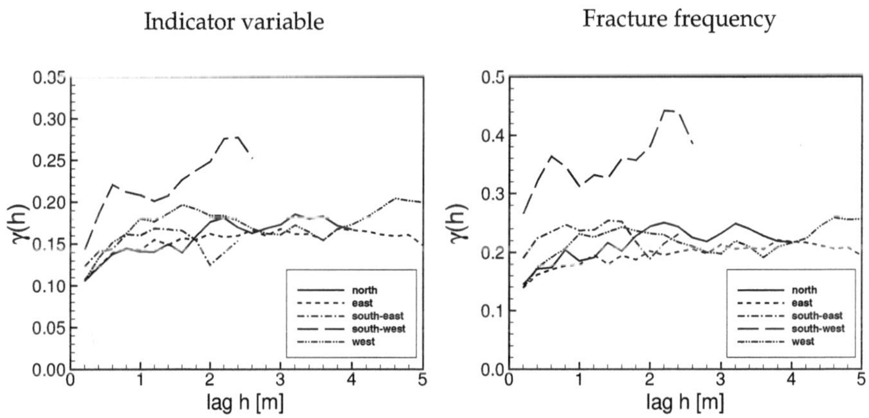

Fig. 5.33. Experimental variogram of the five walls: vector h_i perpendicular to the scanlines, segment length $l = 0.10\,\text{m}$.

5.4.3 Discussion of the Results

A geostatistical analysis of the fracture-trace maps of the five side walls and the top of the field block is conducted. The fracture-trace maps are overlaid by a set of parallel scanlines. The set of scanlines itself is oriented perpendicularly to the main fracture orientation. The scanlines are divided into segments of equal length. The number of intersection points *fracture - scanline segment* leads to the indicator variable (minimum one intersection point existing yes/no) and to the fracture frequency (number of intersection points per scanline segment). The experimental variogram investigations are done in two directions: separation vector h_i parallel to the scanlines, and separation vector h_i perpendicular to the scanlines.

The geoststistical analysis improves the characterization of the field block. It is, however, of great interest to actually implement the knowledge that can be obtained from a geostatistical analysis in discrete numerical simulation models. Currently, the fracture generator FRAC3D (Sect. 2.4.2) is being extended to enable fracture generation based on variograms.

5.4.3.1 Variograms with Separation Vector Parallel to the Scanlines

The variogram investigations parallel to the scanlines focus on the position of the fractures and on the distances of the fractures perpendicular to their main fracture orientation. Some of the variograms show a slight periodic behavior. In the average variogram, the nugget effect dominates the variograms. This means that the spatial arrangement of the fractures is a random process and independent of the neighboring fractures. This is supported by extensive field investigations showing that a Poisson process of the fracture midpoints corresponds to an exponential distribution of the fracture distances (Sect. 2.1). In Sect. 5.3.1, it is shown that the fracture distance distribution at the test site follows an exponential distribution. The assumption that the spatial arrangement of the fractures follows a random process is supported by the exponential distribution of the fracture distances and by the nugget effect of the variograms.

5.4.3.2 Variograms with Separation Vector Perpendicular to the Scanlines

The variogram investigations perpendicular to the scanlines focus on the extension of the fractures along the existing fracture trace. Almost all of the variograms show a steep ascent for small distances $|h_i|$ followed by a flatter ascent. As soon as a threshold value $\gamma(|h_i|) = C$ is reached, the variograms follow an almost horizontal graph. The distance $|h_i|$, at which the threshold value is reached, is called range. Along the direction of the fracture extension, the existence and the extension of a fracture is correlated within this range. This means that the feature *no existence / existence of one or more fractures* at point $x_{i_{P1}}$ depends on the feature at the points $x_{i_0}, x_{i_1}, ..., x_{i_n}$, which are located in the direction h_i at a distance smaller than/equal to the range to the point $x_{i_{P1}}$. A correlation no longer exists for distances greater than the range. This means that the feature at point $x_{i_{P2}}$, which is located at a distance greater than the range from the points $x_{i_0}, x_{i_1}, ..., x_{i_n}$, is independent of the features of points $x_{i_0}, x_{i_1}, ..., x_{i_n}$.

5.4.3.3 Comparing Indicator Variable $I(x_i)$ - Fracture Frequency $FF(x_i)$

The influence of the segment length on the indicator variable and the fracture frequency is pointed out by Fig. 5.34 and Fig. 5.35. Figure 5.34 is based on the variable fields of the vertical fractures which are overlaid by a set of horizontal scanlines. Figure 5.35 shows the results for the variable fields of the horizontal fractures which are overlaid by a set of vertical scanlines. The two figures present the results of the univariate statistic of the indicator and the fracture-frequency fields. The mean value, the variance, and the number of segments per variable field are plotted for the segment lengths $l = 0.04$ m, 0.10 m, 0.20 m.

Here, the difference between the indicator variable and the fracture frequency is discussed for the example of south-west wall for the vertical fractures. For the segment length $l = 0.04$ m, the mean value and the variance are almost equal for the indicator variable fields and the fracture-frequency fields as can be seen in Fig. 5.35 and in Table 5.5. This means that the measurement set with a segment length of $l = 0.04$ m is so fine that almost all fracture traces are detected in an individual segment. Only a few segments comprise a fracture frequency of $FF(x_i) > 1$. The fracture frequency of $FF(x_i) > 1$ points to the following fact: fracture traces which are located very close together (within a segment of $l = 0.04$ m) have to be interpreted as one single natural fracture trace. For the segment length of $l = 0.10$ m, there are clearly higher values for the mean and the variance and smaller values for the number of segments. This trend continues for the analysis with segment length $l = 0.20$ m, the reason being that the number of segments with $I(x_i) = 0$ and $FF(x_i) = 0$ decreases and the number of segments with $I(x_i) = 1$ and $FF(x_i) > 0$ increases for larger segment lengths. This leads to a higher mean value and variance.

As a result of larger segment lengths, the difference between the mean value and the variance gets more significant for the indicator variable and

Table 5.5. South-west wall: mean, variance, number of segments.

Segment length	Indicator variable		Fracture frequency		Number of
	Mean $\bar{x}_I$	Variance s_I^2	Mean $\bar{x}_F$	Variance s_F^2	segments
0.04 m	0.0929	0.0843	0.1084	0.1311	1743
0.10 m	0.1994	0.1596	0.2713	0.3971	682
0.20 m	0.3647	0.2317	0.5500	0.7887	340

the fracture frequency. Considering the mean value

$$\bar{x} = \frac{1}{n} \sum_{i=1}^{n} x_i \tag{5.5}$$

and the variance

$$s^2 = \frac{1}{n-1} \sum_{i=1}^{n} (x_i - \bar{x})^2 \tag{5.6}$$

leads to a better understanding of the differences. A comparison of indicator field and a fracture frequency field based on the same scanline measurement set shows that the total number of segments n and the number of 0-segments ($I(x_i) = 0$, $FF(x_i) = 0$) are equal for the two fields. The maximum value of the indicator field can reach $I(x_i) = 1$. In contrast, the maximum value of the fracture-frequency field can reach $FF(x_i) \geq 1$. The sum $\sum_{i=1}^{n} x_i$ of the fracture-frequency field is at least equal to the sum $\sum_{i=1}^{n} x_i$ of the indicator field. In most cases, the sum of the fracture-frequency field is higher than the sum of the indicator field. Therefore, the mean value $\bar{x}_F$ of the fracture-frequency field is equal to or larger than the mean value $\bar{x}_I$ of the indicator field. The total number n and $n-1$ of the segments is equal for both variable fields. Based on the higher variability of the fracture frequency, the increment $(x_i - \bar{x})$ reaches higher values as in the case of the indicator variable as well. These different facts lead to higher variance values for the fracture-frequency fields. Figure 5.26 clearly shows the different segment values for the indicator field and the fracture-frequency field on the basis of the same scanline measurement set for the south-west wall.

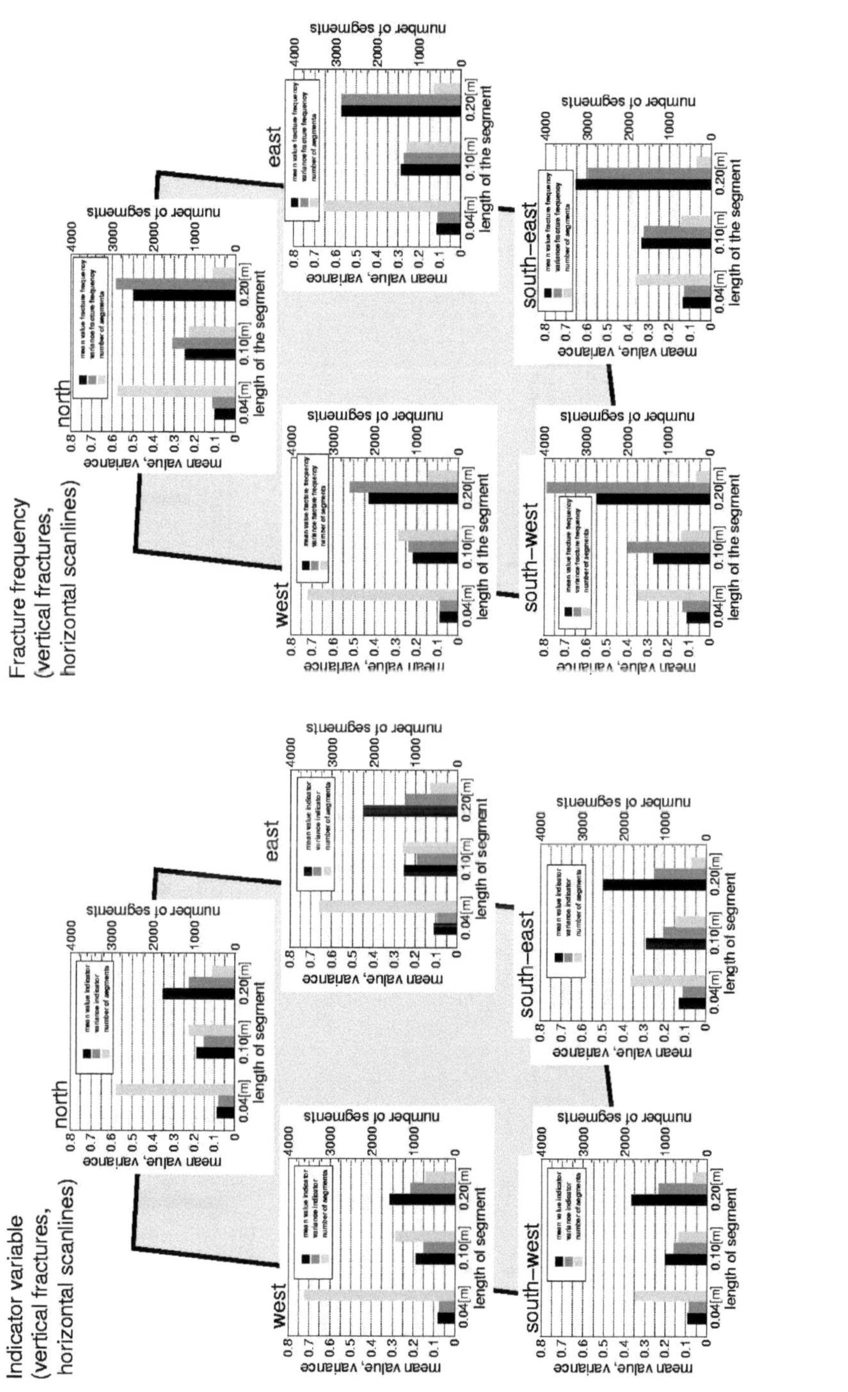

Fig. 5.34. Mean, variance, number of segments of the variable fields (left: indicator variable, right: fracture frequency) for vertical fractures.

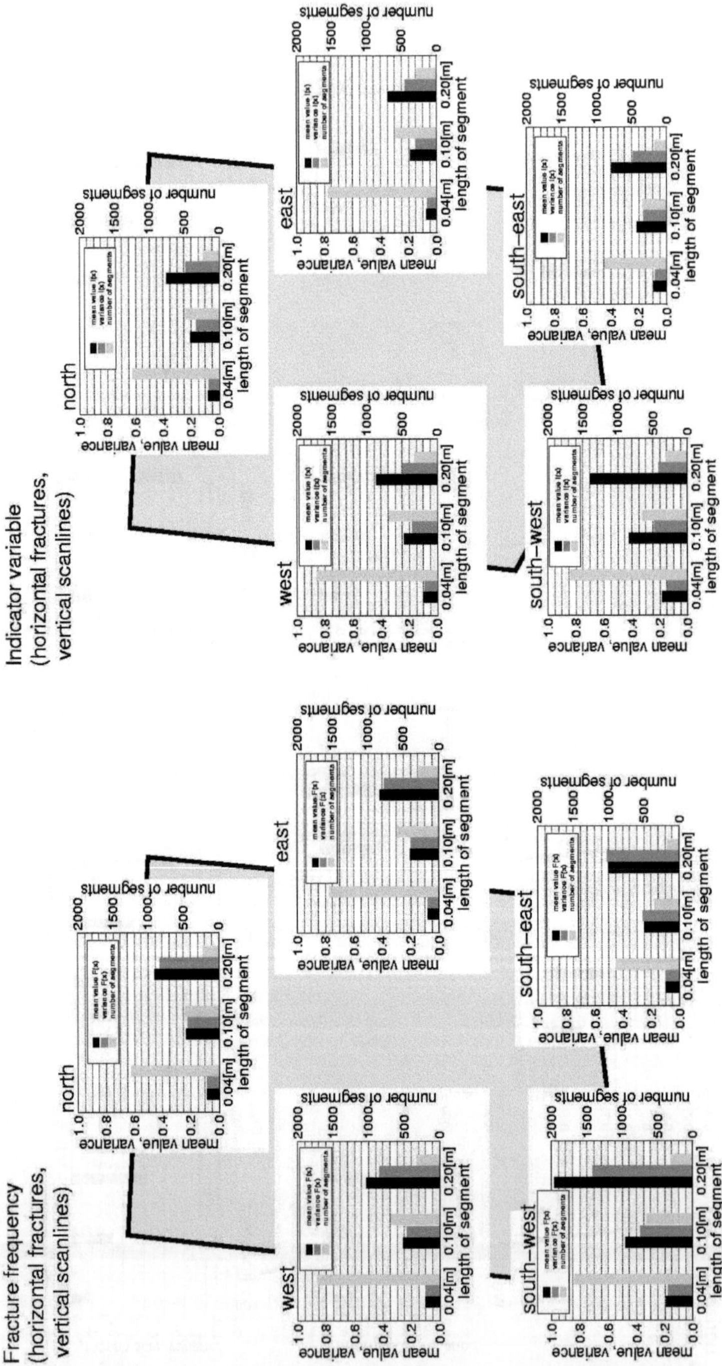

Fig. 5.35. Mean, variance, number of segments of the variable fields (left: indicator variable, right: fracture frequency) for horizontal fractures.

5.4.3.4 Top of the Field Block: Comparing the Variability of the Vertical Fractures

A trend of the variogram values can be observed in the experimental variograms with separation vector $h_i = (0, y)$ of the vertical fractures of segment length $l = 0.04$ m, 0.10 m, 0.20 m. As can be seen in Fig. 5.29, the variogram values of the west wall are generally the lowest, whereas the values of the south-east and the east wall are generally the highest. An additional hint of a possible north/west - south-east/east trend can be gained from a comparison of the mean and the variance of the different side walls in Fig. 5.34. The mean and the variance are mostly higher for the southern and eastern walls than for the western and northern walls. This leads to the assumption that the number of fractures may be higher in the south-east area of the field block than in the north-west area.

In addition to the variograms, the mean and the variance of the different side walls, the top of the field block gives further information about a possible north/west - south-east/east trend. The fractures which can be seen on the top of the field block in Fig. 5.36, left, are fracture traces of the two vertical fracture clusters. The fracture map is overlaid by a mesh of size $0.10\,\text{m} \times 0.10\,\text{m}$. The number of fractures which intersect one cell is recorded in Fig. 5.36, right. Figure 5.36, right, does not give a significant suggestion of a higher fracture frequency for the south-east/east corner of the field block.

However, there is a higher fracture frequency for the cluster $S_t = 56°$ than for the cluster $S_t = 139°$. Comparing the experimental variograms of the vertical fractures of the top plane shows that the variogram values for the cluster $S_t = 56°$ are higher than the variogram values for the second cluster with direction $S_t = 139°$.

5.4.3.5 Comparing the Range and the Segment Length

The influence of the segment length on the range is investigated by assessing the experimental variograms with separation vector h_i perpendicular to the scanlines. Figure 5.37 presents the comparison of the approximated average value of the range of all side walls versus the segment length $l = 0.04$ m, 0.10 m, 0.20 m, 0.30 m. The values of the ranges come from the experimental variograms. Three pairs of graphs are seen in in Fig. 5.37. There is no large difference between the range of the indicator variograms and the range of the fracture-frequency variograms for the single pairs.

The pair of graphs of the vertical fractures (fracture traces on the side walls) does not show a strong dependence of the range and the segment length. The vertical fractures are always recorded as single fracture traces and not as different, disconnected fracture traces. The surface maps, which can be seen in Figs. 5.23 and 5.24, support this fact. Additionally, the direction of the fracture traces is almost parallel to the z-axis. The scanline mesh (set of horizontal scanlines perpendicular to the almost vertical fractures) causes

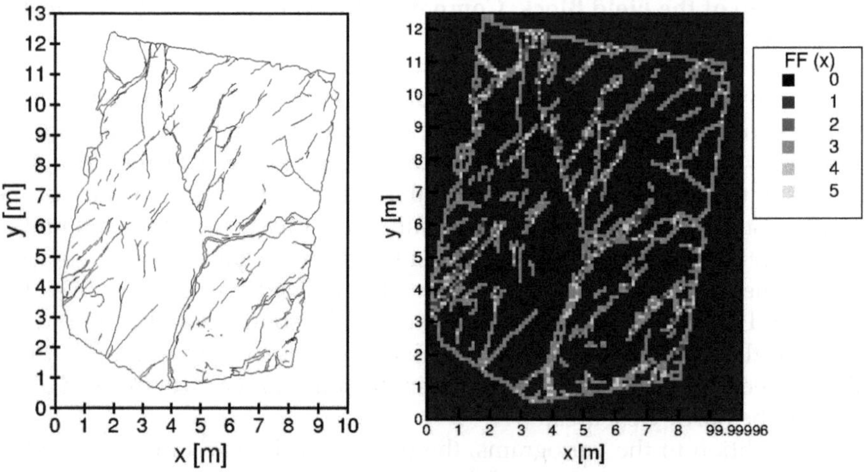

Fig. 5.36. Top of the field block. Left: fracture mapping. Right: fracture frequency per grid element 0.10 m×0.10 m.

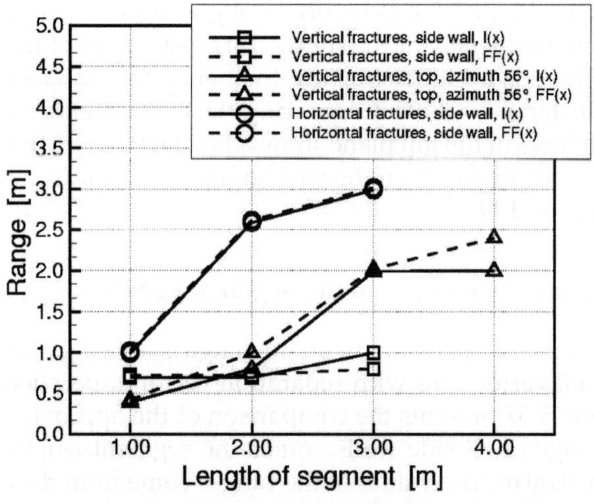

Fig. 5.37. Comparing range versus segment length.

the following effect: the single fracture traces are not recorded in diagonal scanline segments. They are recorded in segments, which lie on a straight line, independently of the segment length used.

The second pair of graphs of vertical fractures from the top of the field block (Fig. 5.37) shows a significant dependence of the range and the segment length. Between the segment lengths $l = 0.04$ m and 0.10 m, the range increases very slowly. This is followed by a strong increase of the range between the segment lengths $l = 0.10$ m and 0.20 m. Between the segment lengths $l = 0.20$ m and 0.30 m the range is almost constant. The reason for the strong increase of the range is that, with the use of a larger segment length, single fractures which lie very close together and almost parallel on a line form a quasi-new longer fracture trace. The scanline mesh runs perpendicularly to the main fracture direction (strike $S_t = 56°$ and azimuth $A_z = 146°$, Table 5.2) but the orientation of the fracture traces varies slightly around the main direction. This is seen in Fig. 5.36, left. Consequently, the scanline mesh does not run exactly perpendicular to the fracture traces. A scanline mesh based on small segment lengths divides a single fracture trace into several "new" shorter fracture traces. A variogram investigation based on such an incorrect variable field (indicator or fracture-frequency field) leads to a smaller range, which does not represent the real natural range of the fracture traces. The choice of a larger segment length reduces or prevents this error.

The pair of graphs of the horizontal fractures shows a relatively steep ascent of the range between the segment length $l = 0.04$ m and 0.10 m. However, the ascent of the range between length $l = 0.10$ m and 0.20 m is very small. The reason for the large ascent is the same as described above. The horizontal fractures are not oriented absolutely parallel to the main direction, the x-axis. The choice of a fine scanline mesh (set of vertical scanlines with small segment lengths) and the more or less horizontal fractures lead to the same error: single, not exactly horizontal fractures are divided in new shorter fractures. By the use of a coarser mesh (larger segment lengths with $l = 0.10$ m, 0.20 m), the error can be avoided. However, using a mesh which is too coarse leads to the following problem: real single fractures are interpreted as new, longer fracture traces. One has to investigate different sizes of mesh (different distances between scanlines, different segment lengths) in order to indicate the jumps within the graphs of range versus segment length.

5.5 Orientating Measurements at the Unsealed Field Block

C. Thüringer, M. Weede, R. Bäumle, H. Hötzl

Before the final shaping and sealing of the field block, some orientating experiments are conducted. On the one hand, information about the fracture connectivity and the range of flow velocities inside the field block are gathered by applying simple connectivity and flow experiments with gas tracers. On the other hand, ground-penetrating radar (Georadar) provides the opportunity, of making structural inhomogeneities inside the field block visible, before it is encapsulated, making this method impossible to use.

5.5.1 Connectivity and Flow Tests

To assess the fracture connectivity inside the field block experimentally, gas-tracer tests are conducted, using carbon dioxide (CO_2) as an artificial tracer. CO_2 is injected into the sandstone through several injection boreholes in a certain area of the test site. The injection occurs by an excess pressure of 1 bar and a flow rate of 5 to 10 m^3/h, using a packer system in order to seal the borehole against external atmospheric influences.

The tracer gas is injected both as a continuous injection over a period of several minutes and as a short impulse of 120 seconds. In this way, not only the horizontal extension of interconnected fractures, but also the effective flow velocity of the gas impulse between points of injection and detection can be determined.

Gas detection occurs at gas measuring gauges. These consist of 20 cm drill holes with embedded polyamide tubes and quick-release couplings. In order to make the measuring gauges airtight, the annulus between tube and drill hole has to be filled with clay and ideally secured with cement. The measuring gauges are installed on the surface of the sandstone field block along the fracture trace lines and, where appropriate, between the fractures directly on the rock matrix.

To determine the CO_2 concentration, an infrared spectrometer is used. Its effective measuring range is from 0.05 to 50 vol.-% CO_2. Data acquisition occurs every 20 seconds.

Figure 5.38 shows a sketch map of the unsealed raw test site with the three horizontally oriented gas-tracer-injection boreholes (GTB1 to GTB3). The thin black lines represent the trace lines of the main fractures, the black dots the 45 gas measuring gauges.

Connectivity tests applied at the test site prove that fractures are well interconnected, in some cases over a horizontal distance of about 5 m. Effective flow velocities of 1 to 4 m/min were determined by impulse tests, using the direct distance between the points of injection and detection.

5.5.2 Electromagnetic Reflection Method

The electromagnetic reflection method (EMR) is also known as Georadar. It is a geophysical technique, based on the reflection of high-frequency electromagnetic waves from 10 MHz up to 4000 MHz at material boundaries with different dielectric properties (Vogelsang (1991); Militzer *et al.* (1986)). Basically, this method is used to identify inhomogeneities in physical parameters in the ground, such as electrical conductivity [S/m] and the relative dielectric constant.

Electrical conductivity affects the depth of intrusion of the electromagnetic wave. The attenuation of the electromagnetic signal and consequently the maximum depth of intrusion is inversely proportional to the electrical

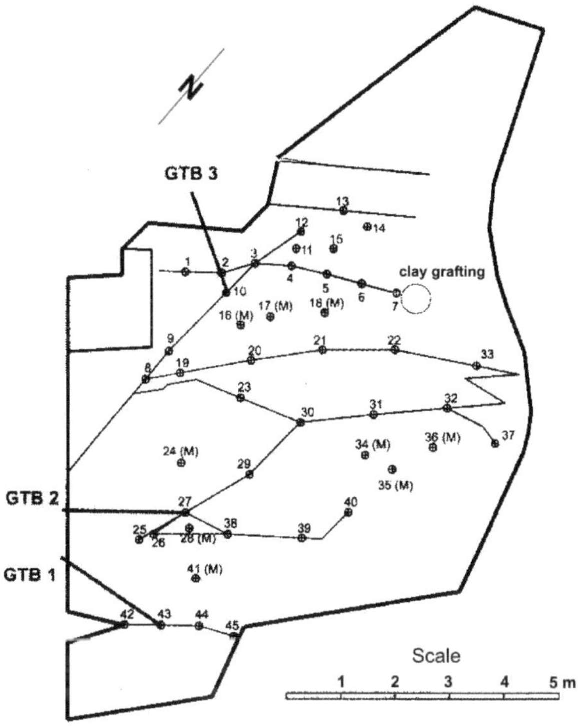

Fig. 5.38. Experimental setup of the fracture connectivity tests at the raw test site with 45 gas measuring gauges and three gas tracer injection boreholes (GTB).

conductivity of the medium. Clay minerals and water-filled pores or fractures normally increase the electrical conductivity and thus decrease the depth of intrusion dramatically.

The velocity of propagation of the emitted and reflected wave and the amount of reflected energy is affected by the non-dimensional relative dielectric constant. In natural materials, this parameter is between 1 (air) and 81 (water). The relative dielectric constant of dry sand, for example, is 3 to 5. With increasing frequency of the electromagnetic wave, the wavelength decreases and the resolution increases, independently of the material under investigation. In contrast to this effect, the depth of intrusion decreases with increasing frequency. The decision on which range of frequency to use depends on the required depth of intrusion and the favored resolution. Thus, EMR measurements should be applied with at least two different frequencies. Which frequency to use is not only a question of the availability of technical equipment, but also of the geological conditions and the structures being investigated. At the *Pliezhausen* test site, EMR measurements were carried out in cooperation with the Department of Geophysics (Karlsruhe University) in September 1997, using SIR-3-equipment with both 900 MHz and 500 MHz antennas. As the data evaluation occurs at the raw field block before the side walls are finally cut, it is also possible to record radargrams along two orthogonal cutting lines marking the east and north faces of the future field block. After the final cuttings, these radargrams can be compared to the created outcrops, making a quantifiable geological interpretation possible.

The whole field block is scanned and the positions and orientations of fractures and sedimentary structures are determined with EMR technology. The measurements were carried out in cooperation with GUS, geophysical services Karlsruhe, in March 1999. The following specifications mainly refer to the GUS report (GUS, 1999).

The measurements are carried out on the surface of the field block along crossing profiles fixed every 25 cm (900 MHz antenna) and every 100 cm (500 MHz antenna). On the side walls, the distance from line to line was 50 cm (900 MHz antenna).

Radargrams are digitally recorded and saved with the dipole direction of the antennas situated vertically to the profile directions. The time window for recording the radargrams is 40 ns (900 MHz antenna) and 60 ns (500 MHz antenna). Before they are analyzed, they are filtered and migrated several times. To locate extensive reflectors, time slices of the radargrams are calculated by averaging over the absolute amplitude in each time and distance interval. Amplitudes of a time slice affect the reflection intensity in the depth and distance range. For 900 MHz measurements, the time interval is chosen in such a way that it represents a depth interval of 5 cm. For this range of frequency (5 cm interval, 0-150 cm depth), 30 time slices are calculated whereas 20 time slices are calculated for 500 MHz measuring (10 cm interval, 0-200 cm depth).

As a bench mark, a geodetic point at the south-east corner of the field block surface is chosen. The coordinates of each reflector, identifiable in at least two adjacent measuring profiles and reaching over more than 1 m, are saved in a separate file. To avoid misinterpretation, smaller reflectors are only recorded at less than 1 m distance to the sidewalls of the test block. The intensity of reflections is dominated by their distance to the measuring profile, fracture filling, saturation, fracture geometry and aperture. Reflections of water- or clay-filled fractures are generally of higher intensity than reflections of air-filled, open fractures. For a reliable detection, the fracture aperture must not be considerably smaller than 1 mm.

The depth of the water-table below the test-site surface is about 5 m during the EMR measurements. Thus, the sandstone block with its thickness of 2 m is completely unsaturated with no afflux at the base. Both in the calculated time slices and in the radargrams, many reflectors, their orientation and their contours can be identified.

The two most dominant, almost horizontally oriented reflectors inside the test block recorded by EMR measuring at the surface are:

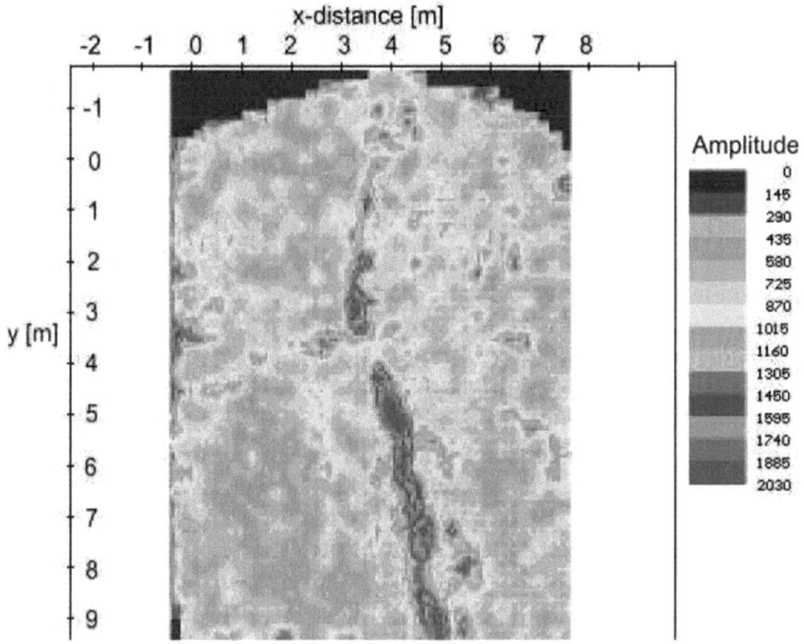

Fig. 5.39. Time slice of depth range 0.95 m to 1.00 m, calculated from unmigrated radargrams with 900 MHz antenna at the field block.

- reflector x = 2.0 to 3.5 m, y = 3.5 to 5.2 m, z = 0.85 to 1.05 m below the surface dipping north-east,
- reflector z = 0.90 to 1.10 m extending almost across the entire field block.

The depth range 0.95 m to 1.00 m below the surface of the block shows a reflection plane oriented north-west to south-east (in Fig. 5.39, north is oriented downwards). It seems to dip steeply and bend to the south in the middle part of the field block. A second, less dominant and orthogonally oriented (south-west to north-east) reflector is visible in the middle left part of Fig. 5.39.

5.6 Flow and Transport Tests at the Sealed Field Block

M. Weede, C. Thüringer, H. Hötzl

With completely encapsulating the whole field block, one fundamental precondition for setting up controlled flow fields within the test site can be achieved. In the following chapter, the technical background as well as the results of the gas-flow and gas-tracer tests applied at the test site *Pliezhausen* are described.

5.6.1 Tracer Injection and Detection Techniques

In order to gain information about the flow and transport parameters of fractured porous media, gas tracer tests are conducted in the gas-saturated field block. For this purpose, a flow field is artificially created over a certain distance via boreholes and gas tracers are injected into the system to investigate.

For each conducted flow and transport experiment two boreholes are used: one borehole is connected to an extraction device to generate low pressure in the borehole and therefore in the field block. The second borehole is opened in order to allow free inflow into the otherwise sealed system. After a certain period of time, a constant flow field will be set up from one borehole to the other inside the fractured porous block. The resulting boundary conditions at the boreholes are constant extraction pressure ($P_{extract}$) and constant inflow pressure, which equals atmospheric pressure ($P_{inflow} = P_{atmos}$). Certain ranges of depth within the boreholes can be tested separately when using packer systems.

After steady flow conditions are reached, impulses of gas tracers (helium (He) and sulfur hexafluoride(SF_6)) are unpressurized injected into the injection borehole while recording the time-depending change of concentration within the air, extracted at the extraction borehole. During the experimental investigation of transport processes by gas tracer tests, the unpressurized injection is an important criterion to fulfill when developing a suitable injection method. By doing so, a significant compression of the gas tracer inside the injection borehole can be avoided.

Tracer injection by a bypass arrangement, as usually applied in aquatic tracer tests, is inappropriate at the test site as the injected gas-tracer mass is insufficient for a significant signal at the detection borehole. Therefore, tracer injection occurs as a direct impulse from the gas bottle. To ensure an unpressurized injection, a pressure equalization box is interposed, in which the temperature of the injected gas can be determined simultaneously (Fig. 5.40). The flow rate of tracer gas out of the bottle and into the box is adjusted to be higher than the passive, unpressurized borehole inflow in order to achieve an injection of undiluted tracer gas into the system.

The injected tracer mass is calculated on the basis of the flow rate and injection time. Actually, this procedure is only correct for incompressible NEWTON fluids as it is based on the continuity equation of the volume. Water, which can be assumed to be incompressible, fulfills this condition in contrast to compressible gases (air, He, SF_6,CO_2...). On the assumption of a non-compressible, constant density flow of air for the time of tracer test, this method can be accepted as sufficiently accurate, provided that the injection is unpressurized and the temperature constant. In addition, investigations

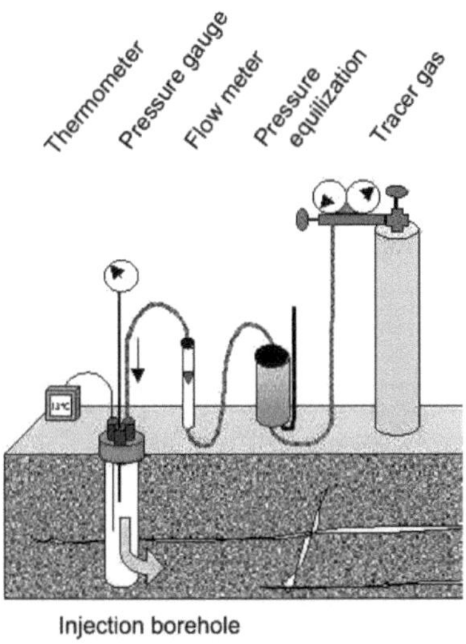

Fig. 5.40. Experimental set-up for gas-tracer injection at test site *Pliezhausen*.

by Jaritz (1999) show that, under steady-state conditions, the error regarding gas as a non-compressible fluid is negligibly small.

The injected tracer gas and the extraction rate are determined by volume flow meters. Pressures at the boreholes are measured by horseshoe-bend manometers with an accuracy of 0.5 mm to 1.0 mm water column (according 5.0 Pa to 10.0 Pa). Rock humidity is selectively measured by a TDR system inside the boreholes.

Tracer gas is detected within the extraction flow at the detection borehole by an online measuring system. To determine helium, a gas detector (HELITEST®) is used. The tracer gas is sampled and analyzed continuously. The measured concentration is recorded every second or every several seconds, depending on the controlling software with a measuring range up to 65 000 ppm.

Sulfur hexafluoride is detected by a field gas chromatograph with an electron collection detector (GC-ECD). When it is in automatic sampling mode, the minimum sampling interval is 120 seconds. This is the period of time needed for the automatic flushing of the measuring chamber and a sampling line with inert gas.

For a more detailed resolution of the tracer breakthrough, an additional manually operated sampling is applied every 20 seconds at the detection borehole. A common measuring range is between 0.02 ppb and 20 ppb.

It should be mentioned that for all conducted experiments, the rate of extraction was always higher than the volume flowing into the field block.

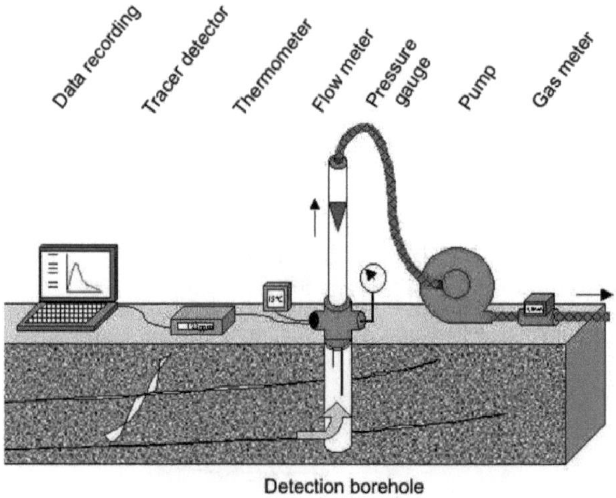

Fig. 5.41. Experimental set-up for gas-tracer detection at test site *Pliezhausen*.

The ratio between extraction rate and flow rate into the block was never balanced. Even an extraction lasting several days with unvaried boundary conditions did not make a change, as expected for a completely sealed system. Although the flow rates and pressure conditions are constant at both boreholes after several hours of testing and steady flow can be expected, the rate of extraction was still a multiple of the flow into the block.

Two reasons can be causal for this observation:

- influence of gas compressibility:
  By the pressure difference at the injection and extraction borehole, differing gas density due to compressibility and therefore differing volumetric gas flow rates can be expected at the boreholes. However, as investigations by Jaritz (1999) show, the error of dealing gas as a non-compressible fluid under steady state conditions in the pressure range of the experiments is negligibly small.
- incomplete sealing of the field block:
  It can be expected that the sealing of the block was not completely gas proof. By different technical arrangements as the afflux of water at the block basement, this problem can partly be equalized.

The permanent air draft within the field block causes a drying of the clay marls which has to be compensated by an extensive drenching of the basal layers in order to prevent them from cracking and leaking. At outcrops existing during the construction phase of the test site, the changing consistency of this material, depending on its water content can be examined. This phenomenon bases on the high content of swellable clay minerals within the basal layers, which get hydrated and expand in contact with water. Drying the base marls causes the risk of fracturing and thus the generation of flow channels for the air outside to intrude into the depressurized block. Earlier investigations indicate the enormous influence of gas flowing through the basement into the field block using high conductivity fractures, when assuming a close contact between sandstone and marl basement.

5.6.2 Measurements at Marginal Boreholes

During the initial experimental phase of this study,exclusively marginally installed boreholes are used for tracer experiments on the field block. This is based on the idea of investigating transport processes at as long experimental distances between two boreholes as possible. At that time, the influences of the field block boundaries, appearing along the defining test site walls, are not regarded in the analysis. However, a detailed discussion of the influence of the field block boundaries is given in Sec. 5.9.

By using the boreholes in the edges of the field block (Fig. 5.11), several variably shaped breakthrough curves can be recorded. As a result of these experiments, information about transport velocities (e.g. maximum velocity, dominating velocity, and mean velocity) and tracer concentration can be gathered. Furthermore, the shape of the curve can give indications for kind and number of the different flow channels, involved in the transport. Narrow and temporally sharp defined peaks can be interpreted as fast, advective dominated transport through well accessible fractures. In contrast, wide and flat peaks indicate a more dispersive-diffusive dominated transport through small ancillary fractures or the rock matrix. Figure 5.42 reviews the different shapes of breakthrough curves, measured in tracer tests between the marginal boreholes in the field block. It accentuates the variable curve shapes and the different ranges of time and concentration scales.

Influence of Thermal Fluctuations
In addition to the type of flow path (main or ancillary fractures, rock matrix) it is found that variations in temperature have great influence on the experimental results when applying gas flow and tracer tests to the field block. This influence of system temperature is exemplarily investigated by the comparison of helium tracer tests applied in winter and in summer time. The most noticeable difference has been recognized during field work in summer: detection of tracer gas at the detection borehole is difficult as the recorded concentration of helium is much smaller than in winter tests. Also the recovery rate of the tracer mass injected, is only a fraction of comparable winter tracer tests. The recovery rate of helium in summer is at an average of about 1 to 5 %. Helium tracer tests in winter show recovery rates of mostly 40 to 80 %, with temperature below zero even up to 90 %. In Fig. 5.43 tracer break-

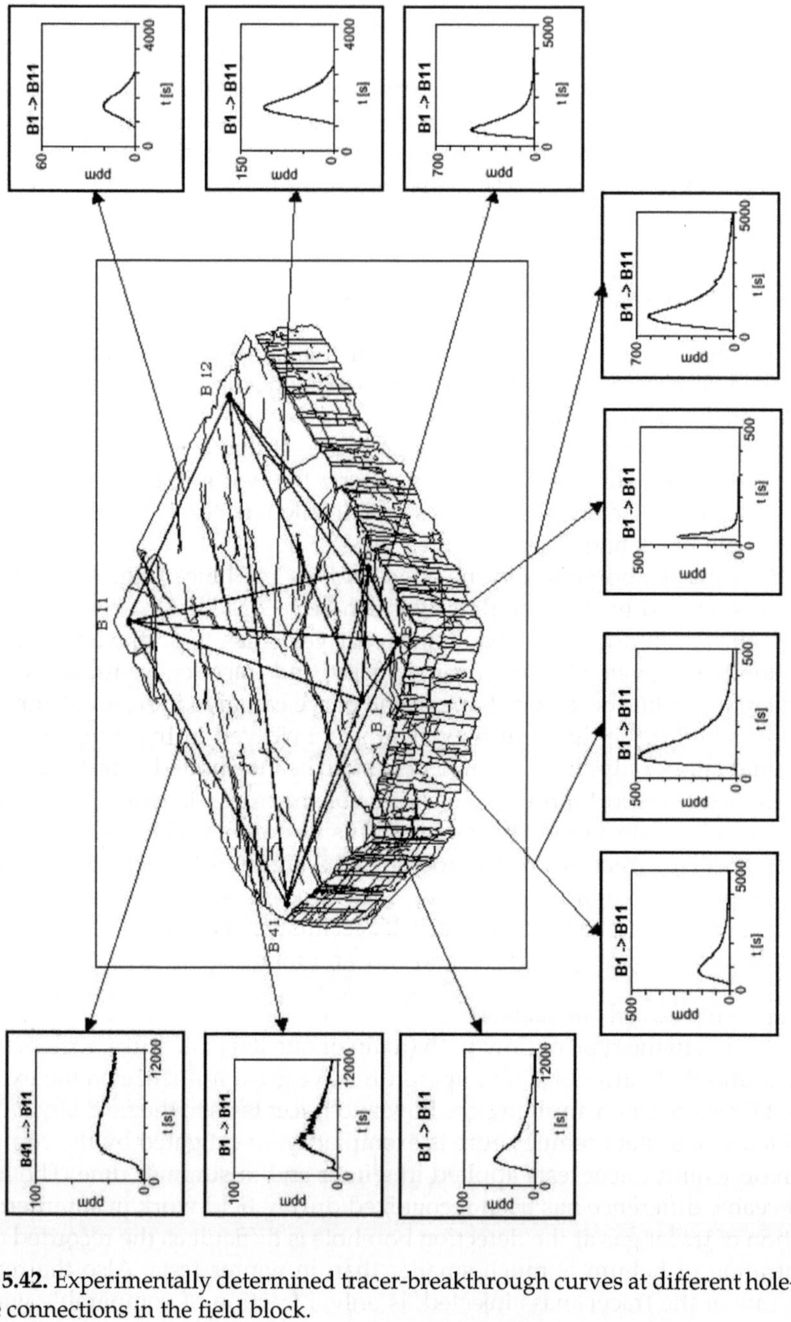

Fig. 5.42. Experimentally determined tracer-breakthrough curves at different hole-to-hole connections in the field block.

through curves of helium-tests in summer and in winter between B31 and B1 are compared. During these experiments, the temperatures at detection boring B1 are -2.1 °C during the winter test and 19.5 °C during the summer test.

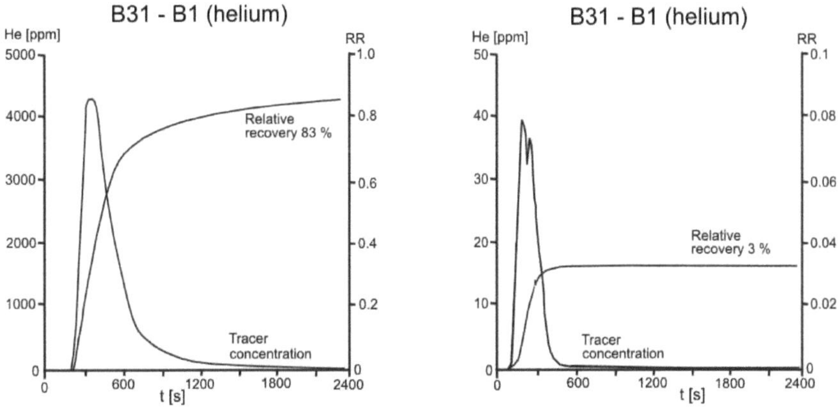

Fig. 5.43. Breakthrough curves and recovery rates of helium tracer tests in winter (left) and summer (right). Note the different scaling between the graphs.

Another significant difference, shown in the diagrams of Fig. 5.43 is the range of concentration, recorded in winter (left) and summer (right). The maximum concentration in winter is 4500 ppm and in summer about one hundredth (42 ppm). This is only partly caused by the higher pressure gradient between extraction borehole B1 (p_{out} = 550 Pa) and injection borehole B31 (p_{in} = 120 Pa) in winter and the 10-times higher tracer input mass M of 327 mg He (28.7 mg He in summer), which is sucked passively into the system (using the same extraction rate Q).

Table 5.6. Experimental parameters of tracer tests with He.

Test type	**T [°C]**	**P_{out} [Pa]**	**P_{in} [Pa]**	**Q [l/sec]**
He-winter	-2.1	-550	120	1.05
He-summer	19.5	-560	40	1.13

Peak shape and number of the helium tracer test breakthrough curves between B31 and B1 differ depending on temperature, too. Whereas in winter there is always one clearly defined single helium peak, in summer in He-tracer tests two peaks are recorded (Fig. 5.43). The single peak, measured in winter tests, is characterized by a quick concentration increase and a relative

Table 5.7. Experimental parameters of tracer tests with He.

Test type	M [mg]	t_1 [sec]	C_{max} [ppm]	RR [%]
He-winter	327.0	160	4500	92.9
He-summer	28.7	124	42	3.2

short but wide tailing. The recovery rate rises up to 93 % at the end of the observation time, 100 min after tracer injection. In summer tests, the recorded time-related concentration rises very quickly to the first of the double peak and then decreases rapidly with a small second peak. The shape of the recovery rate indicates that right after the second peak no more tracer can be recovered, which is only 7 min after the injection. Altogether only about 3 % of the helium can be recovered at the registration borehole.

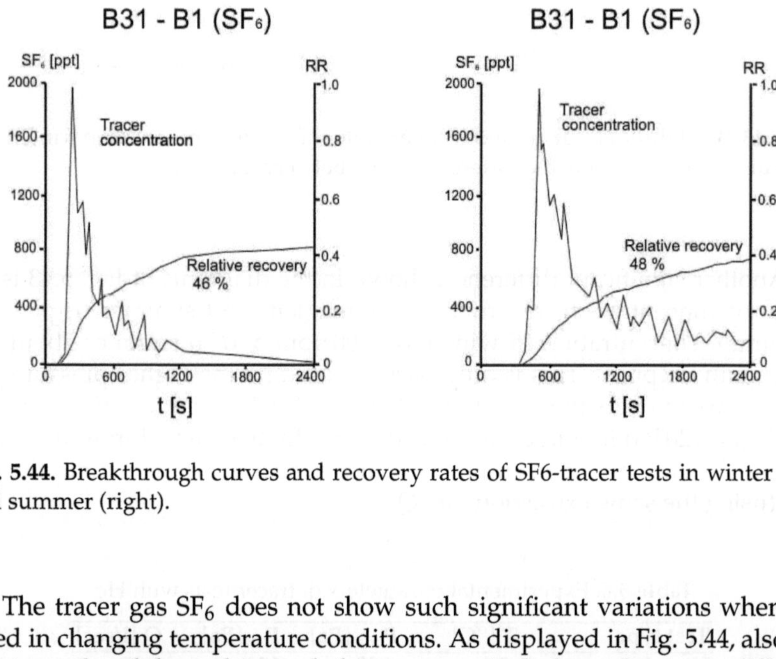

Fig. 5.44. Breakthrough curves and recovery rates of SF6-tracer tests in winter (left) and summer (right).

The tracer gas SF_6 does not show such significant variations when applied in changing temperature conditions. As displayed in Fig. 5.44, also the SF_6 tracer breakthrough recorded in summer includes at least two different peaks and a significant tailing, similar to the He tracer tests in winter. But the recovery rate of the summer test is continuously increasing until the end of observation time after 5400 s, up to 48 % (compared to about 3 % with He in summer). Obviously, the tracer gas SF_6 seems to be less affected by temperature variations.

Preliminary Analytical Parameter Study

As an initial interpretative approach, the measured breakthrough curves are described by simple analytical models in order to gain information about the dominating transport mechanisms.

The recorded tracer breakthrough curves are adapted by the general transport equation (Lever *et al.* (1985) and Tang *et al.* (1981)) for two-dimensional transport of water solute material in a single fracture, in a radial convergent flow field (Sec. 2.3). Each part of this differential equation represents one of the different mechanisms taking place during mass transport. Solutions of the differential equation for the two-dimensional transport in a fracture in a radially convergent flow field are given by Lenda and Zuber (1970) in form of an advection-dispersion-model (ADM) and by Maloszewski and Zuber (1990) in form of a single-fissure-dispersion-model (SFDM). The time dependent tracer concentration at an observation point in transport direction is:

SFDM (Maloszewski and Zuber, 1990):

$$C_f(t) = \frac{aM}{2\pi Q}\sqrt{(Pe \cdot t_0}\int_0^t exp\left(-\frac{Pe(t_0-u)^2}{4ut_0} - \frac{a^2u^2}{t-u}\right)\frac{du}{\sqrt{u(t-u)^3}} \tag{5.7}$$

ADM (Lenda and Zuber, 1970):

$$C(x,t) = \frac{M}{Q}\frac{x}{\sqrt{4\pi D_L t^3}}exp\left(-\frac{(x-v_a t)^2}{4D_L t}\right) \tag{5.8}$$

with:

$$D_L = \frac{v_a x}{Pe}; a = \frac{n_e\sqrt{D_P}}{b}; t_0 = \frac{x}{v_a} \tag{5.9}$$

The application of this approach requires the following simplification of initial and boundary conditions:

- Gas flow primarily occurs at few fractures. By the injection and detection of tracer gas at single boreholes in connection with an orthogonal fracture network this condition can be assumed as approximately fulfilled.
- Transversal dispersion is negligible. Tracer is injected directly into the main flow between injection and detection boring. Thus the effect of transversal dispersion can be prevented particularly at tracer detection boring.
- Tracer is uniformly distributed inside the whole injection borehole at test initiation. This condition has been assumed without the possibility of verification, due to the unavailable recording of tracer injection concentration.

By an iterative adaption of the parameters to fit the experimental tracer breakthrough curves (Fig. 5.42), transport in the field block can be described, using the Peclet number *Pe*, the mean transit time t_0 and the diffusion parameter *a*. Figures 5.45 and 5.46 show that a satisfying adaption to the breakthrough curves, measured at the field block by Thüringer (2002) can be achieved, applying the SFDM.

Table 5.8. Fitted transport parameters of gas tracer tests at test site *Pliezhausen*.

		t_0 [min]	Pe	α_L [m]	a
He	ADM	7.6	13.1	0.15	-
	SFDM	5.9	30.7	0.06	0.011
SF_6	ADM	11.5	4.8	0.40	-
	SFDM	5.4	31	0.06	0.025

This composition clarifies the main differences between ADM and SFDM. As the ADM describes transport only by advective and dispersive processes, the SFDM also implements diffusion-caused retardation by the introduction of the diffusion parameter *a*. Thus the dispersive part of the tailing decreases and the Peclet-number increases, applying the SFDM.

Experiments in fissured aquifers frequently show a tailing in concentration declension which can have many different reasons. This must in no case

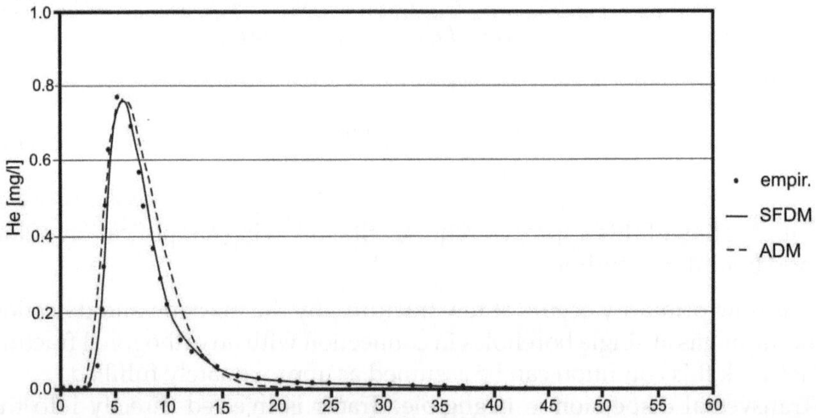

Fig. 5.45. Experimental tracer-breakthrough curve (B31 - B1) and adaption by ADM and SFDM with He as tracer gas.

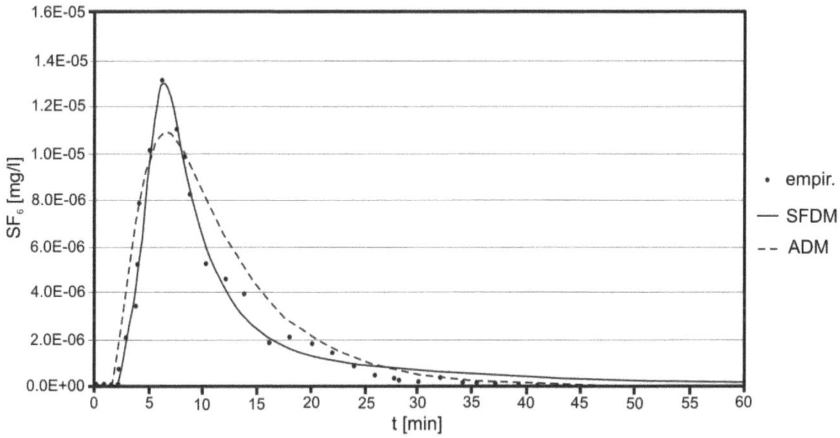

Fig. 5.46. Experimental tracer breakthrough curve (B31 - B1) and adaption by ADM and SFDM with SF_6 as tracer gas.

be interpreted as a proofed sign for diffusive mass transport between fracture and matrix. When discussing the tailing of tracer tests breakthrough curves, investigations by Thüringer (2002) show that also differential advection processes, caused by the use of different and independent flow channels and the technique of tracer injection play a major role. A tailing in tracer breakthrough curves, caused by mixing effects inside the injection boreholes, can hardly be differed from a tailing, caused by diffusion. According to Lever *et al.* (1985) and Tsang *et al.* (1996) a diffusive tailing for an aquatic one-phase-system can be detected in log-log-presentation of the time-concentration-curve. Here a straight line with a slope of -1.5 should appear. The gas tracer tests, done at the unsaturated test site, meet this criteria only between B31 and B1 (Figs. 5.47 and 5.48). Both using He and SF_6, the decrease of concentration follows a straight line with a slope of -1.5 in log-log-presentation. However the number of experiments conducted is too small to deal this result as a proof for the influence of matrix diffusion.

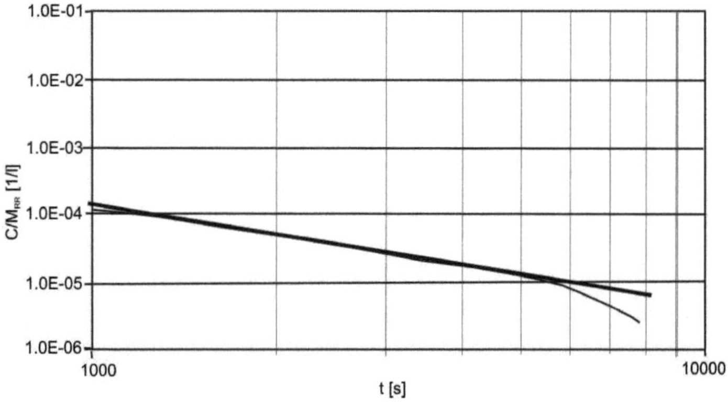

Fig. 5.47. Analysis of tailing behavior with He (B31 - B1), $f(t) = 4t^{-1.5}$.

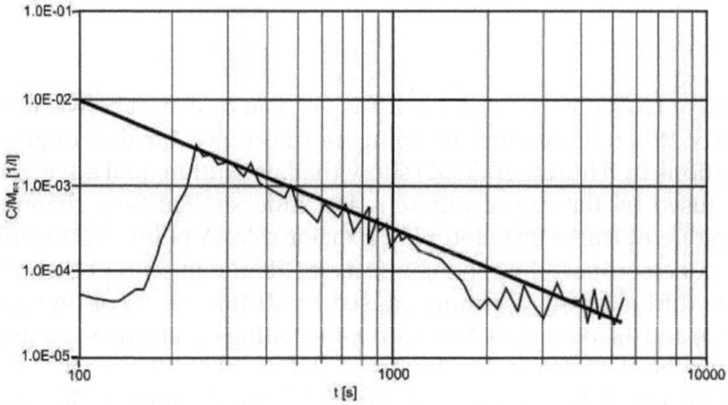

Fig. 5.48. Analysis of tailing behavior with SF_6 (B31 - B1), $f(t) = 10t^{-1.5}$.

5.6.3 Measurements at Central Boreholes

The influence of boundary effects (Sec. 5.9) cannot be neglected at a test site on the field-block scale. After the initial measurements at the marginally installed boreholes (Sec. 5.6.2), experiments were relocated to the central part of the field block. A radial symmetrical arrangement of boreholes (Sec. 5.2.2) ensures that boundary effects are comparable during the experimentation in different directions.

5.6.3.1 Gas-Flow Experiments

In order to quantify the influence of the fracture-system orientation on flow and consequently on solute transport processes within the fractured porous sandstone on the field-block scale, gas flow experiments are applied. A radial symmetrical arrangement of boreholes in the center of the test site with one central borehole and 6 holes in a circle with 2 m radius around it is used. Thus, twelve different flow/transport directions (every 30°) can be reverted to (Fig. 5.50). Every borehole has a diameter of 50 mm and a depth of 1.80 m to 2.20 m through the sandstone bed into the underlying silt and clay.

Pressure gradients are set up between two boreholes of the otherwise completely sealed test block and the air-flow rates measured from hole to hole. In this way, altogether 42 different borehole-to-borehole connections are tested and every direction is represented by three to five different borehole-borehole connections. Without the use of packer systems, the entire thickness of the fractured sandstone is accessible to setup a flow field between the two boreholes.

Every experimental leg is run with gradually increasing and decreasing pressure gradients. A particular air flow rate can be assigned to every pressure gradient (Fig. 5.49). The mean proportion of flow rate and pressure gradient provides information about conductivity with respect to gas flow for every borehole-to-borehole flow path. Figure 5.49 exemplifies this proportion of flow rate and pressure difference using borehole Z3 as the extraction hole.

As is typical of heterogeneous fractured systems, great differences can be detected among the flow rates per pressure gradient when changing the experimental leg. Besides, the relation between flow and pressure gradient is not linear very often. Initially, a slightly decreasing flow-rate change per change of pressure $\frac{\partial Q}{\partial \Delta P}$ can be observed. But when a certain threshold pressure is exceeded, an excursive increase of flow per pressure gradient occurs in some cases. This phenomenon can be interpreted by the sudden opening of numerous flow paths with rather small cross sectional area, beyond a certain pressure gradient.

In order to correlate relative flow rates to the fracture orientation, the mean values of flow per pressure gradient are determined for every flow

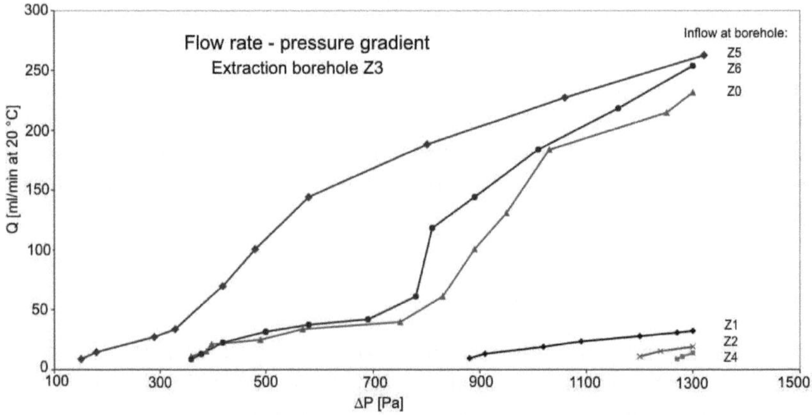

Fig. 5.49. Proportion of air-flow rates and pressure gradients at extraction hole Z3.

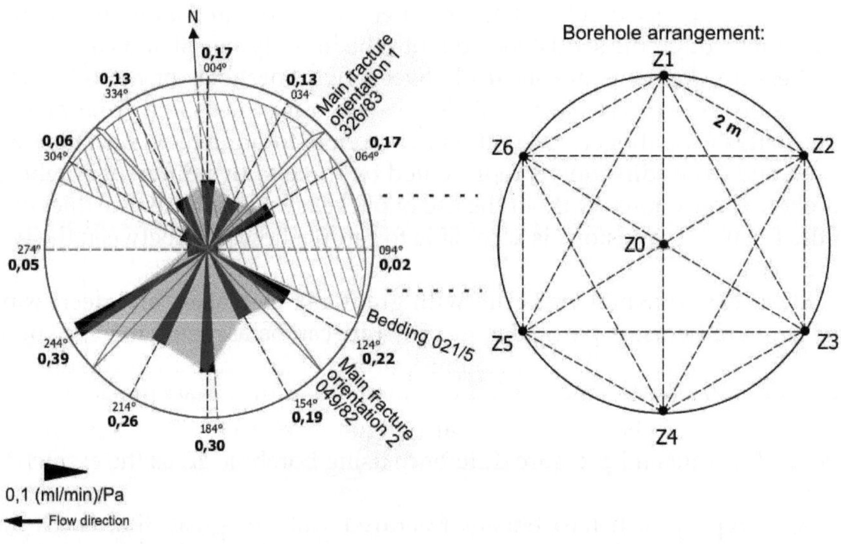

Fig. 5.50. Mean flow rates per pressure gradient [(ml/min)/Pa] (left) for twelve different flow directions (right) compared to the three main discontinuity plains at test site *Pliezhausen*.

direction. Twelve mean proportions of flow rates and applied pressure gradients can be measured, corresponding to the twelve different borehole to borehole connections. The ratios of flow rate per pressure gradient and the three main fracture plains are illustrated in Fig. 5.50.

It is obvious that a direct correlation between flow rates and main fracture plains is not possible. On the one hand, the fractures striking south-west - north-east are well defined at high flow rates in the direction 244° or 64°. On the other hand, there is a high conductivity in the north-south direction at borehole Z0, which has an effect on the mean value of the north-south conductivity of all other boreholes. In order to clarify the influence of discontinuity plains on the gas-flow conductivity, some more detailed investigations are necessary.

5.6.3.2 Dipole Tracer Tests

The borehole annulus in the center of the field block provides the opportunity to apply tracer experiments without the interfering influence of the field-block side walls. With the central borehole (Z0) as the injection or extraction borehole, twelve experiments in six different transport directions are possible. The length of each experimental leg is 2 m. The radial symmetrical arrangement of boreholes allows a direct comparison of the tracer experiments to each other and to the results of the flow experiments (Sec. 5.6.3.1). Furthermore, the influence of the relative orientation of the transport direction to the statistically analyzed fracture system can be assessed.

Experimental Set-Up

In contrast to the tracer tests applied to the field block at the marginal boreholes, the tracer experiments on the central boreholes are carried out exclusively in dipole arrangement. An active extraction from the extraction borehole is used together with an active injection of air into the injection borehole. The rate of extraction is chosen to be the same as the rate of injection during these experiments. In this way, a more uniform flow field can be generated using smaller pressure gradients than during the monopole set-up. This is advantageous with respect to gas compressibility.

For the extraction, the same technical set-up is used as that applied during the monopole tests (Fig. 5.41). A constant air flow of 17 l/min is extracted from the extraction borehole using a side channel compressor. The pressure inside the extraction borehole is monitored and the point at which steady-state conditions are achieved is ascertained. The injection occurs by means of a second compressor, which injects a constant air flow of 17 l/min into the field block. The pressure should be monitored in this borehole in order to make out steady state conditions and to avoid an intensive compression of the gases.

In analogy to the monopole tracer tests, helium is used as the tracer gas. The detection occurs directly from the extracted air by a portable mass spectrometer with a measuring range up to 65 000 ppm and an accuracy of 2 ppm. In contrast to the pressureless tracer injection, as applied by Thüringer (2002), the helium is injected in these experiments as an impulse of 30 s by a three-way valve directly into the injection borehole, with a slight overpressure.

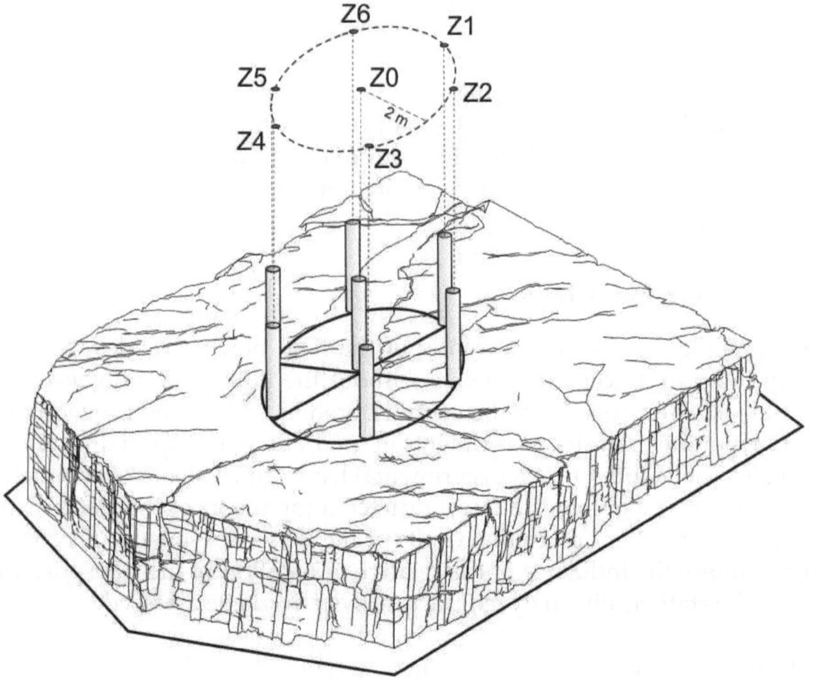

Fig. 5.51. Radial symmetrical borehole annulus in the center of the field block (borehole diameter: 50 mm, borehole depth: 1.80 - 2.20 m, distance between two neighboring boreholes: 2 m).

Thus, an exactly defined mass of helium is introduced into the system, leading to a good reproducibility of the experimental results. To evaluate the injected tracer mass, a gas-volume flow meter is used, which has been calibrated in the run-up by an electronic helium-mass flow meter. During every dipole tracer test, 39.7 mg helium are injected into the field block. Due to the small injected volume of only 0.22 l helium (20°C and 1.0 bar), the temporary disturbance of the steady flow field, caused by the injection at slight overpressure, can be neglected.

Transport to the Central Borehole

In an initial experimental campaign, the tracer transport from the six annulus boreholes (Z1-Z6) to the borehole in the center of the field block (Z0) is investigated. Figure 5.52 illustrates the time-dependent tracer concentration measured in extraction borehole Z0. The helium concentration is displayed in parts per million, in correlation to the experimental time in seconds. All the concentration and time axes of the diagrams are scaled in the same way to facilitate a direct comparison.

It is obvious that significantly different breakthrough curves are recorded during the six transport experiments. Both, single peaks (e.g. Z6Z0) of different spreading and multiple peaks (e.g. Z5Z0) are recorded with different peak concentrations, ranging from 170 ppm to 750 ppm, with recovery rates from about 14 % to 31 %. The registered times of the first arrival of the tracer

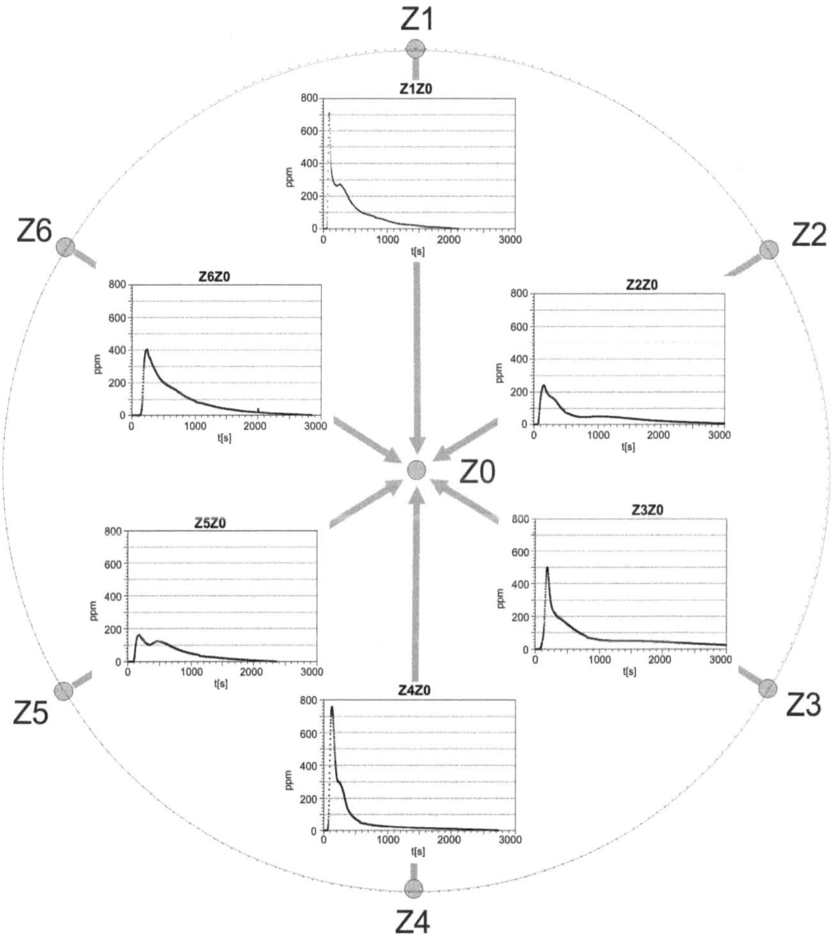

Fig. 5.52. Tracer-breakthrough curves of six helium tracer tests with Z0 as extraction borehole. Distance between two neighboring boreholes: 2 m.

in the central borehole vary from 53 s (injection hole Z1) to 132 s (injection hole Z6).

A comparison of tracer breakthrough curves of the experimental legs lying on a line show noticeable similarities. The breakthrough curves of the injection boreholes Z6 and Z3, Z1 and Z4, as well as Z5 and Z2 represent pairs of curves with comparable maximum concentration, peak number, and peak spreading. Each of the three main directions of the diagonals seems to be characterized by a certain curve shape.

Inversion of Transport Directions and Discussion of Curve Shapes

In a second experimental series, the transport direction on every transportation leg is inverted. Flow field and tracer are now injected at the central borehole (Z0) and extracted successively at one of the six surrounding boreholes (Z1 to Z6). Within these experiments, only five breakthrough curves can be measured and analyzed. Without a stronger side channel compressor and without a significantly lower pressure within the extraction hole, it is impossible to extract a constant air flow of 17 l/min at borehole Z2. Thus, Z2 seems to be only poorly connected to Z0.

When Figs. 5.52 and 5.53 are compared it becomes evident that the inversion of the transport direction between two boreholes only leads to a inconsiderable change in the shape of the breakthrough curves. The shapes, defined by maximum concentration, time of first arrival, peak spreading, and the time for the complete tracer transit, also fit into the shape scheme of the tracer tests to the central borehole.

All the breakthrough curves, measured on the Z1-Z0-Z4 leg feature the earliest times of first arrival of less than one minute, the highest maximum concentration of approximately 700 ppm to 800 ppm, and a sharply defined, dominating concentration peak with a noticeable shoulder or a weakly developed second peak at about 300 ppm. The transport process on this leg seems to be the most significantly fracture-dominated transport mechanism of all these tracer tests. Apparently, there are two independent flow paths with varying participation of the different transport processes. The measurements on the Z6-Z0-Z3 leg feature one dominating concentration peak as well. But this one shows a significantly stronger spreading and tailing and a lower maximum concentration of about 400 ppm to 600 ppm. The time for the complete tracer transit on leg Z3-Z0 is 4800 s and thus more than twice the time for the tracer transit from Z1 to Z0 (2100 s). These breakthrough curves also seem to reflect the coaction of several, variably characterized flow paths. One relatively sharply defined peak is followed by a widely spread tailing, with the decreasing concentration staying on one level over several minutes. At this point, the influence of sparsely dominating auxiliary fractures and the sandstone matrix with its porosity of 20 % (Vol.) defines the shape of the breakthrough curve. The breakthrough curves between Z5 and Z2 indicate an interaction of several, partly poorly conductive small fractures and

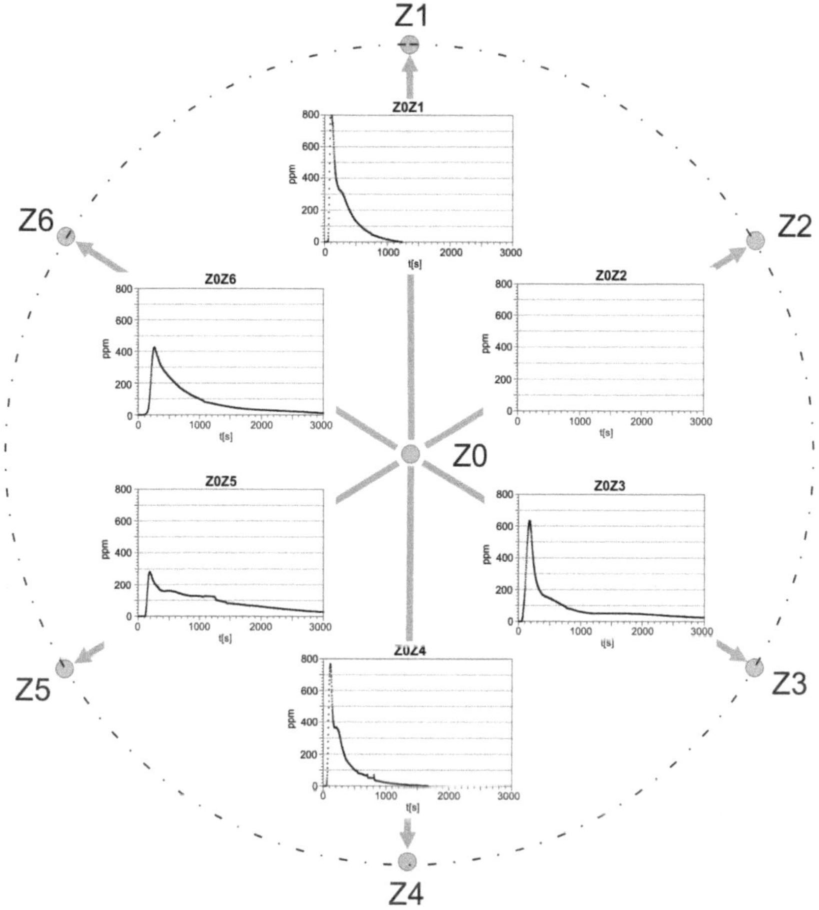

Fig. 5.53. Tracer-breakthrough curves of helium tracer tests with Z0 as injection borehole. Distance between two neighboring boreholes: 2 m.

a probably substantial participation of the highly porous sandstone matrix in the transport process. The measured peaks are significantly spread and show maximum concentrations of only 200 ppm to 300 ppm. Interestingly, however, the first arrival of Z2-Z0 is only inconsiderably later than the first arrival of the mainly fracture-dominated Z1-Z0 leg. Thus, the permeability of at least one of the involved flow paths cannot be considerably lower than that of the dominating main fracture of the Z1-Z0 leg.

Transport Direction and Fracture System

A correlation of the twelve tracer tests described here with the statistically determined fracture system is hardly possible. The significantly dominating fracture between Z1 and Z4 does not fit into the orthogonal fracture system with fractures striking north-west - south-east and south-west - north-east (Fig. 5.50). The EMR measurements described in Sec. 5.5.2 make this dominating discontinuity plain visible (Fig. 5.39).

The results of the flow experiments, conducted at the central boreholes, reflect the position of this fracture roughly. In this case, the most dominating conductivities around borehole Z0 are measured along the north-south direction. Figure 5.51 shows several fractures around the borehole annulus which are oriented in that direction. These fractures do not play a role in the fracture statistics of the whole field block, but they have great influence on the measured transport processes around the central borehole Z0. They mainly cause an early time of first arrival, a high maximum concentration, a quick concentration declension, and a short tailing in tracer tests in north-south or south-north direction.

5.6.4 Conclusions

In a sandstone quarry in Southern Germany, the sealed in-situ hardrock test site *Pliezhausen* is setup, where experiments to evaluate flow and transport parameters of fractured porous media are applied on the field scale. With different experimental set-ups as well as different tracers, the influence of changing boundary conditions and tracer-specific characteristics can be investigated.

The field and laboratory tests, described in this chapter, show that a basic characterization of transport parameters of an unsaturated porous fractured hard rock is possible by using the following techniques in combination:

- statistical evaluation of fracture parameters (orientation, length distribution and distance distribution),
- multidirectional gas tracer tests with He and SF_6,
- analysis of porosity, permeability and fracture-fillings in core samples,
- electro-magnetic-reflection measuring in different orientations.

Within the air-saturated (water-unsaturated) hard-rock system, two different tracer gases were used. Helium is significantly lighter than air, and could be called "floater" in analogy to an aquatic system. The other tracer gas is sulfur hexafluoride, which is heavier than air and could be called "sinker". An exact evaluation of the analogue potential in terms of investigation of pollutant transport of floating and sinking hydrocarbons seems to be profitable. The significant temperature-dependence of physical and chemical gas parameters can be used for worst-case studies on transport parameters. Gas-tracer tests conducted at the frozen test site in winter make the evaluation of transport parameters at extremely reduced retardation potentials (diffusion and gas solution) possible. Such investigations are hardly applicable to a saturated system. The unsaturated in-situ hard-rock test site *Pliezhausen* offers the opportunity to carry out basic research on flow and transport processes in fractured porous media.

5.7 Application of the Discrete Model on the Field-Block Scale

A. Silberhorn-Hemminger, R. Helmig

The investigation of flow and transport processes on the field-block scale using the discrete modeling approach (Sect. 2.4) has the aim of improving the understanding of flow and transport processes occurring in complex fractured porous media. The aim is to achieve the best possible agreement between measured and simulated flow and transport results; bearing in mind that the various processes and geometries have a significat influence on the results. Here, the characteristics of the results rather than exact fit is essential. Figure 5.54 shows the surface of the field block. The fracture traces along the side walls and the top of the field block are clearly seen. First, a three-dimensional deterministic fracture model for the south-east/east area of the field block is developed and presented in Sect. 5.7.1. In Sect. 5.7.2, the flow and transport processes are investigated using a two-dimensional model of the vertical plane between two vertical boreholes. In order to include the outer boundaries of the field block and the surrounding area, an extended two-dimensional model is presented in Sect. 5.7.3. The simulated and the experimental results are compared in Sect. 5.7.4.

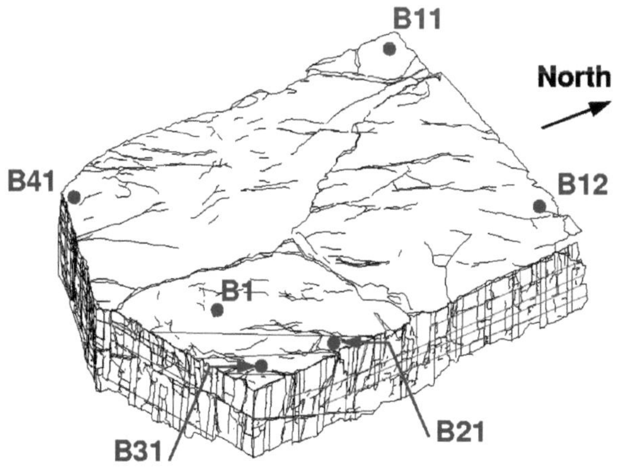

Fig. 5.54. View of the field block from the south-east. Mapping of the fracture traces on the side walls and approximation of the fracture traces.

5.7.1 Deterministic Fracture Model for the South-east/East Area and Boundary Conditions

First, a three-dimensional deterministic fracture model for the south-east/east area of the field block is developed. The relevant and significant fractures on the surface are detected by using stereo-photogrammetric plots (Sect. 5.3.1) and photos of the side walls and of the top. The detected fractures can be seen in Fig. 5.55 a)–c). In order to obtain a three-dimensional fracture model, the fracture traces detected on the different side walls and the top must be linked. The fracture traces of the three vertical fracture planes V1, V2, and V3 are visible on the two side walls (south-east and east wall) and on the top. Hence, the position and extension of the three vertical planes are well known.

However, the detection of the horizontal fracture planes, which are visible on the eastern side wall, are more difficult. The horizontally dominated structure elements represent the boundaries of different geological layers. These boundary layers have a dip to the north-east with an angle of 5°- 10° (Sect. 5.3.1). A first assumption is that the horizontal boundary layers, later called the horizontal fracture planes, extend almost horizontally through the complete field block. On the Basis of the detected vertical and horizontal fracture planes, a three-dimensional deterministic fracture model for the south-east/east corner of the field block is developed. Figure 5.55d) shows the fracture model.

In order to obtain a better understanding of the highly complex flow and transport processes within the field block, the model domain with its material properties, the geological structure and the boundary conditions reduced to a simple but still realistic model. The model is used as a *learning model*, which supports the analysis of the complex process behavior. Therefore, a two-dimensional plane which is located between boreholes B21 and B31 is extracted from the three-dimensional fracture-network model. The boundary conditions around this two-dimensional plane between boreholes B21 and B31 and around the field block are shown in Fig. 5.56. The field block sits on a sandy clay silt layer. There is atmospheric pressure along the outer surface of the silt layer. The whole surface of the field block is sealed. As described in Sect. 5.2.2, the surface can be regarded as almost impermeable. There is a direct contact area between the field block and the sandy clay silt layer along the layer boundary.

5.7.2 Two-Dimensional Case Study: Simulation 1

Tracer measurements between boreholes B21 and B31 show that the injection rate is significantly smaller than the extraction rate (Sect. 5.6.1). Two possible reasons are given: first, in addition to the injection borehole, there could be another inflow area which leads to the large difference between the controlled inflow and outflow rates. Because of the sealed surface of the field

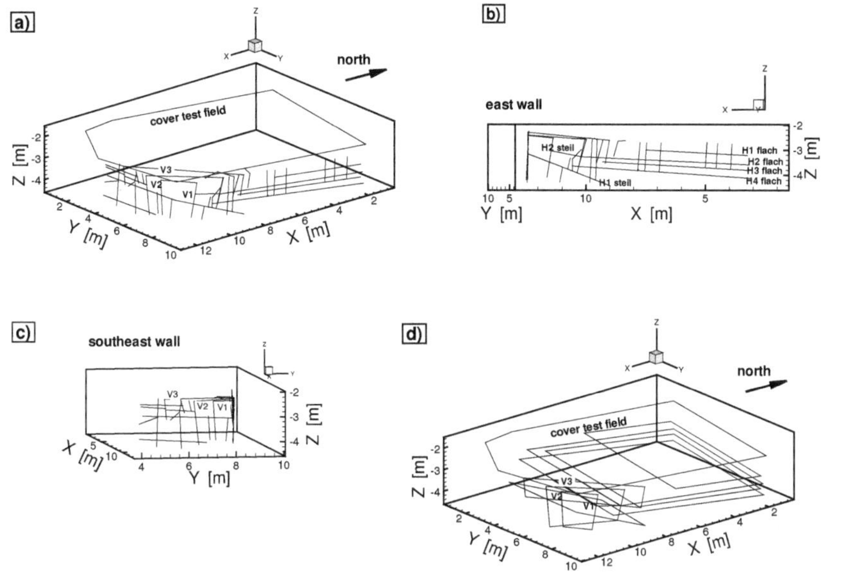

Fig. 5.55. Detected fracture traces (a–c) and the developed three-dimensional deterministic fracture model (d).

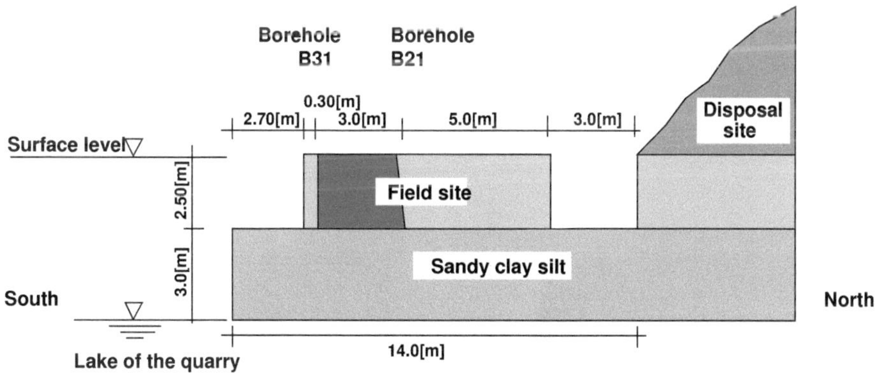

Fig. 5.56. Vertical section of the field block from north to south (year 2000).

block, an additional inflow area must be located at the bottom of the field block. Second, the in- and the outflow discharges can vary due to the compressibility of the gas. Since the imposed pressure ratio between the in- and the outflow boreholes is small (≈ 1.005), the first reason is considered to be the more realistic explanation.

In the first simulation, the additional inflow along the boundary between the field block and the silt layer is taken into account using a pressure boundary condition at the bottom of the simulation domain (see Fig. 5.57). For the two-dimensional fracture matrix model (Fig. 5.58, left), which is extracted from the three-dimensional fracture-network model, one receives only a very small inflow rate along the bottom boundary ($Q_{\text{in,bottom}} = 0.02\% \cdot Q_{\text{out}}$). Because there is no direct connection between the highly permeable fracture network system and the bottom boundary of the simulation domain, representing the boundary to the silt layer, the inflow rate along the bottom boundary is very small.

From the evaluation of the field-block surfaces, information about the statistical distribution of the fractures is available (Section 5.3.1). This additional

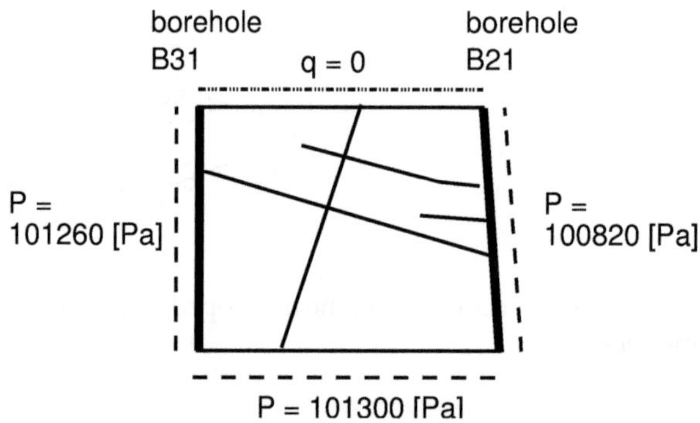

Fig. 5.57. Simulation domain 1: boundary conditions.

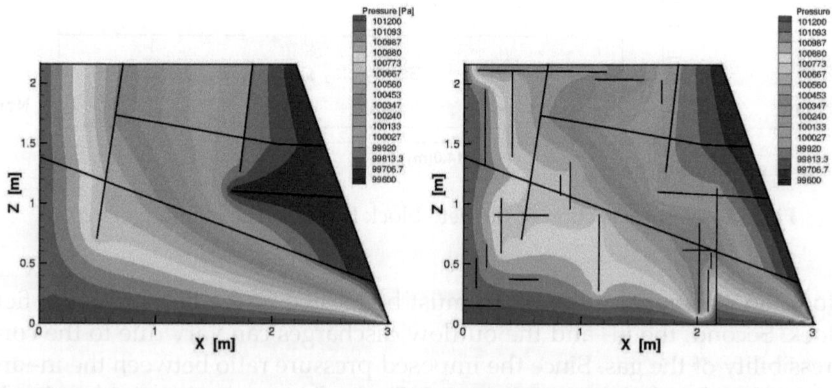

Fig. 5.58. Simulation domain 1, pressure distribution. Left: deterministic fracture network. Right: combination of deterministic and stochastic fracture network.

data is used to generate a stochastic fracture network that is combined with the deterministically detected fractures. The purpose of including the statistical information is to achieve a more realistic model of the actual fracture network. The extracted two-dimensional simulation domain of the fracture matrix system and the pressure distribution are shown in Fig. 5.58, right. There is one fracture which connects the bottom boundary with the highly permeable fracture network. With regard to the inflow and outflow rates, the influence of this fracture is very clear: 68% of the extraction rate Q_{out} flows along the outer boundary into the simulation domain. The simulations lead to the following results:

- There is a significant inflow into the investigation domain along the boundary layer field block / silt layer.
- A purely matrix-dominated connection between the field block and the silt layer does not cause the high inflow rates along the bottom boundary. In order to reach this high inflow rate, the field block and the silt layer must be connected directly by highly permeable fractures.
- The atmospheric-pressure boundary condition along the bottom boundary has the effect of a worst-case condition. However, the simple model clearly shows the relation between the influence of the bottom boundary and the fracture network.

5.7.3 Two-Dimensional Case Study: Simulation 2

In order to obtain a more realistic model, the above model is modified. The sandy clay silt layer is included into the model domain and the deterministic three-dimensional fracture network is updated using more detailed data of the field block and the boreholes themselves. The vertically dominated fractures reach the silt layer. Figure 5.59, left, shows the modified simulation domain and the chosen boundary conditions. To the right, the notation of the fractures as used in Table 5.9 is presented.

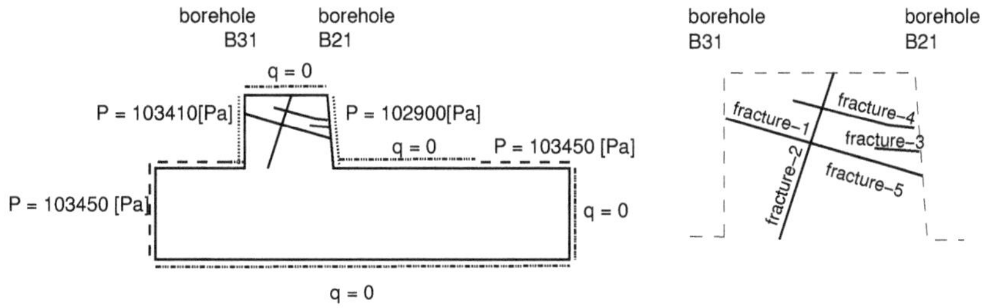

Fig. 5.59. Extended simulation domain. Left: boundary conditions following the selected experiment. Right: notation of the fractures.

5.7.3.1 Flow Simulation

First, the flow model and the experimental results of the selected measurement must correspond. Therefore, the model parameters matrix permeability, permeability of the silt layer, and fracture apertures are varied. Table 5.9 presents the variations of the model parameters and the inflow and outflow rates obtained.

Table 5.9. Variation and calibration of the flow parameters. If no parameter is given, the parameter is the same as in the previous column.

	run01a	run01b	run01c	run01d	run01e	run01f	run01g
Permeability (m^2)							
Matrix	$2.26{\cdot}10^{-15}$			$5.0{\cdot}10^{-13}$			
Clay silt	$1.33{\cdot}10^{-12}$					$1.0{\cdot}10^{-11}$	$7.07{\cdot}10^{-10}$
Aperture (m)							
fract. 1	0.001	0.002	0.0005		0.00045		0.00047
fract. 2	0.001	0.010			0.015	0.02	
fract. 3	0.001	0.002			0.005		
fract. 4	0.001	0.002			0.005		
fract. 5	0.001	0.002			0.010	0.02	
Permeability (m^2)							
fract. 1	$8.33{\cdot}10^{-8}$	$3.33{\cdot}10^{-7}$	$2.08{\cdot}10^{-8}$		$1.69{\cdot}10^{-8}$		$1.84{\cdot}10^{-8}$
fract. 2	$8.33{\cdot}10^{-8}$	$8.33{\cdot}10^{-6}$			$1.88{\cdot}10^{-5}$	$3.33{\cdot}10^{-5}$	
fract. 3	$8.33{\cdot}10^{-8}$	$3.33{\cdot}10^{-7}$			$2.08{\cdot}10^{-6}$		
fract. 4	$8.33{\cdot}10^{-8}$	$3.33{\cdot}10^{-7}$			$2.08{\cdot}10^{-6}$		
fract. 5	$8.33{\cdot}10^{-8}$	$3.33{\cdot}10^{-7}$			$8.3{\cdot}10^{-6}$	$3.33{\cdot}10^{-5}$	
Q_{in} ($m^3\,s^{-1}$)							
Meas. Qin=$1.83{\cdot}10^{-5}$ ($m^3\,s^{-1}$) (at B31)							
Sim.	$1.15{\cdot}10^{-4}$	$9.90{\cdot}10^{-4}$	$2.25{\cdot}10^{-5}$	$2.26{\cdot}10^{-5}$	$1.67{\cdot}10^{-5}$	$1.67{\cdot}10^{-5}$	$1.90{\cdot}10^{-5}$
Q_{out} ($m^3\,s^{-1}$)							
Meas. $Q_{out} = 9.24{\cdot}10^{-4}$ ($m^3\,s^{-1}$) (at B21)							
Sim.	$1.15{\cdot}10^{-4}$	$9.91{\cdot}10^{-4}$	$2.43{\cdot}10^{-5}$	$2.96{\cdot}10^{-5}$	$2.37{\cdot}10^{-5}$	$3.68{\cdot}10^{-5}$	$9.32{\cdot}10^{-4}$

A comparison of case *run01a* and case *run01b* shows that the larger fracture aperture of case *run01b* leads to a significant higher flow rate. The simulated outflow rate is similar to the experimental outflow rate. However, the simulated inflow rate along borehole B31 is too high. In a next step, the fracture between boreholes B21 and B31 is divided in two fracture segments (*fracture-1* and *fracture-5*) with different material properties (see Fig. 5.59). A very small aperture for *fracture-1* minimizes the inflow rate along borehole B31 (see case *run01c*). But the small outflow rate obtained is not desired. As shown in case *run01d*, the fracture network dominates the flow rates strongly, while the sandstone matrix contributes only to very small changes of the flow rates. The chosen permeability of the sandstone matrix $k_M = 5.0 \cdot 10^{-13}\,\mathrm{m}^2$ represents an effective permeability. The effective permeability combines the sandstone matrix and additional fractures which are not included in the pure fracture network.

The experimental measurements and the simulation results presented in Sect. 5.7.2 show that the main inflow into the investigation domain occurs along the boundary layer field block / silt layer. To represent this behavior in the model, a larger aperture of the vertical fracture *fracture-2* is chosen (see case *run01b, run01e, run01f*). Additionally, a larger permeability for the silt layer is chosen (case *run01f*). The inflow and outflow rates obtained are very close to the measured rates. The flow parameters of case *run01f* are the basis of the following transport simulations.

Figure 5.60 shows the pressure distribution for the flow simulation of case *run01a* and *run01f*. Comparing the two cases, one notices the larger reduction of the pressure along the silt layer as well as along the vertical fracture *fracture-2* in case *run01f*. This larger reduction in case *run01f* is based on the higher permeability of the silt layer, the larger aperture of the vertical fracture *fracture-2*, and the very small aperture of fracture *fracture-1* in case *run01f*.

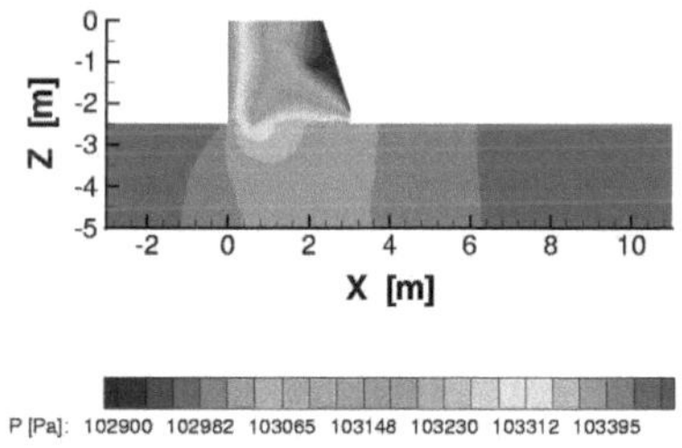

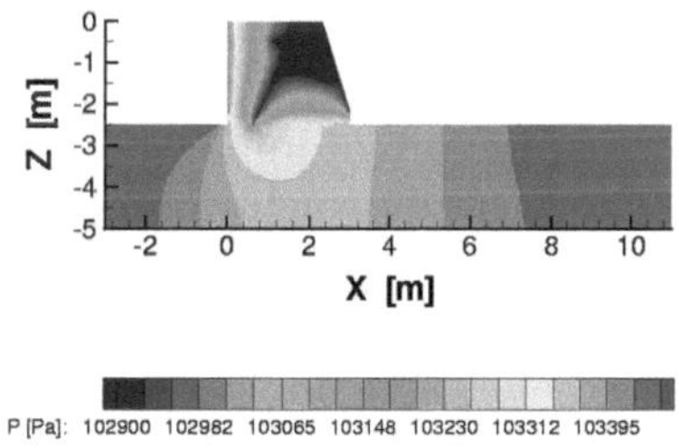

Fig. 5.60. Pressure distribution: case *run01a* and *run01f*.

5.7.3.2 Transport Simulation

The following boundary conditions are chosen for the transport simulations. At the injection borehole, a constant concentration $c/c_o = 1.0$ is injected over an injection time of $t_{in} = 60$ s. At the extraction borehole, a free-flow boundary is defined. Figure 5.63 shows the tracer-breakthrough curves for the various transport simulations presented in Table 5.10. If no parameter is given, the parameter is the same as the previous column. The right plot of Fig. 5.63 does not include the results of case *run03*. Furthermore, the scale of the y-axis is different from the left plot.

Table 5.10. Variation of the transport parameters.

	run01	run02	run03	run04	run05	run06
Porosity (-)						
Matrix	0.30					
Silt	0.30					
Fractures 1 - 5	0.30			0.40	0.30	
Dispersivity (m)						
Matrix longitudinal	0.002	0.01	0.001		0.01	
Matrix transversal	0.002	0.01	0.001		0.01	
Fractures 1 - 5	0.001				0.01	0.01
Molecular						
Diffusion ($m^2\,s^{-1}$)	$6.12{\cdot}10^{-5}$		0.0	$6.12{\cdot}10^{-5}$	0.0	$6.12{\cdot}10^{-5}$

The concentration distributions for the two simulations case *run01* and case *run03* are shown in Figs. 5.61 and 5.62 for three time steps. Note that Figs. 5.61 and 5.62 show only the area of the field block and not the complete simulation domain. The simulation results of case *run01* (Fig. 5.61) show that the fractures still dominate the transport, but the matrix has an additional important influence. Tracer infiltrates into the matrix and flows slowly towards the fracture network. The main tracer transport occurs in the fracture-network system.

The transport behavior in case *run03* (Fig. 5.62) is characterized by a small hydrodynamic dispersion which leads to an advectively dominated transport. As can be seen in Fig. 5.62, tracer mainly flows through the fracture network ($t = 60$ s, $t = 180$ s). At the intersection point of fracture *fracture-1, fracture-2,* and *fracture-5,* the tracer concentration is significantly reduced because of the zero concentration in the lower segment of *fracture-2.* After the tracer injection stops, a large amount of tracer remains in the matrix area around *fracture-1.*

All five curves in Fig. 5.63, left, show a very early first arrival time. Curve *run03* ascends very steeply and, after reaching the maximum peak, it descends very fast. A second, much smaller peak indicates a second flow path within the fracture network. This second flow path can be seen clearly in Fig. 5.62 ($t = 180$ s) along *fracture-4.* The advectively dominated transport is due to the small hydrodynamic dispersion: the molecular diffusion is zero and

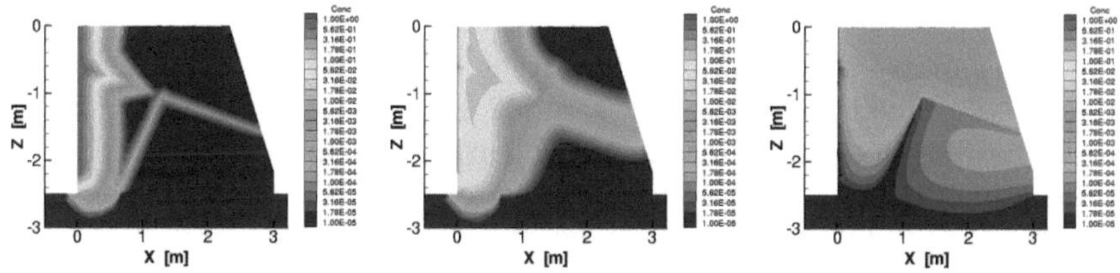

Fig. 5.61. *run01*: concentration c/c_0 after $t = 60\,\mathrm{s}, t = 180\,\mathrm{s}, t = 3000\,\mathrm{s}$.

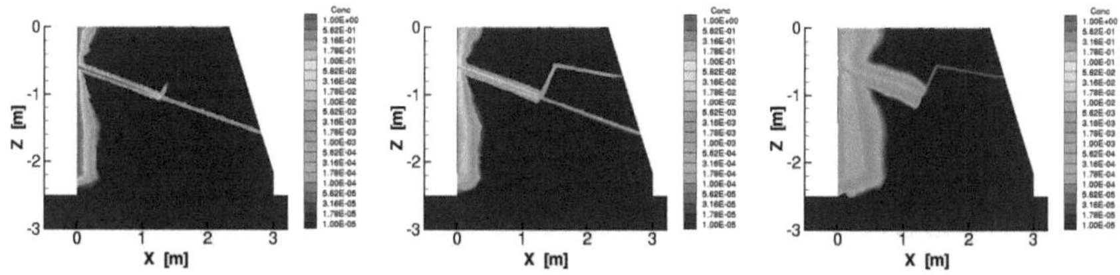

Fig. 5.62. *run03*: concentration c/c_0 after $t = 60\ \mathrm{s}, t = 180\,\mathrm{s}, t = 3000\,\mathrm{s}$.

the dispersivity of the fractures and the sandstone matrix is very small. The breakthrough curves of the simulations *run01, run02, run04,* and *run05* (see also Fig. 5.63, right) are very similar. They ascend steeply as well. However, the maximum peak is smaller than in case *run03*. The observed tailing indicates that the influence of the matrix is not negligible. Curve *run05* reaches a higher maximum than curves *run01, run02,* and *run04*. As in case *run03*, the molecular diffusion is zero, but the high dispersion of case *run05* prevents the development of a strong advectively dominated transport process.

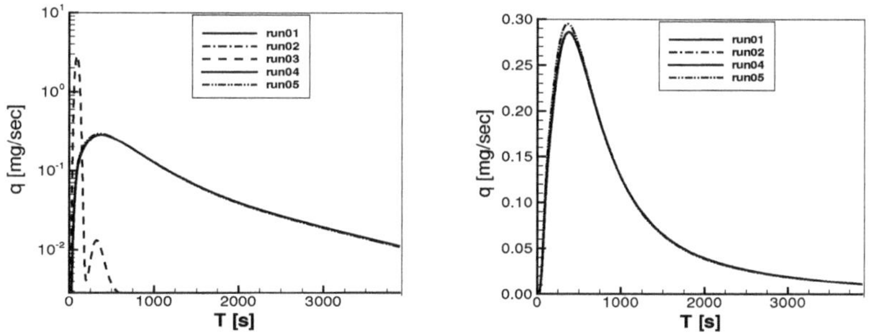

Fig. 5.63. Breakthrough curves for simulations listed in Table 5.10. Left: *run 01 – 05*, logarithmic tracer mass flux scale. Only *run03* differs from the other curves. Right: *run 01, 02, 04, 05*, linear tracer mass flux scale. All curves are very similar.

5.7.4 Comparing Measured and Numerical Results

Figure 5.64 shows the breakthrough curves of the selected experiment and the results of the simulations *run01, run02, run04,* and *run05.* It can be observed that the first arrival occurs much earlier for the simulation than for the experiment. Just shortly after the start of the tracer input at the injection boundary, the tracer plume reaches the extraction boundary in the simulations. In contrast, the measured breakthrough curve reaches the extraction borehole after a couple of minutes. The simulated and the measured curves differ in their maximum peak; however, the characteristics (increase, decrease, tailing) are similar. The tailing of the curves indicates the influence of the matrix.

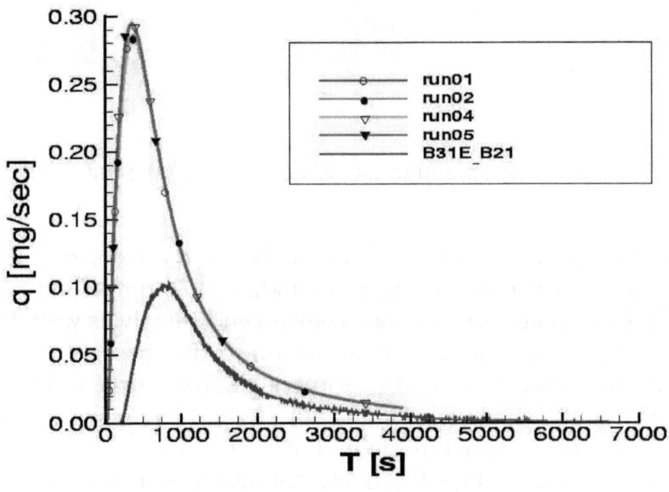

Fig. 5.64. Experimental and simulated breakthrough curves.

When the experimental and the simulated results are compared, it must be kept in mind, that the simulation domain does not represent the natural system of the field block. The simulations are based on a two-dimensional model, which cannot represent the strong three-dimensional flow and transport behavior within a fractured porous aquifer system. The same conclusion is drawn for the cylindrical bench-scale sample (Sect. 4.3.2.2)

The fractures are represented by parallel plates with a constant aperture. However, observations show that the fractures are partly filled with clay or sand. This heterogeneous structure of the fractures probably leads to channeling effects within the fracture planes, which cannot be reproduced by the parallel-plate concept.

Further improvement is expected if the deterministic fracture model is combined with stochastic fracture models which are based on recorded field data. The combined fracture models include more information on the field block than the single deterministic fracture model. Stochastically generated fracture systems are only single realizations of an infinite number of possible realizations. Therefore, the use of stochastic data requires a large number of simulations in order to achieve an average behavior that can be compared to measured data. Since one single measured curve is likely to deviate from the average behavior, several measurements at more that one location are a pre-requisite for achieving useful results.

In Chapt. 10, an approach for characterizing tracer breakthrough curves, measured at different locations within a domain, is presented. This approach is based on the classification of breakthrough curves using multivariate statistical methods. The method is tested on the bench scale, on $60 \times 60 \times 60\,cm^3$-block which does not contain very significant fractures. It is planned to conduct measurements which would allow this characterization method to be applied to the field block as well. Such multiple measurements would also allow simulations of stochastically generated fractured porous domains to be carried out.

5.8 Integral Transport Behavior on the Field-Block Scale

T. Vogel, D. Jansen, J. Köngeter

In the following study, a hypothetical aquifer system (based on the parameters of the field experiments) is used to present the capabilities of the developed multi-continuum model STRAFE (cf. Sec. 2.5) for three identified hydraulic components. This study is only a conceptual investigation and is used to gain an understanding of the processes.

5.8.1 Model Area and Boundary Conditions

In the study, a simplification of the sandstone block (near Tuebingen, Germany) is used. The horizontal circumference is approx. $10.0\,x\,7.0\,m^2$, the thickness approx. 2.0 m in the three-dimensional case study. For the one-dimensional case study, a transport distance of 10.0 m is considered.

The boundary conditions for the one-dimensional case study to identify the integral transport behavior are shown in Fig. 5.65. The tracer is infiltrated locally at one point. At the beginning ($t = t_0$), the whole area is free of tracer and the piezometric water level is 3.0 m. The boundary conditions are kept constant during the simulation. A Dirichlet boundary condition is chosen for the transport simulation, with a relative concentration of 1.0 at the inlet. The

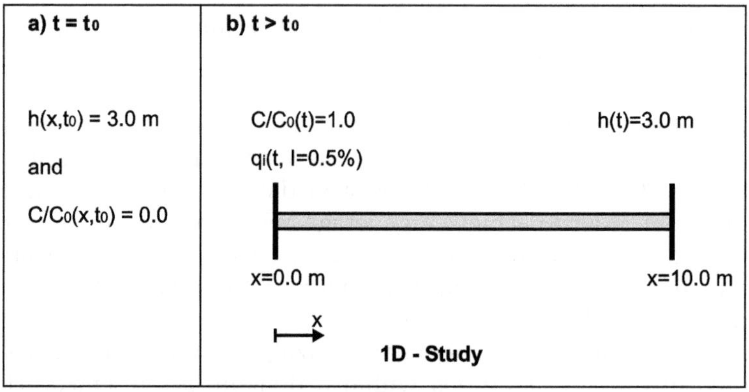

Fig. 5.65. Boundary conditions of the one-dimensional case study.

flow field is stationary, with a hydraulic gradient of 0.5 %. The outlet is on the side opposite to the inlet.

In the three-dimensional case study, a regional flow from one of the smaller face surfaces to the other is considered. The other surfaces are designed to be impermeable. It is very complicated to find an arrangement of integral boundary conditions on the face surfaces for an experiment. For this reason, punctiform boundary conditions are created (cf. Fig. 5.66). For the three-dimensional simulations, the z- and y-dimensions are reduced to half width for reasons of symmetry. This results in a modeling area of $10.0\,x\,3.5\,x\,1.0\,\text{m}^3$. At the beginning ($t = t_0$), the whole area is free of tracer and the piezometric water level is 3.0 m. At the point in time $t > t_0$, the hydraulic gradient between in- and outlet is 0.5 %. The point of the inlet or the outlet is situated at the middle of the respective face surface.

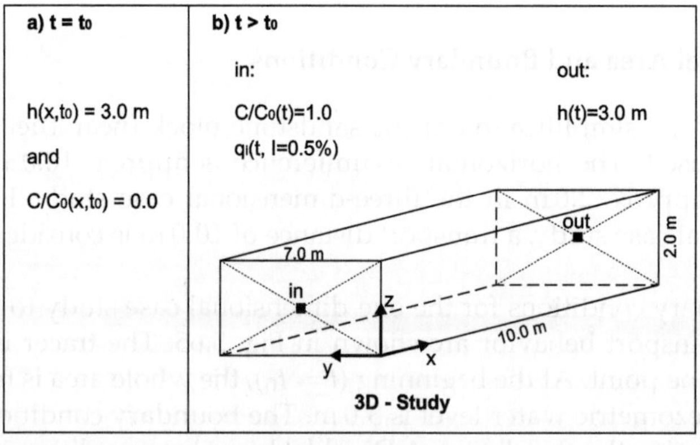

Fig. 5.66. Boundary conditions of the three-dimensional case study.

5.8.2 Aquifer Properties

During the laboratory and field experiments, three hydraulically active components could be identified, including a macro-fracture system, a micro-fracture system and the host matrix. The orientation of the fractures is assumed to be orthogonal, implying that the matrix blocks are rectangular. The equivalent and exchange parameters used for the basic configuration of the multi-continuum model are summarized in Tables 5.11 and 5.12.

Table 5.11. Equivalent parameters for the basic configuration of the multi-continuum model.

Equivalent parameter			Macro (f1)	Micro (f2)	Matrix (m)
Permeability	$\bar{k}$	(m^2)	$5.00 \cdot 10^{-12}$	$5.00 \cdot 10^{-13}$	$5.00 \cdot 10^{-14}$
Specific storativity	$\bar{S}_s$	(m^{-1})	$1.00 \cdot 10^{-5}$	$1.00 \cdot 10^{-5}$	$1.00 \cdot 10^{-5}$
Porosity	$\bar{n}_e$	(-)	$1.00 \cdot 10^{-3}$	$5.00 \cdot 10^{-3}$	$1.00 \cdot 10^{-1}$
Dispersivities:		(m)			
- longitudinal	$\bar{\alpha}_l$		$5.00 \cdot 10^{-2}$	$1.00 \cdot 10^{-1}$	$2.00 \cdot 10^{-1}$
- transversal horizontal	$\bar{\alpha}_{th}$		$5.00 \cdot 10^{-3}$	$1.00 \cdot 10^{-2}$	$2.00 \cdot 10^{-2}$
- transversal vertical	$\bar{\alpha}_{tv}$		$2.00 \cdot 10^{-3}$	$5.00 \cdot 10^{-3}$	$1.00 \cdot 10^{-2}$
Coefficient of Diffusions	$\bar{D}_m$	(m^2s^{-1})	$8.00 \cdot 10^{-10}$	$6.00 \cdot 10^{-10}$	$4.00 \cdot 10^{-10}$
Relative volume	Ω	(-)	1.00	1.00	$9.94 \cdot 10^{-1}$

Table 5.12. Exchange parameters for the basic configuration of the multi-continuum model.

Exchange parameter			Macro / Micro f1f2	Macro / Matrix f1m	Micro / Matrix f2m
Shape of blocks			cube	cube	cube
Penetration depth	$\bar{s}_{max}$	(m)	$5.00 \cdot 10^{-1}$	$5.00 \cdot 10^{-1}$	$2.50 \cdot 10^{-1}$
Specific surface	$\bar{\Omega}_0$	(m^{-1})	6.00	6.00	$1.20 \cdot 10^{+1}$
Specific fracture surface	$\bar{\Omega}_W$	(m^{-1})	2.00	2.00	4.00
Geometric factor	ϵ	(-)	5/3	5/3	5/3
Exchange parameter:					
- flow	$\bar{\alpha}_Q$	(ms^{-1})	$8.33 \cdot 10^{-6}$	$8.33 \cdot 10^{-7}$	$8.33 \cdot 10^{-7}$
- transport	$\bar{\alpha}_c$	(m^2s^{-1})	$5.00 \cdot 10^{-12}$	$6.67 \cdot 10^{-11}$	$6.67 \cdot 10^{-11}$
Interface function	$\bar{A}(s)/\bar{A}_0$	(-)	$1 - (2 - \sqrt{\bar{s}_{max}})s + (1/\bar{s}_{max}^2)s^2$		

For the basic configuration, all components of the system influence the regional transport behavior. In accordance with the conceptual idea of the study homogeneous properties are assumed. A high connectivity is then assumed for the chosen transport parameter.

5.8.3 One-Dimensional Case Study

The results for the basic configuration of the triple-continuum model, compared to the single double-continuum models, are presented in Fig. 5.67. These simulations are performed with a one-dimensional model. The coupling chosen for the basic configuration is the parallel coupling method, i.e. all components are coupled with each other.

In the basic configuration, a consistent temporal development for the transport phenomenon is observed (cf. Fig. 5.67). This development cannot be explained simply with a double-permeability model. The three phases distinguished for the transport behavior are drawn schematically in Fig. 5.67. The breakthrough curve f1f2m (BTC) exhibits a "fast breakthrough" at the beginning, and this phase is defined as phase I.

Within the second phase, which is clearly dominated by the interaction of all components, two subphases can be distinguished (the 1st part and the 2nd part of phase II). When the double-continuum model results are compared with the triple-continuum result, it is evident that there are two significant changes in inclination for the TPTP model. The first change indicates the beginning of phase I, resulting from the interaction of the macro-fracture system with the micro-fracture system (f1f2). The second change in inclination is induced by the interaction of the fracture systems with the matrix, because the point in time corresponds with the point in time for simulation f1f2, where the gradient between the components can no longer be observed.

A comparison of the coupling methods and their influence on the transport behavior is presented in Fig. 5.68. The system reacts differently for the selective coupling method than for a parallel or serial coupling method. This is because of the missing interaction of the micro-fracture system and the host matrix which results in a faster equilibrium in concentration between the macro- and micro-fracture system. This can be observed at the breakthrough curve for the first half of the interaction. The slope of the breakthrough curve for selective coupling is less steep than for serial or parallel coupling within the second half of the interaction phase. This behavior can be explained by the fact that, for these points in time, mass transfer only takes place between the host matrix and the macro-fracture system. This transport mechanism results in a slower regional transport.

5.8.4 Three-Dimensional Case Study

The numerical experiment of the conceptual three-dimensional case study allows an evaluation of the transport behaviour when tracer tests are performed. Because of the isotropic, homogeneous material parameters of the

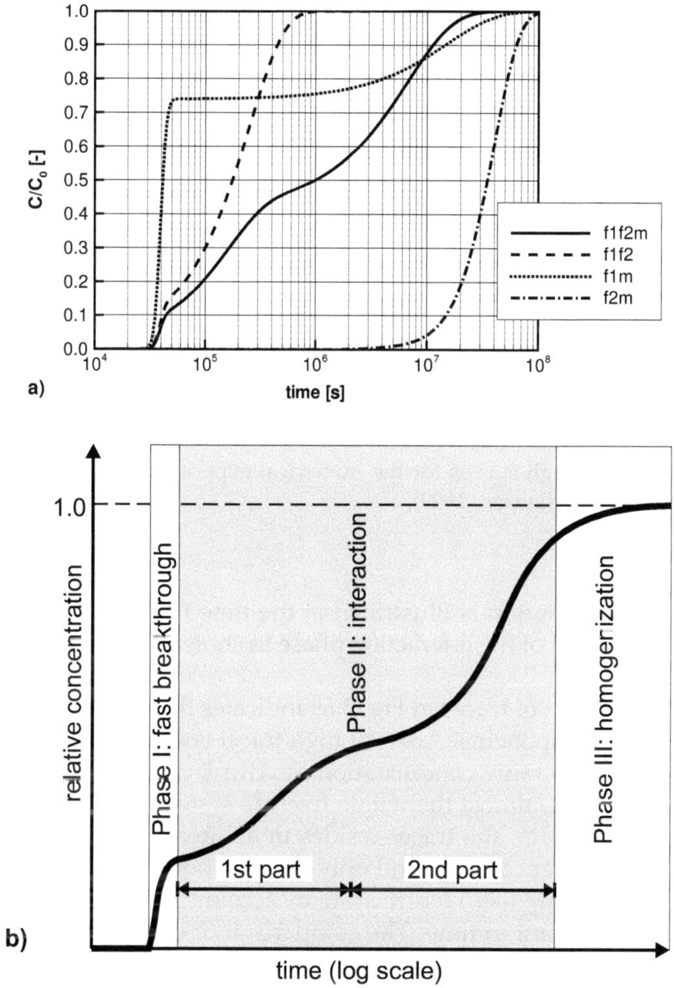

Fig. 5.67. Breakthrough curves for the numerical experiments: (a) comparison of the TPTP model with the different double-continuum models; (b) typing of the integral transport behavior of the TPTP system (Jansen, 1999).

basic configuration and the symmetric boundary conditions (cf. Fig. 5.66), the model area is reduced to a quarter of the original cuboid. The origin of the coordinate system lies at the point of the tracer discharge. The x-axis connects the point of the tracer discharge on the front surface of the cuboid with the measuring point at the back surface. The dimension of the reduced model area is $L_x/L_y/L_z = 10.0/3.5/1.0$ [m].

Figure 5.69 shows the three-dimensional distribution of tracer on the surface of the model area in all components (f1+f2+m) of the multi-continuum

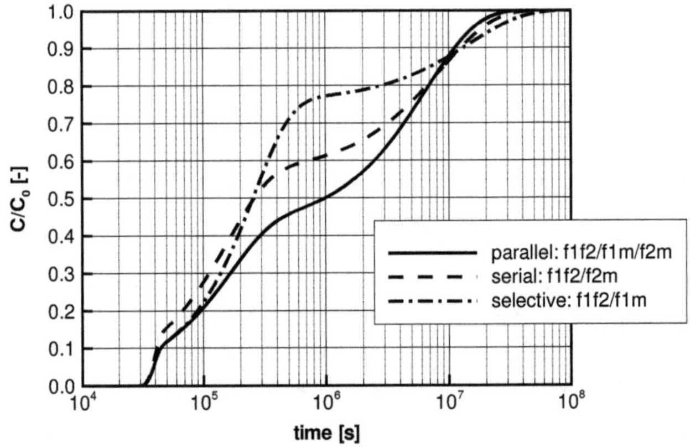

Fig. 5.68. Breakthrough curves for the numerical experiments and a comparison of the coupling models (Jansen, 1999).

system. The distribution is illustrated at the time $t = 2.00 \cdot 10^{+05}$ s. This is during the first half of the interaction phase as shown in the one-dimensional case study.

The distribution of tracer in Fig. 5.69 indicates the different transport behaviour of the components. Areas of high tracer concentration are shown in gray, areas of low tracer concentration are displayed in black. At the time of $t = 2.00 \cdot 10^{+5}$ s, almost the whole fracture system is full of tracer. In the micro-fracture system, the tracer resides in an area $x < 0.5$ m whereas, in the matrix, the tracer can be found only near the point of the tracer injection. This means that only the fracture systems account for the transport properties at this early point in time. The comparison of concentration values at a certain point within the domain shows the different transport velocities in the continua.

5.8.4.1 Summary

This conceptual study on the field-block scale, including two hierarchical fracture categories and a host matrix in a triple-continuum model, illustrates a characteristic transport behavior which may not be explained by a homogeneous single- or double-continuum model. The transport behavior is divided into three characteristic phases.

Further investigations are presented by Lagendijk (2003), in which tracer breakthrough curves in fractured porous media are analyzed. A method is developed that allows tracer pulses to be used as boundary conditions with respect to transport and the given signal (breakthrough curve) of the effluent

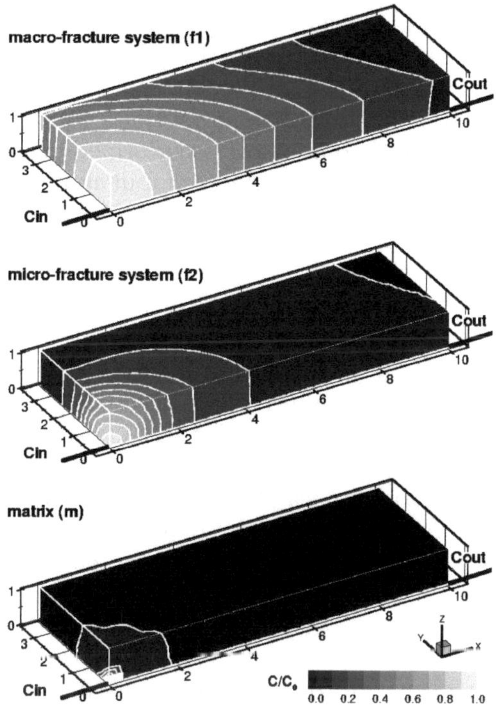

Fig. 5.69. Concentration distributions in the three-dimensional model area; discharge of the tracer in all components occurred at the point in time $t = 2.00 \cdot 10^{+5}$ s.

to be transformed to the signal provided by infinite tracer injection. A variety of studies has been performed for those breakthrough curves and the number of components relevant for flow and transport in fractured porous media identified.

It is proposed that these approaches be applied to experimental data in order to evaluate their capabilities.

5.9 A Study Concerning Boundary Effects on the Field-Block Scale

D. Bachmann, T. Vogel, J. Köngeter

Several boreholes are positioned on the field block (cf. Sect. 5.1) to carry out flow and transport studies. By dint of numerical simulations with the multi-continuum model STRAFE, the extent to which the model boundary affects the shape of the breakthrough curves is analyzed. Because of the fractures in the sandstone block, the flow velocities observed in reality are noticeably larger than those modelled in the context of this study. The influence of the model boundary should be smaller in reality than predicted by the model. Therefore, the simulations can be regarded as a *worst-case-scenario* study. The simulations are performed by using a homogeneous two-dimensional model composed of two coupled media (cf. Sect. 5.9.1).

First, a preliminary study of the flow is performed to calibrate the model and test the boundary conditions used at the input and output ports. Then, a transport calculation is conducted to determine the boundary effects on the tracer-breakthrough curves.

5.9.1 Model Design

The model of the sandstone block is set up as a two-dimensional model with a two-dimensional subdomain, which represents the layer of poor clay found underneath the sandstone block on the test site. It also allows for the model to take into account all of the leaks of the sandstone block, thus simulating flow rates comparable to the ones observed by tracer tests on the test site (cf. Sect. 9.1.1). Both media are coupled by one-dimensional line elements. Figure 5.70 shows a top view of the model area, the section in the middle representing the field block. The outer boundary defines the model area of the subdomain, which exceeds that of the sandstone block by two meters on all sides, to represent natural conditions better. The inner part of the mesh for the subdomain, however, corresponds to the one for the sandstone block. The two lines crossing each other represent the paths on which the in- and output ports are moved for the flow and transport calculations (cf. Sect. 5.9.3). Figure 5.71 presents a profile of the model, illustrating different configurations of the boundary conditions (cf. Sect. 5.9.3).

The boundary of the sandstone layer is impervious, but the system can aspirate air via the subdomain, where the pressure at the boundary is set to air pressure.

5.9.2 Material Properties

Table 5.13 shows the material properties of the model, those of the fluid (air) and the tracers (helium and SF_6) used for the transport simulations. The orders of magnitude of these parameters are taken from Jaritz (1999) and from Sect. 5.6. The dynamic viscosity η, the density ρ and the diffusion coefficients D_{m} are given at a temperature of 20°C and a pressure of 101.30 kPa. The material properties of the subdomain are the same as those of the sandstone block, except for the permeabilities k_{xx} and k_{yy}, which are the values used for calibration of the model (cf. Sect. 5.9.3).

The effective diffusion coefficient $D_{\mathrm{m,e}}$ in porous media is calculated by means of

$$D_{\mathrm{m,e}} = D_{\mathrm{m}}\omega \,, \tag{5.10}$$

where ω is an empirical coefficient that takes into account the effects of the matrix on diffusion and ranges from 0.01 to 0.50 (Freeze and Cherry, 1979).

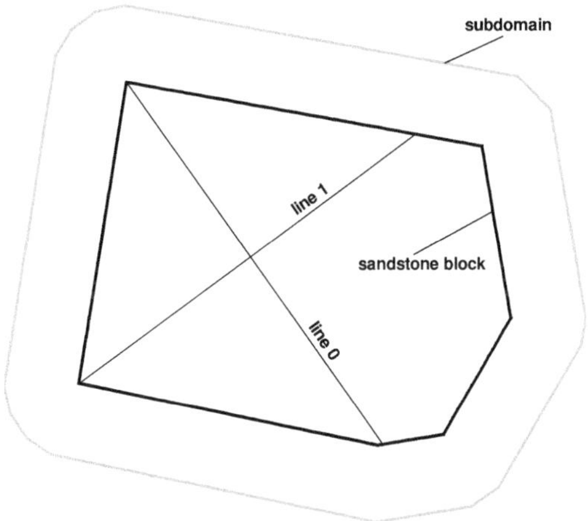

Fig. 5.70. Top view of the model area.

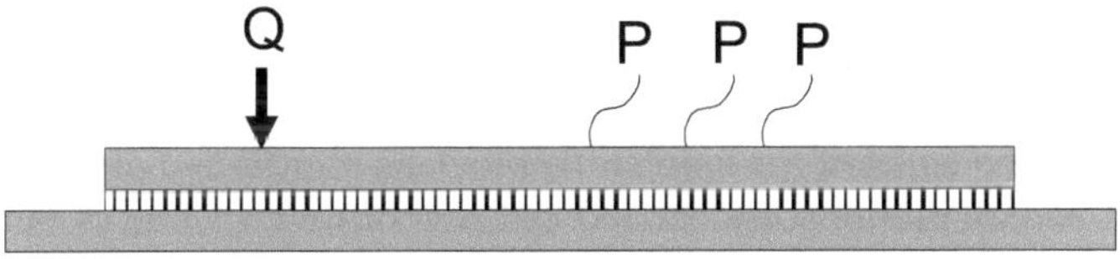

Fig. 5.71. Profile of the model area.

For this study, a value of $\omega = 0.10$ is chosen. The values for the molecular diffusion coefficients D_{m} are taken from Thüringer (2002).

5.9.3 Flow Simulation

The dual-media model is calibrated on the basis of the tracer tests conducted on the sandstone block. The boundary conditions at the input and output are varied between Dirichlet pressure and Neumann flow. Different values for the hydraulic conductivity of the poor clay subdomain are tested.

The variant Neumann - Dirichlet (Q - P) (cf. Fig. 5.71) provides the best results in comparison to the field experiments, so the transport calculation is based on the related boundary conditions. At the output port, a pressure of $p = -420\,\mathrm{Pa}$ is fixed, while the discharge is set to $Q = 1.3 \cdot 10^{-5}\,\mathrm{m^3 s^{-1}}$ at the input port.

Table 5.13. Material properties of the sandstone block, fluid properties and tracer properties.

Material properties of the sandstone block		
n_e	(-)	0.18
k_{xx}	(m^2)	$4.00 \cdot 10^{-14}$
k_{yy}	(m^2)	$4.00 \cdot 10^{-14}$
α_l	(m)	$1.00 \cdot 10^{-2}$
α_t	(m)	$2.00 \cdot 10^{-3}$
Fluid properties		
ρ	($kg\,m^{-3}$)	1.00
μ	($kg\,m^{-1}s^{-1}$)	$1.81 \cdot 10^{-5}$
Tracer properties		
$D_{m,He}$	(m^2s^{-1})	$5.80 \cdot 10^{-6}$
$D_{m,SF6}$	(m^2s^{-1})	$0.83 \cdot 10^{-6}$
ω	(-)	0.10

The value for the permeabilities of the subdomain is set to $k_{xx,sub} = k_{yy,sub} = 10^{-11}\,m^2$. This is a rather high permeability for a clay, but it is in accordance with the observation at the test site that the layer underneath the sandstone block has dried out and as a consequence fractures appeared, increasing the permeability of the clay. Moreover, this permeability value takes into account the leaks of the sandstone block as mentioned above.

The flow calculations are carried out for various distances between the input and output ports (cf. Fig. 5.71). The results are summarized in Table 5.14. When compared to the results of the tracer tests made at the block, they exhibit the same order of magnitude. For a distance of 2 m between the input and output ports, a pressure of -100 Pa and a discharge of $1.3 \cdot 10^{-5}\,m^3s^{-1}$ are measured at the input port, and -420 Pa and $-9.0 \cdot 10^{-4}\,m^3s^{-1}$ at the output port (Hötzl *et al.*, 2000).

Different configurations of the input and output ports are tested. The distance between the ports is varied from 1.00 m to 5.00 m in 1.00 m steps. Furthermore, the ports are moved towards the center of the model area, also in 1.00 m steps, on the crossing lines (cf. Fig. 5.70). The input port at a distance of 1.00 m from the boundary is referred to as *position 1*, the input port at 2.00 m from the boundary as *position 2*, and so on. This is illustrated in Fig. 5.72, which shows the results of the flow simulations for two different configurations: in Fig. 5.72 (a), the input node is in the middle of the model area (*position 7*) while in Fig. 5.72 (b), it is in the upper left-hand corner (*position 1*). In both configurations, the distance between the input and output ports equals 1.00 m. The tracer is injected continuously throughout the whole simulation.

Table 5.14. Results of the flow simulations.

Distance between input and output port (m)	Pressure at the input port (Pa)	Discharge at the output port (m^3s^{-1})
1	−109	$-9.0 \cdot 10^{-4}$
3	−36	$-8.4 \cdot 10^{-4}$
5	−15	$-7.9 \cdot 10^{-4}$

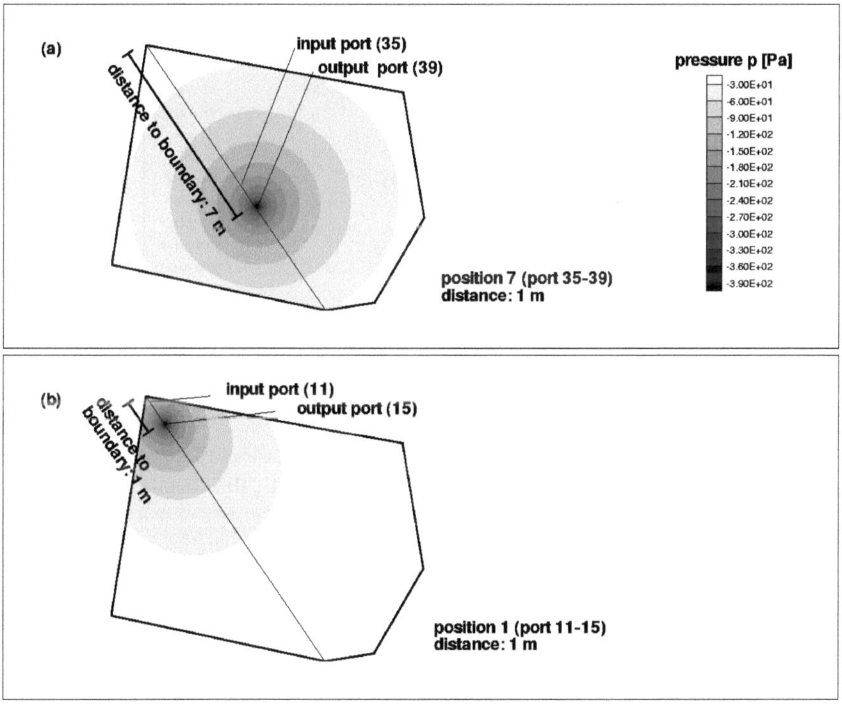

Fig. 5.72. Results of the flow simulation for two different configurations: (a) *position 7*, (b) *position 1*.

5.9.4 Transport Simulation

The results of the transport simulations illustrate the influence of the model boundary on the breakthrough curves. Figure 5.73 presents the breakthrough curves with helium as the tracer, at a distance of 2.00 m between the input and output ports. It can be observed that moving the ports towards the center of the model area, away from the boundary, evokes correspondingly fewer differences in the breakthrough curves. Thus, the boundary has a significant influence on the shape of the breakthrough curves for configurations close to the boundary. From a certain distance onwards, there is also a noticeable influence on the breakthrough curves, but the shape and tailing of the breakthrough curves are affected in a similar way.

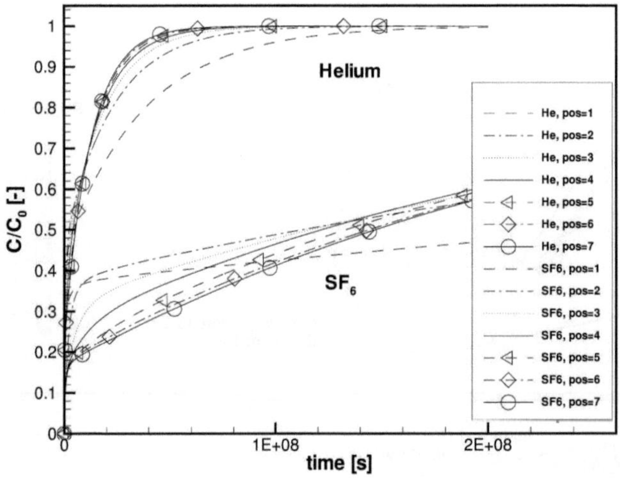

Fig. 5.73. Breakthrough curves for helium and SF_6, with the distance between the input and output ports set at 1 m.

For the present choice of boundary conditions, another phenomenon is observed. Since the boundary of the sandstone block is impervious, once the concentration front arrives at the boundary, the tracer is reflected. So the model starts to fill up with tracer until a concentration of 100% is reached at all nodes. Figures 5.74 (a) and (b) illustrate the distribution of the relative concentration of two different port configurations (*position 7* and *1*) for the same point in time, which is near the intersection point of these breakthrough curves (cf. Fig. 5.73). The reflection of the concentration front is demonstrated by the greater concentration for a port configuration in the center of the block (cf. Fig. 5.74 (a)) than for one close to the boundary (cf. Fig. 5.74 (b)).

The reflection of the concentration front can also be observed in the breakthrough curves (cf. Fig. 5.73): while in position 1, a concentration of 100 % is achieved latest, the breakthrough curves get steeper as the ports are moved

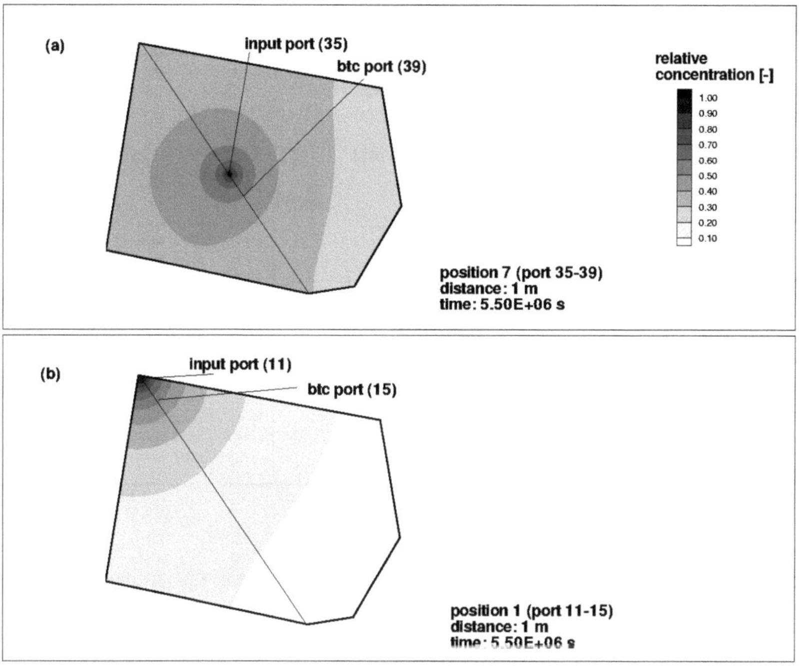

Fig. 5.74. Distribution of the relative concentration (helium) in the sandstone block for two different configurations: (a) *position 7*, (b) *position 1*.

toward the center of the block. At the more central positions, the boundaries are closer to the ports on all sides, so the reflected concentration fronts arrive at the output port earlier. To compare the influence of diffusion on the result, the simulations are carried out for two different tracers, helium and SF_6, the diffusion coefficient of the latter being much smaller than that of helium. Since the flow velocities in the porous medium are relatively small, the dominant transport phenomenon is diffusion. Therefore, the less important diffusion of SF_6 makes the concentration front move more slowly, so that the breakthrough curves are flatter than those with helium as a tracer (cf. Fig. 5.73). The results for the different distances between the input and output ports are essentially the same.

In Fig. 5.75, the breakthrough curves for helium are compared to those of a so-called dipole test, where the sandstone block is represented by a two-dimensional single-medium model, without the subdomain. The boundary conditions at the input and output ports are the same as those of the dual-media model (cf. Sect. 5.9.3).

Due to the impervious boundary, the discharge into the model at the input port must be equal to the one taken out of the model at the output port. No air is aspirated into the sandstone block, so that, in contrast to the simula-

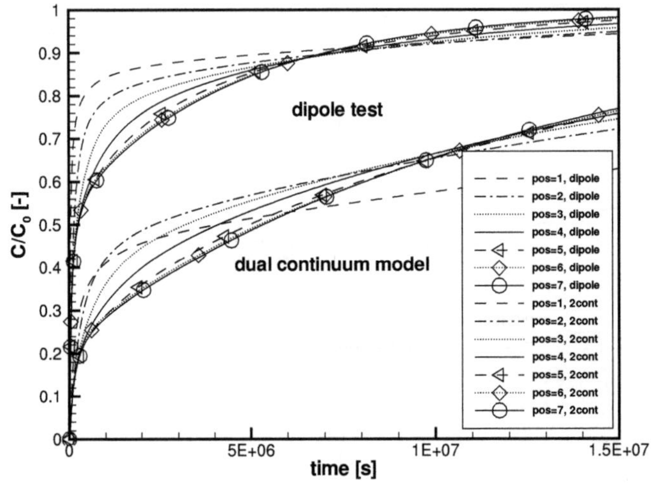

Fig. 5.75. Breakthrough curves of the dipole test compared to the dual medium model (distance between input and output ports = 1 m; helium tracer).

tions with the dual media model, no dilution occurs. Thus, the breakthrough curves of the dipole test are steeper.

Finally, the boundary conditions are changed to simulate pulse injections of a tracer. Figure 5.76 shows the resulting breakthrough curves for configurations with a distance of 3.00 m between the input and output ports, and the curves for the other configurations are essentially the same. The influence of the boundary is obvious as well. The shape of the part of the breakthrough curves before the peak is similar already from a distance of 1.00 *m* from the boundary onwards. The maximum and the tailing of the curves after the peak do not change significantly for configurations that place the input node more than four meters away from the boundary (from position 5 onwards). Again, the reflection of the concentration front is mirrored in the breakthrough curves, here by the bump in the descending part of the curves, which is especially pronounced for the configurations in the centre of the model area.

5.9.5 Conclusions

The investigation concerning boundary effects on the field-block scale demonstrates that the boundary effects are significant if the input and output ports

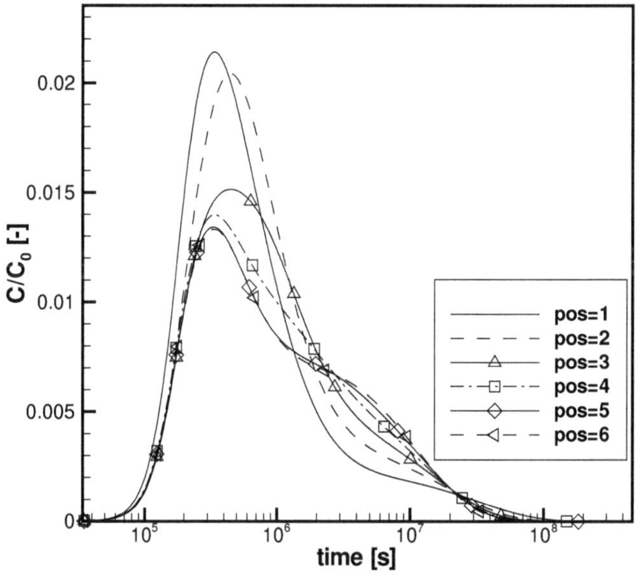

Fig. 5.76. Breakthrough curves for pulse injection (distance between input and output ports = 3 m; Helium tracer).

are positioned near the center of the sandstone block (pos. 4-pos. 6). However, the fact that the breakthrough curves calculated in the center of the block have almost the same shape and tailing (illustrating that the curves converge) shows that breakthrough curves from a test configuration at some distance to the boundary are influenced in a similar way by the boundary effects. This allows for an easier interpretation and makes them more comparable to other field-block scale studies.

The impacts of the boundaries are almost independent of the properties of the tracer (diffusion coefficient), the approach of the model (single-medium model or dual-media model) and the duration of the tracer injection (pulse injection or continuous infiltration).

The results of the numerical investigation of the boundary effects can be transferred to field experiments. They show that tracer experiments on the field block lead to more significant and comparable results when a port configuration in the center of the block is chosen. Therefore, the position of the ports should be chosen as described in Sect. 5.9.4.

Part III

Scale-Independent Approaches and Investigations

6

The Multi-shell Model - A Conceptual Model Approach

C.I. M^c^Dermott, R. Liedl, M. Sauter, G. Teutsch

On a small (discrete) laboratory scale (typically of the order of 10 cm), individual processes and parameters defining the physical conditions of single fractures may be well defined and modelled. However, on a larger scale, where fracture networks dominate the flow and transport characteristics, the interaction of the various processes controlling flow and transport in the fracture system coupled in some cases with a porous medium cannot be easily defined. Here, a conceptual stream-tube model is presented where the stream-tube geometry is defined by the geometry of the flow in the system. This model was used to investigate the three-dimensional flow systems in bench scale laboratory samples (sample diameter 30 cm, length 40 cm) containing fracture networks in porous material. The model avoids the definition of the individual processes and concentrates on the integral flow and transport signal dependent on the geometry of the flow systems developed during gas tomographical flow and transport investigation. The use of the model to analyze the tomographical data allows the definition of three-dimensional anisotropic tensors characterizing the flow and transport characteristics of the fractured systems based on experimental results.

The model is based on the expected flow patterns within the dipole flow fields generated in the experimental cell. The geometry of the stream tubes comprising the model is controlled by the geometry of the flow field, which in turn led to the model being described as a multi-shell model. In principle, the flow field is represented by a series of one dimensional stream tubes which geometrically correspond to flow shells around the center line/plane joining the dipole source and sink, hence the term multi-shell. These stream tubes are combined to give the three-dimensional flow field signal measured in the experimental work.

Not only the multi-shell model provide an approximation of the flow and transport signal to be expected from the various geometries of the experimental investigation (see Chapt. 4) and thereby allow the deviation from this expected signal to be derived, but it also provides a clear conceptual understanding of the geometrical factors effecting flow and transport from a small laboratory scale through to the field scale. By an examination of the deviation of the expected signal, the effects caused by the fracture networks can be investigated.

The subsequent chapter starts with a discussion of the model principle, followed by the mathematical development of the model and a consideration of the effects of boundary conditions. Once the model has been described, it is use to understand the different effects of flow fields where the pressure is allowed to vary in one, two or three dimensions, i.e. a one-, two- or three-dimensional flow field. This is demonstrated on hand of experimental results gained from the Multiple Input Output Jacket discussed in Chapt. 4. The effects are then examined further by considering the transport signal and interpreting its form, again with reference to the development of the flow field. Finally, this conceptual approach is shown to provide a practical understanding of the distribution of tracer in the flow fields developed.

6.1 Model Principle

The principle behind this modeling approach is apparent from Fig. 6.1. The flow field is divided up into a series of shells around the center flow line / plane, i.e. the direct connection between input and output port. If the flow is three-dimensional, the shells have the form of a ball or onion (Fig. 6.1b) and if the flow is two dimensional, the shells have a cylindrically curved form (Fig. 6.1a). The flow in each individual shell is calculated, which allows the multi-dimensional flow to be determined as a combination of the flow occurring in the individual shells. For simplicity, two such shells are illustrated in Fig. 6.2. The shell in Fig. 6.2a illustrates the case where the distance h_n from the center of flow to the shell is less than the distance x_0, i.e. half the distance between the input and output positions. The shell in Fig. 6.2b illustrates the case where the distance h_n from the center of flow to the shell is greater than x_0.

An examination of Fig. 6.2 shows that the flow inside the experimental cell must be a subset of the unlimited flow conditions where no boundaries are present (Fig. 6.2b). Here, the unlimited flow conditions (i.e. flow without the boundaries of the cell walls) in a homogenous medium are considered firstly and then the limited boundary conditions of the experimental cell.

In the case of the two-dimensional flow system, an approach similar to a stratified two well system was applied by Güven *et al.* (1986): the flow field was approximated by a number of thin crescents which matched the pattern of the flow lines between the wells.

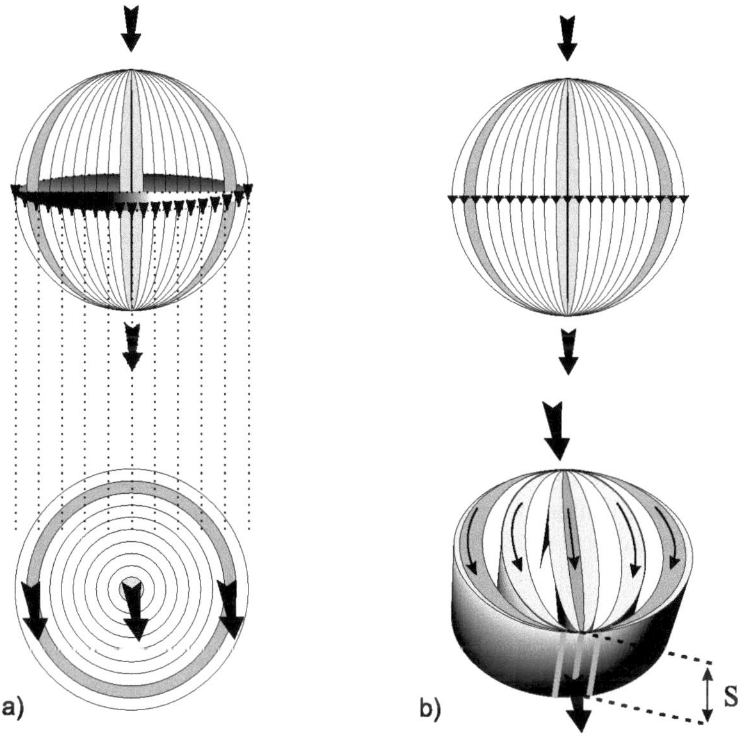

Fig. 6.1. Stream-tube model geometry based on the geometry of the flow field. a) Two-dimensional flow field. b) Three-dimensional flow field.

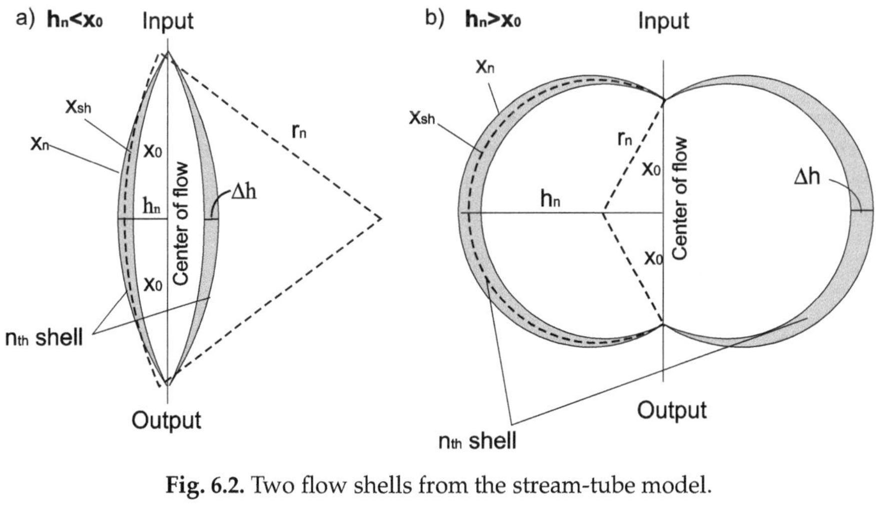

Fig. 6.2. Two flow shells from the stream-tube model.

6.2 Developing the Model

The flow in the system depends on the geometry of each shell. To determine the incremental volume contained within each individual shell, the integral volume of the $(n-1)^{\text{th}}$ shell is subtracted from that of the nth shell. Here, the term "integral volume" refers to the volume which is contained within the outer edge of the shell. The subscript (n) in the following equations is used to represent the integral parameters of the nth shell and the subscript (sh) to represent the incremental parameters of the nth shell.

From Fig. 6.2, the length along the outer edge of each shell from the input position to the output position (x_n) is given by

$$x_n = 2r_n \cos^{-1} \frac{r_n - h_n}{r_n} \quad , \tag{6.1}$$

where r_n is the radius of the nth shell and h_n is the maximum distance from the shell to the center of flow. For the consideration of the flow fields, the value h_n is incrementally increased. The radius r_n can be obtained from

$$r_n^2 = x_0^2 + (r_n - h_n)^2 \quad , \tag{6.2}$$

resulting in

$$r_n = \frac{h_n}{2} + \frac{x_0^2}{2h_n} \quad . \tag{6.3}$$

The integral volume contained within each shell for a two-dimensional flow system (V_{2dn}) is given by

$$V_{2dn} = \left[r_n^2 \cos^- 1 \frac{r_n - h_n}{r_n} - (r_n - h_n)\sqrt{2r_n h_n - h_n^2} \right] \cdot S \quad , \tag{6.4}$$

where S is the depth of the flow system (Fig. 6.1). The integral volume contained within each shell for a three-dimensional flow system (V_{3dn}) is given by rotating the shell (Fig. 6.2) around the center of flow. According to Bronstein and Semedjajew (1977), V_{3dn} is given by

$$V_{3dn} = 4\pi \left[\frac{1}{3} \sqrt{(r_n^2 - (h_n - r_n)^2)^3} + \right.$$
$$\left. \frac{h_n - r_n}{2} \left(r_n^2 \frac{\pi}{2} + (h_n - r_n)\sqrt{r_n^2 - (h_n - r_n)^2} - r_n^2 \sin^{-1} \frac{r_n - h_n}{r_n} \right) \right] \tag{6.5}$$

The relationship of the flow path length, the integral volume of the two-dimensional shells and the integral volume of the three-dimensional shells versus the distance x_0 is presented in Fig. 6.3. Once the integral volume of each shell is known, the volume increment for each shell (V_{2dsh} or V_{3dsh}) can be calculated by subtracting the volume of shell $(n-1)$ from that of shell (n) as

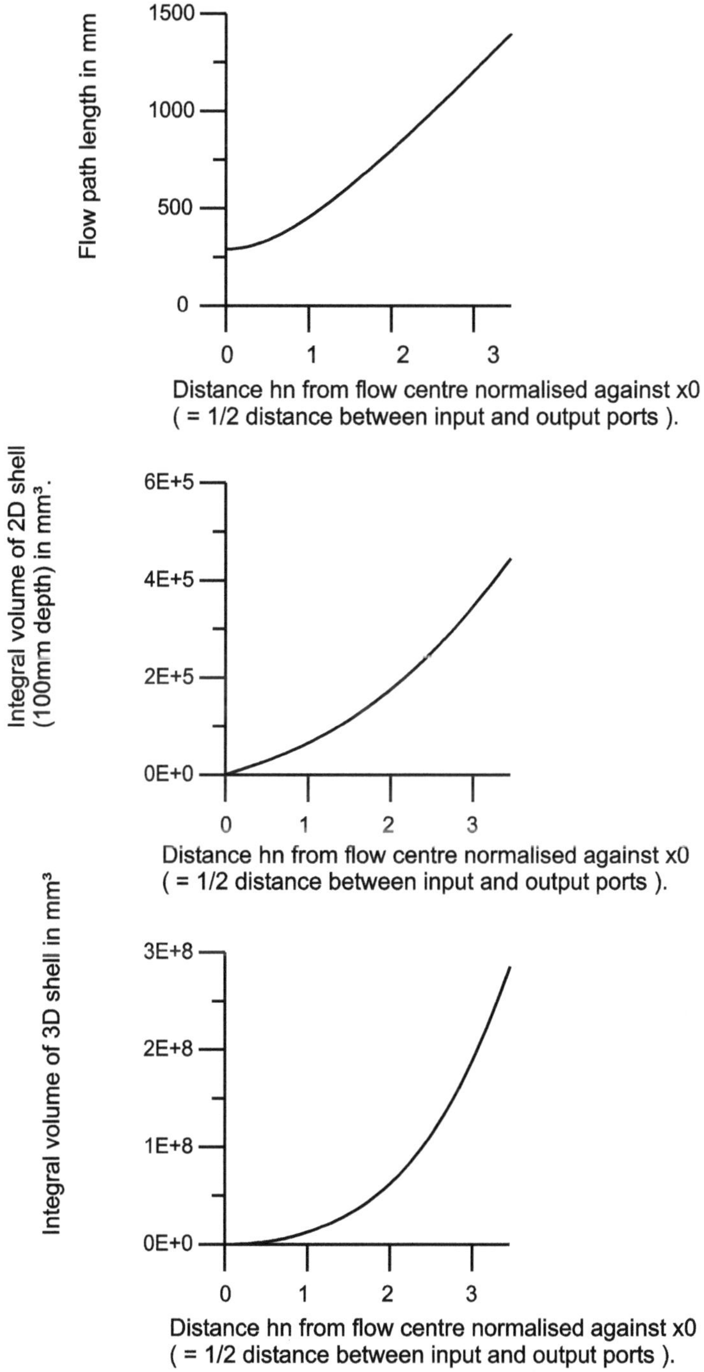

Fig. 6.3. Geometrical relationships of the 2D and 3D multi-shell system.

$$V_{2dn} = V_{2d(n)} - V_{2d(n-1)} \quad \text{or}$$
$$V_{3dn} = V_{3d(n)} - V_{3d(n-1)} \quad . \tag{6.6}$$

Likewise, the central path length for the individual shells (x_{sh}), (Fig. 6.2), can be calculated for each shell as follows

$$x_{sh} = \frac{x_n + x_{n-1}}{2} \quad . \tag{6.7}$$

The average cross sectional flow area (A_{2dsh} or A_{3dsh}) available in each shell to the flow is then derived by dividing the volume of each individual shell (V_{2dsh} or V_{3dsh}) by the central path length (x_{sh}) of that shell

$$A_{2dsh} = V_{2dsh}/x_{sh} \quad \text{or}$$
$$A_{3dsh} = V_{3dsh}/x_{sh} \quad . \tag{6.8}$$

Once the geometrical properties of each shell have been determined as described above, it is possible to calculate the flow characteristics of the system. Each shell is represented mathematically as a one-dimensional pipe of length equivalent to the central path length x_{sh} for that shell and of cross sectional area equivalent to the average cross sectional flow area (A_{2dsh} or A_{3dsh}) for that shell. The pressure difference from one end of the shell, p_1, to the other end, p_2, is constant and known for all pipes. Allowing compressive flow, the average pressure gradient (i_{sh}) in each shell can be calculated as

$$i_{sh} = \frac{p_1^2 - p_2^2}{2 p_2 x_{sh}} \quad . \tag{6.9}$$

In such a manner, the contribution of each shell to flow and transport in the system can be calculated using DARCY's law and combined to provide the total flow in the system Q_{total}:

$$Q_{total} = \sum_{n=1}^{n_{sh}} Q_{sh} \quad \text{with} \quad Q_{sh} = A_{sh} i_{sh} \frac{k}{\mu_a} \quad . \tag{6.10}$$

Here, n_{sh} represents the number of shells considered and the dimensionality of the system is reflected in the equation for the average cross section areas (6.8).

6.3 Boundary Conditions

The initial discussion of the principles behind the multi-shell model is based on an unbounded infinite system. For the application of the model to the boundary conditions found in the experimental cell, a geometric correction is made to the flow path lengths. This is illustrated in Fig. 6.4 for comparison. The unbounded system is presented in Fig. 6.4a and the MIOJ bounded

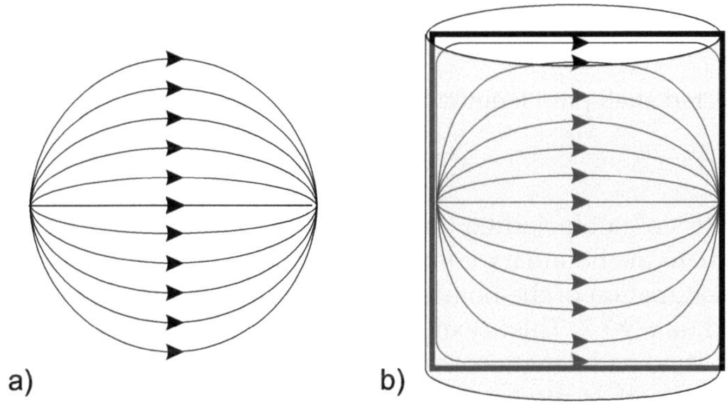

Fig. 6.4. Three-dimensional flow systems. a) Unbounded 3D flow system, flow paths form a sphere. b) Bounded 3D flow system, flow paths influenced by boundary geometry.

system (see Sect. 4.2) in Fig. 6.4b. The influence of the cell walls is apparent in that the outer flow paths bend to follow the shape of the walls.

A simplified assumption is applied to the flow path lengths of the shells to account for this increase in length, illustrated in Fig. 6.5. The maximum correction is given by

$$Maximum\ correction = MCor = \frac{a+b}{\sqrt{a^2+b^2}} \quad . \tag{6.11}$$

Fig. 6.5. Derivation of the correction factor applied to accommodate the boundary conditions of the experimental cell.

The lengths a and b are illustrated in Fig. 6.5. This correction is then applied linearly assuming that the innermost flow path is not affected by the boundary conditions, whereas the largest effect can be expected along the outer flow path

$$Shell\ correction = SCor = \frac{h_n}{a} \cdot MCor \quad . \tag{6.12}$$

The corrected shell path lengths are then given by

$$x'_{sh} = x_{sh} \cdot SCor \quad . \tag{6.13}$$

The effect of not including the path length correction in the model calculations where the boundary conditions indicate that it should be applied is an underestimation of the homogeneous hydraulic conductivity by a factor of approximately 1.7. This is explained by considering the outermost shell in the MIOJ. If no path length correction is applied, the flow gradient is higher and the flow path is shorter than if it is applied. The effect of increasing the path length is an increased travel time of the tracer and a reduction of the flow rate in the shell. Therefore, the path length correction leads to reduced flow rates and a larger amount of tailing. If no path length correction is applied in fitting the model to the experimental results, then the tailing and higher flow rates are compensated for by a reduced hydraulic conductivity of the model.

6.4 Comparison with One-Dimensional Flow Model

The multi-shell model makes it possible to predict the effect of analyzing a multi-dimensional flow system using a simplified one-dimensional model. A one-dimensional approach leads to an apparent increase in hydraulic conductivity with increasing distance between the input and output ports. This is presented in Fig. 6.6 for two- and three-dimensional flow fields in a homogeneous medium if analyzed as a one-dimensional flow field. The term "apparent permeability"refers to the permeability derived by applying a one-dimensional flow model in a multi-dimensional flow system, and is not the actual permeability of the system. Fig. 6.7 presents measured data analyzed as a one dimensional system with the two- and three-dimensional flow field envelopes modelled as described above for a homogeneous medium with a hydraulic conductivity of $2.78 \cdot 10^{-7}$ m/s. In this case, the flow field is modelled with limited boundary conditions representing the experimental cell. In the case of the two-dimensional section, the depth of the section is related to the size of the input ports (port vertical length) and the length of the sample ((vertical length of ports + length of sample)/2). Fig. 6.7 demonstrates clearly that the dimensionality of the flow field is a parameter which cannot be ignored in considering the flow and transport processes operating in the fractured porous rock. The interpretation of Fig. 6.7 is given as the system having been experimentally investigated by the application of one-, two- and three-dimensional flow fields. The initial interpretation of these results has come from a one-dimensional flow field applying DARCY's law to derive the apparent hydraulic conductivity, plotted against the distance between

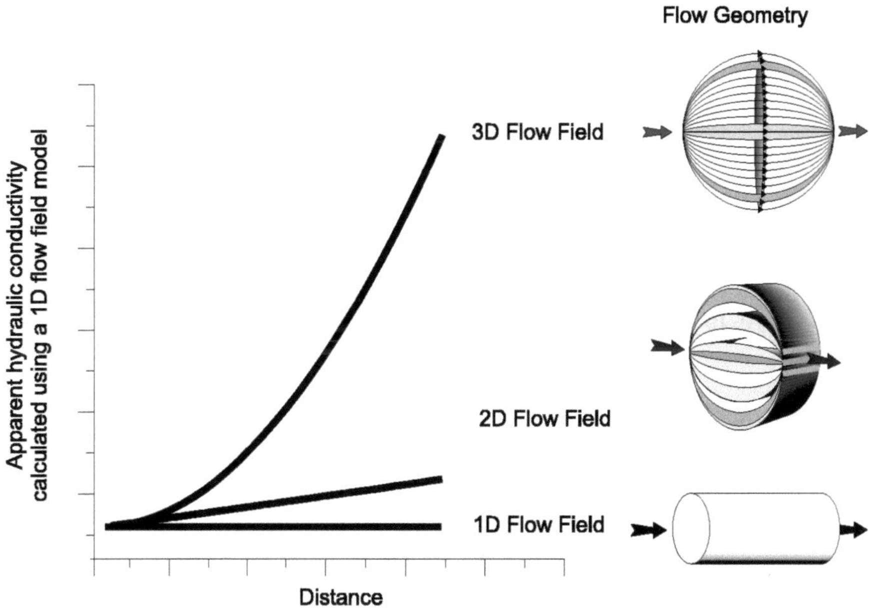

Fig. 6.6. Effect of multi-dimensional flow conditions when analyzed as one dimensional flow on apparent permeability as a function of distance between injection point and extraction point.

the input and output ports. The scatter of the points between the two- and three-dimensional flow field envelopes predicted by the multi-shell model is illustrated. Points lying above the three-dimensional envelope are considered to have been influenced by flow through a fracture.

A one-dimensional model can only be used to provide comparative results if the flow field geometries of the measurements are identical. Again, it should be stressed that an apparent value and not an actual value is gained in this case.

6.5 Calculation of the Tracer Breakthrough Curves

When the multi-shell model is use to simulate the transport of a diffusive substance in a porous medium the advective transport of the tracer is considered, coupled with longitudinal diffusion along the flow path. With respect to diffusion, the flow is assumed to be incompressible and no other physical processes such as dispersion are taken into account. In the experiments, low pressure gradients (150 mbar/flow path length) were applied, allowing such assumptions about the compressibility of the flow. The diffusion within the system is considered to be one-dimensional along the path length of the shells and the diffusion coefficient of the tracer gas in air in the

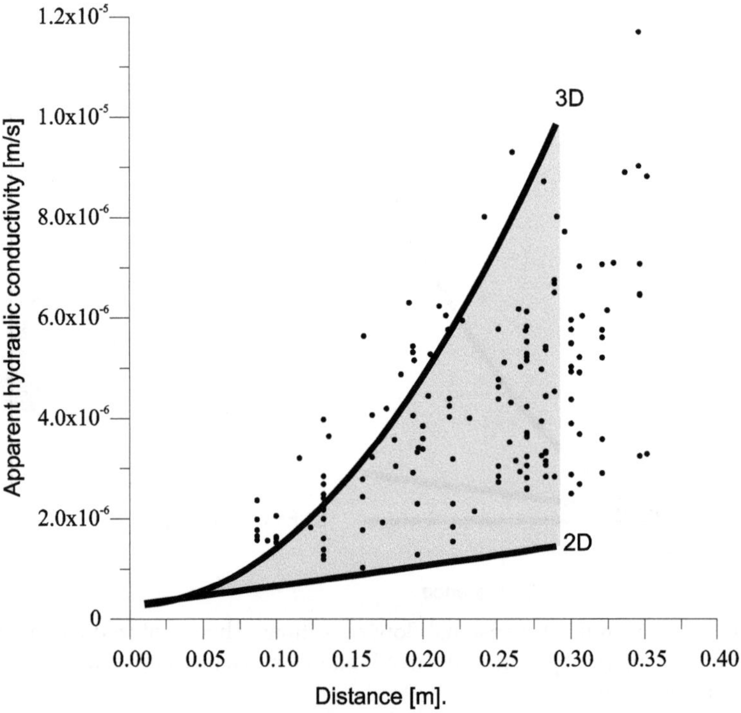

Fig. 6.7. Data from the experimental cell presented with two-dimensional and three-dimensional flow envelopes.

porous medium is used to describe the diffusion of the gas in the system. A DIRACpulse is used to derive the diffusive transport parameters. Each shell with a central path length x_{sh} makes a contribution C_{sh} to the breakthrough curve at the output port after a time t as follows:

$$C_{sh}(x_{sh}, t) = \frac{1}{2\sqrt{D^* \pi t}} \exp\left(\frac{-(x_{sh} - v_{sh} t)}{4D^*}\right) . \tag{6.14}$$

Here, D^* is the pore diffusion coefficient after Grathwohl (1998), given by

$$D^* = \frac{D_{air}}{\tau_f} , \tag{6.15}$$

where D_{air} is the diffusion coefficient of the tracer gas in air, the tracer is non-reactive and τ_f is the tortuosity. The tortuosity can be approximated by

$$\tau_f = \frac{1}{n_e} , \tag{6.16}$$

where n_e is the effective porosity of the matrix. The total concentration at a time t at the output port is then

$$C_{total}(t) = \sum_{n=1}^{n_{sh}} C_{sh}(x_{sh}, t) \cdot \frac{Q_{total}}{Q_{sh}} \quad . \tag{6.17}$$

The normalized breakthrough curve, used for the interpretation of the results, is given by

$$\overline{C}(t) = \frac{C_{total}(t)}{C_{total}^{max}} \quad , \tag{6.18}$$

with C_{total}^{max} as the maximum of $C_{total}(t)$ for the duration of the experiment. From the normalized breakthrough curve, it is then possible to derive the normalized mass flux in the system by integrating the concentration over time and then normalizing the concentration at a particular time with respect to this integral. This allows the percentage of the mass of the initial tracer input migrating through the system per second at a particular time ($F(t)$) to be derived as follows:

$$F(t) = \frac{\overline{C}(t)}{\int_0^{+\infty} \overline{C}(t)dt} \cdot 100 \quad . \tag{6.19}$$

The integral in equation (6.19) is approximated by applying the trapezoidal rule. Breakthrough curves derived in such a manner are presented in Fig. 6.8 along with the results of advection-only transport for comparison. The increase in tailing due to diffusion is clearly visible.

6.6 Experimental and Numerical Confirmation of Transport Modeling Using the Multi-shell Model

The flow within the MIOJ can be forced into a certain dimensionality according to the input and output ports chosen (see Chapt. 4). The curves in Fig. 6.9 present just such a case where two- and three-dimensional flow scenarios were induced. The measured tracer breakthrough curves are modelled using the homogeneous multiple shell approach and a finite element approach. The agreement between the modelled curves and the measured curves is quite clear. The permeability of the sample is used as a fitting parameter when the multi-shell model is calibrated. The very slight difference between the permeability of the multi-shell model and that of the finite element model is considered to be due to geometrical aspects involved in fitting a spherical modelling system (3D flow) into a cylinder (MIOJ). The multi-shell model can be seen to be both experimentally and numerically confirmed.

6.7 Tracer Distribution in a Two-Dimensional and Three Dimensional Flow System

A more detailed analysis of the breakthrough curves obtained by the multi-shell modelling approach leads to interesting results with respect to the dif-

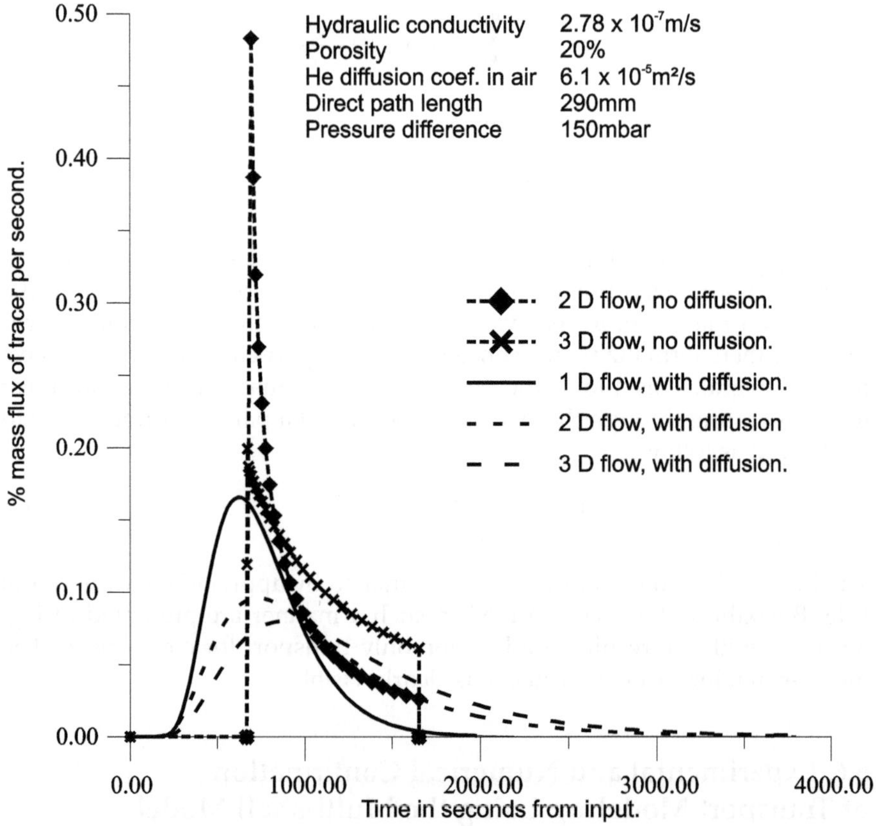

Fig. 6.8. Advective transport only compared to advective and diffusive transport.

ference in the spatial distribution of the tracer according to the dimensionality of the system. The mass of tracer carried in each shell is directly related to the proportion of flow carried in each shell. This is presented in Fig. 6.10 as a percentage contribution of tracer mass against the maximum distance of the shell from the flow center. It can be seen from Fig. 4.31 that, for three-dimensional flow configurations (Fig. 6.1a), the main portion of tracer mass is advected within shells with a distance from the flow center approximately equal to half the distance between the injection and extraction points, x_0, while for two-dimensional flow patterns (Fig. 6.1b), tracer transport occurs predominantly in the center shell. This occurs as a result of the relationship between the available cross sectional flow area and the flow gradient within the system.

In the three-dimensional system, although the gradient driving the flow from the input port to the output port is steepest along the axis joining these two points, the available cross sectional area for flow is at a minimum. Increasing the distance away from this center of flow reduces the flow gradient

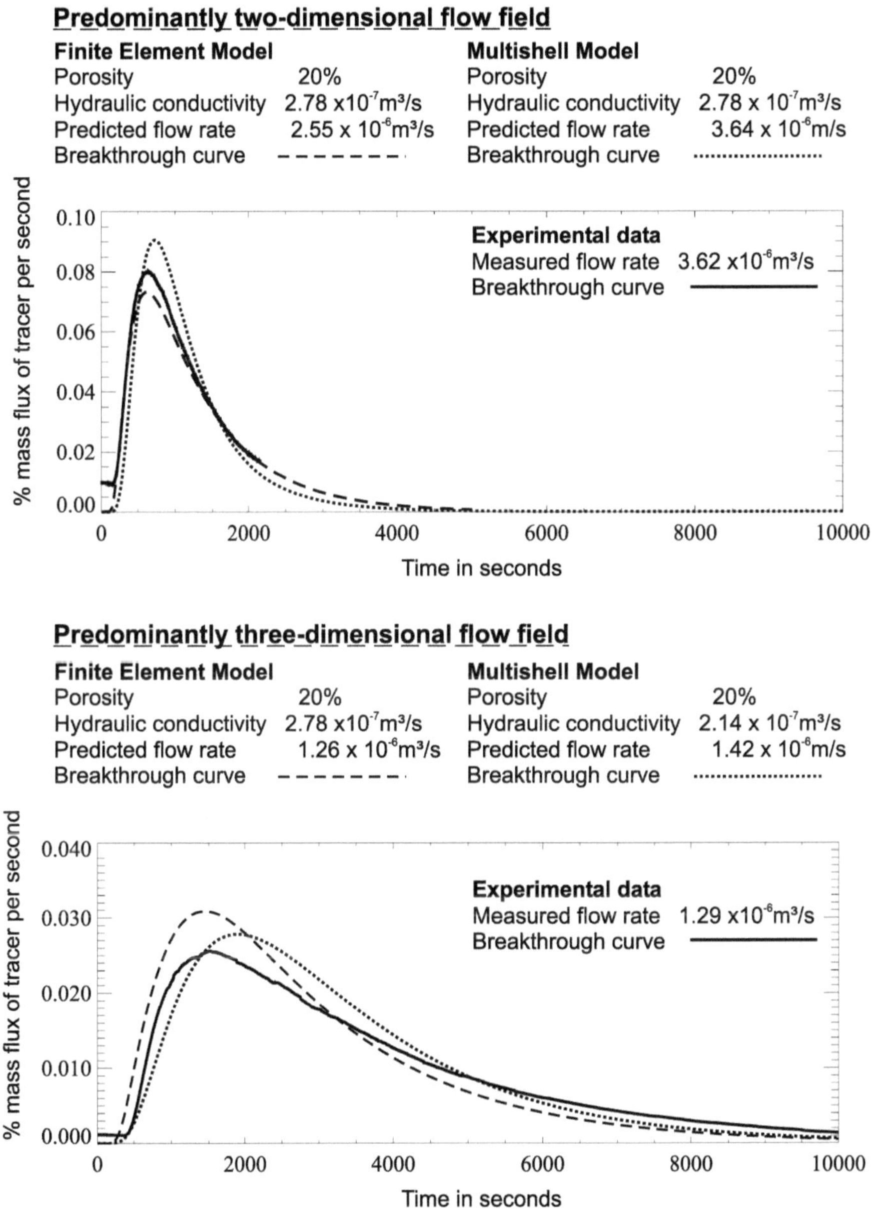

Fig. 6.9. Comparison of experimental results with finite element and multi-shell modelling predictions for two- and three-dimensional flow fields.

due to the increased flow path distance but increases the available cross sectional area to such a degree that the total flow through the shell increases. As stated, at a point approximately x_0 from the flow center, the effect of the

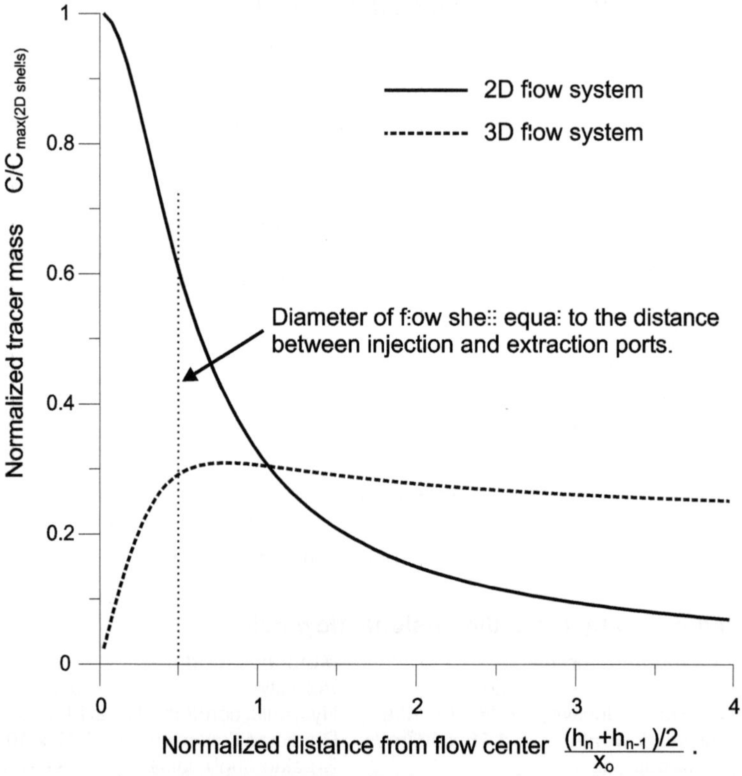

Fig. 6.10. Normalized tracer mass flux as a function of distance from the flow center for two- and three-dimensional flow systems.

reducing flow gradient then becomes larger than the effect of the increasing cross sectional flow area, leading to a reduction of flow in the system with increasing distance from the flow center.

Likewise, in the two-dimensional system, the gradient driving flow is steepest along the axis joining the input and output ports and reduces with increasing distance from the flow center, as in the three-dimensional case. However, unlike the three-dimensional case, the cross sectional area available to flow through the shells does not significantly increase with distance from the flow center. Therefore, the flow in the system is highest along the axis joining the input and output ports and reduces with increasing distance from the flow center.

6.8 Application of Multi-shell Model to Investigate the Anisotropic Nature of the Fractured Porous Samples

A predicted arrival signal for the experimental system is determined using the multi-shell model for all the port configurations used in the experimental investigation. The standard homogenous conductivity used in the model is defined by considering the spread of results as given in Fig. 6.7, fitting the predicted results to those measured and in some cases from independent measurements using conventional one dimensional permeability equipment. Similarly, in the case of the evaluation of the transport characteristics, porosity was both fitted and in most cases determined independently for comparison.

The results actually gained from the experimental system can be compared with the "standard"response predicted by the model. Deviations from this expected result are principally a result of the flow and transport variations in the system. In Fig. 6.11, the "model standard"response is plotted in a plane and the flow and tracer arrival response of the system is presented. With a similar approach, Fig. 6.12 presents the same information, but in three dimensions for a fractured sample. Examining Fig. 6.11 shows the

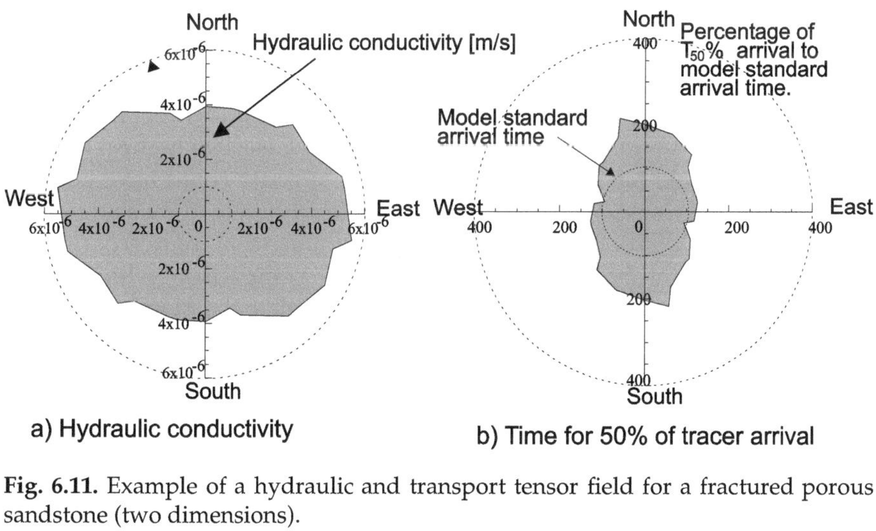

Fig. 6.11. Example of a hydraulic and transport tensor field for a fractured porous sandstone (two dimensions).

multi-shell model provides a standard model response for the homogeneous system, and the heterogeneities within the system can be clearly seen. The hydraulic conductivity is larger in the east-west direction, and this is also reflected in a reduction of the transport times in the east-west direction. The standard model provides the basis for comparison in each case, allowing a quantitative assessment of the differences in the hydraulic and transport parameters. The standard model is derived by taking as ideal a geometrical

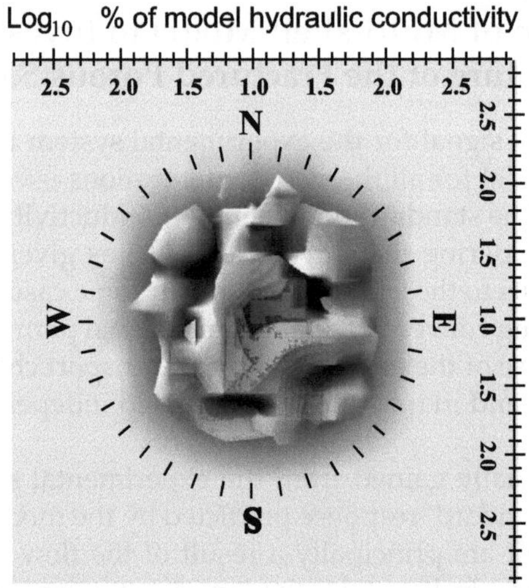

Fig. 6.12. Example of a three dimensional irregular hydraulic conductivity distribution derived by comparison with the stream-tube model prediction.

set-up as possible across the whole sample and fitting the flow and transport information, plus any independently determined data (i.e. porosity), as accurately as possible. The parameters determined in this way are then compared with the rest of the tomographical measurements (see Chapt. 4). The sample presented here is extremely fractured and as a result it is difficult to assess exactly what the standard model values should be. Clearly, a higher standard permeability could have been chosen, and the porosity could have been reduced to fit the transport data more exactly for the overall measurements. However this, would only have made the diagrams seem more accurate but would have had no bearing on the quantitative comparison, which would have remained the same. What this suggests is that the standard model, chosen principally because of the geometry of the measurements, is derived from results which lie outside the sample norm. This demonstrates the power of this technique in eliminating potential irregularities in the measurements of the sample parameters caused by multi-dimensional flow fields, and allows a comparison of the results against a known model value. This comparison can then be quantitatively represented as hydraulic and transport tensor fields.

Fig. 6.11 presents this information in two dimensions, Fig. 6.12 presents the same sample in three dimensions.

6.9 Précis of the Development of the Multi-shell Model

Originating from the aquifer-analogue principle, according to which various scales of investigation are examined to provide an analogue aquifer, a new modelling approach has been presented which has already provided interesting results with respect to the scale-dependent three-dimensional characterization of fractured porous media. Applying the multi-shell model has provided new insights into the flow and transport processes dominating the response of the fractured porous system. What is most important in determining the relevant parameters for use in the construction of an aquifer analogue is the dimensionality of the flow field. This not only affects the evaluation of flow measurements, but also determines to a major degree the shape of the breakthrough curves. In any study using an induced flow field with a higher dimensionality, this is of critical importance for the later accurate determination of the aquifer parameters. These findings are of particular importance for any type of tracer testing done under induced hydraulic or pneumatic gradients. Not only are the flow conditions dependent on the system geometry but, possibly of most importance, the tracer distribution within the system is influenced by the flow geometry. The multi-shell model particularly helps in the understanding of the distribution of tracer and the flow patterns within the system. In induced three-dimensional flow fields in homogeneous media, the largest contribution to the overall breakthrough of the tracer occurs in a flow shell with approximately the same diameter as the distance between the injection and extraction points. In a two-dimensional flow system, the largest contribution occurs along the axis joining the injection and extraction points. The multi-shell model allows the flow field to be divided up into the component stream tubes controlled by geometry of the system. The result of the differing tracer distribution in the flow field is a consequence of the relationship between the flow area available to the stream tubes and the pressure gradient along the stream tubes controlled by the length of the stream tubes. The model is easy to apply and does not require much computational power.

7

The Sensitivity Coefficient Approach

C. Leven, P. Dietrich

7.1 General Considerations

The suitability of subsurface investigation methods for the characterization and the capability to determine variations in the distribution of hydraulic parameters is primarily controlled by the intrinsic characteristics of the measuring configuration. Therefore, concepts have to be utilized which can contribute to understanding the effects of heterogeneities on an observed stimulus response. For this purpose, the **S**ensitivity **C**oefficient **A**pproach (SCA) is a useful tool which allows potential measurements to be analyzed and evaluated, leading to an understanding of the influence of heterogeneities. In this chapter, the theory of the Sensitivity Coefficient Approach will be introduced followed by a detailed analysis of sensitivity coefficients for a classical pumping test configuration as well as for other commonly used hydraulic testing methods.

7.1.1 Governing equations

The relationship between the distribution of a parameter k, the potential u and a source function q can be described for steady-state conditions by

$$-\sum_{i=1}^{n} \frac{\partial}{\partial x_i}\left(K(x)\frac{\partial u}{\partial x_i}\right) = q(x) \tag{7.1}$$

with $x \in \Omega$, where Ω represents the domain of interest and n is the number of considered dimensions. To quantify the potential u for a given parameter distribution k the specification of appropriate boundary conditions is necessary to make the solution of (7.1) unique.

If the potential u changes with time, (7.1) has to be rewritten including the time coordinate t and the storage term S

$$S(x)\frac{\partial u(x,t)}{\partial t} - \sum_{i=1}^{n} \frac{\partial}{\partial x_i}\left(K(x)\frac{\partial u}{\partial x_i}\right) = q(x,t) \tag{7.2}$$

with $x \in \Omega$ and $t \in \{0, T\}$. For a unique determination of the potential u the specification of appropriate initial and boundary conditions is necessary.

7.1.2 Governing Sensitivity Equation

The effect of a particular deviation in the parameter distribution on the observation within a distinct domain can be expressed by

$$\Delta u = \sum_{j=1}^{m} I_j \cdot \Delta k_j \tag{7.3}$$

where Δk_j corresponds to a change in the parameter distribution in the subdomain j causing a change in the observation u; m is the total number of subdomains which potentially contain parameter changes. The coefficient I_j can be considered a sensitivity coefficient, i.e. it is a measure of the effect of a deviation in the parameter distribution on the measured quantity. While changes in the parameter distribution in some subdomain are supposed to be small compared to the absolute value of the parameter in the subdomain, the sensitivity coefficient can be approximated by

$$I_j \approx \frac{\partial u}{\partial k_j} \tag{7.4}$$

which relates a change of the parameter k in the domain j to a particular change in the observation u. In groundwater hydraulics, I quantifies the change of the hydraulic head as result of a particular heterogeneity in the distribution of the hydraulic conductivity k. This meaning of sensitivity coefficients corresponds to the definition used in inverse problem theory for parameter identification. Frequently, the GAUSS-NEWTON algorithm is used, where the elements of the JACOBIAN matrix consist of the partial derivatives of the observation with respect to the considered parameter. Generally, three different methods are commonly used for computing the JACOBIAN matrix, i.e. calculating the sensitivity coefficients (Yeh (1986); Sun (1994)).

7.2 Calculation of the Parameter Derivative $\partial u/\partial k$

In this section, the main approaches to calculating the parameter derivative $\partial u/\partial k_j$, i.e. the sensitivity coefficients, are briefly sketched. All the methods introduced allow the transformation from transient to the steady-state conditions without any restriction; thus, only the formalisms for the transient state are given. More detailed work is cited in the following section and can also be found in Yeh (1986) and Sun (1994), among others.

7.2.1 Influence Coefficient Method

The method of influence coefficients is based on the concept of parameter perturbation (Becker and Yeh, 1972). This approach corresponds to the discrete description of the parameter derivative

$$\frac{\partial u}{\partial k_j} = \frac{u(k+\Delta k_j) - u(k)}{\Delta k_j} \tag{7.5}$$

The values of $u(k)$ and $u(k+\Delta k_j)$ are obtained by solving the governing equations (7.1) and (7.2)respectively, subject to the imposed initial and boundary conditions. This method requires perturbing each parameter once. For m subdomains, the governing equation has to be solved $(m+1)$ times.

7.2.2 Sensitivity Equation Method

With this method, the parameter derivative $\partial u/\partial k_j$ is obtained by taking the partial derivatives with respect to each parameter in the governing equation (7.1) and (7.2)and initial and boundary conditions (Yeh, 1986). This yields the following set of sensitivity equation

$$S\frac{\partial \frac{\partial u}{\partial k_j}}{dt} - \sum_{i=1}^{n} \frac{\partial}{\partial x_i}\left(\frac{k\frac{\partial u}{\partial k_j}}{\partial x_i}\right) = \sum_{i=1}^{n} \frac{\partial}{\partial x_i}\left(\frac{\partial k}{\partial k_j}\frac{\partial u}{\partial x_i}\right) \tag{7.6}$$

where k_j is the parameter in the subdomain j. The number of simulations required to compute the sensitivity coefficients is $(m+1)$, which is the same as for the influence coefficient method.

7.2.3 Adjoint-State Method

7.2.3.1 General Formulation

The adjoint-state method is based on the variational theory, and is applied in various fields, such as parameter estimation, reliability estimates, observation design, and sensitivity analysis (Sun, 1994). While this method requires only one more simulation run than observation points are included, it is obviously the most advantageous approach in terms of the numerical effort, if the number of observation points is small compared to the total number of subdomains m. In literature, various concepts are discussed based on the variational theory (e.g. Carter *et al.* (1974), Chavent *et al.* (1975), Kravaris and Seinfeld (1985), Dietrich (1992); among others). Following Carter *et al.* (1974) and Carter *et al.* (1982), who extended the pioneering work of Jacquard (1964) and Jacquard and Jain (1965), the sensitivity coefficients with respect to the hydraulic conductivity can be calculated by

$$I_K(\Omega_j,t) = \left.\frac{\partial u(x,t)}{\partial K}\right|_{\Omega_j} = -\int_{\Omega_j}\int_0^t \sum_{i=1}^n \frac{\partial \dot{v}(x,t-\tau)}{\partial x_i}\frac{\partial u(x,\tau)}{\partial x_i}d\tau d\omega \tag{7.7}$$

where I_K is the sensitivity coefficient with respect to hydraulic conductivity. The sensitivity coefficient I_S for the storage can be calculated by

$$I_S(\Omega_j,t) = \left.\frac{\partial u(x,t)}{\partial S}\right|_{\Omega_j} = -\int_{\Omega_j}\int_0^t \sum_{i=1}^n \dot{v}(x,t-\tau)\frac{\partial u(x,\tau)}{\partial x_i}d\tau d\omega \tag{7.8}$$

where the sensitivity coefficients I_K and I_S are expressed as functions of time t and spatial extension Ω_j of the subdomain j, in which a change in the parameter distribution is expected. The variable $\dot{v}(x,t)$ is the temporal derivative of $v(x,t)$ as the solution of the adjoint equation

$$S(x)\frac{\partial v(x,t)}{\partial t} - \sum_{i=1}^n \frac{\partial}{\partial x_i}\left(K(x)\frac{\partial v}{\partial x_i}\right) = G(x)H(t) \tag{7.9}$$

The initial and boundary conditions for the calculation of $v(x)$ correspond to those imposed for the calculation of u. $G(x)$ is the continuous source function at the location of the observation. Outside the specified location $G(x) = 0$ and

$$\begin{aligned} H(t) &= 0 \ \text{for} \ t \le 0 \\ H(t) &= 1 \ \text{for} \ t > 0 \end{aligned} \tag{7.10}$$

It is obvious that the adjoint equation for solving $v(x,t)$ is of the same form as (7.2) allowing the same numerical routine for calculating u and v. Practically, equation (7.2) is solved once for the calculation of u and (7.9) is solved once for each observation point. The advantage of the utilized method becomes evident: for a simple pumping test configuration with one pumping and one observation well, only two computations are necessary for the whole model domain. Carter *et al.* (1982) gave an interpretation of the adjoint equation, where the solution of (7.9) is given by the pressure responses to a constant withdrawal, which corresponds to the HEAVISIDE function $H(t)$. Compared to other solutions for the sensitivity coefficients using the adjoint equation method (cf. Chavent *et al.* (1975), Kravaris and Seinfeld (1985), Dietrich (1992), among others), it should be noted that, in the cited approaches, the temporal derivative of the solution for the adjoint equation is missing. This is because their solution can be interpreted as the pressure responses to a single pulse, a DIRAC function-like withdrawal, where the DIRAC function $\delta(t)$ equals the derivative of $H(t)$, at the time $\tau = t$ and the time runs backward from t to zero. Furthermore, the adjoint-state method introduced by Carter *et al.* (1974) is based on particular properties of equation (7.1) and (7.2):

1. Reciprocity, which exists if the pumping and observation well in a hydraulic test were interchanged and the same observation made as in the original configuration (Fig. 7.1).
2. The solutions can be described by convolution integrals.

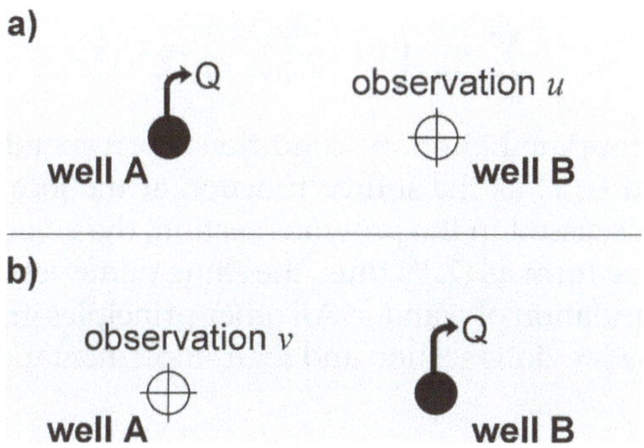

Fig. 7.1. Principle of reciprocity. (a) Initial positioning: pumping (Q = pumping rate) at well A leads to the observation u at well B. (b) Situation with interchanged well positions: pumping at well B leads to the observation v at well A. Due to reciprocity, observation v equals observation u from the initial configuration.

In the present chapter, the adjoint-state method is used to compute the derivatives $\partial u/\partial K$ and $\partial u/\partial S$ in order to gain an understanding of the effect of heterogeneous parameter distributions on hydraulic measurements. The adjoint-state method is advantageous for this purpose, because it can be used to compute the sensitivity coefficients for arbitrary parameter distributions, as well as hydrogeological systems with particular boundary conditions. Furthermore, this approach allows the computation of sensitivity coefficients with analytical solutions of (7.1) and (7.2)for known hydrogeologic situations.

7.2.3.2 The Steady-State Case

With $\partial u/\partial t = 0$, the time-dependence (7.2) is transferred to (7.1) describing the steady-state condition. Consequently, the dependence of the observation of u on storage S will vanish and a reformulation of the parameter derivative $\partial u/\partial I_K$ for steady-state conditions has to be given. Basically, all the approaches discussed above for calculating the parameter derivative $\partial u/\partial I_K$ can be used directly for steady-state conditions. Dietrich (1992) derives a solution for calculating $\partial u/\partial I_K$ for steady-state conditions, also based on the adjoint-state method. Accordingly, the relation between the deviations in hydraulic head due to a deviation in the parameter distribution is given by

$$I_K(\Omega_j) = \left.\frac{\partial u}{\partial K}\right|_{\Omega_j} = -\int_{\Omega_j} \sum_{i=1}^{n} \frac{\partial u(x)}{\partial x_i}\frac{\partial v(x)}{\partial x_i} d\omega \tag{7.11}$$

where the hydraulic head u is a solution of (7.1) and v is the solution of the adjoint equation

$$-\sum_{i=1}^{n} \frac{\partial}{\partial x_i}\left(K(x)\frac{\partial v(x)}{\partial x_i}\right) = G(x) \tag{7.12}$$

with the appropriate boundary conditions corresponding to the conditions of (7.1), and $G(x)$ as the source function at the location of the observation well. As discussed in the previous section, the adjoint equation (7.12) for v is of the same form as (7.1); thus, the same numerical algorithm can be used for the computation of u and v. All other principles given in the general formulation of the previous section and their implementation and realization are valid.

7.3 Performance of the Sensitivity Coefficient Approach

Generally, with the adjoint-state method, sensitivity coefficients can be derived from the superposition of two independent potential fields, i.e. two independent hydraulic tests with the same points of measurement at a given time t or for the steady-state case. The actual pumping test leads to the head distribution u. To gain the solution v of the adjoint equation, a second pumping test is simulated, where the original observation well is used as pumping well. As discussed previously, this is based on the reciprocity of hydraulic tests (e.g. Fig. 7.1) and can also derived from (7.7) and (7.11). Practically, the calculation of the sensitivity coefficients can be implemented numerically, i.e. by using numerically simulated potential distributions with consecutive numerical differentiation and integration, or utilizing analytical solutions for known hydrogeologic conditions.

7.3.1 Numerical Implementation

Numerical methods have to be used to compute the sensitivity coefficients for arbitrary parameter distributions, since appropriate analytical solutions are only known for specific hydrogeologic situations. The flow chart in Fig. 7.2 illustrates the consecutive steps in computing sensitivity coefficients for a numerical implementation.

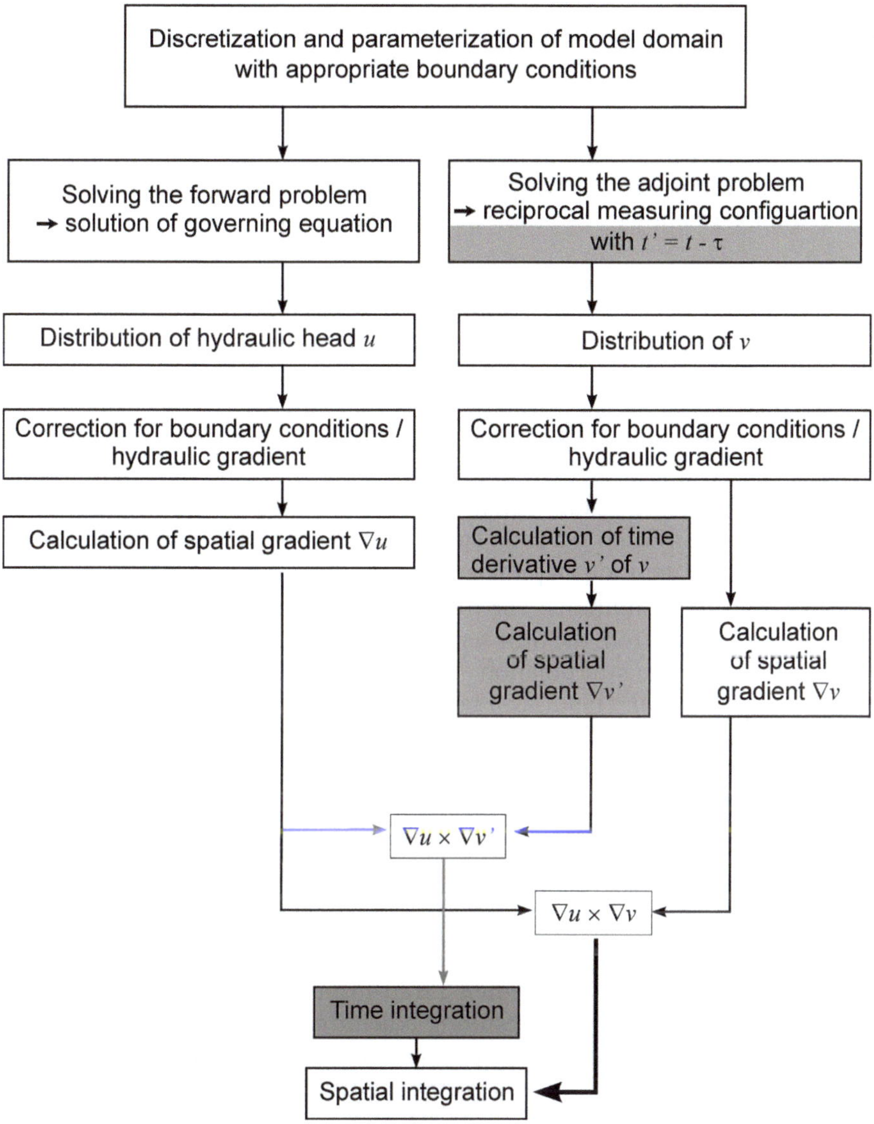

Fig. 7.2. Flow chart of the consecutive steps for the computation of the sensitivity coefficients. Boxes in gray indicate steps necessary for computations of transient-state conditions.

1. *Discretization and parameterization.* In many practical hydrogeological situations, information is available about general or specific boundary conditions and / or certain parameter distributions, e.g. preferential flow directions or macro-heterogeneities based on geophysical investigations. In practice, this information can be implemented most easily by applying of finite difference methods, since they allow a simple discretization and parameterization by appropriate zonation.

2. *Solution of the forward problem and the adjoint equation.* The analysis of equations (7.7) and (7.8) for transient conditions, and (7.11) for steady-state conditions shows that the parameter derivatives $\partial u/\partial K$ and $\partial u/\partial S$ not only depend on the parameters K and S respectively, and the hydraulic head u as the response to the source term $q(x)$, but also on the response v as a result of a hydraulic stimulation $G(x)$ at the location of the observation well. Hence, in order to calculate the sensitivity coefficients $\partial u/\partial K$ and $\partial u/\partial S$, equations (7.1) and (7.2)respectively have to be solved once in each iteration step, and (7.9) or (7.12) have to be solved once in each iteration step for each observation point.
   i. Transient conditions: For transient cases, only a limited number of data points from the time series are available for each grid cell. Therefore, a spline interpolation algorithm (e.g. cubic spline interpolation, cf. Press *et al.* (1996)) can be used to calculate the solution with an adequate time discretization in the numerical realization. The calculation of the coefficients of the spline interpolation requires the computation of the temporal derivatives at the beginning and the end of the interval $(0, t)$. Because the distribution of the hydraulic head at $t = 0$ can be regarded as the steady-state condition, the derivatives equal zero. At the time $t = \tau$, the derivative is calculated by the difference quotient between $t = \tau$ and $t = (\tau - 1)$. In addition, to calculate (7.2) and (7.7), the solution of the adjoint equation is required with the following transformation of the time coordinate $t' = t - \tau$. Since the resulting adjoint equation has the same form as the governing equation of the forward problem, it is possible to use the same numerical algorithm. Furthermore, as reciprocity exists and the solution of (7.7) can be considered physically as the head response to a constant withdrawal at the observation well for the source term $G(x)$, the value $q(x)$ from the solution of the original equation is used. As described above, the numerical implementation is realized using spline approximations for the solution of the adjoint equation.
   ii. Steady-state conditions: For the calculation of the potential distribution under steady-state conditions, only two simulation runs of the finite difference model are required to solve (7.1) and (7.11).

3. *Correction for boundary conditions / hydraulic gradients.* It can be shown that head changes caused by parameter deviations are independent of the hydraulic gradient due to distinct boundary conditions. Hence, for the calculation of sensitivity coefficients, the potential field imposed by specific boundary conditions can be removed, e.g. using the method of singularity removal (e.g. Lowry *et al.* (1989)). The potential distribution without hydraulic disturbance, i.e. pumping well (singularity), can be expressed following (7.2) by

$$S(x)\frac{\partial u(x,t)}{\partial t} - \sum_{i=1}^{n} \frac{\partial}{\partial x_i}\left(k(x)\frac{\partial u(x)}{\partial x_i}\right) = 0 \tag{7.13}$$

where u is the hydraulic head. The potential distribution resulting from pumping can be described by

$$S(x)\frac{\partial u'(x,t)}{\partial t} - \sum_{i=1}^{n} \frac{\partial}{\partial x_i}\left(k(x)\frac{\partial u'(x)}{\partial x_i}\right) = q \tag{7.14}$$

with $u' = (u+s)$, which is the sum of the initial head u and the pumping induced head disturbance s, i.e. the quantity of drawdown. The difference between (7.13) and (7.14) is therefore appropriate for the calculation of the response due to a disturbance at a single spatial position and can be expressed by

$$S(x)\frac{\partial s(x,t)}{\partial t} - \sum_{i-1}^{n} \frac{\partial}{\partial x_i}\left(k(x)\frac{\partial s(x)}{\partial x_i}\right) = q \tag{7.15}$$

For steady-state conditions, as the time derivative of the head, i.e. $\partial u/\partial t$ or $\partial s/\partial t$ equals zero, the term on the left-hand side quantifying the storage S will vanish. In the same way, the boundary conditions can be specified for the domain with and without hydraulic disturbance due to pumping.

4. *Calculation of temporal derivatives and gradients.*
   i. Transient-state: Following (7.7) and (7.8), the calculation of the temporal derivative $\dot{v}$ of the solution of the adjoint equation is required. As mentioned above, the introduction of the temporal derivative results from the kind of source term in (7.9). In practice, the same algorithm - with appropriate conversions - can be used as for the interpolation of the time series of v. In addition, it is necessary to compute the spatial gradients of the solution u of the forward problem and the temporal derivatives $\dot{v}$ of the solution of the adjoint equation, whereby simple numerical algorithms can be used.
   ii. Steady-state conditions: For steady-state conditions the spatial gradients of the solutions u and v are calculated using simple numerical algorithms.

5. *Calculation of the coefficients $\partial u/\partial k$ and $\partial u/\partial S$.*
   i. Transient state: To calculate the parameter derivatives $\partial u/\partial k$ and $\partial u/\partial S$, the calculated potential gradients are used with equations (7.7) and (7.8). For increased accuracy, it is beneficial to calculate first the integral over the time domain $(0, t)$ before carrying out the spatial numerical integration. For this purpose, a ROMBERG integration (cf. Press *et al.* (1996)) is one of the most suitable and powerful approximations of the integral in the time domain $(0, t)$.
   ii. Steady-state conditions: To calculate the parameter derivative $\partial u/\partial k$, the calculated potential gradients are used with equation (7.11) performing a spatial integration.

7.3.2 Application of Analytical Solutions

For certain hydrogeological conditions, analytical solutions of (7.1) or (7.2)can be derived. In this section, the solution of (7.7) and (7.8) for the case of a transient pumping test in a two-dimensional, homogeneous and isotropic confined aquifer is given. Essential considerations with respect to effects of parameter distributions can be derived from this case. The solution representing the specified hydrogeological conditions is given by Theis (1935):

$$s = \frac{Q}{4\pi T}\int_a^\infty \frac{e^{-z}}{z}dz \tag{7.16}$$

with

$$a = \frac{r^2 S}{4Tt}$$

valid for a domain of infinite extent. With the parameter derivatives (7.7) and (7.8) given by the adjoint-state method in combination with (7.16) and after the replacement of the hydraulic conductivity by the transmissivity T where

$$T = K \cdot b$$

with b = aquifer thickness, the sensitivity coefficients for the observation with respect to a local change in the transmissivity distribution can be calculated by

$$I_T(\Omega_j, t) = \left.\frac{\partial s(t)}{\partial T}\right|_{\Omega_j} = -\int_{\Omega_j}\int_0^t \frac{Q^2 S}{16\pi^2 T^3}\frac{e^{-\frac{S}{4T}\left(\frac{r_{ow}^2}{t-\tau}+\frac{r_{pw}^2}{\tau}\right)}}{(t-\tau)^2 r_{pw}^2}\,d\tau\, d\omega \tag{7.17}$$

Here Q is the pumping rate and x_{pw},y_{pw}, x_{ow}, y_{pw} are the coordinates of the pumping (PW) and observation well (OW) in x- and y-directions, respectively. The distance to the observation and pumping well is expressed by r_{ow} and r_{pw}.

The sensitivity coefficient with respect to storage can be calculated accordingly by

$$I_S(\Omega_j, t) = \left.\frac{\partial s(t)}{\partial S}\right|_{\Omega_j} = -\int\limits_{\Omega_j}\int\limits_0^t \frac{Q^2}{16\pi^2 T^2} \frac{e^{-\frac{S}{4T}\left(\frac{r_{ow}^2}{t-\tau}+\frac{r_{pw}^2}{\tau}\right)}}{(t-\tau)\tau} \, d\tau \, d\omega \tag{7.18}$$

7.4 Analysis of Sensitivity Coefficient Distributions

On the basis of the derivation of sensitivity coefficients, an analysis of the sensitivity distribution with respect to variations in the governing parameters is given in this section. In order to evaluate the effects of parameter deviations on the hydraulic measurements it is most appropriate to analyze the distribution of sensitivity coefficients for a homogeneous parameter distribution. In this case, particular characteristics of the distribution of sensitivity coefficients depend solely on the measuring configuration itself. In the following, the distribution of sensitivity coefficients for the configuration of a pumping test with one pumping and one observation well is analyzed on the basis of the analytical solutions (7.17) and (7.18) given in the previous section.

7.4.1 Some General Considerations

A detailed analysis of the governing equations for the calculation of sensitivity coefficients and their solutions, e.g. (7.17) and (7.18), reveals some essential correlations that are independent of the actual measuring configuration: The values of sensitivity coefficients depend on

i. the parameter distribution of the domain of interest,
ii. the location of the points of measurement, i.e. pumping and observation wells, and
iii. the spatial extension and magnitude of a potential discontinuity in the parameter distribution located in the domain of interest.

Furthermore, on the basis of the equations for the calculation of sensitivity coefficients with respect to hydraulic conductivity (e.g. (7.11)), the sensitivity coefficients can be considered as integral over the scalar product of the gradients ∇u and ∇v:

$$\nabla u \cdot \nabla v = |\nabla u| \cdot |\nabla v| \cdot \cos\,\theta \tag{7.19}$$

where θ is the angle between the gradients ∇u and ∇v. If ∇u and ∇v are perpendicular to each other, thus $\theta = 90°$or $\theta = 0°$, the resulting coefficient equals zero. Accordingly, for angles of $0° < \theta < 90°$, the coefficients resulting from (7.11) are negative and, for angles of $90° < \theta < 180°$, the resulting coefficients are positive. From (7.19), it can be derived that the influence on the measured quantity will decrease with decreasing gradients ∇u and ∇v. Hence, the value of the sensitivity coefficient will decrease with increasing distance from the wells. The consequences for the spatial distribution of the sensitivity will be shown in the following.

7.4.2 Sensitivity with Respect to Hydraulic Conductivity

For a detailed analysis of sensitivity coefficients with respect to changes in the hydraulic conductivity, a two-dimensional distribution for a pumping test in a homogeneous, two-dimensional aquifer is given for three time points in Fig. 7.3. In order to obtain values that are independent of the distance between the pumping and observation wells, a normalization is introduced for the considerations with spatially separated pumping and observation positions using the transformation

$$\begin{aligned} x' &= x/r \\ y' &= y/r \end{aligned} \tag{7.20}$$

where r is the distance between the pumping and observation wells. Hence, for the calculation of normalized sensitivity coefficients, the following transformation is useful

$$I'(x', y') = I(x, y) \cdot r \tag{7.21}$$

As the sensitivity distributions of a pumping test for homogeneous conditions under steady-state is of the same form as those for the transient state at very late times, the steady-state case is not explicitly considered in this section. From the spatial distribution of the sensitivity coefficients in Fig. 7.3, it is obvious that two domains of opposite sensitivity characterize the spatial distribution. The transitional zone between these domains is presented by a circle intersecting the pumping and observation wells. Since the values of sensitivity on the transitional zone approach zero, a parameter change on this line will have no impact on the observation. The domain outside the transition zone represents a region with a negative sensitivity on drawdown with respect to a change in the parameter distribution. However, the domain between the pumping and observation wells is characterized by positive sensitivity. Consequently, any heterogeneity of definite spatial extent and quantity could lead to either an increase or a decrease of the observed value u,

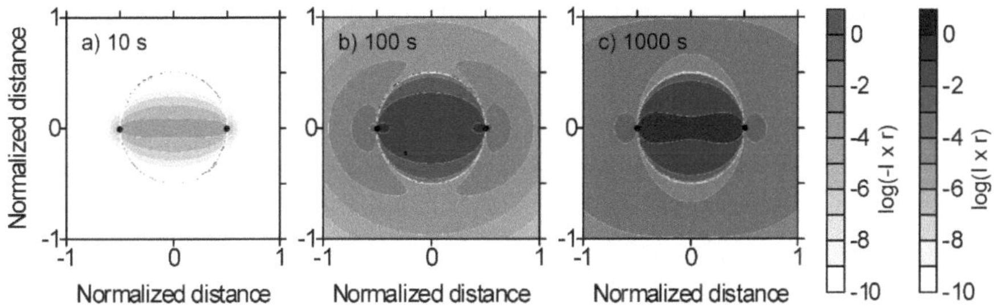

Fig. 7.3. Two-dimensional sensitivity distribution with respect to hydraulic conductivity for the times $t = 10, 100, 1000$ seconds. The black dots indicate the position of the pumping and observation wells ($k = 10^{-4}\,\mathrm{m\,s^{-1}}$, $S = 10^{-4}$ [-], $Q = 3.74 \cdot 10^{-3}\,\mathrm{m^3\,s^{-1}}$)

depending on the position of the heterogeneity relative to the wells. Hence, a particular discontinuity at the same radial distance from the pumping well will not necessarily lead to the same observation. For example, a discontinuity of higher conductivity in the domain of positive sensitivity will result in an increased drawdown, which contradicts the classical understanding of pumping tests, where increased drawdown is assumed to be the result of lowered conductivity.

The highest sensitivities with respect to hydraulic conductivity are in the direct vicinity of the wells and decrease with increasing distance from the points of measurement. Hence, a small spatial discontinuity in the direct vicinity of one well can lead to dramatic effects on the observed potential. However, although the domain of highest influence is located in the vicinity of the wells, not every variation in hydraulic conductivity necessarily has a noticeable influence on the observation u, e.g. if it lies near the line of zero influence, as discussed above. The shape of the sensitivity distribution does not change for differing values of hydraulic conductivity and the symmetric form of the sensitivity distribution reflects reciprocity of the hydraulic test configuration. Considering the temporal evolution of the sensitivity coefficients, it can be derived from (7.17), and consequently from Fig. 7.3, that the distribution described above with two domains of opposite influence persists for all times. However, the absolute values of sensitivity are subject of change over time within the domains. The highest values of sensitivity are situated in the direct vicinity of both wells for all times (Fig. 7.4 (a)). However, these areas with highest but opposite (positive and negative) sensitivity are adjacent. At early times (Fig. 7.3 (a), the domain of positive sensitivity takes on high sensitivity values in contrast to the outer negative domain. Thus, for early times, the drawdown is most sensitive to conductivity changes in between the wells (Fig. 7.3 (a)); however, as discussed below, its relative influence decreases with elapsing time.

The absolute values of sensitivity rise in the domain of negative influence with elapsing time, while an increasing portion of the domain is covered

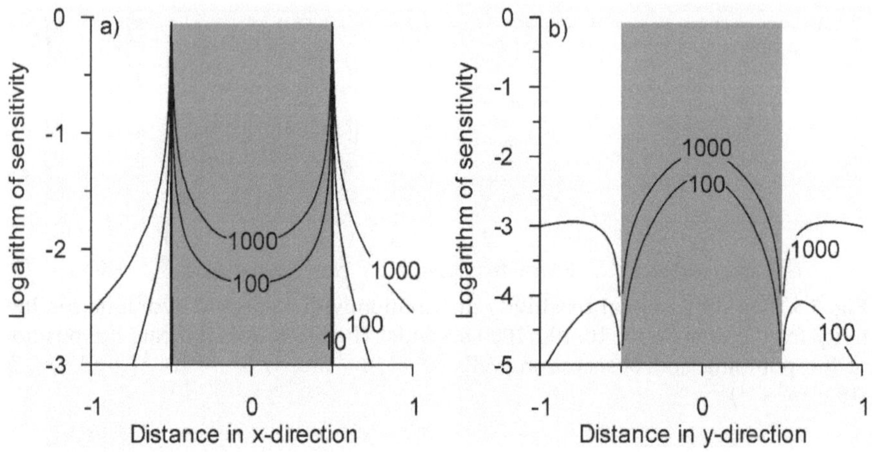

Fig. 7.4. Sections through the two-dimensional distribution of sensitivity coefficients (a) at $y' = 0$ intersecting the pumping and observation well, and (b) at $x' = 0$. The numbers indicate the corresponding times in seconds. The section with gray background indicates positive values of sensitivity. Note that the axes of ordinate have different scaling in (a) and (b) and the sensitivity coefficients at $t = 10\,\text{s}$ are only plotted in (a) since the absolute values are very small compared to those at later times (cf. Fig. 7.3 (a).

with sensitivities of increasing value. In the distribution shown in Figs. 7.3 and 7.5 (a), the sensitivity converges to constant values for the considered domain after $t = 1000$ s, i.e. the absolute contribution of this domain of stationary sensitivity remains constant as time proceeds.

Figures 7.4 and 7.5 (a) show the spatial and temporal distribution of sensitivity coefficients with respect to hydraulic conductivity for different one-dimensional sections. It is obvious that the sensitivity decreases with increasing distance from the wells. Both graphs in Fig. 7.4 show the relatively increasing influence of more distant areas with elapsing time. However, the domain of highest sensitivities remains in the vicinity of the wells.

Due to the characteristics of sensitivity distributions described above, various issues become relevant for the interpretation of hydraulic tests and for a reliable characterization of aquifer heterogeneities. The interpretation of drawdown data, however, is non-unique as different heterogeneities might lead to identical drawdown behavior. For example, a high conductivity zone in the positive domain can lead to the same observation as a low conductivity zone in the negative domain. Other problems associated with the ambiguity of results originate from the reciprocity and the symmetry of sensitivity distributions of hydraulic tests (Fig. 7.3). As discussed previously, the sensitivity coefficients reach constant values as the area of stationary sensitivity coefficients increases with time (e.g. Figs. 7.3 (a) through (c) and 7.5 (a)); thus, the relative influence of parameter changes on the observation in those subdo-

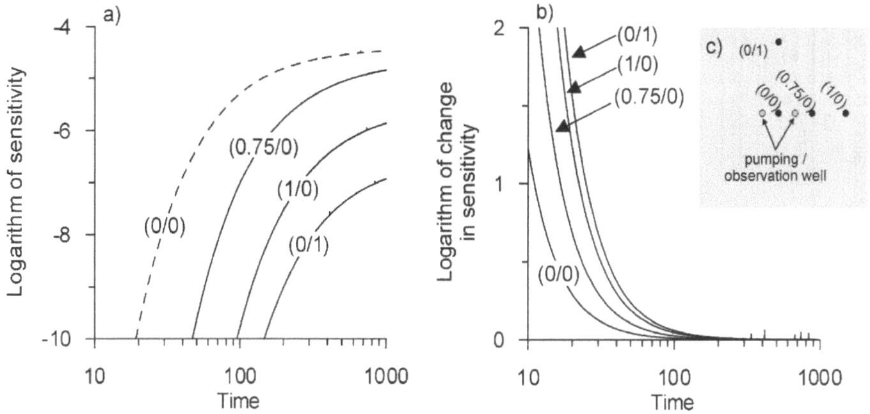

Fig. 7.5. (a) Evolution of sensitivity coefficients with respect to hydraulic conductivity at different spatial positions illustrated in (c). (b) Time series of the relative change of sensitivity coefficients for the spatial positions given in (c).

mains decreases. To account for this, it can be useful to consider the relative change of sensitivity between consecutive time steps. This allows a detailed assessment of results from hydraulic measurements as well as the possibility of a reliable interpretation based on sensitivity coefficients. The relative change of sensitivity coefficients is expressed by

$$\Delta I = \frac{I(t_k)}{I(t_{k+1})} \tag{7.22}$$

where $I(t_k)$ and $I(t_{k+1})$ are the sensitivity coefficients at two consecutive time steps. Figure 7.5 (b) gives the time series of the relative changes in sensitivity for different spatial positions. It is evident that the relative change of sensitivity decreases with elapsing time and converges to the relative change of one, i.e. the logarithm of change equals zero. Mathematically, this corresponds to the transition of the governing equation from the parabolic to the elliptic form. Furthermore, with increasing distance from the center of the measuring configuration, i.e. from the midpoint between pumping and observation wells, the value of relative change increases at a definite time step. This is also obvious from Fig. 7.6. For early times the coefficients of relative change are elliptically distributed and tend to a radially symmetric distribution for later times (Fig. 7.6 (c). From this distribution, three distinct domains of relative sensitivity can be distinguished: 1) a domain where the sensitivity remains constant between the particular time steps, 2) a domain where a considerable change in the sensitivity distribution occurs, and 3) a domain where changes in the sensitivity distribution occur, but the overall sensitivities are too small to affect the observation.

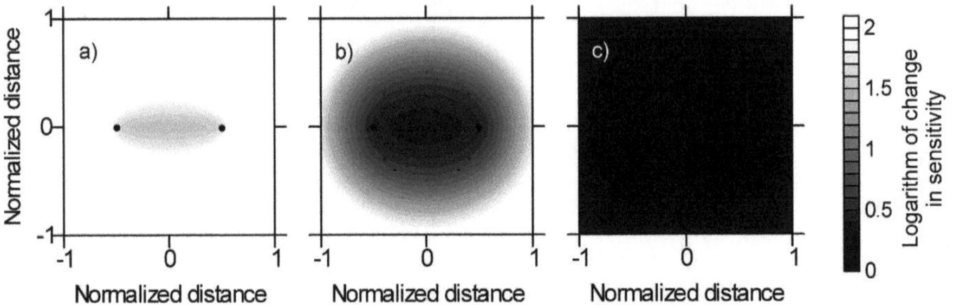

Fig. 7.6. Logarithm of relative change in sensitivity between consecutive logarithmic time steps (a) between $t = 10$ and 100 s, (b) between $t = 100$ and 1000 s and (c) between $t = 1000$ and $10{,}000$ s.

1. The domain where the sensitivity remains constant within a definite time frame is represented by coefficients of relative change of $\log_{10}(\Delta I) \rightarrow 0$. In Fig. 7.6 (b) and (c), this domain corresponds to the innermost part of the domain with the lowest values of relative change and, in Fig. 7.5 (b), it is indicated by the section of the time curve parallel to the abscissa. As time passes, no further information is available from this area of the aquifer during the hydraulic test.
2. A ring embracing the domain described above represents the portion of the aquifer where a considerable change in the sensitivity distribution occurs. In Fig. 7.6, this area is characterized by coefficients of relative change of $\log_{10}(\Delta I) < 2$. In Fig. 7.5 (b), this corresponds to the section of the curve where the slope flattens from high to low relative changes. With elapsing time, any changes in drawdown can be assigned to this portion of the aquifer. The chosen value of $\log_{10}(\Delta I) < 2$ given for the relative change is based on the fact that, if the values of absolute sensitivity are less than two orders of magnitude for the time step t_{k-1}, it can be concluded that this area will not have a significant influence during the given time interval.
3. This is the domain where a negligible or non-measurable influence of parameter discontinuities on the drawdown can be expected, although considerable changes in the sensitivity distribution occur ($\log_{10}(\Delta I) > 2$). This is due to the fact that the overall sensitivities in this domain are too small to affect the observation.

7.4.3 Sensitivity with Respect to Storage

On the basis of the transient nature of the governing differential equation, it is possible to determine the storage and therefore to compute the sensitivity coefficients with respect to storage. Figure 7.7 shows the two-dimensional sensitivity distribution for three consecutive log-cycles. For all time steps, the entire domain is characterized by a negative sensitivity distribution. Consequently, any variation in storage has a reverse influence on the observation;

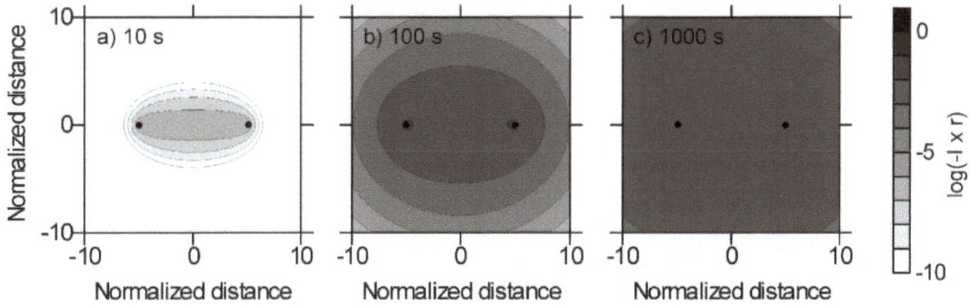

Fig. 7.7. Two-dimensional distribution of sensitivity coefficients with respect to changes in storage for the three times. The black dots indicate the position of the pumping and observation wells ($K = 10^{-4}\,\mathrm{m\,s^{-1}}$, $S = 10^{-4}$ [-], $Q = 3.74 \cdot 10^{-3}\,\mathrm{m^3\,s^{-1}}$).

i.e. heterogeneity of increased storage leads to a decreased drawdown, and vice versa, irrespective of the position of the parameter variation relative to the well locations.

The distribution of the sensitivity coefficients is initially of elliptical form with foci at the wells, and approaches a circular distribution at late times. With increasing sensitivities towards the wells, the highest influence for a change in storage is directly in the vicinity of the points of measurement. The analysis of the temporal evolution reveals that, with elapsing time, the sensitivity increases at each point until a time is reached where

$$\frac{\partial}{\partial t}\left(I_S(\Omega, t)\right) = 0 \tag{7.23}$$

After the maximum is reached, the sensitivity with respect to changes in storage decreases. This is also obvious from both graphs in Fig. 7.8, which show the spatial and temporal distribution of sensitivity coefficients. The domain of highest sensitivities remains in the vicinity of the wells for all times. However, the overall sensitivity of the considered domain increases (Fig. 7.8 (b)), the relative influence of the domain in the vicinity of the wells thus decreases.

7.4.4 Sensitivity Distribution for Different Hydraulic Test Configurations

Following the principle of superposition (Kruseman and de Ridder, 1994), any arbitrary hydraulic measuring configuration can be derived from a summation of individual pumping test configurations with one pumping and one observation well. Consequently, the sensitivity distribution for any hydraulic measuring configuration can also be derived from a summation of sensitivity distributions of individual pumping test configurations. Figure

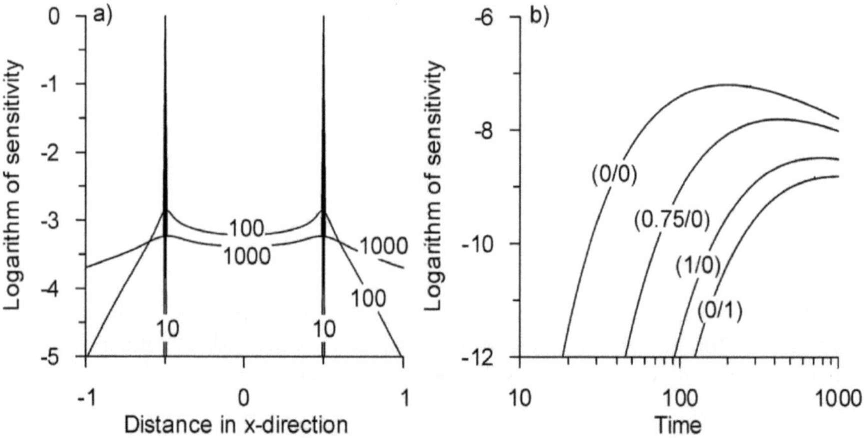

Fig. 7.8. (a) Section through the two-dimensional sensitivity distribution of Fig. 7.7 at $y' = 0$ intersecting the wells. (b) Evolution of sensitivity coefficients at different spatial positions that can be gathered from Fig. 7.5 (c).

7.9 presents examples of two-dimensional sensitivity distributions for different measuring configurations, which are used in hydrogeological practice. Configuration (a) illustrates the case of a doublet consisting of a pumping and injection well and an observation in the injection well. However, hydraulic tests in aquifers with high values of hydraulic conductivity or high values of storativity would lead to very small potential changes. Thus, this injection-extraction method requires precise measuring devices and is more suitable in aquifers of low conductivity.

The configuration given in Fig. 7.9(d) has the same points of measurement as Fig. 7.9(a) except that two extraction wells are used. Configuration (g) is a classical pumping test with two observation wells, while the difference in head is observed between the observation wells. The configurations illustrated in Figs. 7.9(j), (m), and (p) are two-well pumping tests with one observation well at different spatial positions. The configuration shown in (m) corresponds to the DC-geoelectric configuration of the type AMA' (Dietrich, 1999). Applications of the cited configurations can be found in Molz *et al.* (1986), Lebbe *et al.* (1995), Zlotnik and Ledder (1996), Clement *et al.* (1997), for example, and many others.

Following the analysis of the sensitivity distribution for the different hydraulic measurement configurations, some general statements can be made:

a. The conclusions concerning the sensitivity distribution of a standard pumping test (e.g. Fig. 7.3) are reflected in the distributions of Fig. 7.9: For the sensitivity with respect to changes in hydraulic conductivity the domain is divided in subregions of positive and negative sensitivity.

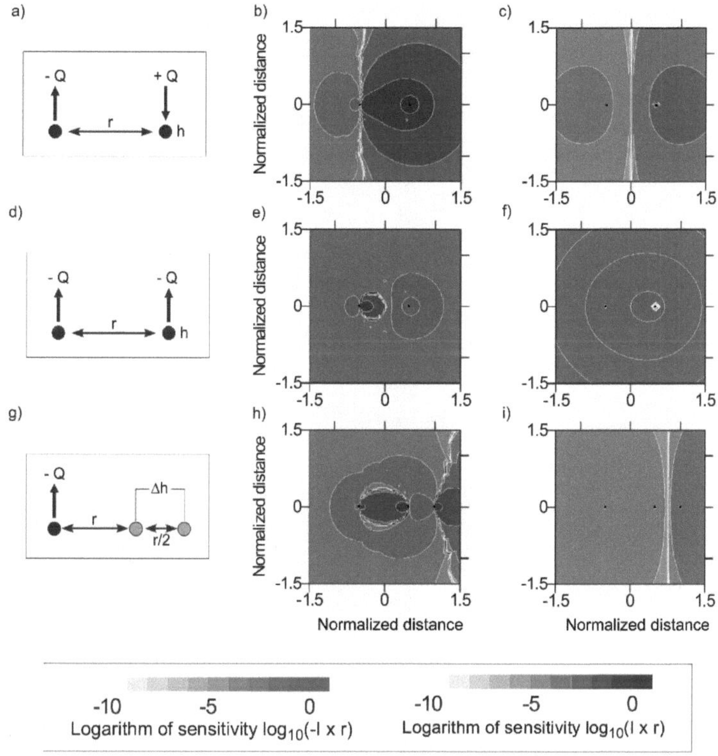

Fig. 7.9. Two-dimensional sensitivity distribution at $t = 1000$ s for different hydraulic measuring configurations (left column) with respect to transmissivity (middle column) and with respect to storage (right column): (a) - (c) Hydraulic dipole with one pumping ($-Q$) and one injection well ($+Q$), changes in the head h are observed in the injection well. (d) - (f) Two-well pumping test with observation in one pumping well. (g) - (i) Pumping test with two observation wells at which the hydraulic gradient (Δh) is observed. (j) - (l) Two-well pumping test with arbitrarily located observation well. (m) - (o) Two-well pumping test with observation well between the pumping wells. (p) - (r) Two-well pumping test with an observation well located on the axis perpendicular to the axis of the pumping locations. The black dots in the sensitivity plots mark the positions of the wells.

The points of measurement lie on the transitional zone between positive and negative sensitivity except for the cases where the observation takes place in the pumping or injection well.

b. For the case of an observation in the pumping well, the well is surrounded by a radially symmetric negative-sensitivity distribution with respect to hydraulic conductivity (Fig. 7.9 (e). By contrast, if the observation occurs in an injection well, positive sensitivity is present around the corresponding well (Fig. 7.9 (b)).

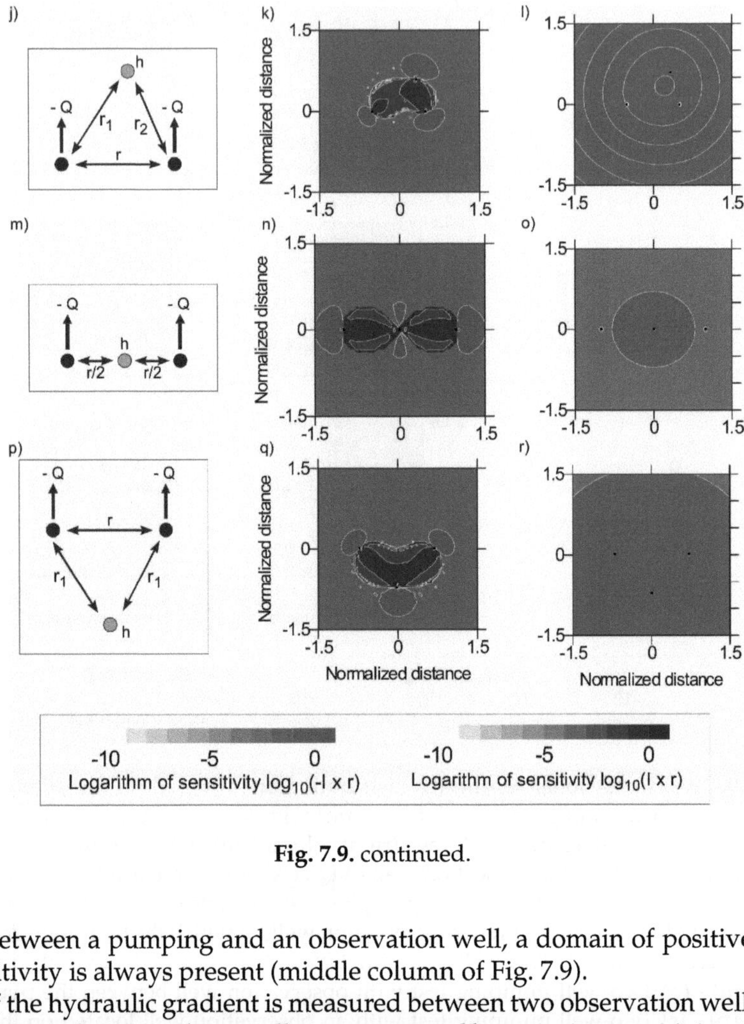

Fig. 7.9. continued.

c. Between a pumping and an observation well, a domain of positive sensitivity is always present (middle column of Fig. 7.9).
d. If the hydraulic gradient is measured between two observation wells, the domain between these wells is represented by negative sensitivity. As the line of zero influence intersects the points of measurement, a line of zero influence also intersects the second observation well at which the head difference is measured.
e. The considered domain has an overall negative sensitivity with respect to changes in storage for the case of one or more pumping wells and a single observation well (Fig. 7.9 (f), (l), (o), and (r)).
f. For a doublet test with pumping and injection well (Fig. 7.9 (a), or if the hydraulic gradient is measured between two observation wells (Fig. 7.9 (g)), the sensitivity distribution with respect to storage is divided into a domain of positive and negative influence (Fig. 7.9 (c) and (i). The transitional zone of positive and negative sensitivity intersects the line between the respective wells halfway.

In Fig. 7.10, an example of the temporal development of the sensitivity distribution with respect to hydraulic conductivity and storage is given based on the hydraulic measuring configuration of Fig. 7.9 (g). It is evident for the sensitivity distribution with respect to hydraulic conductivity that the transitional zone between positive and negative sensitivity persists over time only for the closed line intersecting the pumping and the adjacent observation well. The line of zero influence intersecting the second observation well at which the head difference is measured does not persist with time. Thus, any changes in hydraulic conductivity in the area over which this second line of zero influence moves with time can have an opposite influence on the observation at different times. For the change of coefficients with time within the distinct domains of sensitivity, the findings of Sect. 7.4.2 are applicable. From the sensitivity distribution with respect to storage shown in Fig. 7.10 (b), (d), and (f), it is obvious that the division in the distribution persists for all times. However, the values of sensitivity vary over time, as discussed in the previous section 7.4.3.

7.5 Summary

Based on the derivation of sensitivity coefficients, an analysis of the sensitivity distribution with respect to variations in the governing parameters is given in this chapter. In order to evaluate the effects of parameter deviations on the hydraulic measurements, the distribution of sensitivity coefficients for a homogeneous parameter distribution is analyzed in detail. This analysis reveals the intrinsic characteristics of hydraulic testing configurations in terms of sensitivity distribution. Generally, sensitivity distributions of hydraulic tests with spatially separated stimulation and observation locations are characterized by a division of the domain of interest in two regions of opposite sensitivity separated by a line of zero influence. The sensitivity with respect to changes in the parameter distribution is highest in the direct vicinity of the wells. From the investigation presented in this chapter some general conclusions can be given:

For any hydraulic test, which aims on high-resolving investigations, the wells should be positioned as close as possible to the domain of interest, which should be covered by as many different permutations of sensitivity distributions as possible for a reduction of ambiguity and for an improvement of the resolution of parameter discontinuities. For the characterization of a definite region in terms of effective parameters, several tests, spatially different positioned, should be performed in the way of getting a nearly homogeneous coverage of the domain of interest with average sensitivity values. In hydrogeologic practice, the characterization of aquifers is mostly performed using classical test methods, such as constant rate pumping tests, and by an evaluation of hydraulic parameters by means of homogeneous interpretation models. This can lead to misinterpretations, as these methods

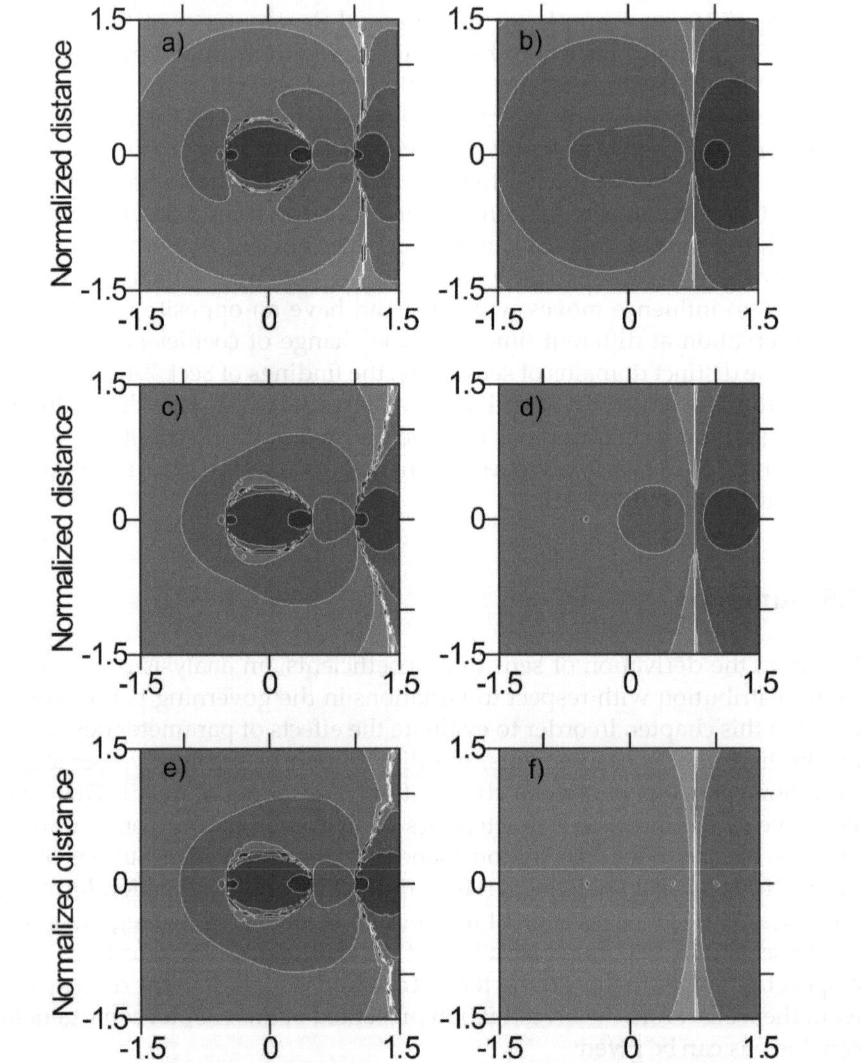

Fig. 7.10. Sensitivity distribution for the hydraulic test configuration illustrated in Fig. 7.9 (g) for three consecutive logarithmic time steps. (a) and (b) at $t = 10$ s, (c) and (d) at $t = 100$ s, (e) and (f) at $t = 1000$ s with respect to changes in hydraulic conductivity (left column) and with respect to storage (right column).

are assumed to give parameters averaged over a domain of uncertain extent and unknown weighting as revealed by the SCA. As a consequence, various issues may arise by the use of classical test and evaluation methods. For example, the intrinsic characteristics of sensitivity distributions discussed in this chapter can lead to the problem of how a definite parameter estimate can be related to a certain parameter distribution. Considering transient groundwater flow, the problem is extended, as the time-dependent information, i.e. distinct intervals in the time-drawdown curve, can be linked to definite spatial information, i.e. spatially varying hydraulic parameters.

The interested reader is referred to Leven (2002) for an illustration of further details and some applications of the approach presented in this chapter.

8

Diffusivity Measurements

R. Brauchler, P. Dietrich

In order to be able to analysize of engineering, geotechnical and hydrogeological problems within the context of water resources management, a knowledge of the spatial position of preferential flow paths is of particular importance. Conventional aquifer investigation methods like pumping tests lead to integral information averaged over a large volume. This kind of information is insufficient to develop groundwater models, which require detailed information about the spatial distribution of preferential flow paths. To circumvent this problem, we have developed a new investigation method based on the relationship between the travel times of a transient pressure signal and the spatial diffusivity distribution of the geological medium under investigation. First, we introduce the concept and the application possibilities of diffusivity measurements. Second, we propose an inversion technique which is derived from seismic travel tomography and is based on a travel time integral relating the diffusivity and the travel time of a transient pressure signal. By means of the inversion, a two- or three-dimensional diffusivity distribution can be determined. The effectiveness of the proposed approach is demonstrated by means of a cylindrical fractured porous sandstone sample on laboratory scale.

8.1 Concept of the Diffusivity Approach

Diffusivity is the quotient of hydraulic conductivity and storage. It is a suitable parameter for characterizing differences in the flow and migration paths because it accentuates the contrast between the fracture- and matrix-dominated system. Usually, fractures have larger diffusivity due to their higher conductivity and lower storage capacity, leading to shorter response times to hydraulic stimulations.

Diffusivity measurements are conducted between two or more boreholes and the travel time of signal between pumping well (source) and observation well (receiver) is determined. With the travel time and distance between source and receiver, the effective diffusivity between the wells can be calculated. A typical set-up of this kind of test is presented in Fig. 8.1. The main advantages of diffusivity measurements are discussed briefly in the following:

- The effective diffusivity is determined directly between one pumping well and one or more observation wells. The evaluation of conventional steady state pumping tests in an anisotropic and heterogeneous medium is much more difficult as an exact assignment of the estimated parameters is still uncertain. A detailed investigation into the relation between parameter distribution and observed signal for conventional pumping tests is presented in Chap. 7.
- The design and the performance of diffusivity measurements are very variable. It is possible to adapt the type of pumping signal to the geological medium. The diffusivity measurements can be conducted with a short impulse (DIRACsignal), a steady state signal (HEAVISIDEsignal) or a short term drawdown (boxcar signal) (Fig. 8.2). The high variability means that the most appropriate pumping signal can be chosen in order to diminish long pumping times and thus lower costs. The most appropriate signal in fractured porous media is the boxcar function, as the pumping time can easily be adapted to the permeability of the geological medium and the distance between pumping and observation well.

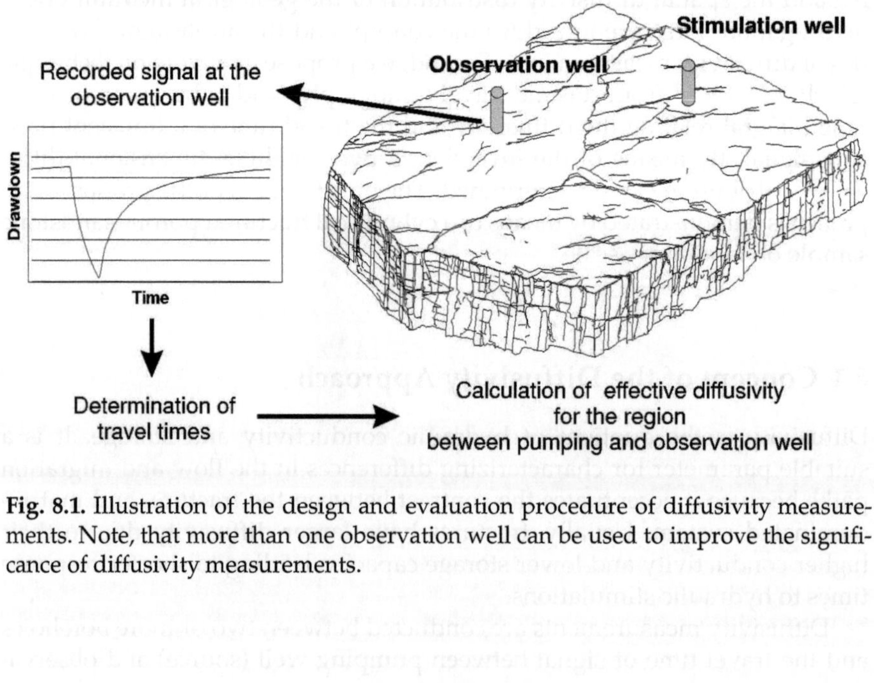

Fig. 8.1. Illustration of the design and evaluation procedure of diffusivity measurements. Note, that more than one observation well can be used to improve the significance of diffusivity measurements.

- As the diffusivity measurements consist of performaning of a series of short term pumping tests, it is possible to vary the position of the sources (pumps) and the receivers (pressure transducers) and to isolate them with packers (Fig. 8.3). This test design allows to produce a streamline pattern which can be compared with the crossed ray paths of a typical seismic tomography experiment (Butler Jr. *et al.*, 1999). The travel times of the recorded transient pressure pulse can be inverted in order to produce a two- or three-dimensional diffusivity distribution of the investigated geological medium. The inversion technique is proposed in the following.

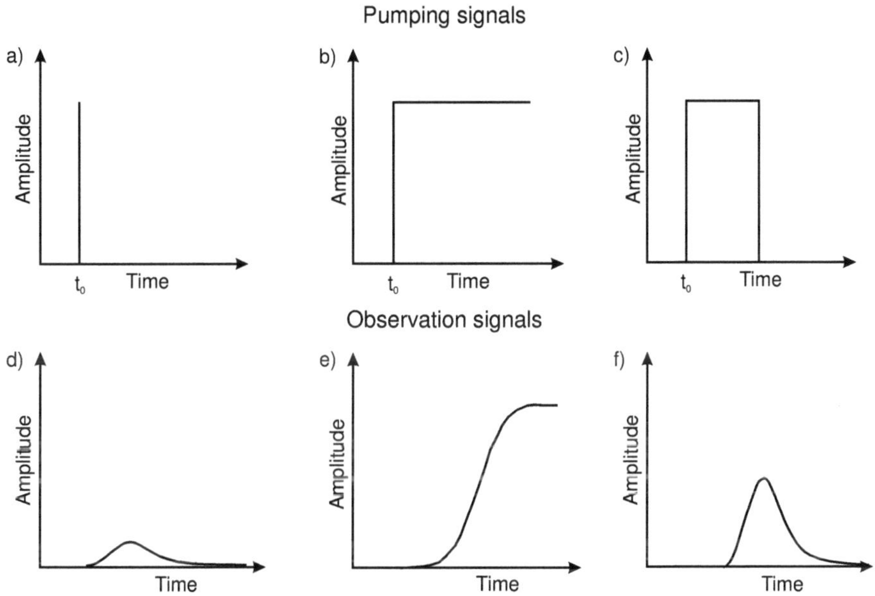

Fig. 8.2. Different kinds of pumping signal and related observations. a) & d) DIRAC signal / related observation; b) & e) HEAVISIDE source / related observation; c) & f) Boxcar source / related observation.

8.2 Inversion Approach

The propagation of a pressure pulse in the three-dimensional subsurface can be described by a line integral relating the arrival time of a "hydraulic signal"to the reciprocal value of diffusivity according to Vasco *et al.* (2000) and Kulkarni *et al.* (2000):

$$\sqrt{t_{\text{peak}}(x_2)} = \frac{1}{\sqrt{6}} \int_{x_1}^{x_2} \frac{ds}{\sqrt{D(s)}} \quad , \tag{8.1}$$

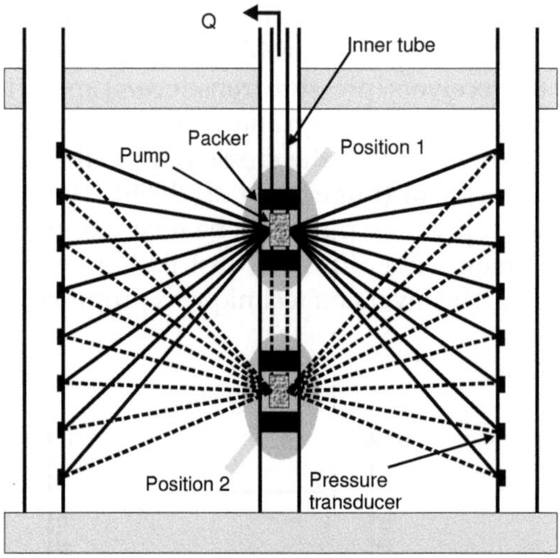

Fig. 8.3. Schematic cross section of a hydraulic tomography experiment. Two separate tests are illustrated; the pump and the packers are moved from position 1 to position 2 after the first test in order to perform the second test.

where t_{peak} is the travel time of the peak of a signal from the point x_1 (source) to the observation point x_2 (receiver) and D is the diffusivity as a function of propagation path s.

The propagation of a seismic signal can be described similarly by a line integral relating the travel time t to the velocity distribution v along the path of signal propagation:

$$t(x_2) = \int_{x_1}^{x_2} \frac{ds}{\sqrt{v(s)}} \quad . \tag{8.2}$$

In our approach, we utilize the similarity between (8.1) and (8.2). This similarity allows the same inversion software to be used. The inversion is conducted with the commercial software package GeoTom 3D, which is based on the last Bureau of Mines tomography program 3DTOM (Jackson and Tweeton, 1996). The program uses a least square solution of a linear inverse problem. The integral defined in (8.1) can be converted to a discrete inverse problem by approximating the integral as a summation as follows:

$$\sqrt{6t_i} = \sum_{j=1}^{n} \frac{1}{\sqrt{D_j}} d_{ij} \ , \tag{8.3}$$

where d_{ij} is the distance along trajectory i in voxel j and D_j is the average diffusivity within voxel j. The goal of the inversion is to choose a diffusivity distribution which minimizes the least squares between the predicted and the measured travel times. In according with this, the function F can be defined:

$$F = \sum_{i=1}^{n} (\sqrt{t_i^m} - \sqrt{t_i^e})^2 \ , \tag{8.4}$$

with t_i^m as the measured travel time, t_i^e as the estimated travel time, and o as the number of measurements. Additionally, a minimum criterion can be defined, allowing the determination of a unique solution:

$$f\left(\sqrt{D^{-1}}\right) = \sum_{j=1}^{n} \left(\sqrt{D_j^{-1}}\right)^2 \rightarrow \min . \tag{8.5}$$

After Dines and Lytle (1979), (8.5) corresponds to a minimum energy correction. Minimizing (8.4) subject to (8.5) leads for each trajectory to an expression for the improvement of diffusivity, $\triangle\sqrt{D_j^{-1}}$:

$$\triangle\sqrt{D_j^{-1}} = \frac{\sqrt{\triangle t_i 6 d_{ij}}}{\sum_{j=1}^{n} d_{ij}^2} \ . \tag{8.6}$$

As the incremental diffusivity correction $\triangle\sqrt{D_j^{-1}}$ is accumulated for all trajectories before being applied to the voxels, the algorithm used is called SIRT, Simultaneous Iterative Reconstruction Technique (Gilbert, 1972). As the propagation of a transient pressure signal and seismic waves follows the same paths, we can use seismic ray tracing techniques.

The proposed inversion technique is based on the relationship between the peak travel time of a recorded transient pressure curve with a DIRAC signal at the origin and the diffusivity of the geological medium. This situation is not very satisfying because much of the information contained in the entire received signal is lost. Therefore, we have developed a transformation factor, allowing our approach be applied to several travel times characterizing each signal. For technical reasons, a HEAVISIDE signal is easier to put in practice, especially for the realization of an adequate signal strength needed for long distances. In order to avoid the differentiation of each recorded curve, a conversion factor relating HEAVISIDEand DIRACsignals is developed. Interested readers are referred to Brauchler *et al.* (2003).

8.3 Application

The methodology presented is applied to data from a set of cross side pneumatic tests conducted in a sandstone cylinder with a height of 34 cm and a diameter of 31 cm. For the recovery of the cylinder, a *Stubensandstein* formation which is quarried in the southern part of Germany was chosen. This formation is part of a continental alluvial depositional system (Hornung and Aigner, 1999) and is mainly composed of arkose sandstone. The sample was situated in facies dominated a bedload channels. The sandstone cylinder was chosen from a series of similar samples because of its simple structural composition with a single fracture embedded in a more or less homogeneous matrix. Figure 8.4 shows this fractured sandstone cylinder prior to preparation for the experiments. The potential position of the fracture is mapped and illustrated in Fig. 8.5. The porosity is approximately 8% and the fracture aperture ranges from 0.1 mm to 0.3 mm.

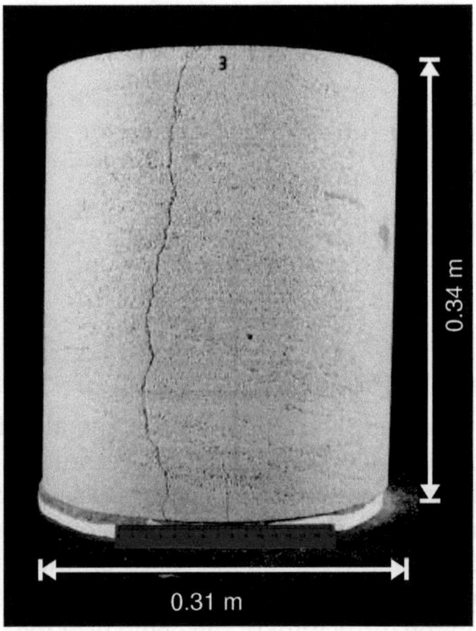

Fig. 8.4. Photo of the unsaturated fractured sandstone cylinder prior to resin coating (Brauchler *et al.*, 2003).

In order to obtain fully controllable boundary conditions, the cylinder is sealed with epoxy resin of approximately 5 mm thickness. To enable access to the sample, 32 ports which can use for injection and measuring were attached on the cylinder in a regular grid. The ports consist of circular metal plugs of 3 cm diameter defining controlled input / output areas. In the center of each plastic plug, a valve is attached allowing a tube to be plugged in. In Sect. 4.1, the recovery and preparation of the cylindrical sandstone sample is described in detail.

In order to receive a large amount of experimental data in a short time, pneumatic tests are conducted instead of hydraulic tests. For the pneumatic experiments, a fully automated multi-purpose measuring device with online data acquisition was developed (Leven, 2002). The experimental set-up is shown in Fig. 8.6. At the injection port, which can be described as a point source,a HEAVISIDEsignal was applied. Compressed air is injected into the

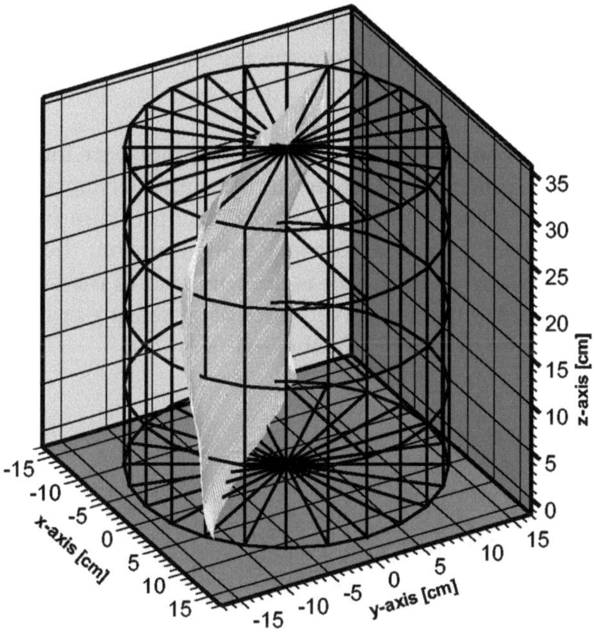

Fig. 8.5. Potential position of the fracture derived from a surface mapping of the cylinder (Brauchler *et al.*, 2003).

sandstone block and the flow rate is recorded at the measuring port. All other ports are opened, allowing the release of the injected air. The flow rate is recorded until a stationary flow field is established between the injection

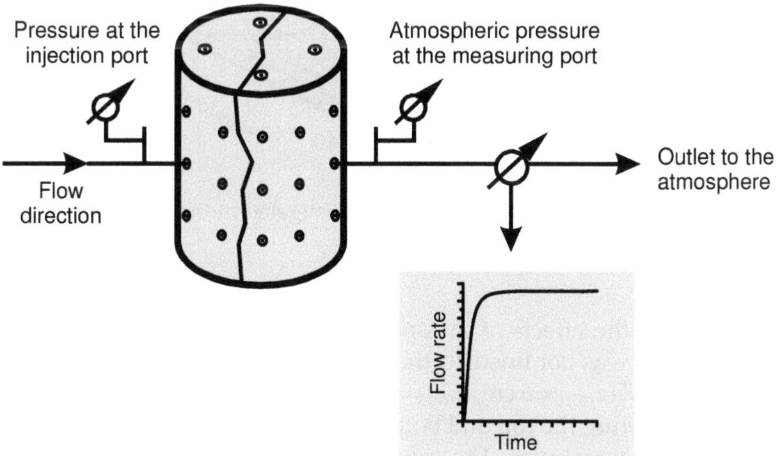

Fig. 8.6. Experimental set-up after Leven (2002).

and the measuring ports. It takes about 20 seconds to reach steady-state conditions. The pressure difference between the injection port and the measuring port is kept constant during the experiment by applying 0.5 bar at the injection port.

In Fig. 8.7, all 487 measured port-port connections are indicated by white lines. Due to the experimental set-up (infinitesimal distance between inside and outside of the cylinder), the gradient of the pressure is proportional to the absolute value of pressure. Consequently, not only the gradient of the pressure but also the pressure itself is proportional to the flow rate. For this reason, it is possible to use the travel times of a pressure curve as well as the travel times of a flow rate curve for the inversion.

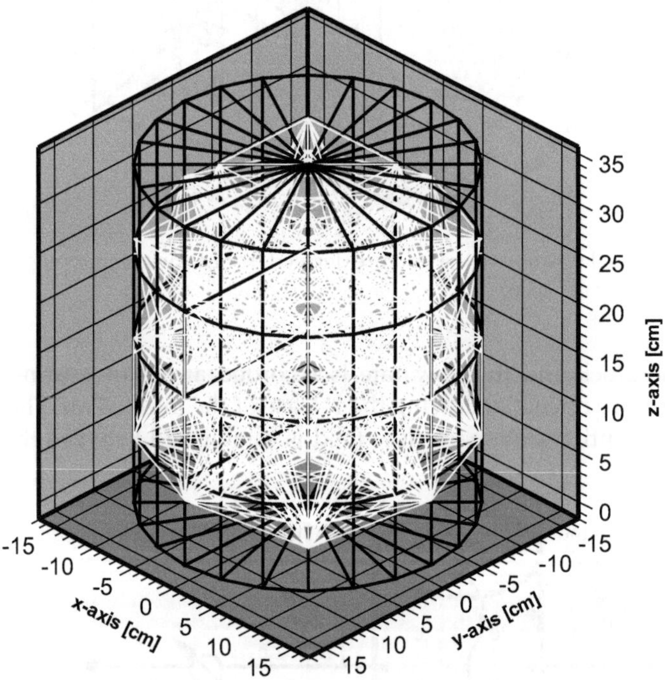

Fig. 8.7. Port-port connections (white lines) investigated in the experiments (Brauchler *et al.*, 2003).

To minimize the effects of the space-dependent trajectory density, te investigation area was confined to the central part of the cylindrical sample shown in Fig. 8.7 ($z_{min} = 6$ cm, $z_{max} = 27$ cm). The reason is the limited angular aperture covering the top and bottom. The variations in trajectory density are least in the central part. The central part was discretized by 500 voxels, 10 in x-direction, 10 in y-direction and 5 in z-direction.

The depicted slices of different depths of the cylinder show the three-dimensional spatial diffusivity distribution (Fig. 8.8). All slices show that significant variations in diffusivity are present, whereby the largest diffusivity values coincide perfectly with the potential position of the fracture. The diffusivity distribution agrees with the well known fact that fractures usually have larger diffusivity due to their higher conductivity and lower ability of storage. The inversion is conducted with ten curved ray iteration steps whereby the initial fit to the measured data is based on a uniform preliminary. As the images are found to be highly reliable and robust, this approach appears to be applicable on the field scale.

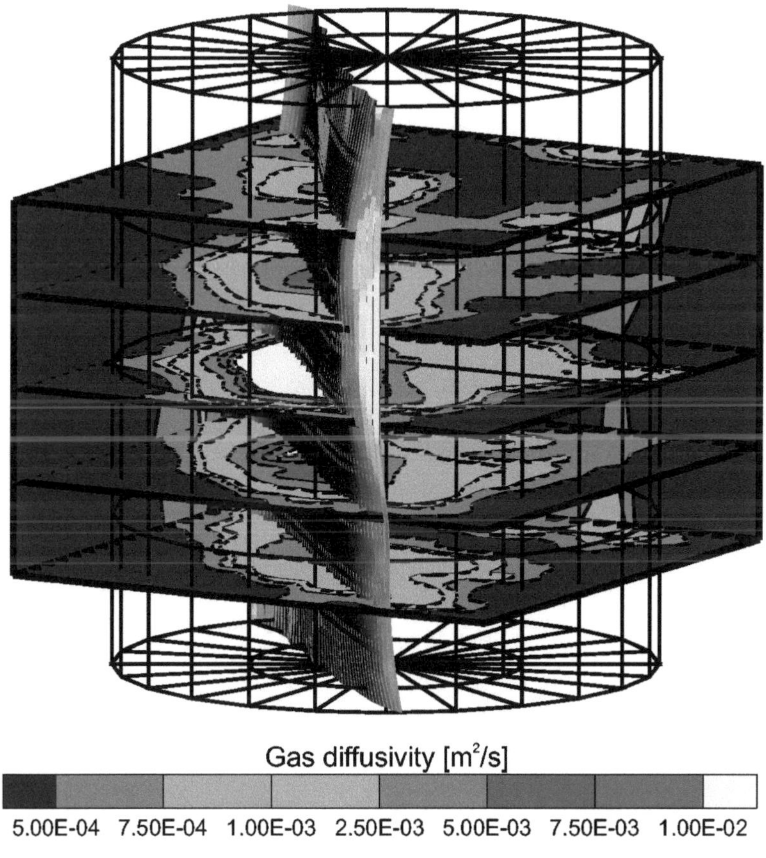

Fig. 8.8. Reconstructed three-dimensional image of the diffusivity distribution of the sandstone cylinder.

8.4 Conclusions

Diffusivity measurements are an alternative to conventional pumping or slug tests, especially in porous fractured media, as it is possible to determine the spatial position of preferential flow paths. The high variability of the test design allows to minimize the pumping time and the arising allocated costs. The similarity between hydraulic travel time tomography and seismic travel time tomography allows to use the same software packages for the inversion of the travel times. As the inversion results are encouraging, applications of this approach in the field seem to be feasible.

9

Analysis of the Influence of Boundaries

M. Süß, H. Eichel, M. Imig, R. Helmig

The aim of conducting flow and transport experiments is to attain a database that can be used for the determination of process behavior and relevant parameters. For practical reasons, domain samples are often taken and measurements conducted under controlled conditions in the laboratory. The results of measurement evaluations of one or more samples are then transfered to the domain from which the samples were taken. It is a well-known fact that the behavior and the parameters of the samples might differ from what one would obtain for the original, larger, domain due to scale effects. Other aspects to be considered when interpreting sample measurement results are, for example, the disturbance of the sample during extraction and transport or the influence of the measurement itself on the flow and transport behavior.

This chapter deals with the fact that samples are always limited in space. The results of flow and transport measurements are therefore always, to a certain extent, influenced by the sample boundaries. Depending on the experimental set-up, this effect is more or less significant and may lead to incorrect measurement interpretation. The significance of the boundary influence is obviously dependent on the distance to a boundary, as illustrated in Fig. 9.1. Other circumstances determining the influence of the boundaries are the choice of boundary conditions and the physical properties of the sample.

The aim of this chapter is to analyze the influence of the distance to boundaries on flow and in particular on tracer-breakthrough curves. The investigation is based on the experimental set-up that was used for tracer measurements on the $60 \times 60 \times 60\,\text{cm}^3$-block (Sect. 4.4.2.2) in the laboratory. The tracer-breakthrough curves obtained were statistically evaluated (Sect. 10.1) and used to determine the structure distribution within the block and their permeabilities (Sect. 10.2). For the quality of a statistical evaluation, it is essential to have a sufficient number of comparable measurements. However, due to the difference in boundary influence, the measurements had to be divided into groups with equal distance to the boundaries first, in order to ensure comparability. From this, the question arises whether it is possible

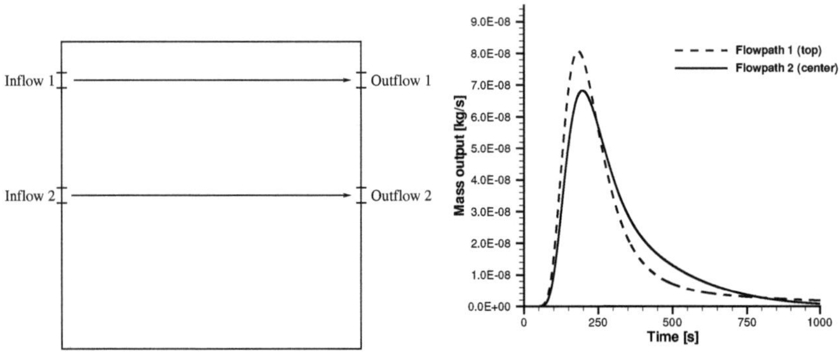

Fig. 9.1. Breakthrough curves for two different port injection / extraction zone configurations.

to normalize a tracer-breakthrough curve so that the influence of the boundaries is eliminated.

This chapter begins with the presentation of basic qualitative investigations to demonstrate the principle influence of the boundaries (Sect. 9.1). In Sect. 9.2, an approach for normalizing tracer-breakthrough curves is introduced. A system of superpositioned wells and image wells is used to obtain a quasi-analytical correction of the flow field.

9.1 Qualitative Analysis of the Boundary Influence

In this section, the results of qualitative investigations of the boundary influence are presented. A model set-up, reflecting the actual experimental set-up of the $60 \times 60 \times 60\,\mathrm{cm}^3$-block, serves as a basis for the discussion of the influence on the parameters: peak concentration c_{peak}, peak arrival time t_{peak}, initial arrival time t_{init} and the behavior of the curve tailing.

9.1.1 Model Set-Up

The geometry as well as the initial and boundary conditions of this principle two-dimensional model reflect the real experimental set-up used for the measurements on the $60 \times 60 \times 60\,\mathrm{cm}^3$-block. A two-dimensional model is chosen for simplicity's sake and to reduce computing time. It is assumed that possible shortcomings due to the missing third dimension are not essential to the principle and qualitative character of the investigation. The geometry of the model is shown in Fig. 9.2. The ports, with a diameter of 3 cm, are named according to the ports in the real three-dimensional set-up. The borders of

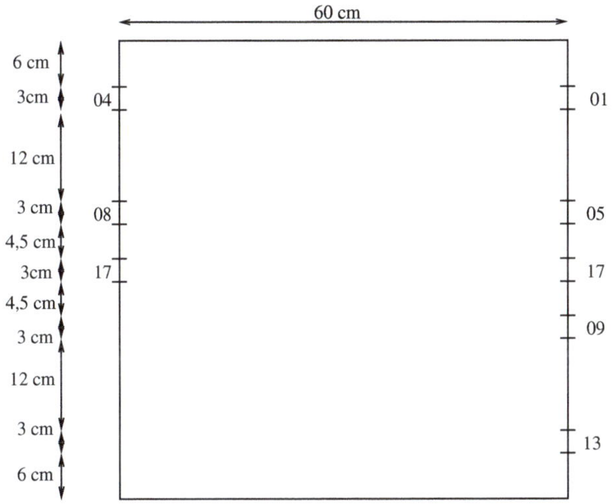

Fig. 9.2. Two-dimensional model configuration.

the model measure 60x60 cm and the depth of the two-dimensional model is set to 60 cm.

The spatial discretization of the domain is realized using a triangulated mesh created by means of the mesh generator ART (Almost Regular Triangulation) (Fuchs, 1999). The various model runs are all based on the same mesh; solely the boundary conditions applied to the ports changes. Using an identical mesh minimizes deviations due to differences in the element geometry.

In each model run, a pressure gradient between two open ports is imposed by assigning a constant pressure to each of them. The other ports and the walls are impervious to flow and transport. The tracer is injected as a pulse at the inflow port, defined by a time-dependent Neumann condition, and the tracer-mass flux is observed at the outflow port.

In total, nine model runs are conducted. The ports with the number 17 are regarded separately from the others as it is the only symmetric configuration and therefore also serves as a reference. The eight other simulations connect the remaining ports on the left side (04 and 08) with the ones on the right side (01, 05, 09, and 13). All configurations are listed in Table 9.1.

9.1.2 Analysis of the Breakthrough Curves

A comparative analysis of the curves shows the principle influence of the boundaries on the simulation result. In order to facilitate the comparison between curves by keeping certain variables constant, the nine curves are classified into groups. One curve can belong to several groups. The first group consists of curves with a common inflow port. The other groups are created by unifying curves with equal distance between the in- and the outflow port.

In Fig. 9.3, all curves of the inflow port 04 are plotted. The tracer-breakthrough curves are plotted as the tracer-mass flux q_M (tracer concentration multiplied by discharge (kg s^{-1})) over time t. As the discharge is not constant for the different configurations, plotting only the concentration would yield misleading results.

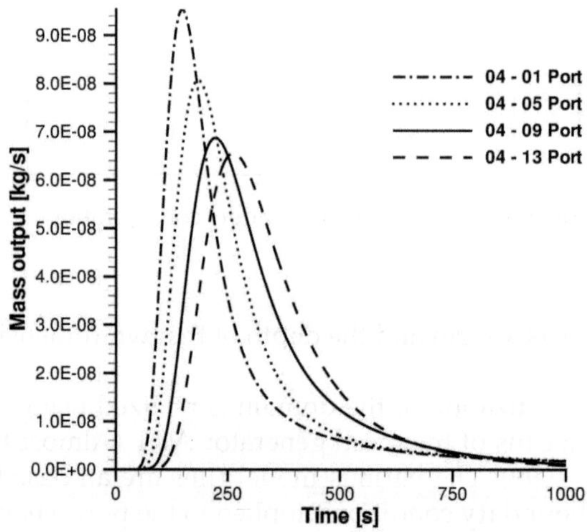

Fig. 9.3. All breakthrough curves with the inflow at port 4.

The four breakthrough curves show a very regular behavior. As the port-port distance increases, the travel time increases and the magnitude of the peak decreases. With increasing port-port distance, the outflow port is shifted further away from the upper boundary and closer to the lower boundary, as in model configuration 04 - 13. To what extent the distance from the walls plays a role is, however, not discernible from this figure. It could be argued that the distance from the upper boundary enhances the effect of the path length, apart from configuration 04 - 13, where the lower boundary influences the flow. This would explain why the peak of this curve, which is the one crossing the block diagonally, is not very far from the peak of the curve of port configuration 04 - 09.

The breakthrough curve that exhibits the highest peak (04 - 01) also shows augmented tailing. This can be explained by the wall influence of this configuration. One part of the tracer travels very fast close to the upper wall, bundling the tracer and accounting for the high peak. The other part of the tracer travels towards the lower wall where the velocities are very low, i.e. the difference in arrival time between the first and the last tracer to arrive is very large. If the main part of the flow occurs in the middle of the block, the travel time difference is smaller, i.e. tailing is not so significant. In general, the velocities are higher close to the in- and outflow port and decrease towards the corners.

In short, Fig. 9.3 shows that the combination of short path lengths and the proximity of the outflow port to the boundary results in very high peaks and strong tailing as these two effects enhance each other. However, the extent of the contibution of the two effects cannot be discerned.

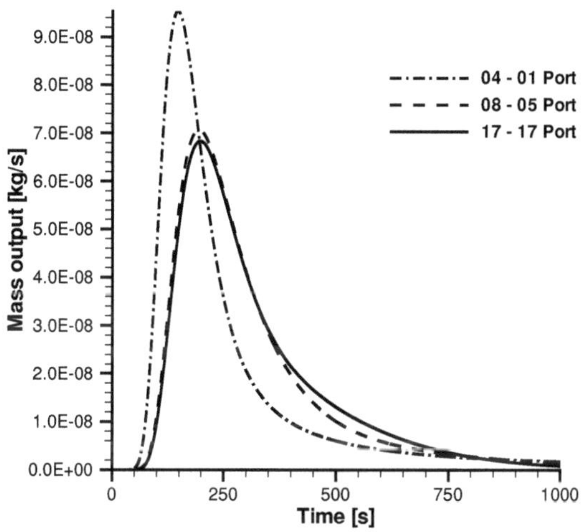

Fig. 9.4. Breakthrough curves from ports directly opposite each other.

In Fig. 9.4, all breakthrough curves with ports directly opposite each other are shown. Here, the influence of the boundaries is very obvious. The closer to the wall the two opposite ports are located, the earlier, the higher, and the narrower the peak and, as mentioned above, the stronger the tailing as well. The distance between the ports is equal and therefore only the boundary influence is relevant.

In Fig. 9.5, which shows all simulations with the port 08 as inflow port, various comparisons can be made. It is interesting that the highest (configuration 08 - 01) and the lowest peak (configuration 08 - 09) stem from two configurations where the in- and the outflow ports are equally far apart. Thus, the only effect that can make the difference must be the proximity of the outflow port to the upper flow boundary. Another intriguing fact is that the highest peak occurs when the outflow port is closest to the wall and not when the path length between the ports is the shortest (configuration 08 - 05). Table

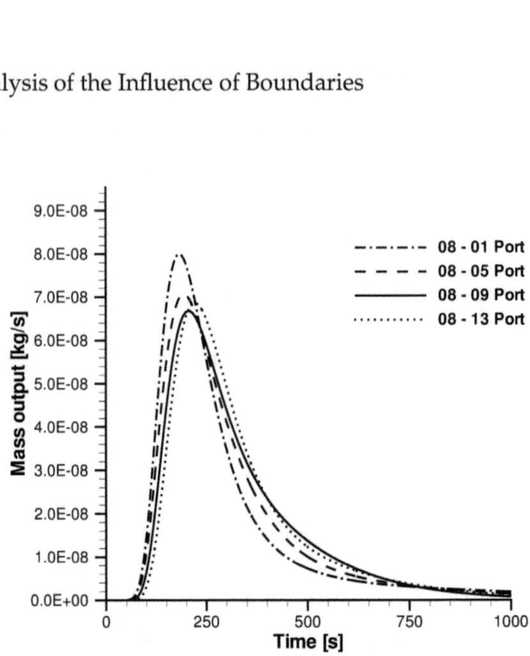

Fig. 9.5. All breakthrough curves with the inflow at port 8.

9.1 shows that the path length of configuration 08 - 01 is 3% longer than the path length for configuration 04 - 05, but the peak arrives in 94% of the time. The same is true for the influence of the lower wall. Here, the peak at outflow port 13 is higher, although not faster, than at outflow port 09. All these observations are evidence of the fact that, for this set-up, the boundary influence has a stronger effect than the path length.

Some of the characteristic parameters that are obtained for different port-port configurations are listed in Table 9.1. The stress is laid on the peak parameters, i.e. the time that elapses until the peak arrives and the magnitude of the peak, although other parameters such as the first response, t_{10} or t_{50} are conceivable. The peak has the obvious advantage that it is identifiable with the naked eye.

In the first column, the open ports are named with the flow always from the left to the right. The second and the third columns display the time of the peak breakthrough and the peak magnitude respectively. The time of peak arrival varies between 146.2 s (04 - 01) and 262.5 s (04 - 13), which is an increase of 80% compared to an increase in distance of only 25%. The most extreme peak magnitudes also stem from these two configurations and are $9.53 \cdot 10^{-8}\,\mathrm{kg\,s^{-1}}$ and $6.55 \cdot 10^{-8}\,\mathrm{kg\,s^{-1}}$ respectively, which is a decrease of 41%. In general, the peaks are higher if they arrive earlier. The fourth column shows the distance between the ports that varies between 60 cm and 75 cm. The last column states the area ratio. The area ratio is defined as the partial area over the total area, where the partial area is the area above the line connecting the two open ports, as illustrated in Fig. 9.6. The area ratio is strongly

Table 9.1. Arrival times, peak magnitudes, distances, and area ratios.

Config.	Peak arrival (s)	Peak magn. ($kg\,s^{-1}$)	Port dist. (m)	Area ratio (-)
04 - 01	146.2	$9.53 \cdot 10^{-8}$	0.600	0.125
04 - 05	182.5	$8.07 \cdot 10^{-8}$	0.619	0.250
04 - 09	221.5	$6.87 \cdot 10^{-8}$	0.677	0.375
04 - 13	262.5	$6.55 \cdot 10^{-8}$	0.750	0.500
08 - 01	182.7	$7.99 \cdot 10^{-8}$	0.619	0.250
08 - 05	193.8	$7.06 \cdot 10^{-8}$	0.600	0.375
08 - 09	206.3	$6.68 \cdot 10^{-8}$	0.619	0.500
08 - 13	221.7	$6.86 \cdot 10^{-8}$	0.677	0.375
17 - 17	197.0	$6.83 \cdot 10^{-8}$	0.600	0.500

related to the peak magnitude. For all cases there holds: the lower the area ratio, the higher the peak. The relationship between other combinations of characteristic parameters is not as unique, as discussed in the following.

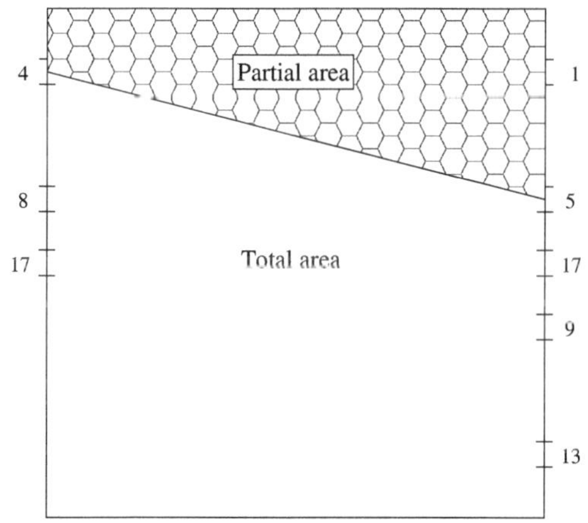

Fig. 9.6. Definition of the area ratio.

9.1.3 Analysis of Characteristic Parameters

Even in a simple case such as this, competing influences are present that make it difficult to decide which feature accounts to what extent for the shape of the breakthrough curves. In order to filter the influences better, the data is evaluated further using only the entries in Table 9.1. This facilitates the identification of trends and dependencies.

An interesting parameter is the area ratio. The connection between two ports can, *as a first approximation*, be looked upon as a dividing streamline.

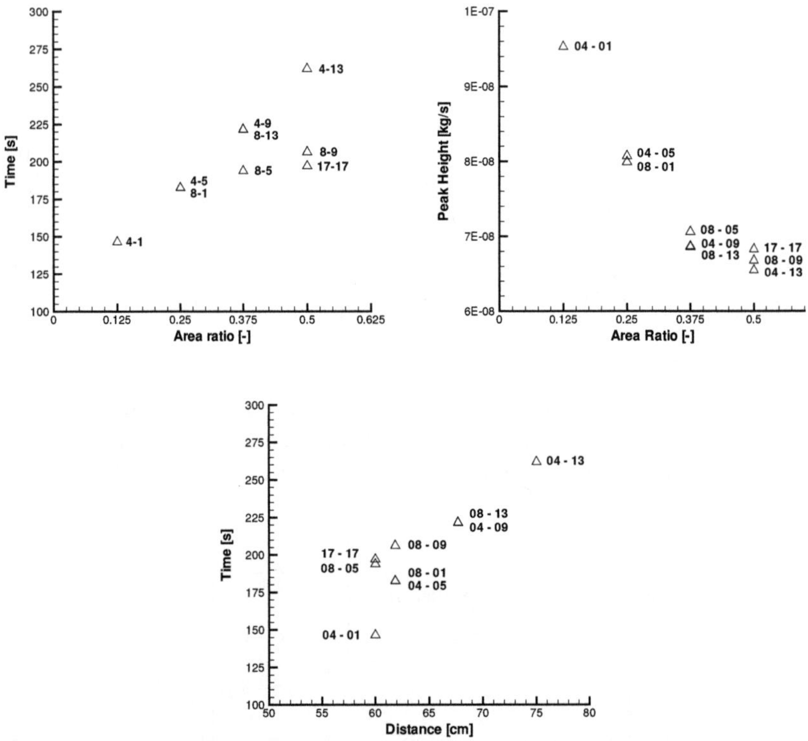

Fig. 9.7. Top, left: Peak time over the area ratio. Top, right: Peak magnitude over the area ratio. Bottom: Peak time over the distance.

Moreover, it can be assumed that half of the flow and thus half of the tracer flows on each side of that line. In configuration 04 - 01, the cross-sectional area above the imaginary line is very small, inferring that half the discharge runs only through a small slice, leading to high velocities. This explains why, in this configuration, the peak is the fastest (Fig. 9.7, top, left) and the highest (Fig. 9.7, top, right). The area-ratio concept also explains the strong tailing. Below the imaginary line, a large cross-sectional area yields low velocities.

This is just a first-order approximation, as the streamlines will bulge away from the closest wall, thus reducing the area differences. The actual area ratios are therefore greater than the assumed values; nevertheless, the concept is valid as an approximation.

If the peak time is plotted over the distance (Fig. 9.7, bottom), a trend towards longer response times for longer pathways can be observed. However, the variations for equal distances are also very high, indicating that another parameter has a strong influence on the curves. The difference between the passing of the fastest peak (04 - 01) and the slowest peak (17 - 17) for a distance of 60 cm has the same order of magnitude as the difference between the slowest peak (17 -17) for 60 cm and the peak for 75 cm (04 -13), showing that the path length is not the dominating parameter.

9.2 Normalization of Tracer-Breakthrough Curves

This section presents an approach for normalizing transport measurement results that are biased by varying boundary influence in such way that the differences due to the boundaries are eliminated. This aim is motivated by the need to compare when a set of data is analyzed. The presented work is considered to be a first step of the development of a normalization method. It is shown, how tracer-recovery curves from advective transport with varying influence of impervious boundaries in a homogeneous domain can be transformed to a reference curve.

The choice to normalize the tracer-recovery curve instead of the breakthrough curve is justified by the fact that it is easier to handle a monotonous curve mathematically. The limitation to advective transport yields a clear problem with a reduced number of parameters. This allows a more specific analysis, without the influence of a large number of possibly competing effects. This is also the reason for restricting the concept development to a two-dimensional homogeneous domain.

In order to exclude possible numerical inaccuracy, a semi-analytical concept for the transport simulation is developed. By superposing simple analytical solutions for the radial flow to wells and image wells, the influence of boundaries on the flow field is considered. The achieved solution is not strictly analytical, since it is an iterative approach which terminates according to certain stopping criteria.

A two-dimensional domain, based on the $60 \times 60 \times 60\,cm^3$-block, is the basis for the investigation. Only configurations with two ports directly opposite each other are considered. The central, symmetrical configuration serves as the reference case to which other configurations are normalized. The curves from port configurations with a distance $b > 0$ are denoted "asymmetric curves".

In Sect. 9.2.1, the development of the normalization concept is presented. This includes (1) the determination of a transfer function as a function of the distance to the center (Sect. 9.2.1), and (2) a discussion of the normalization results achieved (Sect. 9.2.2).

9.2.1 Development of the Normalization Concept

The normalization concept consists of three steps, illustrated in Fig. 9.8. In all plots, the arrival time is made dimensionless by expressing it as the number of pore volumes pv that has flowed through the domain according to the equation:

$$pv = \frac{tQ}{Vn_e} \,. \tag{9.1}$$

Here, t is the arrival time, Q is the discharge, V is the domain volume and n_e is the effective porosity. pv_b denotes the number of pore volumes for asymmetric curves ($b > 0$) and pv_{ref} represents the pore volumes of the reference curve ($b = 0$).

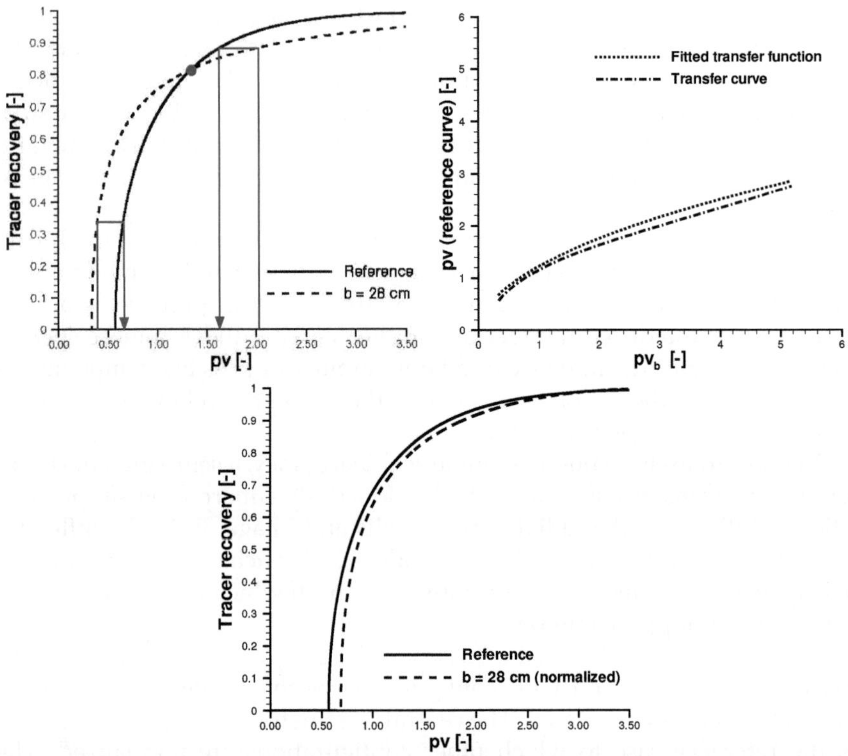

Fig. 9.8. Top, left: The corresponding value of the normalized travel time is determined for each interval of the curve with $b > 0$. Top, right: A transfer function is fitted to the transfer curve. Bottom: The transfer function is applied to the tracer-recovery curve with $b > 0$.

First, the asymmetric curve is sampled equidistantly on the x-axis (pv_b) and the corresponding values of the normalized arrival time (pv_{ref}) of the reference curve are determined (Fig. 9.8, top, left). The obtained pairs of pv values are then plotted. This curve is referred to as the *transfer curve*. Since the goal is to achieve a functional relationship between the two tracer-recovery curves, a mathematical function, the *transfer function*, is fitted to the transfer curve (Fig. 9.8, top, right). The transfer function is then applied to the asymmetric tracer-recovery curve in order to normalize it to the reference curve (Fig. 9.8, bottom). For each distance from the center b, the transfer curve has a different shape. However, it is the aim to have one transfer function only. Consequently, the transfer function is determined as a function of both pv_b and b.

The visualization of the set of transfer curves (Fig. 9.9) reveals that, for small values of pv_b, a root function behavior is expected, whereas for larger values, the curve is almost linear.

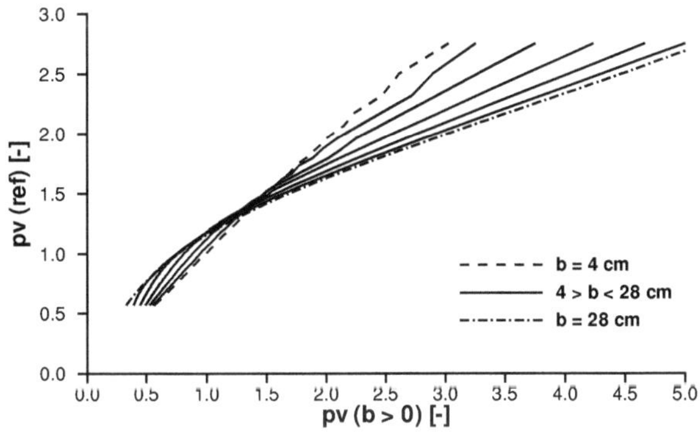

Fig. 9.9. Transfer curves of all port configurations.

With careful regard for the physical plausibility and characteristics of the plotted transfer curves, initially two different transfer functions yielding a very good normalization to the reference curve are found:

$$pv_{\text{norm}}(pv_b) = a_1 + \sqrt[3]{a_2 + a_3 \cdot pv_b + a_4 \cdot pv_b{}^2} \tag{9.2}$$

and

$$pv_{\text{norm}}(pv_b) = a_1 + a_2 \cdot pv_b{}^{\frac{1}{3}} + a_3 \cdot pv_b{}^{\frac{1}{2}} + a_4 \cdot pv_b \,. \tag{9.3}$$

The coefficients a_i are functions of the distance b, which must be fitted as well. Despite the accurate normalization result, with (9.2) and (9.3), another function is finally chosen. This is due to the fact that the coefficients a_i in (9.2) and (9.3) cannot be fitted satisfactorily as their dependence of b is very irregular. The chosen function is a square root function with three coefficients a_i:

$$pv_{\text{norm}}(pv_b) = \frac{-a_2 + \sqrt{4a_1 \cdot pv_b - 4a_1\, a_3 + a_2{}^2}}{2a_1}, \tag{9.4}$$

with

$$a_1(b) = -11 \cdot 10^{-5} \cdot b^3 + 39 \cdot 10^{-4} \cdot b^2 - 61 \cdot 10^{-5} \cdot b - 39 \cdot 10^{-4}$$

$$a_2(b) = +16 \cdot 10^{-5} \cdot b^3 - 46 \cdot 10^{-4} \cdot b^2 - 31 \cdot 10^{-3} \cdot b + 1$$
$$a_3(b) = -96 \cdot 10^{-6} \cdot b^3 + 27 \cdot 10^{-4} \cdot b^2 + 16 \cdot 10^{-5} \cdot b - 15 \cdot 10^{-4}.$$

Equation (9.4) is only valid for a defined range of the variables pv_b and b. Two criteria define the range of validity:

- The numerator $2a_1(b) \neq 0$
  Figure 9.10, left, shows that the value of the coefficient a_1 is close to zero only for very small values of b. Since $a_1(b = 0) < 0$, it is clear that negative values of a_1 exist for $b > 0$. Solving the equation for a_1, shows that $a_1 = 0$ for $b \approx 0.4863$ cm. The exclusion of this value is not a significant limitation of the determined normalization function.
- The radicand $4a_1 \cdot pv_b - 4a_1 a_3 + {a_2}^2 \geq 0$
  This criterion ensures that only real numbers are obtained from the normalization function. The range of validity is defined by both pv_b and b as shown in Fig. 9.10, right. The zero isoline defines the border of the validity range. For $b = 18$ cm, the critical $pv_b \approx 0.32$. For both smaller and larger values of b, the critical pv_b decreases. For the current set of simulated tracer-recovery curves, even the fastest particles do not arrive at times close to the critical value.

It can be concluded that the validity range is sufficient for the homogeneous domain, neglecting minor limitation caused by the first criterion.

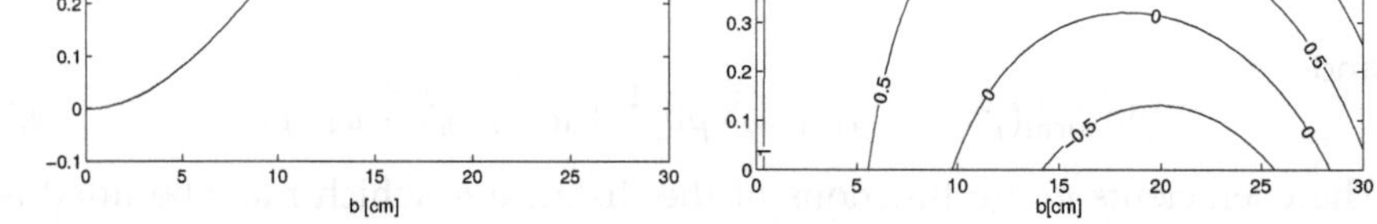

Fig. 9.10. The coefficient a_1 vs. the distance from the center b (left). Contour plot of the radicand of (9.4) as a function of pv_b and b (right).

In Fig. 9.11, the results of the normalization with the largest and the smallest deviations are shown. In general, the result is very satisfactory. For most curves, the largest deviations from the reference curve are observed for the initial arrival of the tracer. The normalized curves generally arrive too late. The latter part of the curves, the tailing, is better reproduced and the obtained deviations are negligible here. There is no trend of decreasing or increasing quality of the normalization for varying distances b.

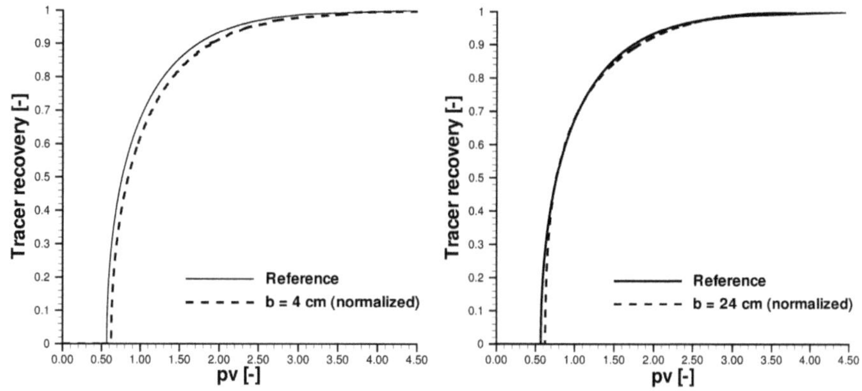

Fig. 9.11. The normalization of the accumulation curve of $b = 4$ cm yields the largest deviations from the reference curve (left). The best result is obtained for $b = 24$ cm (right).

9.2.2 Application to Heterogeneous Domains

The presented normalization function is determined for homogeneous domains. In nature or in laboratory investigations, homogeneity is often assumed, despite the fact that real domains are always, to a certain extent, heterogeneous. In order to test the applicability of the determined normalization function for real domains, two test cases are defined with geostatistically distributed permeabilities with certain variances. In both cases, an exponential variogram model with an isotropic correlation length of 10% of the side length of the domain is used to generate the permeability fields. The exponential variogram model is mathematically defined as:

$$\gamma(h) = s^2 \left[1 - e^{-\frac{h}{a_r}}\right] , \tag{9.5}$$

where s^2 is the variance of the investigated variable, h is the distance between two points and a_r is the length parameter. For exponential variograms, the range is defined as $3a_r$. This type of variogram is often chosen for the description of permeability fields and has proven to yield feasible results (Schafmeister, 1999). The variances of the logarithmic permeability $\sigma^2_{ln\,k}$ of the two cases are 1.0 and 0.25. The first corresponds to the expected variance of a natural system. The latter variance is typical for a very homogeneous natural domain or for an artificially packed homogeneous laboratory set-up.

The semi-analytical model, based on a superposing of sources and sinks, is not applicable to heterogeneous domains due to the irregular perturbation of the flow field. It is, however, desirable to use a model concept which yields transport simulation results that are free of numerical dispersion. Consequently, a particle-tracking algorithm according to POLLOCK (Pollock, 1988) is used in combination with an integral Finite Difference flow model for a structured rectangular mesh. The tool for the geostatistical generation of permeability fields as well as the flow and transport model are made available by O. CIRPKA from the Institute of Hydraulic Engineering of the University of Stuttgart, Germany.

At first, the general influence of heterogeneities on the tracer-recovery curve is demonstrated. In Fig. 9.12 (left), a few samples of tracer-recovery curves for the case with $\sigma^2_{ln\,k_f} = 1.0$ are plotted together with the results of a homogeneous domain. The arithmetic mean of the heterogeneous permeability is simular to the homogenous one. However, the plot shows that the heterogeneities cause both an earlier arrival of the first particles as well as a delay of the last ones. In order to allow a generalization of this statement, a large number of geostatistically equal permeability fields are generated and transport simulations are performed. From the results, an average tracer-recovery curve is determined. The average tracer-recovery curves plotted in Fig. 9.12 (right) are based on 100 realizations for each of the two cases $\sigma^2_{ln\,k} = 1.0$ and $\sigma^2_{ln\,k} = 0.25$. This plot confirms the conclusion drawn for the exemplary curves. It also shows that the effects of the heterogeneities, i.e. earlier initial arrival and longer tailing, become stronger the larger the variance is. The observed effects are due to the fact that the heterogeneous permeability field contains flow paths of higher as well as lower permeability than the homogeneously distributed mean permeability. This phenomenon is referred to as macro-dispersion (Sect. 2.3.3.2).

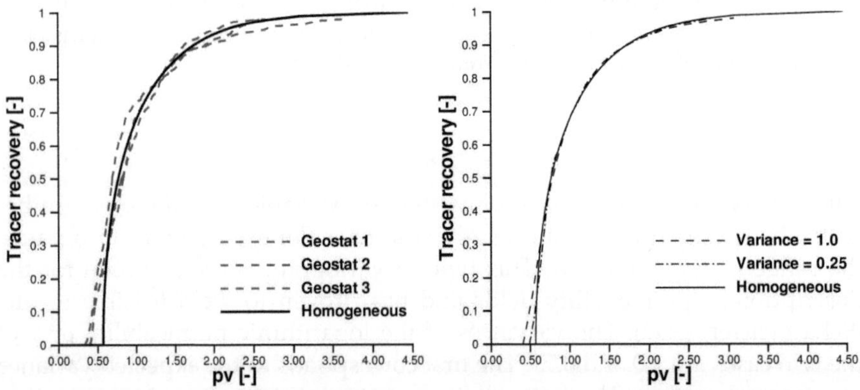

Fig. 9.12. Left: Comparison of exemplary tracer-recovery curves of heterogeneous domains with the curve of the homogeneous domain. Right: Averaged tracer-recovery curves from simulations of different permeability variance $^2_{ln\,k} = 1.0$, 0.25 and 0.0 (homogeneous).

Now that the general influence of this type of heterogeneous permeability distributions on the tracer-recovery curve is known, the applicability of the determined transfer function for the normalization of asymmetric curves to the centered reference curve for such permeability fields is tested. For this purpose, 100 tracer-recovery curves are simulated for each port configuration, corresponding to the 100 realizations of the permeability field. This is done for both $\sigma^2_{ln\,k} = 1.0$ and $\sigma^2_{ln\,k} = 0.25$. The assumption that 100 realizations are sufficient to describe the mean behavior of the tracer transport is justified by the fact that the changes of the average curves are negligible even for fewer than 100 realizations.

The quality of the normalization is determined by calculating the root-mean-square error between the reference curve and the asymmetric curve according to the following equation:

$$e_{\mathrm{RMSE}} = \sqrt{\frac{1}{n_{\mathrm{pv}}} \sum_{i=1}^{n_{\mathrm{pv}}} \left(pv_{\mathrm{ref}}(tr_i) - pv_{\mathrm{norm}}(tr_i)\right)^2}\,, \tag{9.6}$$

where e_{RMSE} is the root-mean-square error, n_{pv} is the number of sampling points, pv_{ref} is the number of pore volumes of the reference curve, pv_{norm} is the number of pore volumes of the normalized curve, and tr_i is the sampling point for a certain fraction of tracer recovery.

The minimum, average and maximum root-mean-square error of the various port configurations for the two permeability variances are presented in Table 9.2. As expected, the deviations are generally larger for the larger variance. There are, however, two exceptions – the minimum deviation of $\sigma^2_{ln\,k_f} = 1.0$ for 20 and 24 cm is smaller than for the same distances of $\sigma^2_{ln\,k_f} = 0.25$. This is an indication of the randomness of the permeability fields, where extreme realizations may occur. An interesting aspect is that there is no clear dependence of the magnitude of the deviation on the distance b, i.e. there is, for example, no increase in the deviations for increasing

Table 9.2. Deviations for the two variances for varying distance b from the center.

	Port position b (cm)						
$^2_{ln\,k} = 0.25$	4	8	12	16	20	24	28
Minimum deviation	0.071	0.076	0.073	0.061	0.072	0.071	0.068
Mean deviation	0.149	0.165	0.159	0.161	0.170	0.167	0.155
Maximum deviation	0.229	0.310	0.300	0.376	0.385	0.381	0.338
$^2_{ln\,k} = 1.0$	4	8	12	16	20	24	28
Minimum deviation	0.088	0.090	0.075	0.068	0.054	0.065	0.078
Mean deviation	0.189	0.235	0.251	0.256	0.288	0.275	0.257
Maximum deviation	0.454	0.525	0.668	0.593	0.689	0.679	0.621

b. It is, however, observed that, for both cases, the smallest mean deviation is obtained for $b = 4$ cm. A possible explanation is that this port configuration falls within the range of the correlation length (6 cm) from the reference curve. It is therefore expected that the local characteristics of the permeability fields for this configuration are similar to the ones of the reference curves. For a general statement, however, the amount of available data is not sufficiently large.

Visualizing examples of the normalized curves gives an impression of the quality of the normalization. In Fig. 9.13, the tracer-recovery curves of port configuration $b = 28$ cm with the minimum, mean and maximum deviation for $\sigma^2_{ln\,k} = 0.25$ and $\sigma^2_{ln\,k} = 1$ are plotted together with the corresponding reference curve.

Figures 9.13(a) and 9.13(b) show the normalized curves with the minimum deviation from the reference curve for the two variances. In both cases, the tailing is accurately reproduced. The initial arrival of the normalized curves occurs slightly too late, an effect which is observed for the homogeneous case as well. The conclusion is that for some heterogeneous realizations, the normalization delivers satisfying results.

Tracer-recovery curves with the mean deviation are shown in Figs. 9.13(c) and 9.13(d). The deviations due to the heterogeneities are clearly seen, especially the effect on the tailing. The arrival time of 98% of the tracer deviates by approximately $-0.5\,pv$ and $+1.0\,pv$ for the small and the large variance respectively. The normalization of the initial arrival is relatively accurate. The large deviations of the tailing are the consequence of the strong influence that the heterogeneities have on this part of the curve. Figure 9.12 shows this very clearly for exemplary curves (left) as well as for the averaged behavior (right).

Finally, in Figs. 9.13(e) and 9.13(f), the curves with the largest deviations are plotted. The observations made for the mean deviation occur here as well; however, the effects are exaggerated. The deviation of the arrival time of 98% of the tracer deviates by approximately $-0.75\,pv$ and $+1.5\,pv$ for the small and the large variance respectively. Again, a relatively accurate normalization of the initial arrival time is achieved.

Valid for all examples shown is the late initial arrival of the normalized curves, following the trend of the homogeneous normalization. The tailing, however, is observed to be both over- and underestimated, excluding a systematic error.

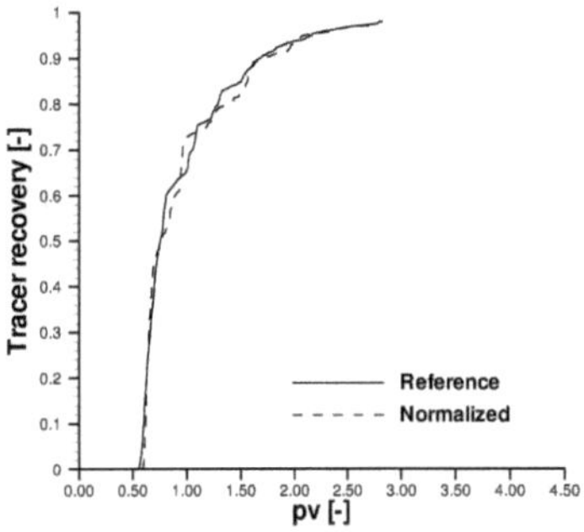

(a) $\sigma^2_{ln\,k} = 0.25$, min. deviation.

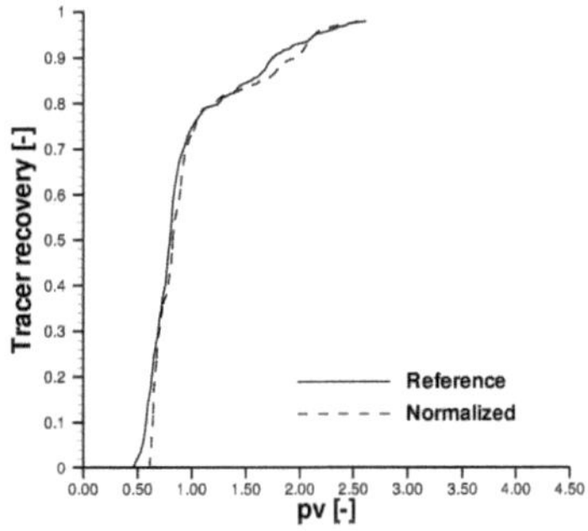

(b) $\sigma^2_{ln\,k} = 1$, min. deviation.

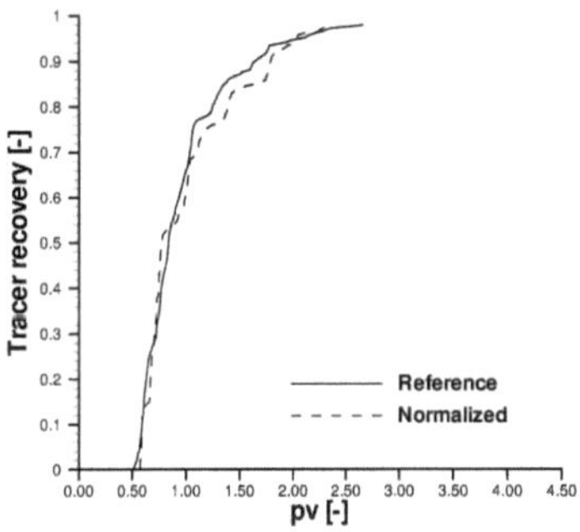

(c) $\sigma^2_{ln\,k} = 0.25$, mean deviation.

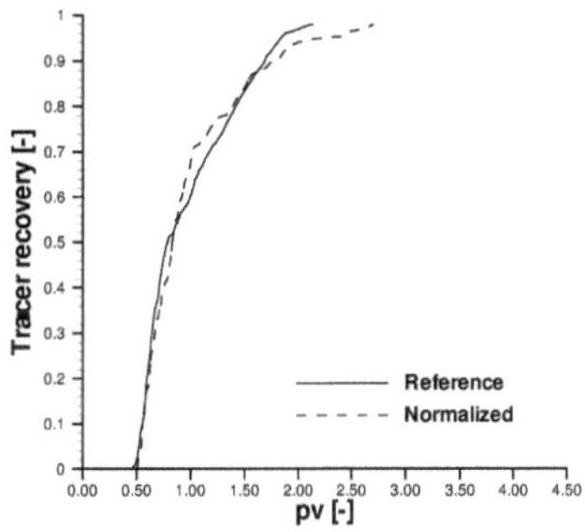

(d) $\sigma^2_{ln\,k} = 1$, mean deviation.

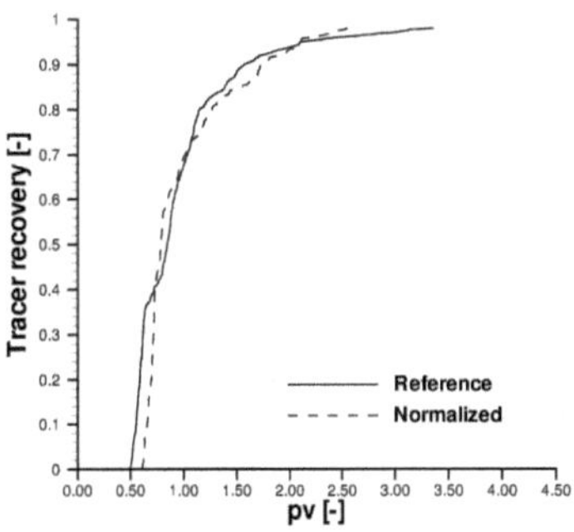

(e) $\sigma^2_{ln\,k} = 0.25$, max. deviation.

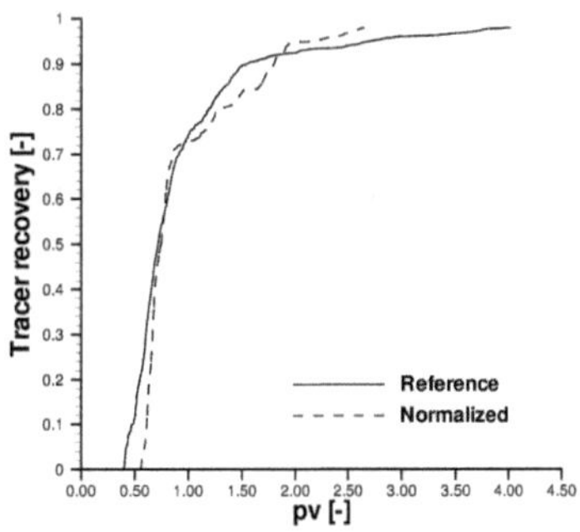

(f) $\sigma^2_{ln\,k} = 1$, max. deviation.

Fig. 9.13. Normalization of the tracer-recovery curves of configuration $b = 28\,\text{cm}$ for heterogeneous domains.

9.3 Summary and Conclusions

The discussions presented in this chapter have te aim to show the importance of considering the influence of boundaries when designing an exprimental set-up, setting up a numerical model or when interpreting measured or simulated data.

First, this chapter presents a discussion of the principle influence of impervious boundaries to flow and transport behaviour on the laboratory scale. Results of numerical simulations of port-port set-ups with varying distances to the model boundaries show that the boundaries influence the discharge and the shape of the tracer-breakthrough curve to an extent that cannot be neglected when interpreting the data.

Second, a first step in the development of an approach for normalizing tracer-recovery curves in order to eliminate differences due to boundary influence is presented. The normalization of tracer-recovery curves of varying distances b from the reference configuration for homogeneous domains yields very good results, especially concerning the tailing of the curve. Since this part of the curve experiences the strongest influence from the domain boundaries, this is a very satisfactory result. The initial tracer arrival-time is generally slightly overestimated, i.e. the normalized curves arrive too late. This delay is, however, considered negligible. It is concluded that the determined transfer function yields good normalization results for homogeneous domains.

As expected, the result of the normalization of tracer-recovery curves from heterogeneous domains with geostatistically generated permeability fields is less accurate. The stronger the variance of the permeability, the larger the deviations from the reference curve. The most significant deviations are obtained at the tailing of the curves as a logical consequence of the strong influence of the heterogeneities on this part of the curve. For small variances, i.e. as expected for homogeneously packed laboratory experiments, the deviations are acceptable.

The presented normalization approach is merely, a first step in the development of a complete normalization method. The next step would be to investigate if influence due to impervious boundaries are actually eliminated, while effects due to heterogeneities are maintained. By using a model set-up that allows the use of impervious as well as periodic boundary conditions, this can be achieved.

10

A Multivariate Statistical Approach

The determination of the spatial distribution of material properties in strongly heterogeneous media is a challenge due to the complexity of the systems and the lack of measurement techniques that can directly detect structures inside a domain.

A widely used method applied to obtain information about the structure of subsurface systems is the evaluation of tracer tests. A tracer is injected at one point and the tracer concentration is measured at another point. The output from such measurements results from the combination of effects of flow and transport processes occurring within the domain. These processes are dependent on the inner structures of the natural system and their properties, suggesting that it is possible to determine the characteristics of the inner structure of a system by analyzing the tracer-breakthrough curve.

In this chapter, a concept for analysing multiple measured or simulated tracer-breakthrough curves obtained from a real or a model domain is presented and discussed. The concept is based on the classification of the tracer-breakthrough curves in clusters according to certain characteristics, of the shape and appendant discharge. The classified tracer-breakthrough curves are interpreted and conclusions are drawn about the inner structure of the domain (Sect. 10.1). The assumed structure is verified by setting up a numerical model based on the properties of the identified clusters of tracer-breakthrough curves (Sect. 10.2).

Additionally, another application of the classification of tracer-breakthrough curves is presented. Clusters of curves from a fractured domain are used to identify different components of a multi-continuum model (Sect. 10.3).

10.1 A Multivariate Statistical Approach for Evaluating Results of Flow and Transport Experiments in an Unsaturated Fissured Sandstone Block

R. Brauchler, C. Leven, P. Dietrich

In the literature, two different approaches are discussed for interpreting flow and transport data and for the investigating, quantifying, and prediction of the relevant processes in fissured and fractured hard rocks:

- deterministic approaches (Huang *et al.* (1999); Karasaki *et al.* (2000); Long and Billaux (1987)) and
- statistical approaches (Clifton and Neumann (1982); Dagan (1982); Harter and Yeh (1996); Yeh and J. (1996)).

Usually, both approaches are based on flux and transport data, *a priori* information, or assumptions on the investigated system. In the case of the deterministic approach, the aim is to develop a model representing to some extent the complexity of the investigated system. For model development, mostly geological or geophysical a priori information or a combination of both together with a certain amount of recorded flow and transport data are used. In contrast, the statistical approaches are based on the statistics of rock parameters, which are derived from field and laboratory data. With a statistical simulation algorithm and variation of the statistical parameters, different realizations are obtained. In both approaches, the model results are compared to experimental data. Usually, neither of the cited approaches includes an *a priori* investigation of the measured data sets for a differentiation of significant characteristics, inherent in the observed flow and transport information. However, such a differentiation can be appropriate for a reliable characterization of the investigated system,

- due to strong anisotropy and significant contrasts in the hydraulic conductivity in hard rocks, and
- since the measured data sets reflect the spatial variation of the distribution of the hydrogeological parameters (e.g. hydraulic conductivity, dispersivity) as well as the physical and hydrogeological boundary conditions that have an increased influence on the experiments especially in three dimensional, high resolution investigations such as tomographical approaches.

In this section, an approach is introduced that is based on methods of multivariate statistics, which

- allows to account for certain characteristics that are inherent in the data of flow and transport experiments,
- enables a rigorous and objective classification of results from flow and transport experiments,

- enable to define an average system response, independent of boundary conditions,
- allows for the identification of parameter zonation in terms of structure identification.

Multivariate statistics have already been successfully applied in nearly all fields of geology and geophysics (Malmgrem and Haq (1982); Schad and Teutsch (1994); Meng and Maynard (2001); Wang *et al.* (2001)). To classify the flow and transport data using multivariate statistics, the investigated object is handled as a black box; thus, no information on the investigated system is required. Furthermore, the variety of applications and the objectivity of the approach are ensured. However, for a reliable investigation using the multivariate statistic approach, a considerable amount of data is required. The method is applied to a high-resolution data set of flow and transport data obtained from tracer tomographical measurements conducted on the fissured sandstone block, which is presented in Sect. 4.4.

10.1.1 Methodology

The main goal of the presented statistical procedure is to identify and characterize representative data sets, reflecting different zones with quasi-homogenous flow and transport properties within the investigated system. The different steps of the statistical procedure are summarized in a flow chart (Fig. 10.1). The initial point of the statistical analysis is a sufficiently large database including results of comparable experiments. The next steps are the definition and standardization of variables describing the results of the conducted experiments. The standardization reduces the influence of the scale dependency. Uncorrelated variables are a prerequisite for the classification of the flow and transport experiments by means of k-means cluster analysis. In order to identify the number of appropriate variables and the variables themselves describe the results of the flow and transport experiments, a combination of principal component analysis and hierarchical clustering is performed. The final classification of the flow and transport processes is conducted by means of k-means cluster analysis. In order to verify the classification, the mapping of the petrofabric and fissured network prior to the resin coating and the numerical model presented in Sect. 10.2 can be used.

10.1.2 Database

Generally, a sufficient amount of data and the comparability of the experimental results are fundamental prerequisites for reliable statistical investigations. The appropriate amount of data can only be determined by means of the statistical results, since there are no rules stating how many objects are necessary to perform a distinct statistical procedure. As shown later, the illustrated database (Fig. 10.2a) is sufficient as the results of the statistical investigation indicate.

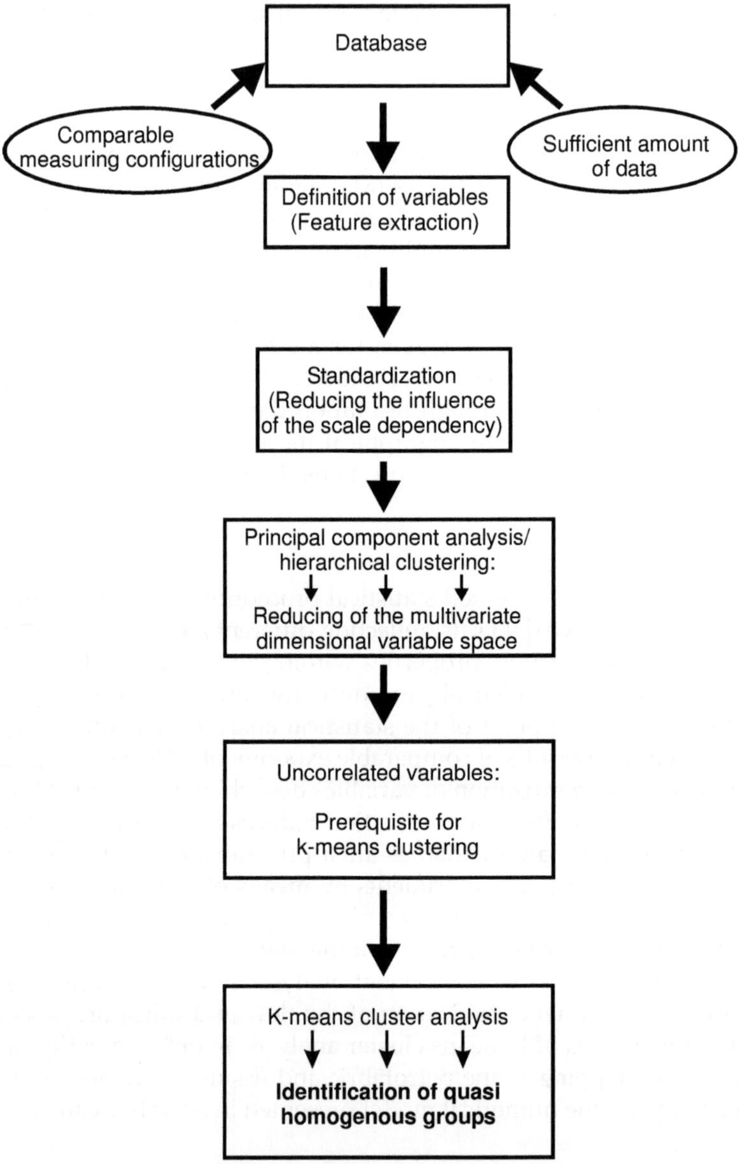

Fig. 10.1. Illustration of the different steps of the statistical procedure.

The conducted experiments are classified by the spatial position of their injection and extraction ports, yielding sets of comparable experimental results with equal distance to the block boundaries.

From an anlysis of Fig. 10.3, the influence of the measuring configuration becomes evident. Without fundamental assumptions about the structure of

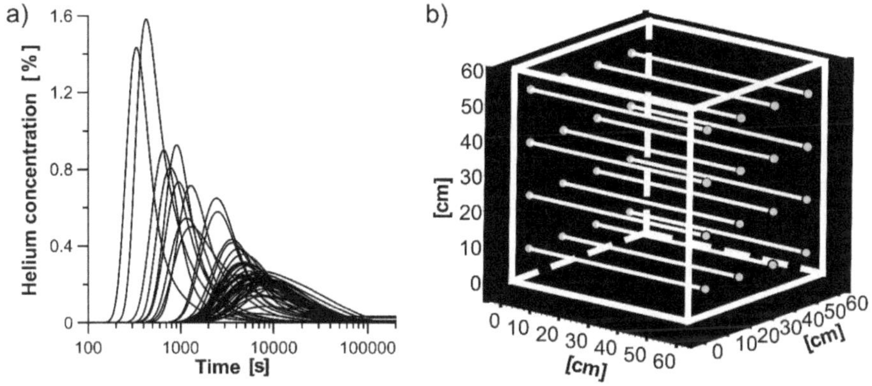

Fig. 10.2. Database for the demonstration of the multivariate statistical approach. a) Recorded breakthrough curves of all 48 tracer experiments conducted on the sandstone block. b) Illustration of the 16 used port-port connections for one of three spatial directions. Note that only one of the shown configurations is used for each flow and transport experiment.

the block, which lead to restricted interpretations, it is not possible to remove the influence of the measuring configurations on the breakthrough of the tracer (Sect. 4.4). To account for this, the presented statistical evaluation is conducted separately for each of the three distinct measuring configurations. The evaluation is presented only for the experiments with measuring configuration I as shown in Fig. 10.3a.

10.1.3 Definition of Variables Characterizing Flow and Transport Processes

To ensure the objectivity of the approach, variables are chosen that can be derived directly from the experimental data, such as definite breakthrough times or measured flow rate. Consequently, the statistical approach described here is based on the measured flow rate and the recorded breakthrough curves. The measured flow rates show a lognormal distribution (not shown here); thus, the logarithm of the flow rates is used in the following statistical investigation. Furthermore, variables are used which describe the run and spread of the curves. In order to describe the run of the curves, the times of the mass-quartiles are chosen. The spread of the curve is described by the quotient of the times of the 60% and 10% decile of the cumulative mass. This coefficient is taken in analogy to sieve curve analysis, where it is used to determine empirically hydraulic conductivities of unconsolidated sediments (Beyer, 1964). Therefore, this quotient is called unconformity coefficient. Additionally, the time of the initial breakthrough (t-initial) and the peak arrival time (t-peak) are used (Fig. 10.4). The variables and their ranges are listed in Table 10.1.

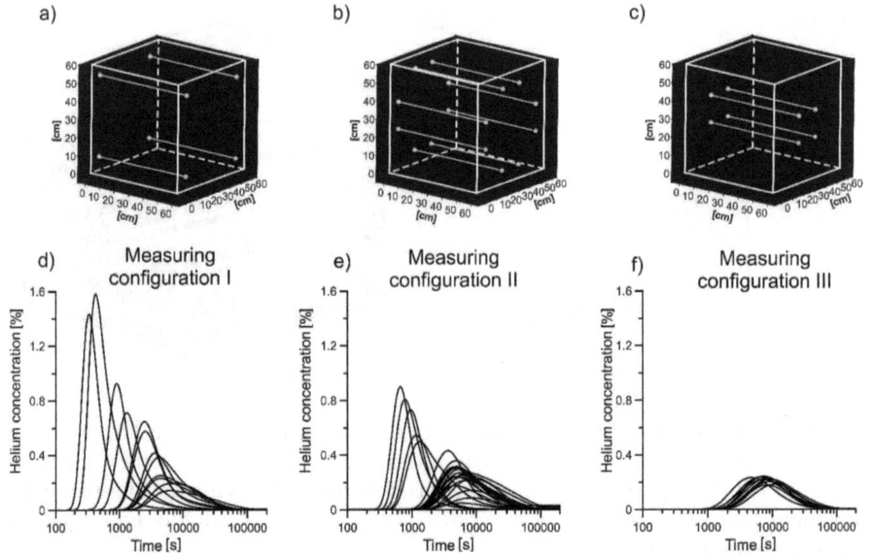

Fig. 10.3. Illustration of the separation of breakthrough curves based on the spatial position of the used port-port configuration. a) – c) indicate the possible spatial positions of the used port-port connections for one of three spatial directions for each group of breakthrough curves. d) - f) illustrate the breakthrough curves classified with respect to the spatial position of the used port-port configuration.

Table 10.1. The minimum and maximum values of each variable used for the statistical investigations. The unconformity coefficient is the quotient of the t-decile (10%) and the t-decile (60%)

Variable	Range
log flow rate (ml/min)	1.6 – 2.9
t-initial (s)	146 – 1252
t-peak (s)	335 – 6572
t-quartile 25% (s)	378 – 10909
t-median (s)	593 – 20537
t-quartile 75% (s)	1809 – 35876
unconformity coefficient (-)	2.7 - 6.3

10.1.4 Reducing the Multidimensional Variable Space

The aim of the first part of the statistical investigation is i) to determine the number of variables that are sufficient to describe the investigated system and ii) to identify uncorrelated variables describing the flow and transport system of the cubic sandstone block. Uncorrelated variables are a prerequisite for many statistical analysis methods such as k-means cluster analysis, which will be described in sect. 10.1.2.5.

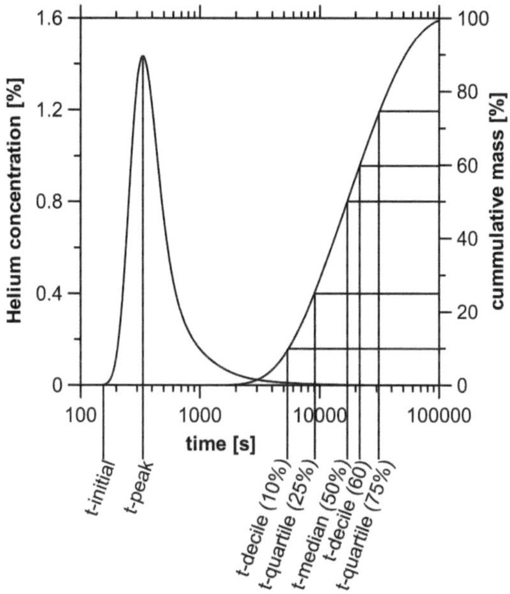

Fig. 10.4. Illustration of the variables derived from the recorded breakthrough curves and used in the multivariate statistical analysis: t-initial, t-peak, t-quartile-(25%, 75%), t-decile-(10%, 60%). An unconformity variable is calculated as the ratio of the t-decile (10%) and the t-decile (60%).

A suitable tool for analyzing the correlation between available variables is Principal Component Analysis (PCA). For the verification of the results of PCA, the application of a hierarchical cluster analysis can be appropriate in some cases in order to identify uncorrelated variables. As shown later, the combination of these two methods allows for an objective identification of uncorrelated variables for the classification of the breakthrough curves. For the statistical procedures, the standardized matrix of the defined variables is used as the starting point. The statistical investigations are carried out using the software package Systat®9.

10.1.4.1 Principal Component Analysis

In the Principal Component Analysis (Jöreskog *et al.*, 1976), each observed variable can be expressed as a linear combination of several components and described by

$$z_{kj} = \sum_{q=1}^{Q} a_{jq} \cdot p_{kq} \quad , \tag{10.1}$$

where z are the observed values of each variable j with respect to each experiment (i.e., breakthrough curve), p are the components and a are the component loadings and Q the number of components. In order to allow a better interpretation, the factors are rotated. In the here presented approach, the so called ""varimax method"according to Kaiser (1958) is used. In this method, the number of variables, that have high loadings on each component, is minimized by an orthogonal rotation of the coordinate system.

10.1.4.2 Hierarchical Cluster Analysis

In hierarchical clustering (Anderberg (1974); Hartigan (1975)), the objects most similar to one another are assembled in a cluster. The next most similar objects are assigned successively to the clusters defined originally. By combining two objects (clusters) P and Q, The distance $D(R, P+Q)$ can be expressed between the new group $(P+Q)$ and any other group R as follows

$$\begin{aligned} D(R,P+Q) &= A\cdot D(R,P)+B\cdot D(R,Q)+E\cdot D(P,Q) \\ &\quad +G\cdot \|D(R,P)-D(R,Q)\| \ . \end{aligned} \tag{10.2}$$

The constants A, B, E and G vary with respect to the algorithm used (Table 10.2). Unfortunately, the literature does not provided an answer as to the most appropriate method for hierarchical clustering. For the presented study, the complete linkage method is chosen, since it yields the most satisfactory results. The complete linkage algorithm uses the most distant pair of objects in two clusters to compute their distances. This method tends to compact globular clusters.

Table 10.2. Values for the constants used with the complete linkage method.

A	B	E	G
0.5	0.5	0.0	0.5

10.1.5 Processing of Data

In order to determine an appropriate number of independent components and variables objectively, the "scree"test method can used. This test uses the component eigenvalues E_j that are arranged in a coordinate system. The eigenvalues are calculated as the sum of the square of the component loadings of one component over all variables and can be expressed as follows:

$$E_j = \sum_j a_{jq}^2 \ . \tag{10.3}$$

The smallest points converging with the x-axis asymptotically are connected. The last point (elbow) on the straight line determines the number of components (Fig. 10.5). In the presented case, three components are chosen.

The eigenvalues and the cumulative variance describing the complete system in dependence on the number of components are listed in Table 10.3. Note that the first three components have much higher eigenvalues than the rest of the list, and they account for 99.44% of the total variance.

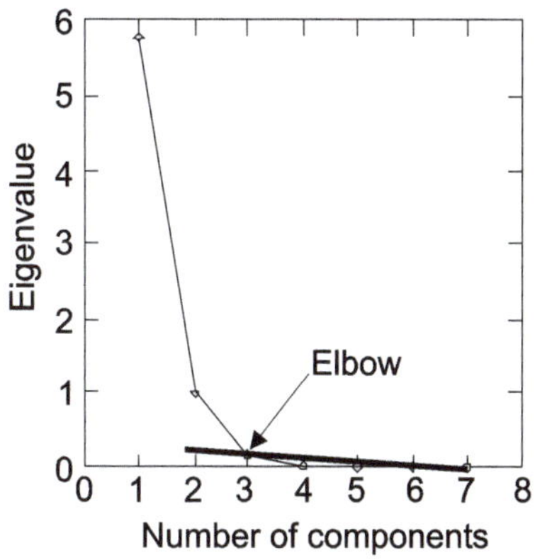

Fig. 10.5. The Screeplot illustrates the eigenvalues of the components. The elbow point indicates the number of appropriate components required for the further analysis of the data set.

Table 10.3. Eigenvalues and cumulative variance of the components. No. is equivalent to the number of components. The components their eigenvalues are illustrated graphically in Fig. 10.5.

No.	Eigenvalue	Cumulative variance
1	5.778	82.537
2	0.99	96.678
3	0.194	99.444
4	0.027	99.828
5	0.007	99.926
6	0.005	99.991
7	0.001	100

The next step is to determine the variables representing the three independent components. The results of the PCA are illustrated in Fig. 10.6a for one measuring configuration (Fig. 10.6b). Because, it is very difficult to determine three uncorrelated variables only by means of PCA, hierarchical cluster analysis was performed using the same database as for PCA. The dendrogram (Fig. 10.6d) resulting from the hierarchical clustering is then separated in the same number of parts as the number of uncorrelated variables. In part

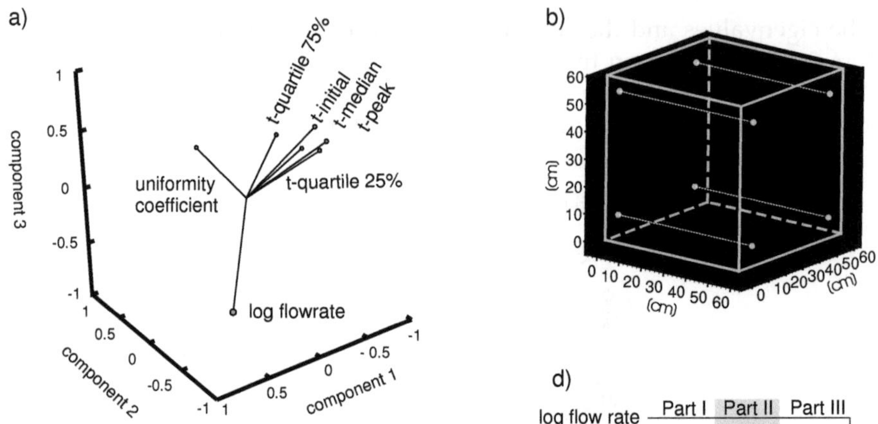

c)

Variables	Component 1	Component 2	Component 3
log flowrate	-0.545	-0.571	-0.613
t-initial	0.901	0.208	0.373
t-peak	0.956	0.109	0.358
t-median	0.912	0.39	0.122
t-quartile 25%	0.966	0.209	0.142
t-quartile 75%	0.769	0.585	0.225
uniformity coefficient	0.143	0.979	0.123

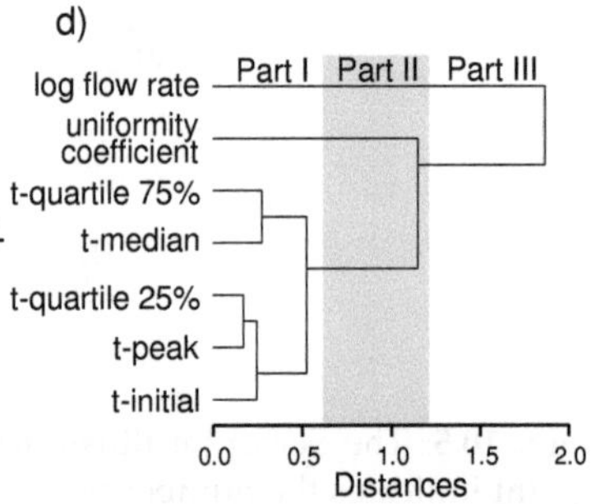

Fig. 10.6. (a) Plot of the results from the principal components analysis for measuring configuration I. (b) Spatial position of the used port-port connections for one of three spatial directions. (c) Rotated loading matrix. (d) Dendrogram of the hierarchical clustering.

I, the cluster contains the five time variables with a distance < 0.62. In part II, the "unconformity coefficient"joins the cluster at a distance of 1.15 and, in part III, the "log flow rate"joins at a distance of 1.86. On the basis of the separation, the variables "unconformity coefficient"and "log flow rate"are chosen representing part II and III, respectively.

The choice of the third variable is much more difficult; thus, the results of the PCA and the hierarchical clustering analysis are combined. The calculated component loadings are given in Figure (Fig. 10.6c and are visualized in Figure (Fig. 10.6a. The rotated loading matrix describes the correlation between the variables and the components. A correlation coefficient of 0 indicates that the variable can not be predicted from the component by using a linear equation, whereas a correlation coefficient of 1 indicates that a variable can be predicted perfectly by a positive linear function of the component. And a value of 1 indicates the same except that the function has a negative sign for the slope.

The rotated loading matrix of the PCA (Fig. 10.6c) shows that component 2 is highly correlated with the variable "unconformity coefficient"and that the variable "log flow rate"is correlated with all three components. Because of this, the third variable should be highly correlated with component 1 and poorly correlated with component 2 and 3. Based on these requirements, the variable t-quartile 25% is chosen.

10.1.6 Classification of the Flow and Transport Data by Using k-Means Cluster Analysis

The aim of the following statistical analysis is to split the breakthrough curves into quasi-homogenous groups using the determined variables which classify the distinct groups. This allows the identification and characterization of representative data sets, reflecting different zones with quasi-homogeneous flow and transport properties within the investigated system.

In the presented approach, k-means clustering (McQueen, 1967) is used for the splitting, since k-means cluster analysis is a powerful tool for identifying natural groupings objectively. Due to the tomographical array of the injection and extraction ports it is possible to identify the spatial distribution of the different flow and transport characteristics.

For k-means clustering, each breakthrough curve is handled as a "case", characterized by the defined uncorrelated variables. The procedure begins by picking "seed cases", one for each cluster, which are separated from the center of all other cases as much as possible. Note, the number of clusters has to be predetermined. Then the cases are allocated to their nearest "seed". Successively, the cases are iteratively rearranged by reducing the total sum of squares within each cluster:

$$\varphi = \sum_{i=1}^{K} \sum_{j=1}^{p} \sum_{m=1}^{n_i} \left(x_{v_{im},j} - \overline{x}_{i,j}\right)^2 \quad , \tag{10.4}$$

where v_{im} denotes the row index of the mth observation (breakthrough curve) in the ith cluster in matrix X (X is a $a \times b$ matrix with a observations and b variables); $\overline{x}_{i,j}$ is the average of the non-missing observations for variable j in cluster i; n_i is the number of rows in X assigned to cluster i; p is the number of variables, and K denotes the number of clusters to be obtained. This processing continues for a given number of clusters, until the sum of squares can be no longer reduced.

10.1.7 Results and Interpretation

The classification starts by splitting up the breakthrough curves into two clusters by using the least correlated variables. Next, the number of clusters is increased and the same procedure is performed with three clusters. To determine the number of clusters needed to describe the system with reasonable certainty, the results of the different classifications are compared. The comparison of classifications (a) and (b) in Fig. 10.7 shows that clusters a/1 and b/1 are identical. This indicates that each of these two clusters represents quasi-homogenous groups of breakthrough curves. Thus, it can be assumed that the clusters characterize different parts of the sandstone block, within the flow and transport parameters are more or less homogenously. The splitting of cluster a/2 leads to a more detailed classification; thus, three clusters can be used to represent the flow and transport processes. The use of four clusters leads to an unreliable result, since cluster c/4 includes breakthrough curves of clusters b/1 and b/3. The mixing of the breakthrough curves of two different clusters would indicate that the previous classification based on only three clusters does not represent homogenous groups. The described classification reveals that at least three different zones with a homogeneous distribution of flow and transport parameters within each zone can then be distinguished. These zones are characterized by the breakthrough curves of clusters b/1, b/2, b/3.

In order to verify the classification, the mapping of the petrofabric and fissured network prior to the resin coating is used. Splitting into clusters a/1 and a/2 fits with the deterministic separation with respect to the texture of the matrix and the bedding. Cluster a/1 contains all the breakthrough curves recorded from direct connections to the upper part of the cubic block which consists of a more coarsely grained matrix (Fig. 10.7). The fifth breakthrough curve of cluster a/1 (dashed line) is measured for a port-port connection in the lower finely grained part of the block. A more detailed mapping of the petrofabric shows that the matrix is coarser in the vicinity of these ports as well. The breakthrough curves of cluster a/2 are recorded either at the lower part of the block, which consists of a more finely grained matrix, or orthogonally to the bedding (recorded from bottom to top).

The shape of the curves within a distinct cluster also corresponds well with the petrofabric derived from the mapping of the block. For example, the breakthrough curves of cluster a/1 are characterized by an initial and dominant breakthrough occurring at significantly earlier times with a sharp concentration increase up to the peak, while the breakthrough curves of cluster a/2 are characterized by a broad and flat shape. This indicates that the upper part is dominated by well interconnected pores with a higher permeability than the lower part.

The classification of the breakthrough curves using three clusters contains more detailed information. It is possible to distinguish the breakthrough curves of cluster a/2 that are recorded orthogonally to the bedding (cluster b/2) from the breakthrough curves recorded with direct connections to the finer matrix (cluster b/3). The experiments conducted orthogonally to the bedding (from bottom to top) are dominated by low flow rates. This agrees with the shape of the breakthrough curves of cluster b/2, because the first initial breakthrough occurs at earlier times than the breakthrough curves of cluster b/3. The classification shows that the matrix properties and the bedding dominate the system while the fissure network has no evident influence of the flow and transport experiments.

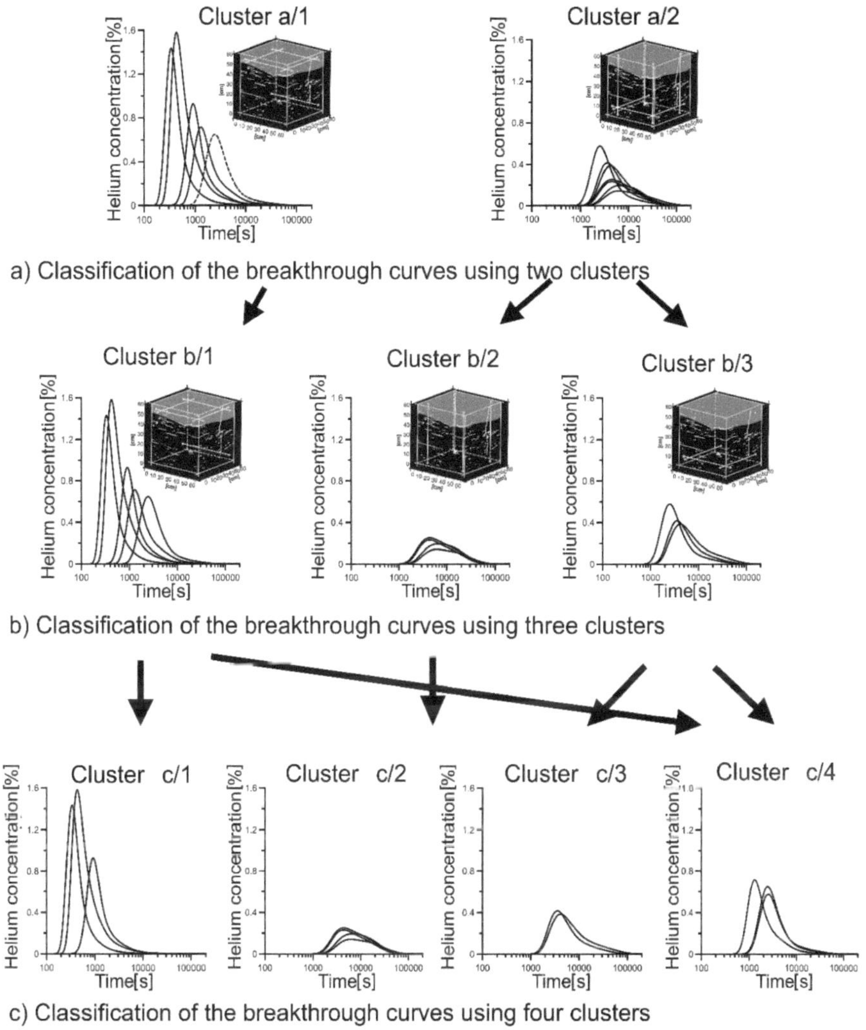

Fig. 10.7. Results of the classification of breakthrough curves using k-means cluster analysis.

A classification of the breakthrough curves into four clusters (c/1 to c/4) is not suitable, as it leads to an unreliable result because cluster c/4 includes breakthrough curves of clusters b/1 and b/3.

10.1.8 Summary and Conclusions

This section presents an approach based on multivariate statistics. In this approach, the investigated fissured sandstone block is handled as a black box, i.e. no a priori information is used for the identification of zones within the block with quasi-homogeneous flow and transport parameters. The approach allows a rigorous, objective classification of results from flow and transport experiments in characteristic, quasi-homogenous groups. These quasi-homogenous groups reflect the different flow and transport properties of the investigated system. Due to the array of the injection and extraction points of the tracer, it is possible to determine the spatial distribution of the flow and transport properties.

The approach is illustrated by analyzing 48 flow and transport experiments which are conducted on a fissured cubic sandstone block on laboratory scale. The cubical shape of the block and the regular array of the injection and extraction ports makes a clear definition of three different measurement configurations possible. The classification of the experiments according to the measurement configurations ensures the comparability of the data.

For the multivariate statistical evaluation, variables are chosen which describe the run and the spread of the recorded breakthrough curves. Additionally, the measured flow rate of each recorded breakthrough curve is used. The classification of the flow and transport experiments is conducted by k-means cluster analysis. Uncorrelated variables are a prerequisite for the application of k-means cluster analysis. In order to identify objectively uncorrelated variables, a procedure is developed which combines the results of principal component analysis and hierarchical cluster analysis.

The detailed geological surface mapping of the investigated sandstone block allows a verification of the classification. It shows, that the statistical classification reflects the textural composition of the matrix of the sandstone block. The conformity between the statistical and the geological classification demonstrates that the multivariate statistical approach presented here is a powerful tool for detecting natural groupings, i.e. zones with a homogeneous distribution of flow and transport parameters within each zone. In some cases, the procedure leads to even more reliable and objective classification results than the regular use of geological a priori information.

The investigations point out that it is useful to conduct measurements under comparable conditions both in the field and in the laboratory in order to identify a parameter zonation in terms of structure identification. Thus, it is possible to characterize and simplify complex systems without loosing essential information about the investigated system. On the basis of the presented classification, a flow and transport model is developed (see next section) . This model enables the reproduction of the recorded breakthrough curves.

10.2 Determination of Domain Properties and Verification of the Results Using Numerical Simulation

M. Süß, R. Helmig

In the previous section (10.1), the results of the application of the multivariate statistical approach on a fissured sandstone block are presented. It is concluded that the system is determined mainly by two quasi-homogeneous layers of different permeability and that there is no significant influence of highly permeable fractures. The object of this section is to verify this statement by setting up a numerical model for simulating the tracer measurements. The model set-up is based on the results from the multivariate statistical analysis of the measured curves. First, it is shown how the measured tracer-breakthrough curves are used to determine the permeability of the assumed structures. Second, the simulation results are compared with the measurements in order to assess the validity of the assumed structure distribution.

10.2.1 Determination of Permeabilities

10.2.1.1 Concept

The approximation of the permeability of the two assumed layers as well as the anisotropy factor between the horizontal and the vertical flow direction is based on the initial tracer-arrival time t_{init}, assuming that the first tracer to arrive has traveled along the shortest flow path from the injection port to the outflow port. For homogeneous conditions, neglecting dispersive and diffusive processes, this is a valid approach. For heterogeneous domains, this is a very strong simplification. First, if, for example, a structure of low permeability is located between the two ports, as illustrated in Fig. 10.8, the tracer may travel faster around the structure than through it, as indicated by the dashed (D') and the solid (D) lines respectively. Second, the discharge of a certain port configuration is always, to a varying degree, the result of the permeability in the total domain, and not only of the permeability between the ports. This is reflected by the fact that a disproportional fraction of the discharge will pass through highly permeable regions, which, if the cross-section is narrow, will lead to increased velocities. Third, it is assumed that the dispersive and diffusive influence on the tracer pulse may be neglected. Despite these simplifications of the actual process, a satisfactory approximation is obtained, as discussed in Sect. 10.2.1.2. In Fig. 10.8, the notations of the equations below are explained. The initial arrival time is represented by the time of arrival of 1% of the tracer mass $t_{1\%}$, since the actual initial arrival time cannot be uniquely defined.

From the equations

$$D = vt_{1\%} \tag{10.5}$$

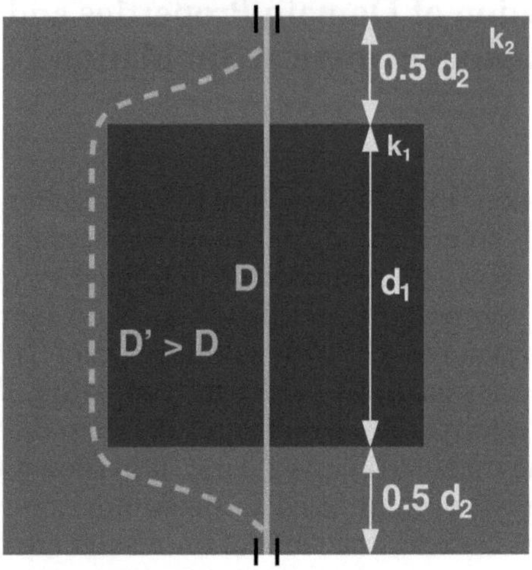

Fig. 10.8. Notations for the determination of the absolute permeabilities.

and

$$v = \frac{k_{\mathrm{eff}}}{n_{\mathrm{e}}\mu} \frac{\Delta p}{D} , \tag{10.6}$$

the following equation for the effective permeability k_{eff} between two ports is obtained:

$$k_{\mathrm{eff}} = \frac{D^2 n_{\mathrm{e}}\mu}{\Delta p t_{1\%}} . \tag{10.7}$$

Here, D is the direct distance between the two ports. If the direct distance between two ports passes through an area of one single permeability, this permeability is assumed to be the same as the effective one. In cases where the direct port connection passes areas with different properties, the effective permeability is assumed to be the harmonic mean of the individual ones:

$$\frac{D}{k_{\mathrm{eff}}} = \sum_{i=1}^{n} \frac{d_i}{k_i} . \tag{10.8}$$

Here, d_i is the length of the corresponding permeability areas in flow direction.

10.2.1.2 Feasibility of the Concept

In order to test the feasibility of the proposed approximation concept, three test cases are defined, as presented in Fig. 10.9. These test cases with block-shaped structures are defined taking the expected character of the sandstone block under investigation into account. It is assumed that simulations in two dimensions are sufficient for this principle investigation.

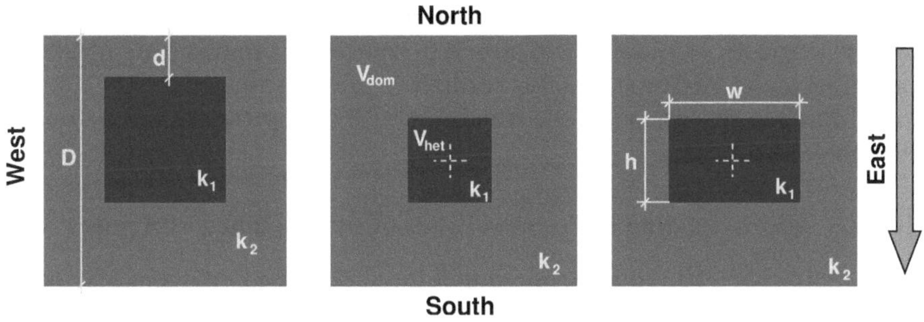

Fig. 10.9. Principle sketches of the three test cases. Left: Case A (distance to the boundary). Center: Case B (relative volume). Right: Case C (shape of the structure).

Common to all test cases is that they contain one block-shaped structure with either lower or higher permeability k_1 than the background permeability k_2. The absolute values of the permeabilities are irrelevant. The general flow direction is always from the north to the south.

The *first test case (A)* is designed to investigate the influence of the distance of the structure to the in- or to the outflow boundary. The structure is gradually shifted away from the northern boundary closer to the southern one, i.e. d is increased. The width of the structure is half of the width of the domain.

In the *second test case (B)*, the relative volume V_r (assuming a domain thickness of 1 m) is varied, while the centered position of the structure is maintained. The relative volume V_r is determined according to the following expression:

$$V_\mathrm{r} = \frac{V_\mathrm{structure}}{V_\mathrm{domain}} \,. \tag{10.9}$$

Finally, the *third test case (C)* deals with the shape of the block structure. The shape is described by the shape ratio P, defined as

$$P = \frac{w \quad (\text{width of the structure})}{h \quad (\text{height of the structure})} \,. \tag{10.10}$$

The results of the feasibility test are not discussed in great detail. It can, however, in general be concluded that it is possible to determine the absolute values of the permeabilities by applying the proposed method. The best results are obtained for the background permeability k_2, because it can be determined directly, without involving the permeability of the structure k_1. The results of the calculation of k_1 prove to be less accurate. The first reason for this behavior is that the flow field in the vicinity of the structure is strongly influenced by the structure and therefore the fastest flow path for the tracer to reach the outflow port, in many cases, is not the direct distance between the ports. The second reason is that the calculations involve the harmonic mean of the two permeabilities, leading to a very high sensitivity to deviations of

k_2 from its true value. In general, the accuracy is better for smaller permeability contrasts. In cases where strong permeability contrasts are obtained, one must always assume that this contrast is underestimated.

A detailed discussion of the feasibility of this approximation method is given in Süß (2004).

10.2.1.3 Approximation of Permeabilities of the Assumed Structures

As stated in Sect. 10.1, it is assumed that the sandstone block consists of two quasi-homogeneous layers, where the upper layer is more permeable than the lower one. Plots of the distribution of the measured discharge (not shown here), as well as a priori knowledge, confirm this assumption and additionally indicate that the boundary between the two layers is slightly inclined. In order to enable slightly different characteristics within the identified lower layer to be considered, a third middle layer is introduced, as indicated in Fig. 10.10.

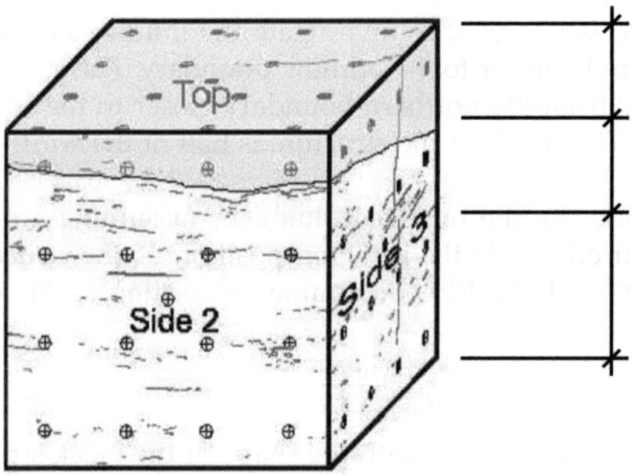

Fig. 10.10. View of the sample, indicating the approximate layer boundaries and the port locations.

The basic concept of the presented multivariate statistical approach is to consider groups of similar curves instead of single ones. In this way, the most significant structures are revealed, at the cost of suppressing minor local variations in the material properties.

The averaged measured tracer-breakthrough curves over each identified cluster are shown in Fig. 10.11. For each measurement configuration (I – III, denoted according to Figs. 10.3a – c), three clusters (1 – 3) are identified. The characteristics and the spatial allocation of these clusters are presented in Table 10.4, referring to the layers indicated in Fig. 10.10.

It is very evident that the curves of the upper layer originate from significantly different flow conditions, compared to all the other averaged curves.

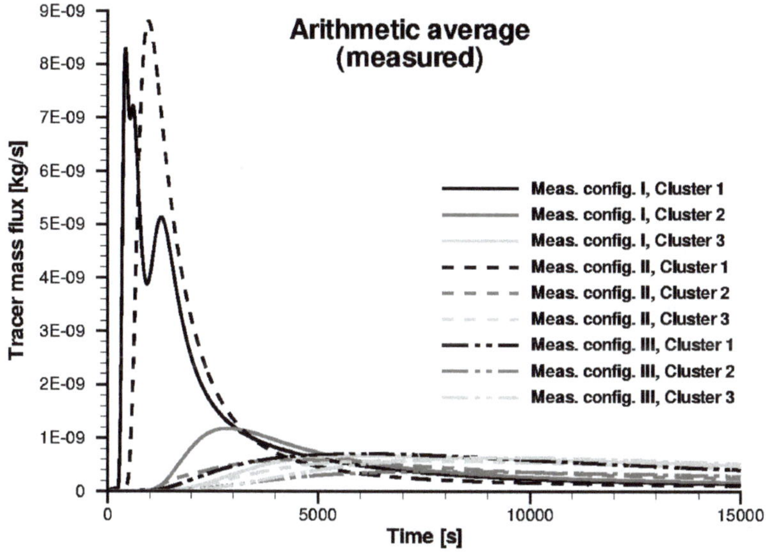

Fig. 10.11. The measured tracer-breakthrough curves, averaged over each cluster (1 – 3) for each of the three possible port configurations (I – III).

Table 10.4. Characteristics of each of the identified clusters and their spatial distribution for each of the measurement configurations.

Cluster	Allocation	Arrival	Peak	Spreading
Measurement Configuration I (corner ports)				
1	Upper layer	very early	clear, high	narrow
2	Lower layer	late	clear, intermediate	wide
3	Bottom-top	very late	flat, low	very wide
Measurement Configuration II (outer middle ports)				
1	Upper layer	early	clear, high	very narrow
2	Middle/lower layer	late	flat, low	wide
3	Bottom-top	very late	flat, low	very wide
Measurement Configuration III (middle ports)				
1	Lower layer	late	flat, low	very wide
2	Middle layer	latest	flat, very low	widest
3	Bottom-top	latest	flat, very low	very wide

Moreover, the curves measured in the direction from bottom to top are generally the slowest and the most widely distributed in time, except for the curve of the middle layer for the middle port configuration, which has a late initial arrival and the widest spreading of all six curves.

Using the average arrival time $t_{1\%}$ of each of the clusters and assumptions for the effective porosities (0.23 for the upper and 0.12 for the middle and the lower layers (Bengelsdorf (1997); Müller (1997))) yields a first approximation of the permeabilities as listed in Table 10.5. Instead of listing the notation of the clusters, the location of the ports on the block is used, e.g. "upper layer". From the approximated permeability values, three observations can be made: (1) the difference in permeability between the upper and the middle/lower layers is just less than one order of magnitude; (2) there seems to be a tendency towards lower approximated permeability further away from the boundaries; and (3) the harmonic mean of the horizontal permeabilities of the layers is higher than the approximated permeability from bottom to top, indicating anisotropy. The first observation indicates a strong contrast in permeability between the layers. The second observation is due to two competing effects occurring as the ports approach a boundary: (1) the total discharge decreases slightly; and (2) the effective velocity increases because the area through which half of the discharge flows decreases. Since the second effect is stronger, the initial arrival time of a tracer-breakthrough curve measured close to a boundary is less than if the boundaries are far away (Sect. 9). The observed anisotropy is reasonable, as the permeability perpendicular to the bedding is usually lower than in the horizontal direction.

Table 10.5. First approximation of the permeabilities (m^2) of each of the identified clusters.

Allocation	Meas. Config. I	Meas. Config. II	Meas. Config. III
Upper layer	$8.9 \cdot 10^{-14}$	$5.0 \cdot 10^{-14}$	
Middle layer		$1.6 \cdot 10^{-14}$	$8.4 \cdot 10^{-15}$
Lower layer	$1.7 \cdot 10^{-14}$		$1.3 \cdot 10^{-14}$
Bottom-top	$1.1 \cdot 10^{-14}$	$1.0 \cdot 10^{-14}$	$8.3 \cdot 10^{-15}$

10.2.2 Set-Up of the Numerical Model

10.2.2.1 Boundary Conditions

At the in- and the outflow port, a constant pressure is imposed, whereas the rest of the block surface is impervious to flow and transport. A tracer pulse is injected by keeping a certain tracer-mass flux constant during one time step. In Fig. 10.12, a mesh for one of the port configurations is shown. The assumed layered structure as well as the strong refinement around the ports can be clearly seen.

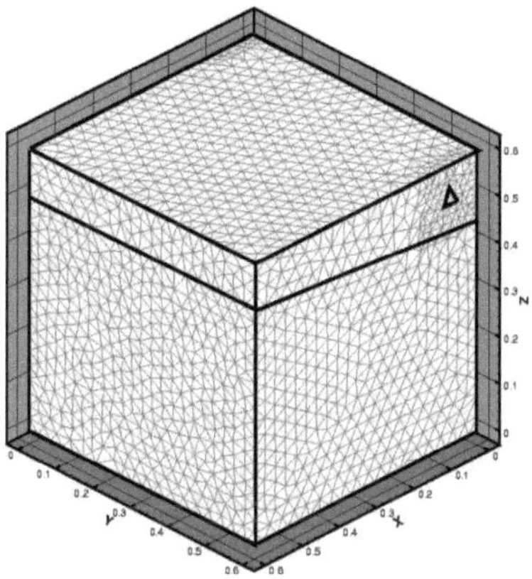

	Fluid Properties		
μ	$1.81 \cdot 10^{-5}$		$\mathrm{kg\,m^{-1}\,s^{-1}}$
D_m	$6.8 \cdot 10^{-6}$		$\mathrm{m^2\,s^{-1}}$
	Matrix Properties		
	Up. layer	*Mid./Low. layer*	
$k_{x,y}$	$5.5 \cdot 10^{-14}$	$8.9 \cdot 10^{-15}$	$\mathrm{m^2}$
k_z	$3.7 \cdot 10^{-14}$	$6.4 \cdot 10^{-15}$	$\mathrm{m^2}$
n_e	0.23	0.12	-
α_l	$1.5 \cdot 10^{-2}$		m
α_t	$1.5 \cdot 10^{-3}$		m

Fig. 10.12. The model set-up. Left: Example of a generated mesh with refinement in the vicinity of a port. Right: Used model fluid and matrix properties of the assumed structures. The anisotropy ratio between the horizontal and the vertical direction is 1.5.

10.2.2.2 Fluid Properties

The dynamic viscosity μ of air at 20° C and 101.3 kPa is $1.81 \cdot 10^{-5}\,\mathrm{kg\,m^{-1}\,s^{-1}}$ and is kept constant during the simulation. The compressibility due to pressure changes is considered in the model by applying the universal gas law for an ideal gas.

The molecular diffusion coefficient $D_\mathrm{m} = 7.0 \cdot 10^{-5}\,\mathrm{m^2\,s^{-1}}$ is valid for helium and air at low pressures and a temperature of 20° C (Reid *et al.*, 1987). There are different approaches to determining the *effective* diffusion coefficient $D_{\mathrm{m,e}}$, considering the reduced diffusive flux due to the porous medium, as discussed by, for example, Cunningham and Williams (1980), Birkhölzer (1994a), and McDermott (1999). Based on these approaches, an effective diffusion coefficient of $D_{\mathrm{m,e}} = 6.8 \cdot 10^{-6}\,\mathrm{m^2\,s^{-1}}$ is used for both layers.

10.2.2.3 Matrix Properties

The approximated permeabilities as presented in Table 10.5 are only to be considered a first but relatively accurate guess of the permeabilities of the assumed structures. Resulting from a few initial flow calibration runs of the model, the values presented in the table to the right in Fig. 10.12 are used for all further simulations.

The effective porosity n_e is 0.23 for the upper and 0.12 for the lower layer, as already mentioned in Sect. 10.2.1.3. The longitudinal and the transversal dispersivities are set to $\alpha_\mathrm{l} = 1.5 \cdot 10^{-2}\,\mathrm{m}$ and $\alpha_\mathrm{t} = 1.5 \cdot 10^{-3}\,\mathrm{m}$.

10.2.3 Discussion of the Results

10.2.3.1 Flow Simulation

In Table 10.6, the ratios between the simulated and the measured discharges are presented. The approximation of the discharges of all corner configurations (configuration I) and of the bottom-top clusters are very accurate (the maximum deviation is 11%). For the other clusters, under- as well as overestimation of the discharges is obtained. The absolute deviations of the middle ports (configuration III) are relatively small in comparison to the deviations of all ports and are therefore not considered significant.

The same accounts for the outer middle ports (configuration II) of the lower layer. The only deviation which cannot be neglected is the one from the outer middle ports (configuration II) of the upper layer. Here, there is a clear underestimation of the measured discharge. Increasing the permeability would, however, lead to an overestimation of the discharge of the upper layer of the corner ports (configuration I). Therefore, the chosen permeability distribution is accepted and used for the transport simulation.

Table 10.6. Ratios of $Q_{\mathrm{sim}}/Q_{\mathrm{meas}}$.

Allocation	Meas. Config. I	Meas. Config. II	Meas. Config. III
Upper layer	1.00	0.76	
Middle layer		1.34	1.41
Lower layer	0.99		0.71
Bottom-top	1.01	0.99	1.11

10.2.3.2 Transport Simulation

The quality of the transport simulations is judged by comparing the measured and the simulated tracer-breakthrough curves as well as the arrival times $t_{1\%}$. In Fig. 10.13, the averaged measured tracer-breakthrough curves are plotted together with the individual curves of each port configuration. The ratios of the arrival times are presented in Table 10.7.

The first impression from Table 10.7 is that the simulated curves in the upper layer arrive too late whereas the other curves arrive too early (except for the lower layer, configuration III). A visual comparison of the average measured curves and the individual simulated ones, however, leads to the conclusion that the initial tracer arrival time in general coincides very well. The reason for the deviations in $t_{1\%}$ are due to the shape of the curves, i.e. the initial arrival time of the simulated curves of the upper layer is very accurate but they are not as steep as the averaged measured ones. In the following, the observed differences and similarities between the simulated curves and

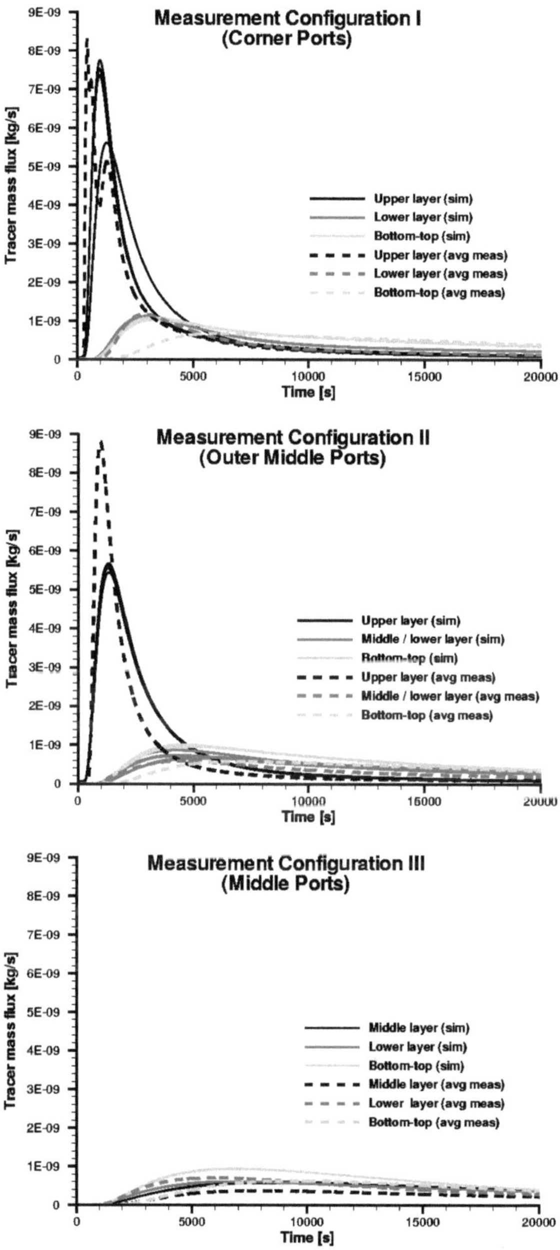

Fig. 10.13. Comparison of average measured tracer breakthrough curves with all individual simulated curves for each measurement configuration.

the averaged measured clusters are presented. These observations are interpreted in Sect. 10.2.3.3.

Table 10.7. Ratios of $t_{1\%}(\text{sim})/t_{1\%}(\text{avg. meas})$.

Allocation	Meas. Config. I	Meas. Config. II	Meas. Config. III
Upper layer	1.41	1.17	
Middle layer		0.88	0.64
Lower layer	0.86		1.07
Bottom-top	0.57	0.64	0.63

Starting with the *corner port configuration (I)*, the curves of the upper layer have a very accurate initial arrival time, but the increase of the tracer-mass flux is not strong enough, leading to a smaller and delayed peak. At later times, the tailing fits well. As regards the curves of the lower layer, the initial arrival comes slightly too early, whereas the shape of the curves is very accurately reproduced. The simulated bottom-top curves definitely arrive too early but, as the simulation continues, the characteristic strong tailing is satisfactorily described.

The characteristics of the *outer middle port curves (II)* are very similar to the ones for the corner ports. The observations are not repeated.

For the *middle port configuration (III)*, there is a satisfactory agreement for the curves of the lower layer. The simulated curves of the other two clusters arrive too early and are steeper than the averaged measured cluster curves. This is most significant for the bottom-top simulations.

10.2.3.3 Interpretation of the Simulation Results

Interpreting the observations made for the simulated discharges and tracer-breakthrough curves in the two previous parts of this section (10.2.3.1 and 10.2.3.2) leads to conclusions concerning the inner structure of the investigated fissured sandstone block.

The simulated corner port curves (configuration I) of the *upper layer* are observed to very accurately reproduce the measured discharge, whereas the discharge of the outer middle ports (configuration II) is underestimated. This suggests that the real layer is not as homogeneous as initially assumed. As regards the shape of the simulated curves, the initial arrival time is very close to the average measured curves, whereas the tracer-mass flux increase is not steep enough. A principle model investigation, illustrated in Fig. 10.14, shows that thin layers of material of low permeability, parallel to the main flow direction, cause only a slight decrease in discharge but significantly steeper tracer-breakthrough curves with higher peaks. It is possible that the upper layer contains structures such as fractures filled with consolidated material or clay, causing this type of behavior.

The fit of the discharges and the curves of the *lower layer* is very good for the corner as well as for the outer middle ports (configurations I and II). For the middle ports (configuration III), the curves of the lower part are

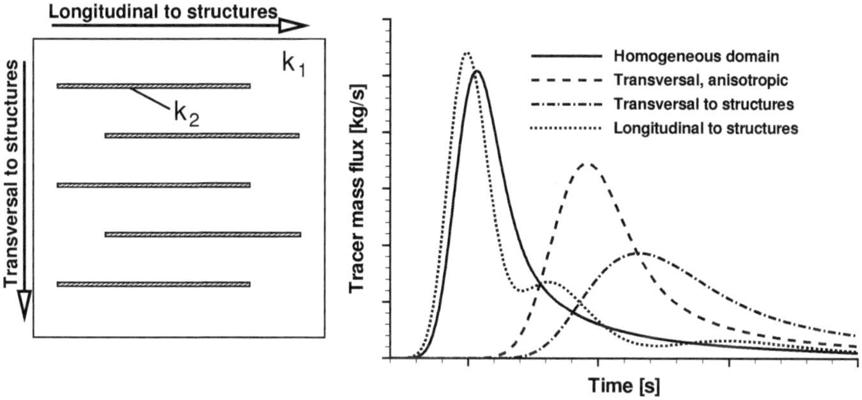

Permeabilities	(m^2)
k_1	$1.0 \cdot 10^{-13}$
k_2	$1.0 \cdot 10^{-14}$
k_z (Hom., anisotr.)	$5.4 \cdot 10^{-14}$

Simulated discharges	($m^3\,s^{-1}$)
Homogeneous domain	$1.09 \cdot 10^{-4}$
Transversal, anisotropic	$0.70 \cdot 10^{-4}$
Transversal to structures	$0.59 \cdot 10^{-4}$
Longitudinal to structures	$1.03 \cdot 10^{-4}$

Fig. 10.14. Results of a principle investigation of the influence of thin layers on flow and transport. The anisotropy factor for the homogeneous, anisotropic case is determined as $Q_{\text{homogeneous}}/Q_{\text{transv. to structures}}$.

very well approximated, whereas the simulations of the middle layer have a slightly overestimated discharge. This leads to an arrival which is too early and peaks which are too high for the latter curves. It is possible that the middle layer represents a region of lower effective permeability, which causes this behavior. Since these deviations are not very significant, the two-layer structure is accepted.

The largest deviations are obtained for the *bottom-top* simulations. The applied anisotropy factor of 1.5 yields a very good approximation of the discharge. However, the simulated tracer-breakthrough curves arrive too early and have peaks which are too high. The principle investigation of the influence of thin layers of low permeability, mentioned for the upper layer (see Fig. 10.14), shows that flow and transport perpendicular to such layers are strongly influenced by the structures. Using an anisotropy factor, determined by the ratio of the measured discharges, to represent the structures reproduces neither the discharge nor the tracer-breakthrough curves perpendicular to the structures correctly. The effect of the real layers is stronger. The implementation of these thin layers of low permeability into the upper layer would improve the shape of the tracer-breakthrough curves in the bottom-top direction. The middle layer may contain such structures as well.

10.2.4 Conclusions

After a numerical model has been set up, based on the results of the multivariate statistical analysis of the measured flow and transport data, and the simulated results compared with the measurements, the assumed layered structure distribution of the fissured sandstone block is verified. Additionally, there are strong and realistic indications of minor heterogeneities in the form of thin horizontal layers of low permeability within the upper layer and also in the upper part of the lower layer.

Apart from the verification of the assumed structure distribution, it is shown that the initial arrival time, represented by $t_{1\%}$, can be used to approximate the permeability between two ports. The feasibility of the approach is successfully tested on various test cases and rules for the interpretation of the approximation results are given.

The discussed procedure is applied here on a laboratory-scale sample with very well defined dimensions and a regular raster of measurement ports. The measurements are conducted under controlled conditions in a laboratory. These circumstances are very advantageous for the procedure. Additionally, due to the characteristics of the sample, i.e. layered structure without significant influence of single permeable fractures, the identified clusters yield a rather clear picture of the sample structure and its properties.

The conditions are not always so favorable, especially if measurements are conducted in the field. At present, an experimental set-up on the field block scale is being prepared. This involves carrying out measurements under less controlled conditions and the existence of significant fractures within the domain. With a 7-spot star-shaped configuration of vertical bore holes and packers within the bore holes, a large set of tracer-breakthrough is obtained. By applying the discussed evaluation procedure on this data set, the feasibility is tested.

Further testing of the approach is presented by Süß (2004). Applying the procedure to various synthetics domains of different character, the applicability is assessed.

10.3 Cluster Analysis to Set Up a Conceptual Multi-continuum Model

T. Vogel, V. Lagendijk, J. Köngeter

The complexity of fractured formations requires a high standard of experimental and modeling techniques to model flow and transport phenomena. As the necessary equivalent parameters are often difficult to determine, the goal is to characterize the experimental results by means of statistical methods. This information could then be used to identify and separate active hydraulic components.

To examine this approach, a simplified two-dimensional model aquifer (2.5 x 2.5 m^2) was chosen for the discrete modelling (Hase, 2002). A cylindrical core (0.3 m) is extracted for discrete (port-to-port) measurements on a small scale. The resulting breakthrough curves are interpreted by both factor and cluster analysis. The hydraulic components of a multi-continuum model are calibrated to the resulting clusters. The parameters that are determined provide a good description of the system behavior, on the scale of the cylindrical core and on the scale of the model aquifer, compared to the discrete and *classical* multi-continuum results.

10.3.1 The Discrete Model

An idealized two-dimensional fractured aquifer, as shown in Fig. 10.15, is considered with respect to flow and transport processes. The model aquifer represents a regularly fractured permeable formation that consists of two bands of fractures inclined at $-45°$ and $45°$. The resulting matrix blocks are assumed to be homogenous and isotropic. These simplifications are applied to reduce the magnitude of the simulation process without influencing the fracture / matrix interactions which are to be analyzed.

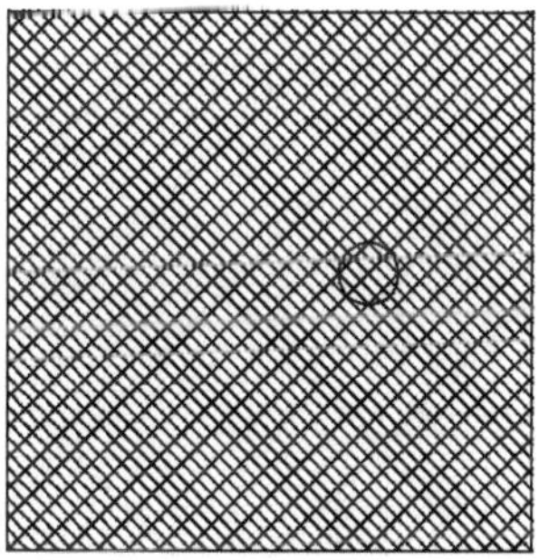

Fig. 10.15. Idealized two-dimensional 2.5 m×2.5 m fractured model area and core location.

10.3.1.1 Discrete Reference Simulation

The matrix blocks of the idealized aquifer are discretized by a triangular finite element mesh, and the fractures are modeled by one-dimensional line elements. The model parameters are summarized in Table 10.8.

In order to eliminate different boundary effects, a cylindrical probe is extracted (cf. Fig. 10.15). This *core* is approximated by a polygon of twenty sides, where the nodes of each side form a measurement port. On this scale

Table 10.8. Model parameters for the discrete reference simulation of the idealized aquifer.

Model parameters			Fracture	Matrix
Permeability	k	(m^2)	$7.200 \cdot 10^{-11}$	$1.132 \cdot 10^{-13}$
Porosity	n_e	(-)	0.300	0.200
Thickness of slice	d_z	(m)	1.000	1.000
aperture	2b	(m)	$8.400 \cdot 10^{-5}$	
Dispersivity				
-longitudinal	α_l	(m)	0.000	$1.000 \cdot 10^{-3}$
-transversal horizontal	α_{th}	(m)	0.000	$1.000 \cdot 10^{-3}$
Molecular				
diffusivity coefficient	D_m			
-water		(m^2s^{-1})	$5.093 \cdot 10^{-9}$	$5.093 \cdot 10^{-9}$
-gas		(m^2s^{-1})	$3.300 \cdot 10^{-5}$	$3.300 \cdot 10^{-5}$

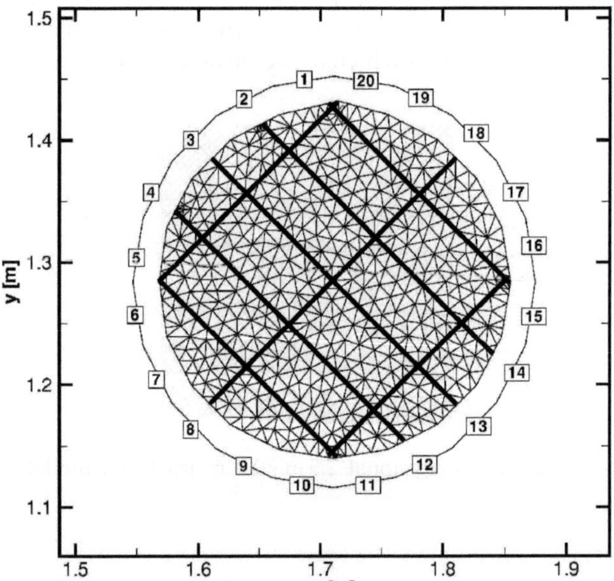

Fig. 10.16. Cylindrical core: finite element mesh and port locations.

(30 cm), the discrete reference simulations are performed by connecting different opposing input and output ports (cf. Fig. 10.16).

In accordance with the experiments performed on the scale of the cylinder (cf. Sect. 4.2), a pressure gradient of 15 000 Pa is applied to the system. The infiltration of the tracer is stopped at 30 ml. Figure 10.17 illustrates the breakthrough curves (BTC) that can be detected at the output ports. A logarithmic time scale is chosen to visualize the different characteristic curves. The BTCs exhibiting a high peak and early first arrival belong to configurations where a direct fracture-fracture connection can be detected (e.g. port 3 - port 13 or port 8 - port 18). Matrix-matrix connections exhibit a shallow peak and late arrival (e.g. port 7 - port 17 or port 9 - port 19). The double peaks can be detected if different flow paths are present.

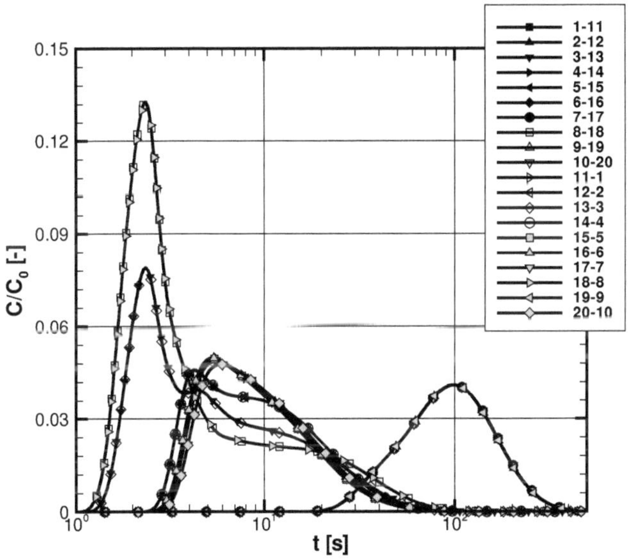

Fig. 10.17. Breakthrough curves for all port-port configurations.

Figure 10.18 illustrates the relative concentration distribution for the port-port connection 3-13 for three points in time. The main portion of the tracer is transported along the direct fracture-fracture connection, while a smaller portion flows into adjacent fractures. This smaller portion is responsible for the second peak in the BTC of configuration 3-13. The tailing of the BTC is caused by the interaction of the fracture-matrix system.

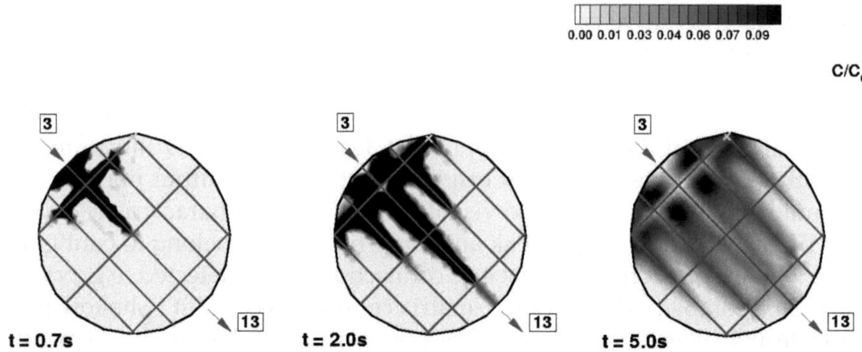

Fig. 10.18. Relative concentration for the simulation 3-13.

10.3.1.2 Statistical Evaluation of Discrete Simulations

The results of the discrete transport simulations are analyzed by means of factor and cluster analysis similar to those described in Sect. 10.1. To compare the different BTCs, the different relative concentrations are scaled by the flow rate.

The following parameters are chosen to characterize the system:

- *t-initial:* first arrival at the output port
- *t-peak:* time of maximum concentration
- *t-median (t(50 %)):* 50 % of tracer have reached the output port
- *t(25 %):* 25 % of tracer have reached the output port
- *t(75 %):* 75 % of tracer have reached the output port
- *unconformity coefficient:* $U = t(60\,\%)/t(10\,\%)$
- *log flow rate:* logarithm of the flow rate
- *C-peak:* value of maximum concentration at the point in time *t-peak*

For the classification of the BTCs, the different times and the unconformity coefficient describe the characteristics of the BTCs sufficiently. In addition, the log flow rate considers the permeabilities of the different port-port-connections. The value of C-peak is a good indicator of the peak characteristic.

The factor analysis is used to select the independent parameters among the parameters listed above. The estimated commonality is supposed to be equal to the highest correlation coefficient of the parameters. The multidimensional space of the variables is reduced and the highly correlated parameters are eliminated, since these parameters would affect the quality of the cluster analysis.

Figure 10.19 illustrates the results of the factor analysis concerning the aforementioned parameters in a three-dimensional space. The appendent

Table 10.9. Eigenvalues and variances of the components (factors) as a result of the factor analysis.

Factor	Eigenvalue	Variance (%)	Cumulative variance (%)
1	6.8168	85.2106	85.2106
2	1.0109	12.6361	97.8467
3	0.1074	1.3429	99.1896
4	0.0068	0.0848	99.2744
5	0.0008	0.0105	99.2849
6	0.0006	0.0080	99.2930
7	0.0004	0.0047	99.2977
8	0.0000	0.0003	99.2980

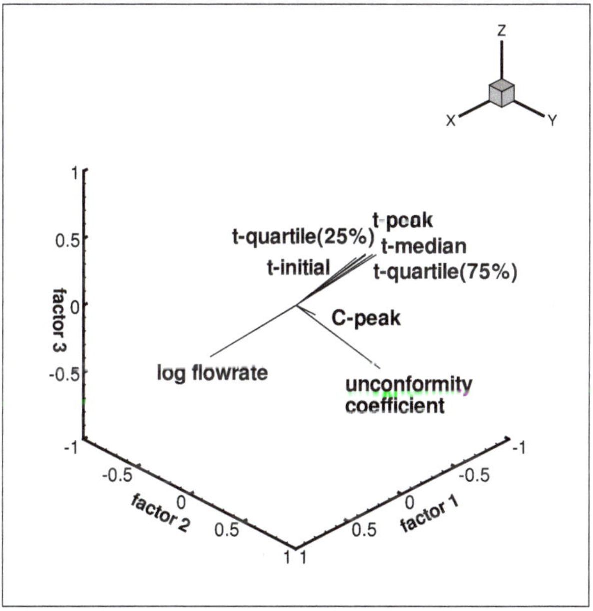

Fig. 10.19. Illustration of the results of the factor analysis for the parameters describing the discrete model results.

eigenvalues and variances, as well as the factor loadings, are summarized in Tables 10.9 and 10.10. The different times form a group of dependent variables, whereas the unconformity coefficient and the log flow rate exhibit a different orientation. Therefore, the *log-flowrate*, *t-median* and the *unconformity coefficient* are chosen to classify the BTCs by cluster analysis.

A separation of the BTCs into three clusters allows a physical classification. The tracer breakthrough curves of the direct fracture-fracture connec-

Table 10.10. Factor loadings as a result of the factor analysis.

	Factor 1	Factor 2	Factor 3
Log flow rate	0.9809	0.1757	0.0895
t-median	-0.9643	-0.2425	-0.1119
U	0.1628	0.9548	-0.0139
t(75 %)	-0.9717	-0.2141	-0.1073
t(25 %)	-0.9490	-0.2926	-0.1221
t-peak	-0.9483	-0.2895	-0.1306
t-initial	-0.9155	-0.3499	-0.1607
C-peak	0.5370	0.7216	0.4428

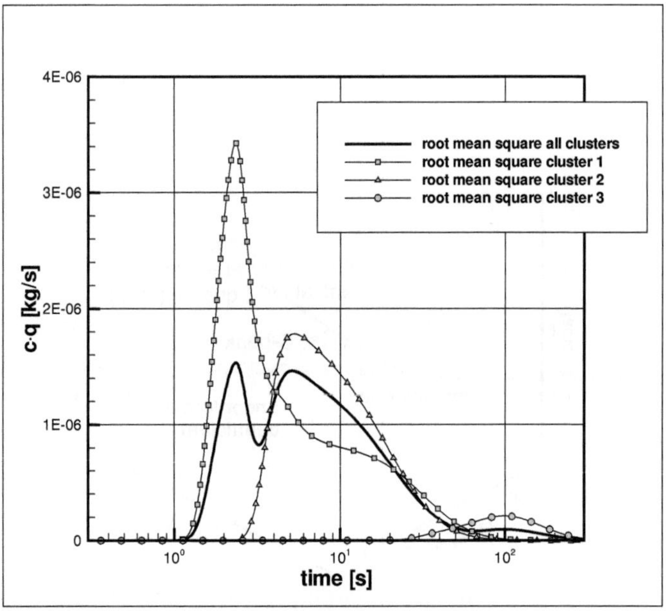

Fig. 10.20. Root-mean-squares of all BTCs and of clusters 1 to 3.

tions form cluster 1, the indirect fracture-fracture connections cluster 2 and the third cluster represents the simulations with matrix ports. Figure 10.20 shows the root-mean-square breakthrough curve of all simulations and the root-mean-square breakthrough curve of the three clusters are identified.

On the basis of this classification, a multi-continuum model is designed as postulated in Sect. 10.3.2.

10.3.2 Multi-continuum Model

On the basis of the simulation results of the discrete model (cf. Sect. 10.3.1), two multi-continuum models are developed. The characteristic clusters are represented by different continua (*new* approach). This model is compared to a multi-continuum model, and the associated parameters are determined by the analysis of the geometric aspects and the hydraulical properties of the fractures and the matrix (*classical* approach).

10.3.2.1 *Classical* Approach

In the following approach, the structure of a *classical* multi-continuum model describing the discrete model of the core scale is illustrated. First an appropriate multi-continuum approach is chosen. Then, the characteristic variables and the equivalent parameters which are incorporated in the continua and the exchange elements as characteristic values of the material are determined.

There are several ways of modeling a rock aquifer with a multi-continuum model. Jansen (1999) distinguishes three models by means of typical tracer-breakthrough curves and the magnitude of the characteristic values (cf. Table 4.8).

The determination of the characteristic values, the mobility number N_M and the loss of identity length L^* is described in detail in Sect. 2.5.5. Three possible models are distinguished, including:

- double-porous / single-permeable (DPSP),
- double-porous / double-permeable (DPDP), and
- single-porous / single-permeable (SPSP).

For the discrete model of the core scale, the mobility number N_M ranges from 0.01 to 0.03. The loss of identity length amounts to about $L^* \approx 35\,\text{m}$. These values suggest the choice of a DPDP approach (compare Table 4.8). Appropriate values for the porosity and the permeability are then assigned to each continuum.

Both continua are represented by a two-dimensional finite element mesh with 585 nodes and 1 064 triangles, which corresponds to the mesh size of the discrete model of the core scale (cf. Sect. 10.3.1). Only the triangular elements are considered, and the one-dimensional line elements are removed. This simplification is made although the aim of multi-continuum modeling is to reduce the complexity of the calculations by using smaller meshes.

Most of the data can be adopted from the discrete calculations. The ports and their denotations are known (cf. Fig. 10.21).

The coupling of the continua is done by 585 one-dimensional line elements that connect the nodes of the two continua. The equivalent parameters for the system with two components are summarized in Table 10.11.

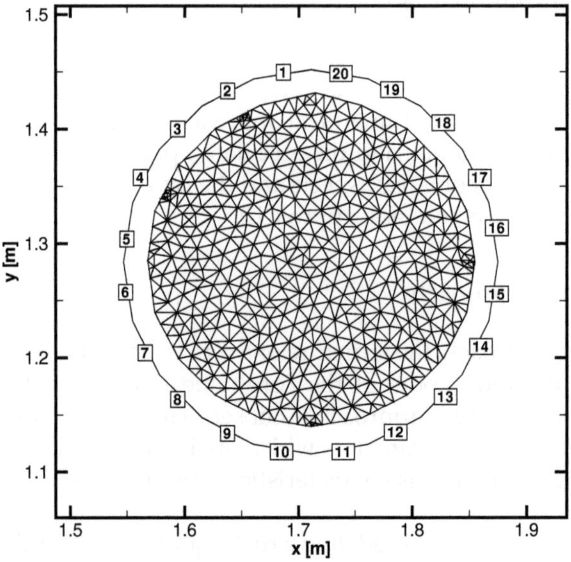

Fig. 10.21. Finite element mesh of a continuum representing the cylindrical core.

Table 10.11. Model parameters for the multi-continuum simulation.

Equivalent parameters		Fracture	Matrix
Permeability	k_{xx} (m^2)	$8.913 \cdot 10^{-14}$	$1.132 \cdot 10^{-13}$
	k_{xy} (m^2)	$-2.599 \cdot 10^{-14}$	0.000
	k_{yy} (m^2)	$8.913 \cdot 10^{-14}$	$1.132 \cdot 10^{-13}$
Porosity	$\bar{n}_e$ (-)	$7.500 \cdot 10^{-4}$	0.200
Dispersivity			
-longitudinal	α_l (m)	$1.000 \cdot 10^{-3}$	$1.000 \cdot 10^{-3}$
-transversal horizontal	α_{th} (m)	$1.000 \cdot 10^{-3}$	$1.000 \cdot 10^{-3}$
Molecular			
diffusion coefficient	D_m (m^2s^{-1})	$3.300 \cdot 10^{-5}$	$3.300 \cdot 10^{-5}$
Volumetric weighting factor	Φ (-)	1.000	$9.975 \cdot 10^{-1}$

The permeability tensor, k_{ij}, and the porosity, $\overline{n}_e$, of the matrix component correspond to the parameters of the discrete model. Instead of being recalculated, the dispersivities α_l and α_{th}, and the molecular diffusion coefficient D_m are also taken from the discrete model.

For the discretely distributed fractures, average values for permeability and porosity are assumed. The two groups of fractures are not equally distributed over the system, which results in different transmissivities for the complete system in the directions of their axes (x_1 and y_1). The fracture axes are rotated by an angle of $\theta_1 = 45°$, with respect to the axes x_i and y_i of the global system (cf. Fig. 10.22). The two fracture systems are perpendicular to each other.

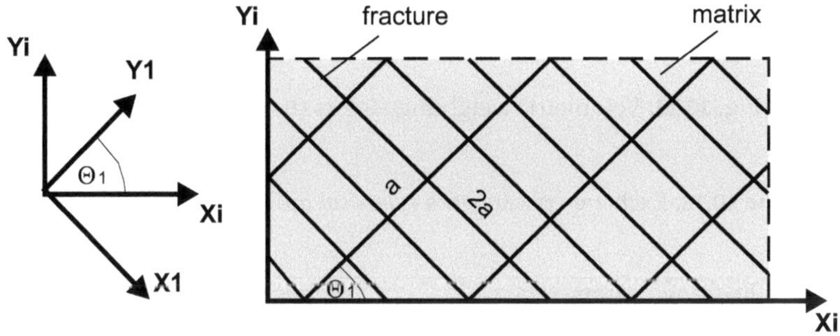

Fig. 10.22. Definition of the axes and a detail of the discrete model.

The volumetric weighting factors of the fracture component, Φ_F, and the matrix component, Φ_M, are illustrated in Fig. 10.23. In a multi-continuum model, the matrix component is represented in the whole model area V_0, although it takes up only a part V_M of the total volume. Therefore, the equations describing the matrix continuum have to be transformed before being applied to the total system.

The volumetric weighting factor, Φ, is defined as the ratio between the volume, V, of a component and the volume of the model area, V_0. For the fracture continuum, Φ is defined as:

$$\Phi_F = 1 \,. \tag{10.11}$$

For the matrix continuum, Φ is defined as:

$$\Phi_M = \frac{V_M}{V_0} \,. \tag{10.12}$$

The parameters required to calculate the exchange processes are listed in Table 10.12. The method used to determine the tensor of the specific fracture area $\Omega_{W,ij}$ is illustrated in Sect. 2.5.3.2.

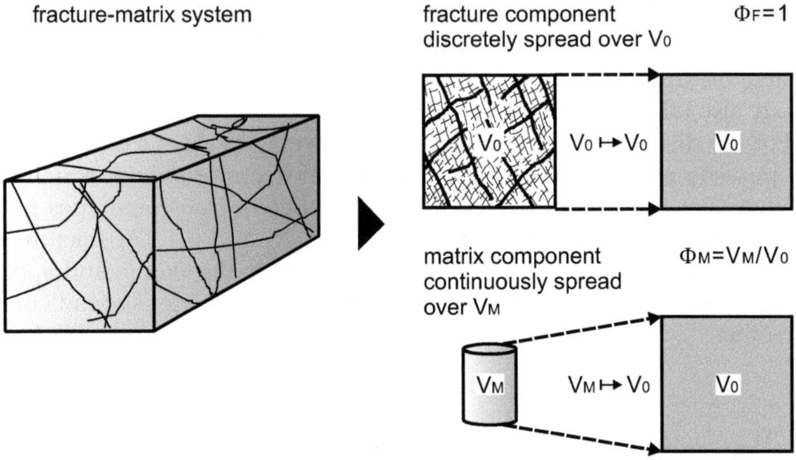

Fig. 10.23. Volumetric weighting factors (from Jansen (1999)).

Table 10.12. Exchange parameters - *classical* multi-continuum system.

Exchange parameters			Fracture component / Matrix component
Specific surface	Ω_0	(m^{-1})	$6.000 \cdot 10^{+1}$
Specific fracture area	$\Omega_{W,xx}$	(m^{-1})	$1.500 \cdot 10^{+1}$
	$\Omega_{W,xy}$	(m^{-1})	$-0.500 \cdot 10^{+1}$
	$\Omega_{W,yy}$	(m^{-1})	$1.500 \cdot 10^{+1}$
Shape factor	ϵ	(-)	1.416
Exchange coefficient -transport	$\overline{\alpha}_c$	(m^2s^{-1})	$9.346 \cdot 10^{-5}$

As a steady-state flow is assumed in the context of this conceptual study, a mass transfer due to local advection does not exist. The relevant exchange processes are regional advection and diffusive mass transfer (cf. Sect. 2.5). The diffusive mass transfer is described by the following equation:

$$W_{D,FM} = \overline{\Omega}^2_{0,M} \epsilon \overline{n}_{e,M} \overline{D}_{m,M} \left(\overline{\overline{c}}_F - \overline{\overline{c}}_M \right) , \tag{10.13}$$

with:

$$\overline{\alpha}_c = \epsilon \overline{n}_{e,M} \overline{D}_{m,M} . \tag{10.14}$$

ϵ is a parameter for calibration. Along with the specific surface, $\overline{\Omega}_{0,M}$, this parameter describes the geometric characteristics of the matrix. The specific surface, $\overline{\Omega}_0$, is the ratio of the surface of the whole interface, $\overline{A}_0$, to the volume, V_0:

$$\overline{\Omega}_0 = \frac{\overline{A}_0}{V_0} . \tag{10.15}$$

Jansen (1999) gives an outline of different expressions that propose a range for ϵ. The product $\epsilon\overline{\Omega}^2_{0,M}$ is taken as a constant of proportionality. By means of the following equations, a value for ϵ can be determined.

For cylindrical matrix blocks with the radius $r = b$ and unit height, Van Genuchten and Dalton (1986) determine:

$$\epsilon\overline{\Omega}^2_0 = \frac{8}{a^2} . \tag{10.16}$$

Equating the cross-section surfaces of a cylinder, $A_\circ = \pi b^2$, and a rectangular matrix block (cf. Fig. 10.22) results in a ratio of the radius b to the side length a of the block, which, inserted in equation (10.15), gives the following expression:

$$\overline{\Omega}_0 \approx \frac{2.377}{b} . \tag{10.17}$$

It is assumed that a cylinder and a rectangular matrix block of unit height have the same specific surfaces if their cross-section surfaces are equal. Inserting equation (10.17) in (10.16) results in:

$$\epsilon \approx 1.416 . \tag{10.18}$$

The multi-continuum model of the *classical* approach is not calibrated. The results achieved with it are to be compared with the discrete model and the multi-continuum model according to the *new* approach. The acquired parameters are transfered to the scale of the discrete conceptual model with a side length of 2.5 m (Fig. 10.15) without any modification. The results of the simulations using this model are also compared to other models in Sect. 10.3.4.

10.3.2.2 *New* Approach

This *new* approach to developping a multi-continuum model using characteristic clusters does not make a clear differentiation between the fracture and matrix properties of the discrete system. The different components of the multi-continuum model are fracture- or matrix-dominated. The clusters described in Sect. 10.3.1.2 represent the characteristic properties of the discrete model. These clusters are represented by different continua which are calibrated to the mean breakthrough curves. To simplify the notation, these continua are named fracture and matrix component in a manner analogous to the *classical* approach.

The discrete model of the cylindrical core is represented by a DPDP model (double-porous - double-permeable), in analogy to the *classical* approach. In this mobility model, the matrix flux is not negligible, and each continuum is given a permeability and a porosity. Both continua participate in flow and transport. Birkhölzer (1994a) provides a detailed description of this approach. The finite element mesh of the described model is identical to the *classical* approach. Figure 10.21 illustrates the mesh with port denotations, which serve to identify the simulation configurations.

The data of the discrete measurements belonging to the cluster of the direct fracture-fracture connections (cluster 1) and the matrix-matrix connections (cluster 3) are each used to obtain parameters for one continuum. These equivalent parameters for each component are determined as described in Sect. 2.5.5 for an initial configuration before calibration. The exchange processes are associated with the cluster of the indirect fracture-fracture connections (cluster 2).

The initial parameters of the exchange component before calibration correspond to those of the *classical* approach (cf. Table 10.12).

The parameters of the two continua and the exchange parameters are adapted iteratively. With the methods of the *classical* approach, the parameters cannot be determined precisely, because the model is not clearly divided into a fracture and a matrix component.

The calibration is first conducted separately for the models of each continuum, and the coupled system is then considered. First, the fracture component is treated, followed by an examination of the matrix component. The initial and boundary conditions of the discrete port-to-port simulation on the core scale are applied to this approach, including:

- pressure at the input: $1.163 \cdot 10^5$ Pa,
- pressure at the output: $1.013 \cdot 10^5$ Pa,
- initial pressure at the other nodes: $1.013 \cdot 10^5$ Pa, and
- quantity of tracer introduced.

At the influx boundary, a unit concentration is introduced until the desired quantity of tracer is reached in the model area.

Tables 10.13 and 10.14 summarize the equivalent parameters and exchange parameters obtained by manual calibration. The results of calibration concerning the different clusters are illustrated in Fig. 10.24. These plots show the root-mean-square breakthrough curves for all clusters and the three clusters in detail. As the multi-continuum model is not capable of representing the effects of certain flow paths but does represent the overall hydraulic behavior, the breakthrough curves for each single port-port connection are averaged (root-mean-square). The integral flow and transport behavior of the system is to be presented by the components and their coupling.

In order to evaluate whether the *new* approach is capable of representing the hydraulic properties of the cylindrical core, further simulations are performed. With both the discrete and the *new* approach, transport simulations

Table 10.13. Equivalent parameters of the multi-continuum model - *new* approach resulting from manual calibration.

Equivalent parameters		Fracture component	Matrix component
Permeability	k_{xx} (m^2)	$5.262 \cdot 10^{-12}$	$2.861 \cdot 10^{-13}$
	k_{xy} (m^2)	$6.190 \cdot 10^{-13}$	$3.016 \cdot 10^{-17}$
	k_{yy} (m^2)	$5.262 \cdot 10^{-12}$	$2.861 \cdot 10^{-13}$
Porosity	$\bar{n}_e$ (-)	0.100	0.300
Dispersivity			
-longitudinal	α_l (m)	$1.000 \cdot 10^{-3}$	$1.000 \cdot 10^{-3}$
-transversal horizontal	α_{th} (m)	$1.000 \cdot 10^{-3}$	$1.000 \cdot 10^{-3}$
Molecular diffusivity coefficient	D_m (m^{2}s^{-1})	$3.300 \cdot 10^{-5}$	$3.300 \cdot 10^{-5}$
Volumetric weighting factor	Φ (-)	0.0025	1.000

Table 10.14. Exchange parameters for the multi-continuum model - *new* approach resulting from manual calibration.

Exchange parameters			Fracture component / Matrix component
Specific surface	Ω_0	(m^{-1})	$1.400 \cdot 10^{+02}$
Specific fracture area	$\Omega_{W,xx}$	(m^{-1})	$1.100 \cdot 10^{+01}$
	$\Omega_{W,xy}$	(m^{-1})	$-0.367 \cdot 10^{+01}$
	$\Omega_{W,yy}$	(m^{-1})	$1.100 \cdot 10^{+01}$
Shape factor	ϵ	(-)	$1.416 \cdot 10^{+00}$
Exchange coefficient			
-transport	$\overline{\alpha}_c$	(m^{2}s^{-1})	$6.600 \cdot 10^{-07}$

are always conducted with four ports at the input side and four ports at the output side of the cylinder.

These simulations are performed for any possible configuration around the cylinder. The root-mean-square breakthrough curves of all configurations are plotted for the discrete and the multi-continuum results. As Fig. 10.25 illustrates, there is a good approximation of the integral behavior of the core system.

10.3.3 Comparison of *Classical* and *New* Approach

When the *classical* approach is used, there is a clear separation between the fracture and matrix components. Parameters of both continua are related to known parameters, such as fracture apartures or the number and orientation

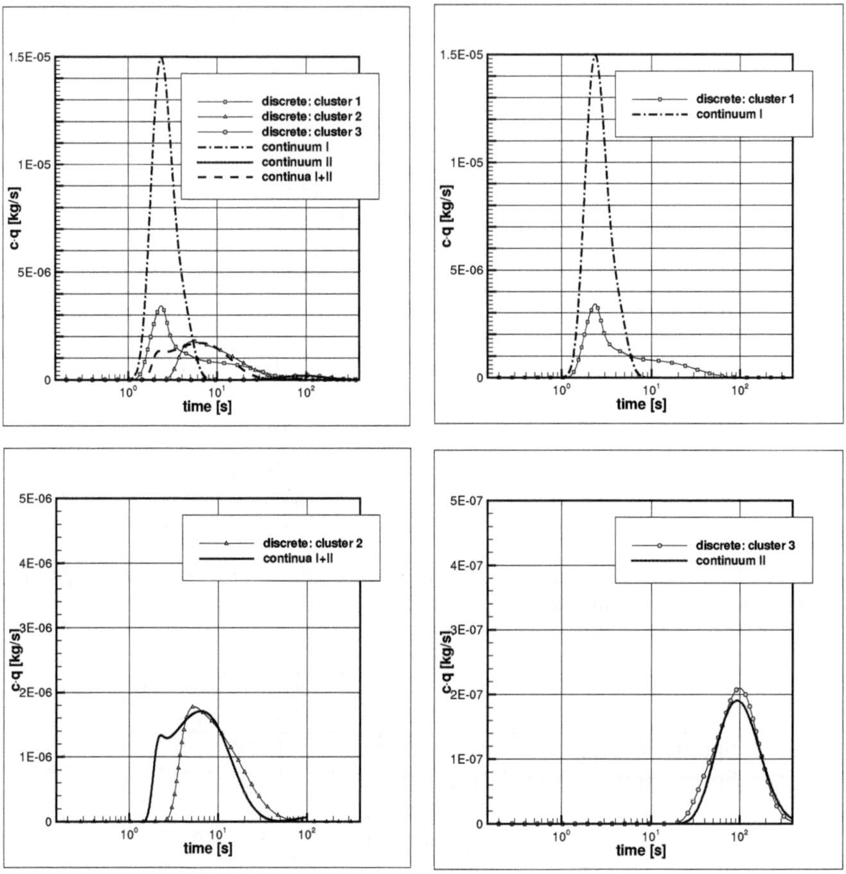

Fig. 10.24. Root-mean-square breakthrough curves for all clusters and the three clusters in detail: discrete versus multi-continuum results.

of the fractures. The exchange processes describe an interaction between the fracture and the matrix component, where, for example, the orientation of the fractures with respect to the flow direction is relevant.

The *new* approach does not strictly distinguish between a fracture and a host matrix component. It is, rather, a mixture of both components, where one component is mostly fracture-dominated and the other one is matrix-dominated. The calibration of the system within the *new* approach leads to a synthetic model of measured breakthrough curves. It does not provide the material properties of the natural system.

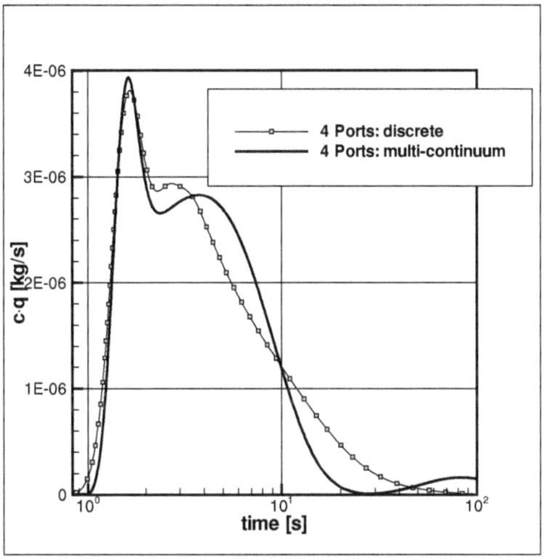

Fig. 10.25. Root-mean-square breakthrough curves of all configurations: discrete versus multi-continuum results.

10.3.4 Upscaling of the Multi-continuum Model

On the basis of the investigations on the bench scale (cf. Sect. 10.3.2), a field block scale model is set up. The length of its sides is 2.5 m, its depth is 1.0 m. Figure 10.26 shows the field-scale model with the fractures and the ports. The field-block scale is represented by different two-dimensional finite element systems, one discrete and two multi-continuum models with different approaches, as described in Sect. 10.3.2.

The mapping of both multi-continuum models by finite elements is identical. Each continuum consists of 33 025 nodes and 65 536 triangles. Both continua are linked by 33 025 one-dimensional line elements. The equivalent parameters of both components and the characteristic variables of the exchange parameters of the *classical* multi-continuum system are listed in Tables 10.11 and 10.12 (*classical* multi-continuum system) and Tables 10.13 and 10.12 (*new* approach). Two different studies, as described in the following paragraphs *Port - Port Study* and *Cross-Section Study*, are performed.

10.3.4.1 Port-to-Port Study

In this study, twelve different configurations are simulated. For each configuration, three ports are coupled as the input or output, for example ports 1-4-7 as the input and 2-6-8 as the output, or 4-12-8 as the input and 2-11-5 as the output. The numbers of the ports are shown in Fig. 10.26. The various port-port configurations are listed in Table 10.15.

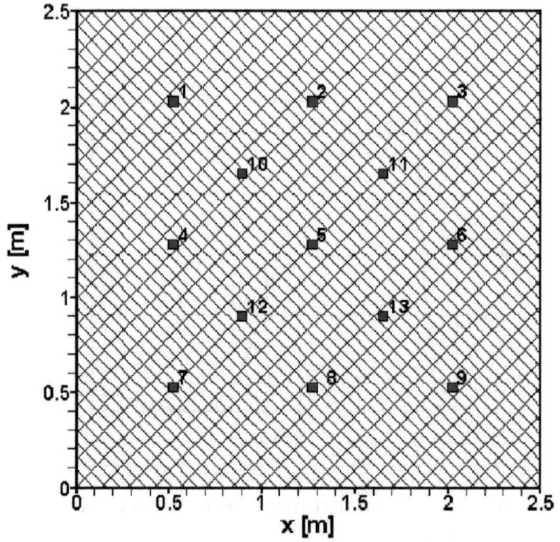

Fig. 10.26. Field-block scale model with fractures and ports for the validation of the multi-continuum approaches.

Table 10.15. Configurations of the port-to-port study for the comparison of the different model approaches.

Configuration no.	Input (port no.)	Output (port no.)
1	1 - 2 - 3	4 - 5 - 6
2	1 - 2 - 3	7 - 8 - 9
3	4 - 5 - 6	7 - 8 - 9
4	1 - 4 - 7	2 - 5 - 8
5	1 - 4 - 7	3 - 6 - 9
6	2 - 5 - 8	3 - 6 - 9
7	2 - 10 - 4	11 - 5 - 12
8	2 - 10 - 4	6 - 13 - 8
9	11 - 5 - 12	6 - 13 - 8
10	2 - 11 - 6	10 - 5 - 13
11	2 - 11 - 6	4 - 12 - 8
12	10 - 5 - 13	4 - 12 - 8

A pressure gradient of 10 000 Pa is applied to the system. After steady state is obtained, the helium tracer is injected and the BTCs are recorded. Figure 10.27 shows the root-mean-square breakthrough curves of all configurations for the three models considered. A good agreement can be observed for the mean square breakthrough curves as well as for the single configurations.

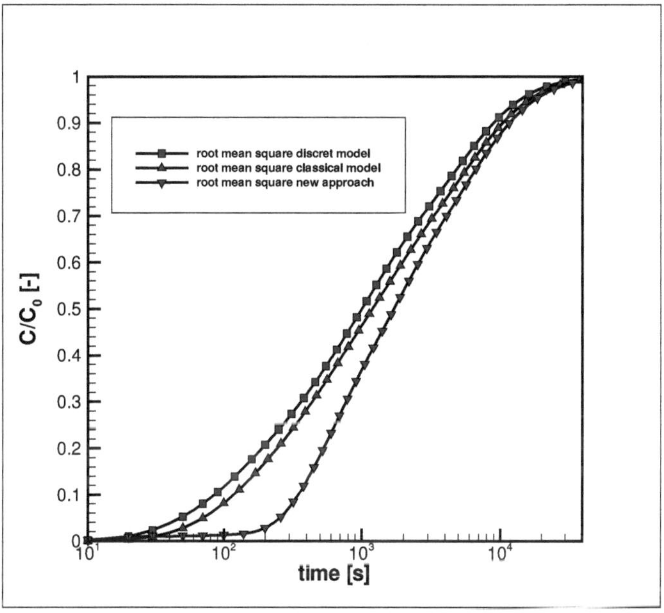

Fig. 10.27. Root-mean-square breakthrough curves of all three models for the port-to-port study.

10.3.4.2 Cross-Section Study

In this study, only one simulation is performed for each of the three models. Ports 1, 4 and 7 are used as input ports, whereas the output is formed by ports 3, 6 and 9. The tracer is recorded every 0.25 m, from $x = 0.75$ m to $x = 2.0$ m in the entire cross-section (cf. Fig. 10.28).

Figure 10.29, which shows the tracer breakthrough at a cross-section 1.5 m from the left boundary, illustrates a good approximation of the discrete results of both models - the *new* approach and the *classical* approach. For early tracer arrival, as well as for late tracer detection, there is a good conformity of the different models.

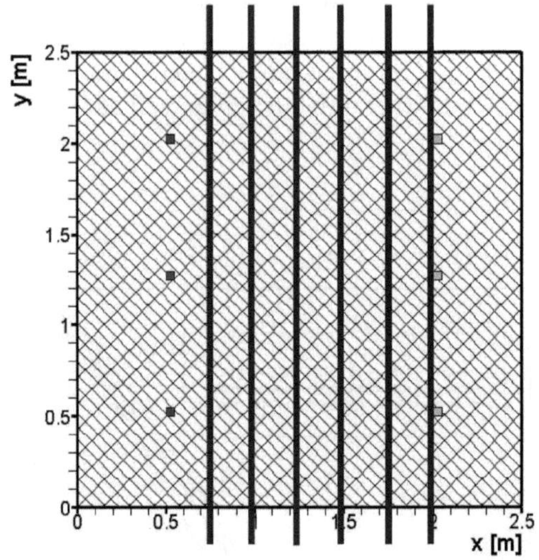

Fig. 10.28. Field scale: cross-sections where tracer-breakthrough curves are determined.

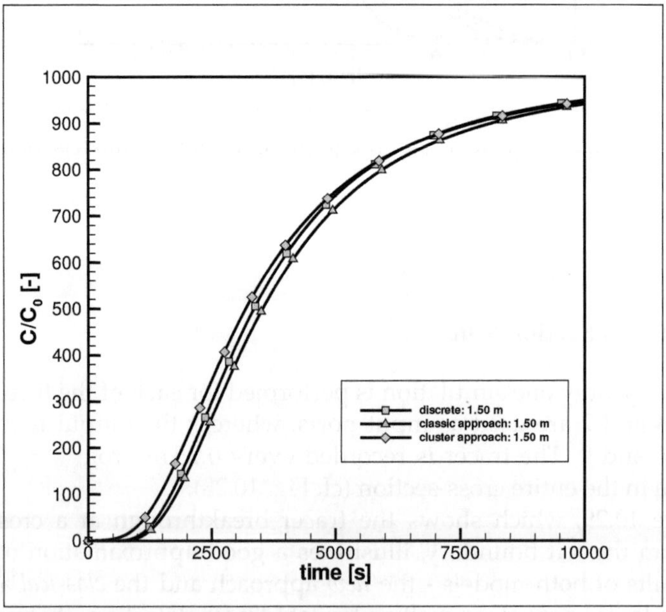

Fig. 10.29. Tracer-breakthrough curve for a cross-section 1.5 m from the input ports.

10.3.5 Conclusions

This conceptual study suggests that a statistical approach to interpreting measured breakthrough curves can be very helpful in supporting the modeling of experimental investigations. In this case, it could even be used to set up a model that is able to reproduce the flow and transport processes in a fractured permeable medium. As this approach has not been evaluated with experimental data, it is suggested that it be considered a tool to assist in the development of a numerical model.

10.3.5 Conclusions

[illegible] ... al ... [illegible] ... e, it could [illegible] ... flow, and transp[illegible] ... [illegible]

References

Abelin, H., Birgersson, L., Gidlund, J., Neretnieks, I., and Tunbrandt, S. (1987). Results from some tracer experiments in crystalline rocks in Sweden. *Academic Press Inc*, **28**, 363–379.

Alexander, W., Frieg, B., Ota, K., and Bossart, P. (1996). Untersuchungen der Retardation von Radionukliden im Wirtgestein. *Nagra informiert*, **27**, 43–55.

Anderberg, M. R. (1974). *Cluster analysis for applications*. Academic Press Inc., New York.

Appelo, C. A. J. and Postma, D. (1993). *Geochemistry, groundwater and pollution*. Balkema, Rotterdam, Netherlands.

Baecher, G., Lanney, N., and Einstein, H. (1977). Statistical description of rock properties and sampling. Proc. of the 18th U.S. Symposium on Rock Mechanics. 5C 1–8.

Baehr, A. and Hult, M. (1991). Evaluation of unsaturated zone air permeability through pneumatic tests. *Water Resources Research*, **27**, 2605–2617.

Baraka-Lokmane, S. (2002a). *Determination of hydraulic conductivities from discrete geometrical characterisation of fractured sandstone cores*. Tübinger Geowissenschaftliche Arbeiten (TGA) C 51, Geowissenschaftliche Fakultät der Eberhard-Karls-Universität Tübingen.

Baraka-Lokmane, S. (2002b). Technical note. A new resin impregnation technique for characterising fracture geometry in sandstone cores. *Int. J. Rock Mech. & Mining Sciences*, **39**, 815–823.

Baraka-Lokmane, S., Teutsch, G., and Main, I. G. (2001). Influence of open and sealed fractures on fluid flow and water saturation in sandstone cores using magnetic resonance imaging,. *Geophys. J. Int.*, **147**, 263–271.

Baraka-Lokmane, S., Liedl, R., and Teutsch, G. (2003). Comparison of measured and modelled hydraulic conductivities of fractured sandstone cores. *Pure & Appl. Geophys.*, **160**, 909–927.

Barenblatt, G., Zheltov, I., and Kochina, I. (1960). Basic concepts in the theory of seepage of homogeneous liquids in fissured rocks (strata). *Journal of Applied Mathematics and Mechanics*, **24**, 1286–1303.

Barker, J. (1988). A generalized radial-flow model for pumping tests in fracured rock. *Water Resources Research*, **24**(10), 1796–1804.

Barlag, C. (1997). *Adaptive Methoden zur Modellierung von Stofftransport im Kluftgestein.* (Institut für Strömungsmechanik und Elektronisches Rechnen im Bauwesen Hannover, Bericht; 52) (Zugl.: Hannover, Univ., Diss., 1997). Eigenverlag, Hannover.

Barton, N. (1973). Review of a new shear-strength criterion for rock joints. *Eng. Geol.*, **7**, 287–332.

Barton, N. K. and Choubey, V. (1977). The shear strength of rock joint in theory and practice. *Rock Mechanics*, **10**, 1–54.

Bastian, P., Chen, Z., Ewing, R., Helmig, R., Jakobs, H., and Reichenberger, V. (1999). Numerical simulation of flow in fractured porous media. In Z. Chen, R. Ewing, and Shi, editors, *Numerical treatment of multiphase flows in porous media*, Lecture notes in physics, pages 1–18. Springer-Verlag, Berlin, Heidelberg.

Bastian, P. T. (1997). UG A flexible software toolbox for solving partial differential equations. *Computing and Visualization in Science*, pages 27–40.

Bear, J.and Tsang, C. F. and de Marsily, G. (1993). *Flow and contaminant transport in fractured rock.* Academic Press Inc., San Diego, California, USA,.

Bear, J. (1972). *Dynamics of Fluids in Porous Media.* (Environmental Science Series). American Elsevier Publ. Comp., New York.

Bear, J. (1993). Modeling flow and contaminant transport in fractured rocks. chapter 1.1.: Introduction. In *Flow and Contaminant Transport in Fractured Rock / Ed. by J.Bear*, pages 1–10. Academic Press, San Diego u.a.

Bear, J. and Bachmat, Y. (1990). *Introduction to Modeling of Transport Phenomena in Porous Media.* (Theory and Applications of Transport in Porous Media; 4). Kluwer, Dordrecht u.a.

Becker, L. and Yeh, W. (1972). Identification of parameters in unsteady open channel flows. *Water Resources Research*, **8**(4), 956–965.

Bengelsdorf, K. (1997). *Fazies- und Reservoirgeologie im Stubensandstein von Pliezhausen-Rübgarten.* Diplomarbeit, Geowissenschaftliche Fakultät der Eberhard-Karls-Universität Tübingen.

Berkowitz, B. (2002). Characterizing flow and transport in fractured geological media: A review. *Advances in Water Resources*, **25**, 861–884.

Berkowitz, B., Bear, J., and Braester, C. (1988). Continuum Models for Contaminant Transport in Fractured Porous Formations. *Water Resources Research*, **24**(8), 1225–1236.

Bertin, H. and Panfiloy, M.;Quintard, M. (2000). Two types of transient phenomena and full relaxation macroscale model for single phase flow through double porosity media. *Transport Porous Media*, **39**(1), 73–96.

Beyer, W. (1964). Zur Bestimmung der Wasserdurchlässigkeit von Sanden und Kiesen. *Z. Wasserwirt.-Wassertech.*, **14**, 165–168.

Billaux, D., Chiles, J., Hestir, K., and Long, J. (1989). Three-dimensional statistical modeling of a fractured rock mass – an example from the Fanay-Augre mine. *International Journal of Rock Mechanics and Mining Sciences and Geomechanics Abstracts*, **26**, 281–299.

Birgersson, N. and Neretnieks, I. (1990). Diffusion in the matrix of granitic rock: field test in the stripa min. *Water Resources Research*, **26**(11), 2833–2842.

Birkhölzer, J. (1994a). *Numerische Untersuchungen zur Mehrkontinuumsmodellierung von Stofftransportvorgängen in Kluftgrundwasserleitern.* (Technische Hochschule Aachen / Lehrstuhl und Institut für Wasserbau und Wasserwirtschaft: Mitteilungen; 93) (Zugl.: Aachen, T.H., Diss.). Eigenverlag, Aachen.

Birkhölzer, J. (1994b). *Programmdokumentation STRAFE4: Ein 3D Finite-Elemente-Programm zur Simulation von Strömungs- und Transportvorgängen in doppelt-permeablen Medien.* (unveröffentlicht). Lehrstuhl und Institut für Wasserbau und Wasserwirtschaft der RWTH, Aachen.

Bitton, G. and Gerba, C. P. (1984). *Microbiological polluants, their survival and transport pattern to groundwater, groundwater pollution microbiology*, pages 65–88. C.P. Gerba, Eds. John Wiley & Sons, New York.

Bloomfield, J. P. and Williams, A. T. (1995). An empirical liquid permeability – gas permeability correlation for use in aquifer properties studies. *Quaterly J. Engng. Geol.*, **28**, 143–150.

Bodvarsson, G., Ho, C., and Robinson, B. (2003). Yucca mountain projekt. *J. of Contam. Hydrology*, **62–63**, Special Issue.

Böttger, M. (1989). Die nutzbaren Gesteine Baden-Württembergs - ein überblick. *Oberrhein. Geol. Abh.*, **35**, 1–4.

Bourke, P. (1987). Channeling of flow through fractures in rock. Proc. GEOVAL-87, Swe. Nuc. Power Insp., Stockholm.

Brauchler, R., Liedl, R., and Dietrich, P. (2003). A travel time-based hydraulic tomographic approach. *Water Resources Research*, **39**(12), 1370 10.1029/2003WR002262.

Brenner, S. and Gudmundsson, A. (2002). Permeability development during hydrofracture propagation in layered reservoirs. Proc. of Nordic Workshop 2001, NGU, 439.

Bronstein, I. and Semedjajew (1977). *Taschenbuch der Mathematik.* Harri Deutsch.

Brown, J. A., Lee, F., and Jackson, J. A. (1981). NMR measurements on Western Gas sand cores, spe/doe 9861. Low Permeability Symp., Denver, Colorado.

Brown, R. J. S. and Fatt, I. (1956). Measurements of fractional wettability of oil field rocks by the Nuclear Magnetic Relaxation method. *Petrol. Trans. AIME*, **207**, 262–264.

Brown, S. R. (1987). Fluid flow through rock joints: The effect of surface roughness. *J. Geophys. Res.*, **92**(B2), 1337–1347.

Brown, S. R. and Scholz, C. H. (1986). Closure of rock joints. *J. Geophys. Res.*, **91**(B5), 4939–4948.

Brusseau, M. (1991). Transport of organic chemicals by gas advection in structured or heterogeneous porous media: development of a model and application to column experiments. *Water Resources Research*, **27**, 3189–3199.

Busch, K.-F., Luckner, L., and Tiemer, K. (1993). *Lehrbuch der Hydrogeologie - Geohydraulik*. Gebrüder Borntraeger, Berlin, Stuttgart.

Butler Jr., J., McElwee, C., and Bohling, G. (1999). Pumping tests in networks of multilevel sampling wells: Motivation and methodology. *Water Resources Research*, **35**(11), 3553–3560.

Cacas, M., Ledoux, E., Marsily, G. d., and Tillie, B. (1990). Modeling fracture flow with stochastic discrete fracture network: calibration and validation. 1: The flow model. *Water Resources Research*, **26**, 479–489.

Carman, P. (1956). *Flow of gases through porous media*. Butterworths scientific publications, London.

Carter, R., Kemp, L., Pierce, A., and Williams, D. (1974). Performance matching with constraints. *Soc. Pet. Eng.*, **14**, 187–196.

Carter, R., Kemp, L., and Pierce, A. (1982). Discussion of comparision of sensitivity coefficient calculation methods in automatic history matching. *Soc. Pet. Eng.*, **22**, 205–208.

Chang, J. and Yortsos, Y. (1990). *Pressure transient analysis of fractal reservoirs*. SPE Formation Evaluation. SPE.

Chardaire, C. and Roussel, J. (1990). *Advances in core evaluation, accuracy and precision in reserves estimation*, chapter NMR imaging of fluid saturation distributions in core samples using a high magnetic field, pages 301–315. Gordon & Breach, Philadelphia, USA.

Chavent, G., Dupuy, M., and Lemmonier, P. (1975). History matching by use of optimal theory. *Society of Petroleum Engineers Journal*, **15**, 74–86.

Chen, G., Illman, W., Thompson, D., Vesselinov, V., and Neuman, S. (2000). Geostatistical, type-curve, and inverse analyses of pneumatic injection tests in unsaturated fractured tuffs at the apache leap research site near superior, arizona. In: Dynamics of Fluids in Fractured Rock. Geoph. Monograph, 122.

Chilès, J.-P. and de Marsily, G. (1993). Stochastic models of fracture systems and their use in flow and transport modeling. In J. Bear, C.-F. Tsang, and G. de Marsily, editors, *Flow and Contaminant Transport in Fractured Rock*, San Diego, California. Academic Press.

Cirpka, O. (1997). *Numerische Methoden zur Simulation des reaktiven Mehrkomponententransports im Grundwasser*. (Mitteilungen / Institut für Wasserbau, Universität Stuttgart; H. 95) (Zugl.: Stuttgart, Univ., Diss., 1997). Eigenverlag, Stuttgart.

Clement, T., Truex, M., and Hooker, B. (1997). Two-well test method for determining hydraulic properties of aquifers. *Ground Water*, **35**(4), 698–703.

Clifton, P. M. and Neumann, S. P. (1982). Effects of kriging and inverse modelling on conditional simulation of the Avra Valley aquifer in southern Arizona. *Water Resources Research*, **18**, 1215–1234.

Closmann, P. (1975). An aquifer model for fissured reservoirs. (4434). *Society of Petroleum Engineers Journal, Vol. 15, No. 5*, pages 385–398.

Cowgill, D. F., Pitman, J. K., and Seevers, D. O. (1981a). NMR and flow estimation. DOE Rep. N 810313, Tulsa.

Cowgill, D. F., Pitman, J. K., and Seevers, D. O. (1981b). NMR determination of porosity and permeability of western tight gas sands, spe/doe 9875. Low Permeability Symp., Denver, Colorado.

Cox, B. L., Pruess, K., and Persoff, P. (1990). A casting and imaging technique for determining void geometry and relative permeability behaviour of a single fracture specimen. Technical Report Int. Rep. 28485, 1–6, University of California, Berkeley, Int. Earth Sciences Division, Lawrence Berkeley Laboratory, University of California, Berkeley, USA.

Croisé, J. (1996). *Extraktion von leichtflchtigen Chemikalien aus porösen ungesättigten Zonen mittels erzwungener Luftströmung*. Ph.D. thesis, Universität Stuttgart, Germany.

Cunningham, R. and Williams, R. (1980). *Diffusion in gases and porous media*. Plenum press, New York and London.

Dagan, G. (1982). Stochastic modeling of groundwater flow by unconditional and conditional probabilities, 2, The solute transport. *Water Resources Research*, **18**, 835–848.

de Marsily, G. (1981). *Hydrogéologie quantitative*. Masson, Paris.

Dechter, J. J., Komoroski, R. A., and Ramaprasad, S. (1989). NMR chemical shift selective imaging of individual fluids in sandstone and dolomite cores. SCA Conf., paper N 8903.

Dershowitz, W. and Einstein, H. (1988). Charaterizing rock joint geometry with joint system models. *Rock Mechanics and Rock Engineering*, **21**, 21 – 51.

Desbarats, A. and Bachu, S. (1994). Geostatistical analysis of aquifer heterogeneity from the core scale to the basin scale, A case study. *Water Resources Research*, **30**(3), 673–684.

Deutsch, C. and Journel, A. (1992). *Geostatistical Software Library and User's Guide (GSLIB)*. Oxford Univ. Press, New York u.a.

Dietrich, P. (1992). *Impedanztomographie: Entwicklungeines Rechnerprogramms zur Bestimmung der Durchlässigkeitsverteilung des Untergrundes mittels hydraulischer Tomographie*. Diplomarbeit, Bergakademie Freiberg, Freiberg, Germany.

Dietrich, P. (1999). *Konzeption und Auswertung gleichstromgeoelektrischer Tracerversuche unter Verwendung von Sensitivitätskoeffizienten*. Tübinger Geowissenschaftliche Arbeiten (TGA) C 50, Geowissenschaftliche Fakultät der Eberhard-Karls-Universität Tübingen.

Dijk, P., Berkowitz, B., and Bendel, P. (1999). Investigation of flow in water-saturated rock fractures using nuclear magnetic resonance imaging (NMRI). *Water Resour. Res.*, **35**(2), 347–360.

Dines, K. and Lytle, R. (1979). Computerized geophysical tomography. *Proc. IEEE*, **67**(7), 1065–1073.

Doe, T. (2001). What do drops do? Surface wetting and network geometry effects on vadose-zone fractured flow. In: Conceptional Models of Flow and Transport in the Fractured Vadose Zone, National Research Council.

Dollinger, H., Bühler, C., and Bossart, P. (1994). Exkavationsprojekt: Epoxidharz-Imprägnations-Experimente (Excavation project: Epoxy resin impregnation experiments). Internal Report 92-24, Nagra.

Domenico, P. and Schwartz, F. (1990). *Physical and chemical hydrogeology*. John Wiley & Sons, New York, Chicester, Brisbane, Toronto, Singapore.

Dragila, M. I. and S. W. Wheatcraft, S. (2001). Free-surface films. In: Conceptional Models of Flow and Transport in the Fractured Vadose Zone, National Research Council.

Dullien, F. A. L. (1991). *Porous media: Fluid transport and pore structure*. Academic, New York, 2 edition.

Earlougher, R. (1977). Advances in well test analysis. Society of Petroleum Engineers, Dallas, Texas.

Einsele, G. and Agster, G. (1986). *Überblick zur Geologie und Morphologie des Schönbuchs*. Deutsche Forschungsgemeinschaft, Weinheim.

Essaid, H., Herkelrath, W., and Hess, K. (1993). Simulation of fluid distributions observed at a crude oil spill site incorporating hysteresis, oil entrapment, and spatial variability of hydraulic properties. *Water Resources Research*, **29**(6), 1753–1770.

Evans, D. and Nicholson, T. (1987). *Flow and Transport through Unsaturated Fractured Rock*, volume 42 of *Geophys. Monograph*. AGU, Washington D.C., 1 edition.

Evans, D., Nicholson, T., and Rasmussen, T. (2001). *Flow and Transport through Unsaturated Fractured Rock*, volume 42 of *Geophys. Monograph*. AGU, Washington D.C., 2 edition.

Everitt, R., Martin, C., and Thompson, P. (1994). An approach to the underground characterization of a disposal vault in granite. Technical Report AECL-10560, Atomic Energy of Canada Ltd., Whiteshell Laboratories, Pinawa, Manitoba, Canada.

Faybishenko, B., Witherspoon, P., and Benson, S. e. (2000). *Dynamics of Fluids in Fractured Rocks*, volume 122 of *Geophys. Monograph*. AGU, Washington D.C.

Fisher, N., Lewis, T., and Embleton, B. (1993). *Statistical analysis of spherical data*. Cambridge University Press, 2 edition.

Flint, S. and Bryant, I. e. (1993). *The Geological Modelling of Hydrocarbon Reservoirs and Outcrop Analogues*, volume 15 of *Spec. Publs.* Int. Ass. Sediment.

Flury, M. (1996). Experimental evidence of transport of pesticides through field soils – a review. *J. Env. Qual.*, **25**, 25–45.

Fondeur, C. (1964). Etude petrographique detaillee d' un gres a structure en feuillets. (a petrographic study of a sandstone with layered structure). *Revue de l'Institut Francais du Petrole*, **19**(7–8), 901–920.

Freeze, R. and Cherry, J. (1979). *Groundwater*. Prentice Hall, Englewood Cliffs, NJ.

Freig, B., Alexander, W., Dollinger, H., Buehler, C., Haag, P., Moeri, A., and Ota, K. (1998). In situ resin impregnation for investigating radionuclide retardation in fractured repository host rocks. *Journal of Contaminant Hydrology*, **35**(1–3), 115–130.

Frind, E., Sudicky, E., and Schellenberg, S. (1988). Micro-scale modelling in the study of plume evolution in heterogeneous media. In: Custodio, E. et al. (eds.), Groundwater FIow and Quality Modelling. Reidel, Dordrecht.

Fuchs, A. (1999). *Optimierte Delaunay-Triangulierungen zur Vernetzung getrimmter NURBS-Körper*. Ph.D. thesis, Universität.

Gale, J. E. (1987). Comparison of coupled fracture deformation and fluid flow models with direct measurements of fracture pore structure and stress-flow properties. Proc. 28th U.S. Symp. on Rock Mechanics.

Gallegos, D. P. and Smith, D. M. (1988). A NMR technique for the analysis of pore structure: Determination of continuous pore size distributions. *J. Colloid Interface Sci.*, **122**(1), 144–153.

Gallegos, D. P., Munn, K., Smith, D., and Stermer, D. (1987). A NMR technique for the analysis of pore structure: Application to materials with well-defined pore structure. *J. Colloid Interface Sci.*, **119**(1), 127–140.

Gentier, S. (1986). *Morphologie et comportement hydromécanique d' une fracture naturelle dans un granite sous contrainte normale*. Ph.D. thesis, Université d' Orléans, Orlans, France.

Gentier, S. and Billaux, D.and van Vliet, L. (1989). Laboratory testing of the voids of a fracture. *Rock Mech. Rock Engng.*, **22**, 149–157.

Gentier, S., Hopkins, D., and Riss, J. (2000). Role of fracture geometry in the evolution of flow paths under stress. In: Dynamics of Fluids in Fractured Rock. Geoph. Monograph, 122.

Gilbert, P. (1972). Iterative methods for three-dimensional reconstruction of an object from projections. *J. theor. Biol.*, **36**, 105–117.

Grader, A., Balzarini, M., Radaelli, F., Capasso, G., and Pellegrino, A. (2000). Fracture-matrix flow: Quantification and visualization using x-ray computerized tomography. In: Dynamics of Fluids in Fractured Rock. Geoph. Monograph, 122.

Grathwohl, P. (1998). *Diffusion in natural porous media: Contaminant Transport, Sorption/Desorption and dissolution kinetics*. Kluwer Academic Publishers.

Grimm, W.-D. (1990). *Bildatlas wichtiger Denkmalgesteine der Bundesrepublik Deutschland*. Bayerisches Landesamt für Denkmalpflege.

Gringarten, A. (1982). Flow-test evaluation of fractured reservoirs. In: T.N. Narasimhan, Recent trends in hydrogeology, GSA, Special Paper, 189.

Gudmundsson, A., Fjeldskaar, I., and Gjesdal, O. (2002). Fracture-generated permeability and groundwater yield in norway. Proc. of Nordic Workshop 2001, NGU, 439.

Gudmundsson, A., Gjesdal, O., Brenner, S., and Fjeldskaar, I. (2003). Effects of linking up of discontinuities on fracture growth and groundwater transport. *Hydrogeo. J.*, **11**, 84–99.

Gummerson, R. J., Hall, C., and Hoff, W. G. (1979). Unsaturated water flow within porous materials observed by NMR imaging. *Nature*, **281**.

GUS (1999). *Bericht der Georadar-Messungen auf dem Testfeld Pliezhausen*. GUS - Geophysikalische Untersuchungen Karlsruhe, Karlsruhe.

Güven, O., Falta, R., Molz, F., and Melville, J. (1986). A simplified analysis of two well tracer tests in stratified aquifers. *Ground Water*, **24**(1), 63–71.

Guzman, A., Geddis, A., Henrich, M., Lohrstorfer, C., and Neuman, S. (1996). Summary of air permeability data from single-hole injection tests in unsaturated fractured tuffs at the apache leap research site: Results of steady-state test interpretation. NUREG/CR-6360, U.S. Nuclear Regulatory Commission, Washington, D.C.

Gwo, J., Jardine, P., and Wilson, G. e. (1995). A multiple-pore-region concept to modeling mass transfer in subsurface media. *Journal of Hydrology*, **164**, 217–237.

Hagemann, B. (2001). *Untersuchung und Modellierung von Wasserdurchlässigkeit und Transporteigenschaften in gekläuftetem Gestein,*. Diplomarbeit, Geowissenschaftliche Fakultät der Eberhard-Karls-Universität Tübingen.

Hakami, E. (1988). *Water Flow in Single Rock Joints*. Licentiale thesis, Lulea University of Technology, Lulea Sweden.

Harter, T. and Yeh, T.-C. J. (1996). Conditional stochastic analysis of solute transport in heterogenous, variably saturated soils. *Water Resources Research*, **32**(6), 1597–1609.

Hartigan, J. A. (1975). *Clustering algorithms*. John Wiley & Sons. Inc., New York.

Hartung, J. (1995). *Lehr- und Handbuch der angewandten Statistik*. Oldenburg.

Hase, M. (2002). *Festgesteins-Aquiferanalog: Charakterisierung eines diskreten Modells und Entwicklung eines neuen Ansatzes der Mehrkontinuum Modellierung*. (Dipl.-Arbeit) (unveröffentlicht). Lehrstuhl und Institut für Wasserbau und Wasserwirtschaft der RWTH, Aachen.

Headley, L. C. (1973). NMR relaxation of ^{7}Li and ^{1}H in appalachian petroleum reservoir rocks containing licl solutions. *Nature Physical Science*, **242**, 87–88.

Helmig, R. (1993). *Theorie und Numerik der Mehrphasenströmung in geklüftet-porösen Medien*. (Institut für Strömungsmechanik und Elektronisches Rechnen im Bauwesen Hannover, Bericht; 34) (Zugl.: Hannover, Univ., Diss., 1993). Eigenverlag, Hannover.

Helmig, R. (1997). *Multiphase Flow and Transport Processes in the Subsurface*. Springer-Verlag, Heidelberg.

Helmig, R., Class, H., Huber, R., Sheta, H., Ewing, R., Hinkelmann, R., Jakobs, H., and Bastian, P. (1998). Architecture of the modular program system MUFTE UG for simulating multiphase flow and transport processes in heterogeneous porous media. *Mathematische Geologie*, **2**.

Hemminger, A., Neunhäuserer, L., Bardossy, A., and Helmig, R. (1998). Identification of equivalent parameters on different scales in fractured–porous media . In K. Holz, W. Bechteler, S. Wang, and M. Kawahara, editors, *Advances in Hydro–Science and –Engineering*, volume 3, page 205. 3rd International Conference on Hydroscience and Engineering, BTU Cottbus, Germany, Center for Computational Hydroscience and Engineering, The University of Mississippi. cdrom.

Himmelsbach, T. (1993). *Untersuchungen zum Wasser- und Stofftransportverhalten von Störungszonen im Grundgebirge Albtalgranit, Sdschwarzwald).* Ph.D. thesis, Schr. Angew. Geol. Karlsruhe, 23.

Hinedi, Z., Chang, A., Anderson, M., and Borchardt, D. (1997). Quantification of microporosity by nuclear magnetic resonance relaxation of water imbibed in porous media. *Water Resour. Res.*, **33**(12), 2697–2704.

Hirsch, C. (1984). *Numerical computation of internal and external flows*, volume 2: Computational methods for inviscid and viscous flows. John Wiley & Sons Ltd., Chichester, England.

Hofmann, U., Schaller, D., Kottenhan, H., Dammler, I., and Morcos, S. (1976). Die adsorption von methylenblau an kaolin, ton und bentonit. *Giesserei*, **54**(1), 97–101.

Hölting, B. (1980). *Hydrogeologie: Einführung in die allgemeine und angewandte Hydrogeologie*. Enke, Stuttgart.

Hornung, J. (1998). *Dynamische Stratigraphie, Reservoir- und Aquifersedimentologie einer alluvialen Ebene: Der Stubensandstein in Baden-Württemberg.* Tübinger Geowissenschaftliche Arbeiten (TGA), Geowissenschaftliche Fakultät der Eberhard-Karls-Universität Tübingen.

Hornung, J. and Aigner, T. (1999). Reservoir and aquifer characterization of fluvial architectural elements: Stubensandstein, upper triassic, southwest germany. *Sedimentary Geology*, **129**, 215–280.

Horsfield, M. A., Fordham, E. J., Hall, C., and Hall, D. (1989). ^{1}H NMR imaging studies of filtration in colloidal suspensions. *J. Magnetic Resonance*, **81**, 593–596.

Hötzl, H., Bäumle, R., Thüringer, C., and Witthüser, K. (2000). Festgesteins–Aquiferanalog: Experimente und Modellierung, Teilprojekt 1: Feldexperimente zur Wasser– und Stoffausbreitung in Abhängigkeit von Kluft– und Gesteinsdurchlässigkeit . Technical report, Lehrstuhl für Angewandte Geologie , Universität Karlsruhe.

Hsieh, P. (2000). A brief survey of hydraulic tests in fractured rocks. In: Dynamics of Fluids in Fractured Rock. Geoph. Monograph, 122.

Huang, K., Tsang, Y., and Bodvardson, G. (1999). Simultaneous inversion of air-injection tests in fractured unsaturated tuff at Yucca Mountain. *Water Resources Research*, **35**(8), 2375–2386.

Huggenberger, P. and Aigner, T. (1999). Introduction to the special issue on aquifer-sedimentology: problems, perspectives and modern approaches. *Sed. Geol.*, **129**, 179–186.

Huggenberger, P., Siegenthaler, C., and Aigner, T. (1997). Quaternary fluvial sedimentation in northern Switzerland and southern Germany: an integrated approach of sedimentological analysis and ultra-high resolution geophysical techniques. *Gaea Heidelbergensis*, **4**, 141–157.

ICDD (1980). *Mineral Powder Diffraction File: Data Book*. Joint Committee on Powder Diffraction Standards, Swarthmore.

Illman, W., Thompson, D., Vesselinov, V., G., C., and Neuman, S. (1998). Single- and cross-hole pneumatic tests in unsaturated fractured tuffs at the apache leap research site: Phenomenology, spatial variabiity, connectivity and scale. NUREG/CR-5559, U.S. Nuclear Regulatory Commission, Wachington, D.C.

Jackson, M. and Tweeton, D. (1996). 3DTOM: Three-dimensional geophysical tomography. Report of Investigation 9617, United States Department of the Interior, Bureau of Mines.

Jacquard, P. (1964). Théorie de linterprétation des mesures de pression. *Revue IFP*, **Vol. XIX**, 297–338.

Jacquard, P. and Jain, C. (1965). Permeability distribution from field pressure data. *Society of Petroleum Engineers Journal*, **5**, 281–294.

Jakobs, H. (2004). *Simulation nicht-isothermer Gas-Wasser-Prozesse in komplexen Kluft-Matrix-Systemen*. Ph.D. thesis, Institut für Wasserbau, Universität Stuttgart. To be published.

Jansen, D. (1999). *Identifikation des Mehrkontinuum-Modells zur Simulation des Stofftransportes in multiporösen Festgesteinsaquiferen*. (Technische Hochschule Aachen / Lehrstuhl und Institut für Wasserbau und Wasserwirtschaft: Mitteilungen; 118) (Zugl.: Aachen, T.H., Diss.). Eigenverlag, Aachen.

Jansen, D., Birkhölzer, J., and Köngeter, J. (1996). Contaminant transport in fractured porous formations with strongly heterogeneous matrix properties. In *Rock Mechanics: Tools and Techniques; Proceedings of the 2nd North American Rock Mechanics Symposium, NARMS '96, A Regiönal Conference of ISRM, Montral, Qubec, Canada, 19–21 June 1996 / Ed. by M. Aubertin [et al.]*, pages 1421–1428. Balkema, Rotterdam u.a.

Jansen, D., Forkel, C., Lagendijk, V., and Köngeter, J. (1998). Evaluating the capability of scale dependent multi-continuum modeling in fractured permeable formations based on field, laboratory and numerical experiments. In *Computational Methods in Water Resources XII : Twelth International Conference held in Crete, Greece, June 1998 / Ed. by V. N. Burganos. Vol 2.: Computational Methods in Surface and Ground Water Transport*, pages 93–100. Computational Mechanics Publ., Southampton u.a.

Jaritz, R. (1999). *Quantifizierung der Heterogenität einer Sandsteinmatrix am Beispiel des Stubensandsteins.* Tübinger Geowissenschaftliche Arbeiten (TGA) C 48, Geowissenschaftliche Fakultät der Eberhard-Karls-Universität Tübingen.

Jasmund, K. and Lagaly, G. (1993). *Tone und Tonminerale.* Steinkopff Verlag, Darmstadt.

Jöreskog, K. G., Klovan, J. E., and Reyment, R. A. (1976). *Geological Factor Analysis.* Elsevier, Amsterdam.

Jury, W. and Wang, Z. (2000). Unresolved problems in vadose zone hydrology and contaminant transport. In: Dynamics of Fluids in Fractured Rock. Geoph. Monograph, 122.

Kaiser, H. F. (1958). The varimax criterion for analytical rotation in factor analysis. *Psychometrika,* **35**, 187–200.

Karasaki, K., Long, J., and Witherspoon, P. (1988). Analytical models of slug tests. *Water Resources Research,* **24**, 115–126.

Karasaki, K., Freifeld, B., Cohen, A., Grossenbacher, K., Cook, P., and Vasco, D. (2000). A mulitidisciplinary fractured rock characterization study at Raymond field site, Raymond, CA. *Journal of Hydrology,* **236**, 17–34.

Kazemi, H. (1969). Pressure transient analysis of naturally fractured reservoirs with uniform fracture distribution. *Society of Petroleum Engineers Journal, December,* pages 451–462.

Kelley, V., Pickens, J., Reeves, M., and Beauheim, R. (1987). Double-porosity tracer-test analysis for interpretation of the fracture characteristics of a dolomite formation. Proc. Solving Ground Water Problems with models, Denver, Colorado.

Kendorski, F. and Bindokas, A. (1987). Fracture geometry characterisation for use in rock mechanics. Proc. of 28th US Sympos. Rock Mechanics/Tucson.

Kenyon, W. E., Day, P. I., Straley, C., and Willemsen, J. F. (1986). Compact and consistent representation of rock NMR data for permeability estimation, spe 15643. 61st Annual Technical Conf. and Exhibition of the SPE, New Orleans.

Kenyon, W. E., Howard, J. J., Sezginer, A.and Straley, C., Matteson, A., and Horkowitz, K.and Ehrlich, R. (1989). Pore-size distribution and NMR in microporous cherty sandstones. SPWLA Annual Logging Symp.

Kilbury, R., Rasmussen, T., Evans, D., and Warrick, A. (1986). Water and air intake to surface exposed rock fracture in situ. *Water Resources Research,* **22**, 1431–1443.

Kinzelbach, W. (1992). *Numerische Methoden zur Modellierung des Transports von Schadstoffen im Grundwasser.* R. Oldenburg Verlag, München Wien. 2. Auflage.

Kiraly, L. (1969). Statistical analysis of fractures (orientation and density). *Geol. Rundschau,* **59**.

Kleineidam, S., Rügner, H., and Grathwohl, P. (1999). Influence of petrographic composition/organic matter distribution of fluvial aquifer sediments on the sorption of hydrophobic contaminants. *Sed. Geol.*, **129**, 311–325.

Klingbeil, R., Kleineidam, S., Asprion, U., Aigner, T., and Teutsch, G. (1999). Relating lithofacies to hydrofacies: outcrop-based hydrological characterisation of Quaternary gravel deposits. *Sed. Geol.*, **129**, 299–310.

Klinkenberg, L. J. (1941). The permeability of porous media to liquids and gases. *Drill. Prod. Pract.*, pages 200–213.

Kolditz, O. (1997). *Strömung, Stoff– und Wärmetransport im Kluftgestein*. Gebr. Borntraeger, Berlin-Stuttgart.

Kozlov, G. A. and Ivanchuk, A. P. (1982). NMR study of the distribution of a liquid in a porous medium. *Kolloidnyi Zhurnal*, **44**(3), 574–577.

Krasny, J. (2002). Quantitative hardrock hydrogeology in a regional scale. Proc. of Nordic Workshop 2001, NGU, 439.

Krasny, J. and Mls, J. e. (1996). First workshop on "Hardrock hydrogeology of the Bohemian Massif", 1994. *Acta Universitatis Carolinae Geologica*, **40**(2), 115–122.

Kravaris, C. and Seinfeld, J. (1985). Identification of parameters in distributed parameter systems by regularization. *SIAM J. Control and Optimization*, **23**, 217–241.

Kröhn, K. (1991). *Simulation von Transportvorgängen im klüftigen Gestein mit der Methode der Finite Elemente.* (Institut für Strömungsmechanik und Elektronisches Rechnen im Bauwesen Hannover, Bericht; 29) (Zugl.: Hannover, Univ., Diss., 1990). Eigenverlag, Hannover.

Kruseman, G. and de Ridder, N. (1994). Analysis and evaluation of pumping test data. . ILRI publication, 47, Wageningen.

Kulatilake, P., Wathugala, D., and Stephansson, O. (1993). Joint network modelling with a validation exercise in stripa mine, sweden. *International Journal of Rock Mechanics and Mining Sciences*, **30**, 503–526.

Kulkarni, K., Datta-Gupta, A., and Vasco, D. (2000). A streamline approach to integrating transient pressure data into high resolution reservoir models,. Proceedings of the SPE European Petroleum Conference, Paris.

Kulke, H. (1967). *Petrographie und Diagenese des Stubensandsteins (mittlerer Keuper) aus Tiefbohrungen im Raum Memmingen*. Ph.D. thesis, Universität Tübingen.

Kumar, J., Fatt, I., and Saraf, D. N. (1969). Nuclear magnetic relaxation time of water in a porous medium with heterogeneous surface wettability. *J. Appl. Phys.*, **40**(10), 4165–4171.

Kundt, A. and Warburg, E. (1875). Über Reibung und Wärmeleitung verdünnter Gase (On friction and heat conduction of diluted gases). *Ann. Phys.*, **155**, 337–525.

La Pointe, P. (2000). Predicting hydrology of fractured rock masses from geology. In: Dynamics of Fluids in Fractured Rock. Geoph. Monograph, 122.

La Pointe, P. and Hudson, J. (1985). Characterization and Interpretation of Rock Mass Joint Patterns. Special paper 199, Geological Society of America.

Lagendijk, V. (1997). *Mehrkontinuum-Modellierung: Erstellung eines Modells zur Simulation der ungesttigten Strmung in doppelt-permeablen Grundwassersystemen.* (Dipl.-Arbeit) (unveröffentlicht). Lehrstuhl und Institut fr Wasserbau und Wasserwirtschaft der RWTH, Aachen.

Lagendijk, V. (2003). *Stofftransportvorgänge in Festgesteinsaquiferen: Analyse von Tracerdurchbruchskurven zur Identifikation eines geeigneten Mehrkontinuum-Ansatzes.* Electr. Ph.D. thesis. Lehrstuhl und Institut für Wasserbau und Wasserwirtschaft der RWTH, Aachen.

Lebbe, L., Tarhouni, J., Van, H., and De, B. (1995). Results of an artificial recharge test and a double pumping test as preliminary studies for optimizing water supply in the western Belgian coastal plain. *Hydrogeology Journal*, **3**(3), 53–63.

Lee, B. and Tan, T. (1987). Application of a multiple porosity/permeability simulator in fractured reservoir simulation. (SPE 16009). In *SPE Ninth Symposium on Reservoir Simulation held in San Antonio, Texas, February 1-4 1987*, pages 181–192. Society of Petroleum Engineers, Richardson.

Lenda, A. and Zuber, A. (1970). Tracer dispersion in groundwater experiments. *Isotope techniques in groundwater hydrology*, **2**, 619 641.

Leven, C. (2002). *Effects of Heterogeneous Parameter Distributions on Hydraulic Test - Analysis and Assessment.* Tübinger Geowissenschaftliche Arbeiten (TGA) C 65, Geowissenschaftliche Fakultät der Eberhard-Karls-Universität Tübingen.

Leven, C., McDermott, C. I., Baraka-Lokmane, S., Sauter, M.and Liedl, R., and Teutsch, G. (2000). Festgesteins-Aquiferanalog: Experimente und Modellierung – Laborexperimente und Entwicklung neuer Untersuchungsmethoden,. 3. Workshop Kluftaquifere – Gekoppelte Prozesse, Institut fr Strömungsmechanik im Internationalen Zentrum für Computergestützte Ingenieurwissenschaften (ICCES) der Universität Hannover.

Lever, D., Bradbury, M., and Hemingway, S. (1985). The effect of dead-end porosity on rock matrix-diffusion. *J. Hydrol.*, **80**, 45–76.

Long, J. (1983). *Investihation of equivalent porous medium permeability in networks of discontinuous fractures.* Ph.D. thesis, Berkeley.

Long, J. and Billaux, D. (1987). From field data to fracture network modeling: An example incorporating spatial structure. *Water Resources Research*, **23**(7), 1201–1216.

Long, J. and Billaux, D. (1987). From field data to fracture network modeling: An example incorporatingspatial structure. *Water Resources Research*, **23**(7), 1201–1216.

Long, J. and Witherspoon, P. (1985). The relationship of interconnection to permeability in fracture networks. *Journal of Geophysical Research*, **90**, 3087–3098.

Long, J., Remer, J., and Wilson, C. e. (1982). Porous media equivalents for networks of discontinuous fractures. *Water Resources Research*, **18**(3), 645–658.

Lovelock, P. E. R. (1977). Aquifer properties of Permo-Triassic sandstones in the United Kingdom. *Bull. Geol. Survey of Great Britain*, **56**, 1–40.

Lowry, T., Allen, M., and Shive, P. (1989). Singularity removal: A refinement of resistivity modeling techniques. *Geophysics*, **54**(6), 766–774.

Luckner, L. and Schestakow, W. (1986). *Migrationsprozesse im Boden- und Grundwasserbereich*. Verlag für Grundstoffindustrie, Leipzig.

Lunn, R. and Mackay, R. (1996). Geostatistical Analysis of Fractured Rock Characteristics. In A. Soares, J. Gomez-Hernandez, and R. Froidevaux, editors, *geoENV I – Geostatistics for Environmental Applications*, Proceedings of the Geostatistics for Environmental Applications Workshop, Lisabon, Portugal, 18.–19. November 1996. Kluwer Academic Publisher.

Majer, E., Myer, L., Peterson, J., Karasaki, K., Long, J., Martel, S., Blumling, P., and Vomvoris, S. (1990). Joint seismic, hydrogeological and geomechanical investigations of a fracture zone in the grimsel rock laboratory, switzerland. Technical Report NAGRA TR 90–94, Swiss National Cooperative for Radioactive Waste Disposal (NAGRA), Wettingen, Switzerland.

Malmgrem, B. A. and Haq, B. U. (1982). Assessment of quantitatives techniques in paleobiogeography. *Marine Micropaleontology*, **7**, 213–236.

Maloszewski, P. and Zuber, A. (1990). Mathematical modeling of tracer bahavior in short-term experiments in fissured rocks. *Water Resources Research*, **26**(7), 1517–1528.

Matthess, G. and Ubell, K. (1983). *Lehrbuch der Hydrogeologie*. Gebrüder Borntraeger, Berlin, Stuttgart.

Matthews, C. and Russel, D. (1967). Pressure build-up and flow tests in wells. Society of Petroleum Engineers, Dallas, Texas.

McDermott, C. (1999). *New Experimental and Modelling Techniques to Investigate the Fractured Porous System*. Tübinger Geowissenschaftliche Arbeiten (TGA) C 52, Geowissenschaftliche Fakultät der Eberhard-Karls-Universität Tübingen.

McDermott, C., Sauter, M., and Liedl, R. (1998). Investigating fractured-porous systems - The aquifer analogue approach. In Rossmanith, editor, *Mechanics of Jointed and Faulted Rock, Balkema, Rotterdam*, pages 607–612.

McDermott, C., Sauter, M., and Liedl, R. (2003a). New experimental techniques for pneumatic tomographical determination of the flow and transport parameters of highly fractured porous rock samples. *Journal of Hydrology*, **278**(1-4), 51–63.

McDermott, C., Sinclair, B., and Sauter, M. (2003b). Recovery of undisturbed highly fractured bench scale samples (diameter 30cm x 35cm) for laboratory investigation. *Engineering Geology*, **69**(122), 161–170.

McQueen, J. (1967). Some methods for classification and analysis of multivariate observations. *5th Berkeley Symposium on Mathematics, Statistics and Probalility*, **1**, 281–298.

Meier, D. and Kronberg, P. (1989). *Klüftung in Sedimentgesteinen*. Enke Verlag, Stuttgart.

Meng, S. X. and Maynard, J. B. (2001). Use of statistical analysis to formulate conceptual models of geochemical behavior: Water chemical data from the Botucatu aquifer in Sao Paulo state, Brazil. *Journal of Hydrology*, **250**, 78–79.

Mertz, J. P. (1991). Structures de porosité et propriétés de transport dans les grs.

Miall, A. (1985). Architectural element analysis: a new method of facies analysis applied to fluvial deposits. *Earth Sci. Rev.*, **22**, 261–308.

Militzer, H., Schön, J., and Stötzner, U. (1986). *Angewandte Geophysik im Ingenieur- und Bergbau*. Deutscher Verlag für Grundstoffindustrie, Leipzig.

Miller, S., McWilliams, P., and Kerkering, J. (1990). Ambiguities in estimating fractal dimensions of rock fracture. Proc. of 31. Symp. Rock Mech., Golden, CO, Balkema, Rotterdam.

Moench, A. and Ogata, A. (1981). A numerical inversion of the Laplace transform solution to radial dispersion in a porous medium. *Water Resources Research*, **17**(1), 250–252.

Molz, F., Güven, O., Melville, R., Crocker, R., and Matteson, K. (1986). Performance, analysis, and simulation of a two well tracer test at the Mobile site. *Water Resources Research*, **22**(7), 1031–1037.

Moreno, L., Neretnieks, I., and Eriksen, T. (1985). Analysis of some laboratory tracer runs in natural fissures. *Water Resources Research*, **21**, 951–958.

Morin, R., Hess, A., and Paillet, F. (1988). Determining the distribution of hydraulic conductivity in a fractured limestone aquifer by simultaneous injection and geophysical logging. *Ground Water*, **26**, 587–595.

Moyne, C. (1997). Two-equation model for a diffusive process in porous media using the volume averaging method with an unsteady-state closure. *Advances in Water Resources*, **20**(2-3), 63–76.

Müller, T. (1997). Teil B: Bestimmung der Gesteins- und Kluftgeometrie eines Stubensandsteinhorizontes. Diplomarbeit, Geologisches Institut, Lehrstuhl für Angewandte Geologie, Universität Karlsruhe (TH).

Murawski, H. (1998). *Geologisches Wörterbuch*. dtv, Ferdinand Enke Verlag, Stuttgart, 10 edition.

Naceur, B. and Economides, M. (1988). Production from naturally fissured reservoirs intercepted by a vertical hydraulic fracture. Proc. - Society of Petroleum Engineers of AIME. California Regional Meeting.

Narr, W. and Suppe, J. (1991). Joint spacing in sedimentary rocks. *J. Struct. Geol.*, **13**, 1037–1048.

Neretnieks, I. (1985). Transport in fractured rocks. *Mem. Int. Assoc. Hydrogeol.*, **17**, 301–318.

Neretnieks, I., Abelin, H., and Birgersson, L. (1987). Some recent observations of channeling in fractured rocks – its potential impact on radionuclide migration. Proc. of Geostat. Sens. Uncert. Methods Groundwater Flow and Rad. Transport Modelling, Battelle Press.

Neuman, S. and Federico, V. d. (2003). Multifaceted nature of hydrogeologic scaling and its interpretation. *Reviews of Geophysics*, **41**(3), DOI 10.1029/2003RG000130.

Neunhäuserer, L. (2003). *Diskretisierungsansätze zur Modellierung von Strömungs- und Transportprozessen in geklüftet-porösen Medien.* Ph.D. thesis, Institut für Wasserbau, Universität Stuttgart.

Novakowski, K. and Lapcevic, P. (1994). Field measurement of radial solute transport in fractured rock. *Water Resources Research*, **30**, 37–44.

NRC (1996). *National Research Council: Rock Fractures ans Fluid Flow - Contemporary Understanding and Applications.* National Academy Press, Washington D.C.

NRC (2001). *National Research Council: Conceptual Models of Flow and Transport in the Fractured Vadose Zone.* National Academy Press, Washington D.C.

Ochs, S., Hinkelmann, R., Neunh"auserer, L., Suess, M., Helmig, R., Gebauer, S., and Kornhuber, R. (2002). Adaptive methods for the equidimensional modelling of flow and transport processes in fractured aquifers. In *5th International Conference on Hydro-Science & -Engineering*, Warsaw, Poland.

Odling, N. (1995). The development of network properties in natural fracture patterns: An example from the Devonian sandstones of western Norway. In L. Myer, C. Tsang, N. Cook, and R. Goodman, editors, *Fractured and jointed rock masses*, Proceedings of the conference on fractured and jointed rock masses, Lake Tahoe / California / USA / 3–5 June 1992. A.A. Balkema / Rotterdam / Brookfield.

Olofsson, B. (2002). Estimating groundwater resources in hardrock areas – a water balance approach. Proc. of Nordic Workshop 2001, NGU, 439.

Olsson, O. e. ((1992)). Site characterization and validation - final report: Stripa project. Technical Report TR 92–22, Swedish Nuclear Fuel and Waste Management Co., Stockholm.

Ostensen, R. (1998). Tracer test and contaminant transport rates in dual-porosity formations with application to the wipp. *Journal of Hydrology*, **204**(1–4), 197–216.

Paillet, F., Hess, A., Cheng, C., and Hardin, E. (1987). Characterisation of fracture permeability with high resolution vertical flow measurements during borehole pumping. *Ground Water*, **25**, 28–40.

Peters, R. and Klavetter, E. (1988). A continuum model for water movement in an unsaturated fractured rock mass. *Water Resources Research*, **24**, 416–430.

Plummer, C., McGeary, D., and Carlson, D. (2002). *Physical Geology*. McGraw-Hill Science/Engineering/Math, Sacramento.

Pointe, P. and Hudson, J. (1985). *Characterization and interpretation of rock mass joint patterns*, volume 199. Geological Society of America, Wisconsin.

Pollard, D. and Aydin, A. (1988). Progress in understanding jointing over the past century. *Geol. Soc. Am. Bull.*, **100**, 1181–1204.

Pollock, D. (1988). Semianalytical computation of path lines for finite-difference models. *Ground Water*, **26**(6), 743–750.

Press, W. e., Teukolsky, S., Vetterling, W., and Flannery, B. (1996). *Numerical recipes in Fortran 90: The art of parallel scientific computing*. Cambridge University Press, Cambridge.

Priest, S. and Hudson, J. (1976). Discontinuity spacings in rock. *Int. Journal of Rock Mechanics and Mining Sciences*, **18**, 135–148.

Priest, S. e. (1993). *Discontinuity Analysis for Rock Engineering*. Chapman & Hall, London.

Pruess, K. (1999). A mechanistic model for water seepage through thick unsaturated zones in fractured rocks of low matix permeability. *Water Resources Research*, **35**, 1039–1051.

Pruess, K. and Karasaki, K. (1982). Proximity functions for modeling fluid and heat flow in reservoirs with stochastic fracture distributions. In *Proceedings of the Eighth Workshop on Geothermal Reservoir Engineering, held at Stanford, California, December 14–16, 1982*, volume 8 of *Workshop on Geothermal Reservoir Engineering*, pages 219–224. Univ., Stanford, CA.

Pruess, K. and Narasimhan, T. (1985). A practical method for modeling fluid and heat flow in fractured porous media. *Society of Petroleum Engineers Journal*, **25**(1), 14–26.

Pyrak-Nolte, L. J., Myer, L. R. O., Cook, N. G. W., and Witherspoon, P. A. (1987). Hydraulic and mechanical properties of natural fractures in low permeability rock. Proc. 6th Int. Congr. on Rock Mechanics, Montreal, Canada.

Quintard, M. and Whitaker, S. (1996). Transport in chemically and mechanically heterogeneous porous media ii: Theoretical development of region-averaged equations for slightly compressible single-phase flow. *Advances in Water Resources Research*, **10**(1), 29–47.

Rasmussen, T., Evans, D., Sheets, P., and Blanford, J. (1993). Permeability of Apache Leap Tuff: Borehole and core measurements using water and air. *Water Resources Research*, **29**, 1997–2006.

Rats, M. and Chernyashov, S. (1967). Statistical aspects of the problem on the permeability of the jointy rocks. In *Hydrologie des roches fissurées: Hydrology of Fractured Rocks. Dubrovnik Symposium 1965. Organis par l'Unesco. Vol. 1*, (International Association of Scientific Hydrology; 73), pages 227–236. Unesco, Paris.

Raven, K. and Gale, J. (1985). Water flow in a natural rock fracture as a function of stress and sample size. *Int. J. Rock Mech. Min. Sci. & Geomech. Abstr.*, **22**, 251–261.

Rawnsley, K., Rives, T., Petit, J.-P., Hencher, S., and Lumsden, A. (1992). Joint development in perturbed stress fields near faults. *J. Struct. Geol.*, **14**, 939–951.

Reichenberger, V. (2003). *Multiphase flow in fractured porous media*. Ph.D. thesis, Interdisziplinäres Zentrum für Wissenschaftliches Rechnen, Universität Heidelberg.

Reid, R., Prausnitz, J., and Poling, B. (1987). *The properties of gases and liquids*. McGrawHill.

Rives, T., Razack, M., Petit, J.-P., and Rawnsley, K. (1992). Joint spacing: analogue and numerical simulations. *J. Struct. Geol.*, **14**, 925–937.

Rohr-Torp, E. and Roberts, D. (2002). Hardrock hydrogeology - proceedings of a nordic workshop. Oslo August 2001, NGU, 439.

Romm, E. (1966). *Flow characteristics of fractured rocks (in Russian)*. Nedra, Moscow.

Rutqvist, J. and Stephansson, O. (2003). The role of hydromechanical coupling in fractured rock engineering. *Hydrogeo. J.*, **11**, 7–40.

Ryan, T., Farmer, I., and Kimbrell, A. (1987). Laboratory determination of fracture permeability. Proc. of 28th Symp. Rock Mech., Tucson, AZ.

Sachs, L. (1997). *Angewandte Statistik*. Springer-Verlag, Berlin, 8 edition.

Sahimi, M. (1995). *Flow and transport in porous fractured media and fractured rock: From classical methods to modern approaches*. VHC Verlagsgesellschaft, Weinheim.

Sampath, K. and Keighin, C. W. (1982). Factors affecting gas slippage in tight sandstones of Cretaceous age in the Uinta Basin. *J. Petrol. Technol*, pages 2715–2720.

Saraf, D. N. and Fatt, I. (1966). *Measurement of fluid saturations by Nuclear Magnetic Resonance and its application to three-phase relative permeability studies*. Phd dissertation, Univ. of California, Berkeley.

Saraf, D. N. and Fatt, I. (1967). Three-phase relative permeability measurement using a Nuclear Magnetic Resonance technique for estimating fluid saturation. *Soc. Petrol. Eng. J.*, **7**, 235–242.

Schachtschabel, P., Blume, H.-P., Brümmer, G., Hartge, K.-H., and Schwertmann, U. (1992). *Lehrbuch der Bodenkunde*. Enke, Stuttgart.

Schad, H. and Teutsch, G. (1994). Effects of the investigation scale on pumping test results in heterogeneous porous aquifers. *Journal of Hydrology*, **159**, 61–77.

Schafmeister, M. (1999). *Geostatistik für die hydrogeologische Praxis*. Springer.

Scheidegger, A. (1961). General theory of dispersion in porous media. *Journal of Geophysical Research*, **66**(10), 3273–3278.

Schrauf, T. and Evans, D. (1986). Laboratory Studies of Gas Flow Through a Single Natural Fracture. *Water Resources Research*, **22**(7), 1038–1050.

Schwarzenbach, R., Gschwend, P. M., and Imboden, D. M. (1993). *Environmental organic chemistry*. John Wiley & Sons, New York.

Seevers, D. O. (1966). A nuclear magnetic method for determining the permeability of sandstones. Trans. 7th Ann. SPWLA Logging Symp., Tulsa, Oklahoma, USA.

Silberhorn-Hemminger, A. (2002). *Modellierung von Kluftaquifersystemen: Geostatistische Analyse und deterministisch-stochastische Kluftgenerierung*. (Mitteilungen / Institut für Wasserbau, Universität Stuttgart; H. 114) (Zugl.: Stuttgart, Univ., Diss., 2002). Eigenverlag, Stuttgart.

Silliman, S. E. (1989). An interpretation of difference between aperture estimates derived from hydraulic and tracer tests in a single fracture. *Water resour. res.*, **25**(10), 2275–2283.

Skagius, K. and Neretnieks, I. (1986). Porosities and diffusivities of some nonsorbing species in crystalline rocks. *Water Resources Research*, **22**, 389–398.

Snow, D. (1965). *A parallel plate model of fractured permeable media*. Ph.D. thesis, Univ. of California, Berkeley.

Snow, D. (1969). Anisotropic permeability of fractured media. *Water Resources Research*, **5**(6), 1273–1289.

Snow, D. T. (1968). Hydraulic character of fractured metamorphic rocks of the Front Range and implications to the Rocky Mountains Arsenal well. *Colo. Sch. Mines*, **63**(1), 201–244.

Streltsova, T. (1988). *Well testing in heterogeneous formations*. Exxon monographs, John Wiley & Sons,.

Sun, N. (1994). *Inverse problems in groundwater modeling*. Kluwer Acad. Publ., Dordrecht.

Sundaram, P., Watkins, D., and Ralph, W. (1987). Laboratory investigations of coupled stress-deformation-hydraulic flow in a natural rock fracture. Proc. of 28th Symp. Rock Mech., Tucson, AZ.

Süß, M. (2004). *Analysis of the influence of structures and boundaries on flow and transport processes in fractured forous media (preliminary title)*. Ph.D. thesis, Universität Stuttgart, Germany. To be published.

Tang, D., Frind, E., and Sudicky, E. (1981). Contaminant transport in fractured porous media: Analytical solution for a single fracture. *Water Resources Research*, **17**(3), 555–564.

Tang, D., Frind, E., and Sudicky, E. (1981). Contaminant transport in fractured porous media: Analytical solution for a single fracture. *Water Resources Research*, **17**(3), 555–564.

Terzaghi, R. (1965). Sources of error in joint surveys. *Geotechnique*, **15**(3), 287–304.

Teutsch, G. and Sauter, M. (1992). Groundwater modeling in karst terranes: Scale effects, data aquisition and field validation. In *Proceedings of the Third International Conference on Hydrogeology, Ecology, Monitoring and Management of Groundwater in Karst Terranes, December 4–6, 1991 Nashville, USA*, volume 10 of *Ground Water Management*, pages 17–35. Water Well Journal Publishing, Dublin, Ohio.

Theis, C. (1935). The relation between the lowering of the piezometric surface and the rate and duration of discharge of a well using groundwater storage. *Transactions, American Geophysical Union*, **16**, 519–524.

Thielen, A. (2002). *Entwicklung und Verifikation eines Modellansatzes zur numerischen Simulation von Stofftransportvorgängen in ungesättigten doppelt-permeablen Grundwassersystemen*. (Dipl.-Arbeit) (unveröffentlicht). Lehrstuhl und Institut für Wasserbau und Wasserwirtschaft der RWTH, Aachen.

Thorez, J. (1975). *Phyllosilicates and clay minerals*. Dison, Belgium.

Thorez, J. (1976). *Practical identification of clay minerals*. Dison, Belgium.

Thunvik, R. and Braester, C. (1990). Gas migration in discrete fracture networks. *Water Resources Research*, **26**(10), 2425–2434.

Thüringer, C. (2002). *Untersuchungen zum Gastransport im ungesättigten, geklüftet-porösen Festgestein*. (Schriftenreihe Angewandte Geologie Karlsruhe). Lehrstuhl für Angewandte Geologie, Karlsruhe.

Tidwell, V. and Wilson, J. (1997). Laboratory method for investigating permeability upscaling. *Water Resources Research*, **33**(7), 1607–1616.

Timur, A. (1969). Pulsed Nuclear Magnetic Resonance studies of porosity, movable fluid, and permeability of sandstones. *J. Pet. Tech.*, **246**, 775–786.

Tokunaga, T. and Wan, J. (1997). Water film flow along fracture surfaces in porous rock. *Water Resources Research*, **36**, 1287–1295.

Tsang, C., Hufschmied, P., and Hale, F. (1990). Determination of fracture inflow parameters with a borehole fluid conductivity logging method. *Water Resources Research*, **26**(4), 561–578.

Tsang, C., Tsang, Y., and Hale, F. (1991). Tracer Transport in Fractures: Analysis of Field Data Based on a Variable–Aperture Channel Model. *Water Resources Research*, **27**(12), 3095–3106.

Tsang, Y. (1992). Usage of "equivalent" apertures for rock fractures as derived from hydraulic and tracer tests. *Water Resources Research*, **28**(5), 1451–1455.

Tsang, Y. and Tsang, C. (1987). Channel model of flow through fractured media. *Water Resources Research*, **23**(3), 467–479.

Tsang, Y., Tsang, C., Neretnieks, I., and Moreno, L. (1988). Flow and transport in fractured media: a variable aperture channel model and its properties. *Water Resources Research*, **24**(12), 2049–2060.

Tsang, Y., Tsang, C., Hale, F., and Dverstorp, B. (1996). Tracer transport in a stochastic continuum model of fractured media. *Water Resources Research*, **32**(10), 3077–3092.

Tsang, Y. W. and Tsang, C. F. (1987). Channel model of flow through fractured media. *Water Resources Research*, **23**(3), 467–479.

Tsang, Y. W. and Tsang, C. F. (1989). Flow channelling in a single fracture as a two dimensional strongly variable heterogeneous medium. *Water Resources Research*, **25**(9), 2076–2080.

Ufrecht, W. (1987). *Zur Hydrogeologie und Hydrochemie des SAndsteinkeupers in Mittel- und Ostwürttemberg*. Ph.D. thesis, Universität Stuttgart.

Van Genuchten, M. and Dalton, F. (1986). Models for simulating salt movement in aggregated field soils. *Geoderma*, **38**, 165–183.

Vandergraaf, T. T., Drew, J., Archambault, D., and Ticknor, K. (1997). Transport of radionuclides in natural fractures: some aspects of laboratory migration experiments. *Journal of Contaminant Hydrology*, **26**(1–4), 83–95.

Vasco, D., Keers, H., and Karasaki, K. (2000). Estimation of reservoir properties using transient pressure data: An asymptotic approach. *Water Resources Research*, **36**(12), 3447–3465.

Vesselinov, V., Neuman, S., and Illman, W. (2001). Three-dimensional numerical inversion of pneumatic cross hole tests in unsaturated fractured tuff: 1. Methodology and borehole effects. *Water Resources Research*, **37**(12), 3001–3018.

Vogelsang, D. (1991). *Geophysik an Altlasten - Leitfaden für Ingenieure, Naturwissenschaftler und Juristen*. Springer, Berlin, Heidelberg, New York.

Voos, C. and Shotwell, L. (1990). An investigation of the mechanical and hydraulic behavior of tuff fractures under saturated conditions. Proc. of High Level Rad. Waste Managem., Am. Nuc. Soc.

Wallbrecher, E. (1986). *Tektonische und gefügeanalytische Arbeitsweisen*. Enke Verlag, Stuttgart.

Wang, J. (1991). Flow and transport in fractured rocks. In: Review of Geophysics, Supplement, U.S. National Report to International Union of Geodesy and Geophysics (1987-1990), Contributions in Hydrology.

Wang, J. and Narasimhan, T. (1985). Hydrologic mechanism governing fluid flow in saturated, fractured porous media. *Water Resources Research*, **21**, 1861–1874.

Wang, Y. and Dusseault, M. (1991). The effect of quadratic term on the borehole solution in poroelastic media. *Water Resources Research*, **27**, 3215–3223.

Wang, Y., Luo, T., and Z., M. (2001). Geostatistical and geochemical analysis of surface water leakage into groundwater on a regional scale: a case study in the liulin karst system, northwestern china.

Warren, J. and Root, P. (1963). The behaviour of naturally fractured reservoirs. (spej 426). *Society of Petroleum Engineers Journal*, pages 245–255.

Washburn, E. (1941). Note on a method of determining the distribution of pore sizes in porous media. *Nat. Acad. Sci. Proc.*, **7**, 115 pp.

White, F. (1999). *Fluid Mechanics*. McGraw-Hill, USA, 4th edition edition.

Whittaker, J. and Teutsch, G. (1996). The simulation of subsurface characterisation methods applied to a natural aquifer analogue. Proc. of IAHS, Modelcare 96, Golden, Colorado, USA.

Wichter, L. and Gudehus, G. (1976). Ein Verfahren zur Entnahme und Prüfungen von geklüfteten Grobohrkernen. Proc. 2 Nat. Taguung ber Felsmechanik, Aachen.

Wikberg, P. e., Gustafson, G., Rhén, I., and Stanfors, R. (1991). Äspö Hard Rock Laboratory. Evaluation of conceptual modelling based on the pre-investigations 1986-1990. SKB Technical Reports 91–22, Swedish Nuclear Fuel and Waste Management Company, Stockholm.

Williams, C. E. and Fung, B. M. (1982). The determination of wettability by hydrocarbons of small particles by deuteron T1 measurement. *J. Magnetic Resonance*, **50**, 71–80.

Wilson, M. (1987). *A Handbook of Determinative Methods in Clay Mineralogy*. Chapman and Hall, Glasgow - New York.

Wollrath, J. (1990). *Ein Strömungs- und Transportmodell für klüftiges Gestein und Untersuchungen zu homogenen Ersatzsystemen.* (Institut für Strömungsmechanik und Elektronisches Rechnen im Bauwesen Hannover, Bericht; 28) (Zugl.: Hannover, Univ., Diss., 1990). Eigenverlag, Hannover.

Wollrath, J. (1990). Ein Strömungs- und Transportmodell für klüftiges Gestein und Untersuchungen zu homogenen Ersatzsystemen. Technical Report 28, Institut für Strömungsmechanik und Elektronisches Rechnen im Bauwesen, Universität Hannover.

Yeh, T.-C. J. and J., Z. (1996). A geostatistical inverse method for variably saturated flow in the vadose zone. *Water Resources Research*, **32**(9), 2757–2766.

Yeh, W. (1986). Review of parameter identification procedures in groundwater hydrology: The inverse problem. *Water Resources Research*, **22**(2), 95–108.

Zhu, W. and Wong, T. (1996). Permeability reduction in a dilating rock; network modelling of damage and tortuosity. *Geophys. Res. Lett*, **23**(22), 3099–3102.

Zimmermann, R. and Yeo, I.-W. (2000). Fluid flow in rock fractured: From the navier-stokes equations to the cubic law. In: Dynamics of Fluids in Fractured Rock. Geoph. Monograph, 122.

Zimmermann, R., Chen, G., and Hadgu, T. (1993). A numerical dual-porosity model with semianalytical treatment of fracture/matrix flow. *Water Resources Research*, **29**(7), 2127–2137.

Zlotnik, V. and Ledder, G. (1996). Theory of dipole flow in uniform anisotropic aquifers. *Water Resources Research*, **32**(4), 1119–1128.

Nomenclature

Coordinates, indices, conventions

$\overline{\overline{\square}}$	Equivalent parameter after averaging over the REV and averaging over single fractures or matrix blocks	
$\overline{\square}$	Equivalent parameter after averaging over the REV	-
$\square_\alpha$, $\square_\beta$	Parameters of a joint face	
$\square_\alpha$, $\square_\beta$	Number of a continuum for the multi continuum modelling	
$\square_\mathrm{a}$	Property associated with air	
$\square_\mathrm{hydr}$	Derived from hydraulic test	
$\square_\mathrm{i}$	Index of rectangles	
$\square_\mathrm{n}$	Property associated with non wetting phase	
$\square_\mathrm{ow}$	Property associated with observation well	
$\square_\mathrm{pneu}$	Derived from pneumatic test	
$\square_\mathrm{pw}$	Property associated with pumping well	
$\square_\mathrm{sh}$	Property associated with the shell model	
$\square_\mathrm{w}$	Property associated with water/wetting phase	
$\square_\mathrm{z}$	Index of slices	
$\square_F$	Fracture continuum (regarding fracture-matrix-systems)	
$\square_{ij}$	Tensor indices	
$\square_M$	Matrix continuum (regarding fracture-matrix-systems)	
$\square_T$	Parameter of an interface	
s	Local coordinates of the matrix blocks	m
x′, y′	Normalized coordinates	-
x, y, z	Carthesian coordinates	m

Scalar quantities

α	Θ-pole coordinate (colatitude) of the main direction (FISHER distr.)	-
α_Q	Exchange coefficient flow	$\mathrm{m\,s^{-1}}$
α_c	Exchange coefficient transport	$\mathrm{m^2\,s^{-1}}$
α_l	Longitudinal dispersivity	m
α_t	Transversal dispersivity	m

β	ϕ-pole coordinate (longitude) of the main direction (FISHER distr.)	-
δ	DIRAC function	
γ	Angle of flow direction	°
γ	Surface tension	$kg\,m\,s^{-2}\,m^{-1}\,s^{-2}$
κ	Concentration parameter (FISHER distr.)	-
λ	Mean free path of air molecules	m
μ	Dynamic viscosity	$kg\,m^{-1}\,s^{-1}$
ω	Geometry coefficient	-
ω	Spherical aperture(FISHER distr.)	*TODO*
Ω_0	Specific surface	m^{-1}
Ω_j	Subdomain j	
Ω_W	Specific fracture area	m^{-1}
Φ	Relative volume	-
σ^2	Variance	(unit of the variable)^2
τ, ω	Integration variables	
θ	Angle of incidence	°
θ_C	Contact angle	°
ε	Absolute roughness	m
ρ	Density	$kg\,m^{-3}$
A	Cross-sectional area	m^2
a	Diffusion parameter	$s^{-\frac{1}{2}}$
a	Half height of an experimental cell	m
a_r	Length parameter (variogram)	m
A_s	Azimuth	°
b	Aquifer thickness	m
b	Diameter of an experimental cell	m
b	Distance from center port configuration	m
b	Fracture aperture	m
b_k	KLINKENBERG factor	-
C	Proportionality factor	-
c	Solute concentration	$kg\,m^{-3}$
C_0	Nugget effect	(unit of the variable)^2
c_R	Concentration of boundary flow	$kg\,m^{-3}$
D	Diffusivity	$m^2\,s^{-1}$
d	Pore diameter	m
D^*	Pore diffusion coefficient	$m^2\,s^{-1}$
d_1	Linear fracture density	m^{-1}
d_2	Area related fracture density	$m\,m^{-2}$
d_3	Volumetric fracture density	$m^2\,m^{-3}$
D_m	Molecular diffusion coefficient	$m^2\,s^{-1}$
D_P	Pore diffusion coefficient	$m^2\,s^{-1}$
D_p	Dip angle	°
d_p	Effective pore diameter	m
D_l	Longitudinal dispersion coefficient	$m^2\,s^{-1}$

$D_{\mathrm{m,e}}$	Effective diffusion coefficient	$\mathrm{m^2\,s^{-1}}$
D_{t}	Transversal dispersion coefficient	$\mathrm{m^2\,s^{-1}}$
d_{ij}	Distance along trajectory i in voxel j	m
e_{RMSE}	Root mean square error	unit of the variable
e_q	Direction of flow	-
F	Mass flux	$\mathrm{kg\,m^{-2}\,s^{-1}}$
G	Continuous source	
g	Gravitational acceleration	$\mathrm{m\,s^{-2}}$
H	HEAVISIDE function	
h	Distance between two points (variogram)	m
h	Piezometric head	m
h_n	Maximum distance of nth shell to center of flow	m
I	Hydraulic gradient	-
I	Normalized sensitivity coefficient	-
I_j	Sensitivity coefficient of subdomain j	s
K	Hydraulic conductivity (isotropic)	$\mathrm{m\,s^{-1}}$
k	Permeability (isotropic)	$\mathrm{m^2}$
k_{eff}	Effective permeability	$\mathrm{m^2}$
k_r	Permeability ratio	-
L	Typical length scale, length of sample	m
l	Length	m
L^*	Loss of identity length	m
L_z	Thickness of slice z	m
M	Injected tracer mass	kg
m_z	Number of slices	-
$MCor$	Maximum correction in the multi-shell model	-
n	Total porosity	-
N_M	Mobility number	-
n_{e}	Effective porosity	-
p	Pressure	$\mathrm{kg\,m^{-1}\,s^{-2}\,m^{-2}}$
Pe	PECLET number	-
pv	Number of pore volumes	-
pv_b	pv assymmetric port configurations	-
pv_{norm}	pv normalized curve	-
pv_{ref}	pv reference curve	-
Q	Volume discharge	$\mathrm{m^3\,s^{-1}}$
q_{s}	Source and sink term	$\mathrm{kg\,m^{-3}s^{-1}}$
q_{M}	Tracer mass flux	$\mathrm{kg\,s^{-1}}$
q_{m}	Tracer mass source/sink term	$\mathrm{kg\,m^{-2}s^{-1}}$
$Q_{Q,S}$	Volume discharge due to sources/sinks per unit volume	$\mathrm{s^{-1}}$
r	Distance	m
R_a	Core radius	m
r_n	Radius of the nth shell	m
R_r	Relative roughness	-

R_i	Individual gas constant	$\mathrm{J\,kg^{-1}\,K^{-1}}$
Re	REYNOLDs number	-
S	Depth of flow system	m
S^2	Variance (variogram)	(unit of the variable)^2
S^*	Spherical variance	TODO
S_t	Strike angle	$^\circ$
S_S	Specific storage coefficient	$\mathrm{m^{-1}}$
$SCor$	Shell correction in the multi-shell model	-
T	Temperature	K
t	Time	s
T^*	Loading time	s
t_0	Mean transit time	s
t_i^e	Estimated travel time	s
t_i^m	Measured travel time	s
tr	Tracer recovery	-
u	Integration variable	-
u	Solution of groundwater flow equation	m
V	Volume	$\mathrm{m^3}$
v	Absolute velocity value	$\mathrm{m\,s^{-1}}$
v	Solution of adjoint groundwater flow equation	m
v_a	Mean velocity	$\mathrm{m\,s^{-1}}$
W_Al	Mass exchange per unit volume (by local advection)	$\mathrm{kg\,m^{-3}\,s^{-1}}$
W_Ar	Mass exchange per unit volume (by regional advection)	$\mathrm{kg\,m^{-3}\,s^{-1}}$
W_c	Mass exchange per unit volume	$\mathrm{kg\,m^{-3}\,s^{-1}}$
W_D	Mass exchange per unit volume (by diffusive exchange)	$\mathrm{kg\,m^{-3}\,s^{-1}}$
W_Ql	Fluid exchange per unit volume (by local pressure gradient)	$\mathrm{s^{-1}}$
W_Qr	Fluid exchange per unit volume (by regional pressure gradient)	$\mathrm{s^{-1}}$
x	Distance	m
x_n	Length of the outer edge of nth shell	m

Tensor quantities

$\Omega_{W,ij}$	Tensor of specific fracture area	$\mathrm{m^{-1}}$
$D_{\mathrm{d},ij}$	Dispersion tensor	$\mathrm{m^2\,s^{-1}}$
D_{ij}	Hydrodynamic dispersion tensor	$\mathrm{m^2\,s^{-1}}$
$FF(x_i)$	Fracture frequency	-
$I(x_i)$	Indicator variable	-
J_i	Total mass flux vector	$\mathrm{kg\,m^{-2}\,s^{-1}}$
$J_{\mathrm{a},i}$	Advective mass flux vector	$\mathrm{kg\,m^{-2}\,s^{-1}}$
$J_{\mathrm{d},ij}$	Dispersive mass flux tensor	$\mathrm{kg\,m^{-2}\,s^{-1}}$
$J_{\mathrm{hd},i}$	Hydrodynamic dispersion mass flux vector	$\mathrm{kg\,m^{-2}\,s^{-1}}$
$J_{\mathrm{m},i}$	Diffusive mass flux vector	$\mathrm{kg\,m^{-2}\,s^{-1}}$
K_{ij}	Hydraulic conductivity tensor	$\mathrm{m\,s^{-1}}$
k_{ij}	Permeability tensor	$\mathrm{m^2}$

q_i	DARCY velocity vector	$\mathrm{m\,s^{-1}}$
R_i	Main orientation (FISHER distr.)	-
v_i	Seepage velocity vector	$\mathrm{m\,s^{-1}}$

MIX
Papier aus verantwortungsvollen Quellen
Paper from responsible sources
FSC® C105338

If you have any concerns about our products,
you can contact us on
ProductSafety@springernature.com

In case Publisher is established outside the EU,
the EU authorized representative is:
Springer Nature Customer Service Center GmbH
Europaplatz 3, 69115 Heidelberg, Germany

Printed by Libri Plureos GmbH
in Hamburg, Germany